Organic Electronics in Sensors and Biotechnology

Biophotonics Series

Series Editors: Israel Gannot and Joseph Neev

Artificial Tactile Sensing in Biomedical Engineering by Siamak Najarian, Javad Dargahi, and Ali Abouei Mehrizi

Biomedical Applications of Light Scattering by Adam Wax and Vadim Backman

Optofluidics: Fundamentals, Devices, and Applications by Yeshaiahu Fainman, Demetri Psaltis, and Changhuei Yang

Organic Electronics in Sensors and Biotechnology by Ruth Shinar and Joseph Shinar

Organic Electronics in Sensors and Biotechnology

Ruth Shinar

Joseph Shinar

McGraw Hill

New York Chicago San Francisco
Lisbon London Madrid Mexico City
Milan New Delhi San Juan
Seoul Singapore Sydney Toronto

The McGraw-Hill Companies

Library of Congress Cataloging-in-Publication Data

Shinar, Ruth.
Organic electronics in sensors and biotechnology / Ruth Shinar, Joseph Shinar.
p. cm.—(McGraw-Hill biophotonics series)
Includes bibliographical references.
ISBN-13: 978-0-07-159675-6 (alk. paper)
ISBN-10: 0-07-159675-5 (alk. paper)
1. Detectors—Materials. 2. Organic semiconductors. 3. Biotechnology—Equipment and supplies. 4. Organic electronics. 5. Biosensors. I. Shinar, Joseph, date. II. Title.
TK7872.D48S55 2009
681′.2—dc22 2009019107

McGraw-Hill books are available at special quantity discounts to use as premiums and sales promotions, or for use in corporate training programs. To contact a representative please e-mail us at bulksales@mcgraw-hill.com.

Organic Electronics in Sensors and Biotechnology

1 2 3 4 5 6 7 8 9 0 DOC/DOC 0 1 4 3 2 1 0 9

ISBN 978-0-07-159675-6
MHID 0-07-159675-5

The pages within this book were printed on acid-free paper.

Sponsoring Editor
Taisuke Soda

Acquisitions Coordinat
Michael Mulcahy

Editorial Supervisor
David E. Fogarty

Project Manager
Somya Rustagi, International Typesetting and Composition

Copy Editor
Patti Scott

Proofreader
Upendra Prasad, International Typesetting and Composition

Indexer
WordCo Indexing Services, Inc.

Production Supervisor
Pamela A. Pelton

Composition
International Typesetting and Composition

Art Director, Cover
Jeff Weeks

About the Editors

Ruth Shinar is a Senior Scientist at the Microelectronics Research Center of the Institute of Physical Research and Technology and Adjunct Professor of Electrical and Computer Engineering at Iowa State University.

Joseph Shinar a Senior Physicist in the Ames Laboratory, U.S. Department of Energy, and a Professor of Physics and Astronomy and of Electrical and Computer Engineering at Iowa State University.

Contents

Contributors

Siegfried Bauer *Soft Matter Physics, Johannes Kepler University, Linz, Austria* (CHAP. 8)

Magnus Berggren *OBOE—Strategic Center for Organic Bioelectronics Organic Electronics, ITN, Linköpings Universitet, Norrköping, Sweden* (CHAP. 11)

Annalisa Bonfiglio *Department of Electrical and Electronic Engineering, University of Cagliari, Italy, and INFM-CNR S3 Centre for nanoStructures and bioSystems at Surfaces, Modena, Italy* (CHAP. 3)

M. Bruendel *Institut für Mikrostrukturtechnik, Forschungszentrum Karlsruhe GmbH, Karlsruhe, Germany; current affiliation: Robert Bosch, GmbH, Stuttgart* (CHAP. 7)

K. Buchholt *Department of Physics, Chemistry and Biology, Linköping University, Linköping, Sweden* (CHAP. 2)

Yuankun Cai *Ames Laboratory-USDOE, and Department of Physics, and Astronomy, Iowa State University, Ames* (CHAP. 5)

N. Cioffi *Dipartimento di Chimica e Centro di Eccellenza TIRES, Università degli Studi di Bari, Bari, Italy* (CHAP. 2)

L. Colaianni *Dipartimento di Chimica, Università degli Studi di Bari, Bari, Italy* (CHAP. 2)

A. Dell'Aquila *Department of Water Engineering and of Chemistr, Polytechnic of Bari, Bari, Italy* (CHAP. 2)

John deMello *Chemistry Department, Imperial College London, London, UK* (CHAP. 6)

Chetna Dhand *Department of Science and Technology Centre on Biomolecular Electronics, National Physical Laboratory, New Delhi, India* (CHAP. 10)

Ananth Dodabalapur *The University of Texas at Austin* (CHAP. 1)

Daniel H. Fine *The University of Texas at Austin* (CHAP. 1)

Jonas Groten *Institute of Nanostructured Materials and Photonics, Joanneum Research Forschungsgesellchaft, Weiz, Austria* (CHAP. 4)

Jingsong Huang *Molecular Vision Ltd., BioIncubator Unit, London, UK* (CHAP. 6)

E. Ieva *Dipartimento di Chimica, Università degli Studi di Bari, Bari, Italy* (CHAP. 2)

Georg Jakopic *Institute of Nanostructured Materials and Photonics, Joanneum Research Forschungsgesellchaft, Weiz, Austria* (CHAP. 4)

U. Lemmer *Light Technology Institute (LTI), Universität Karlsruhe (TH), Karlsruhe, Germany* (CHAP. 7)

B. D. Malhotra *Department of Science and Technology Centre on Biomolecular Electronics, National Physical Laboratory, New Delhi, India* (CHAP. 10)

Ileana Manunza *Department of Electrical and Electronic Engineering, University of Cagliari, Italy, and INFM-CNR S3 Centre for nanoStructures, and bioSystems at Surfaces, Modena, Italy* (CHAP. 3)

T. Mappes *Institut für Mikrostrukturtechnik, Universität Karlsruhe (TH), Karlsruhe, Germany* (CHAP. 7)

F. Marinelli *Dipartimento di Chimica, Università degli Studi di Bari, Bari, Italy* (CHAP. 2)

P. Mastrorilli *Department of Water Engineering and of Chemistry, Polytechnic of Bari, Bari, Italy* (CHAP. 2)

S. G. Mhaisalkar *School of Materials Science and Engineering, Nanyang Technological University, Singapore* (CHAP. 2)

J. Mohr *Institut für Mikrostrukturtechnik, Forschungszentrum Karlsruhe GmbH, Karlsruhe, Germany* (CHAP. 7)

K. P. R. Nilsson *Department of Chemistry, Linköping University, Linköping, Sweden* (CHAP. 9)

M. Punke *Light Technology Institute (LTI), Universität Karlsruhe (TH), Karlsruhe, Germany* (CHAP. 7)

Agneta Richter-Dahlfors *OBOE—Strategic Center for Organic Bioelectronics, Department of Neuroscience, Karolinska Institutet, Stockholm, Sweden* (CHAP. 11)

L. Sabbatini *Dipartimento di Chimica, e Centro di Eccellenza TIRES, Università degli Studi di Bari, Bari, Italy* (CHAP. 2)

Niyazi Serdar Sariciftci *Linz Institute of Organic Solar Cells (LIOS), Institute of Physical Chemistry, Johannes Kepler University, Linz, Austria* (CHAP. 8)

M. Schelb *Institut für Mikrostrukturtechnik, Universität Karlsruhe (TH), Karlsruhe, Germany* (CHAP. 7)

Helmut Schön *Institute of Nanostructured Materials and Photonics, Joanneum Research Forschungsgesellchaft, Weiz, Austria* (CHAP. 4)

Joseph Shinar *Ames Laboratory-USDOE, and Department of Physics and Astronomy, Iowa State University, Ames* (CHAP. 5)

Ruth Shinar *Microelectronics Research Center, and Department of Electrical and Computer Engineering, Iowa State University, Ames* (CHAP. 5)

Th. Birendra Singh *Linz Institute of Organic Solar Cells (LIOS), Institute of Physical Chemistry, Johannes Kepler University, Linz, Austria; current affiliation: CSIRO Molecular and Health Technology, Ian Work Laboratory, Clayton, Victoria, Australia* (CHAP. 8)

A. Lloyd Spetz *Department of Physics, Chemistry and Biology, Linköping University, Linköping, Sweden* (CHAP. 2)

Barbara Stadlober *Institute of Nanostructured Materials and Photonics, Joanneum Research Forschungsgesellchaft, Weiz, Austria* (CHAP. 4)

M. Stroisch *Light Technology Institute (LTI), Universität Karlsruhe (TH), Karlsruhe, Germany* (CHAP. 7)

G. P. Suranna *Department of Water Engineering and of Chemistry, Polytechnic of Bari, Bari, Italy* (CHAP. 2)

L. Torsi *Dipartimento di Chimica e Centro di Eccellenza TIRES, Università degli Studi di Bari, Bari, Italy* (CHAP. 2)

C. Vannahme *Institut für Mikrostrukturtechnik, Forschungszentrum Karlsruhe GmbH, Karlsruhe, Germany* (CHAP. 7)

Liang Wang *The University of Texas at Austin* (CHAP. 1)

T. Woggon *Light Technology Institute (LTI), Universität Karlsruhe (TH), Karlsruhe, Germany* (CHAP. 7)

Martin Zirkl *Institute of Nanostructured Materials and Photonics, Joanneum Research Forschungsgesellschaft Weiz, Austria* (CHAP. 4)

Preface

Organic electronics, i.e., the field of (opto)electronic devices utilizing organic active layers, has been growing rapidly over the last two decades. Examples of intensive ongoing research and development areas include organic light-emitting diodes (OLEDs), organic field-effect transistors (OFETs), and organic photovoltaics (OPVs). The appeal of organic electronics stems from attributes such as the abundance of synthetic π-conjugated small molecules and polymers, whose photoluminescence and electroluminescence span a broad spectral range; the ease of fabricating organic thin films by well-established techniques such as thermal evaporation, spin coating, and inkjet printing; and the mechanical flexibility and compatibility of organics with substrates such as glass and plastic. As a result, devices are amenable to large-scale fabrication and are expected to be of low cost.

OLEDs have advanced from short-lived dim devices, with a lifetime of less than 1 min in air, to red and green OLEDs and blue OLEDs that can operate continuously, based on accelerated measurements, for over 200,000 h and 100,000 h, with efficiencies of ~60 and ~20 lm/W, respectively, at a brightness of 100 Cd/m^2 (i.e., slightly less than a typical TV or computer monitor). In pulsed operation, OLED brightness values > 10^7 Cd/m^2 have been reported. Indeed, OLEDs are already common in flat-panel displays of, e.g., car stereos, mobile phones, MP3 players, and small TV screens, and their sales are growing. A bright-color 11 in. OLED TV was recently commercialized by Sony. OLEDs are also promising for solid-state lighting applications, and commercialization of bright white OLED panels by, e.g., Matsushita, is expected in the near future. Similarly, OFETs are being developed for applications in flat panel displays and radio-frequency identification smart tags, and organic and hybrid PV is an intensely growing research field with currently reported power efficiencies approaching 6%.

The growing activity and progress in organic electronics have led to emerging R&D in the field of organic electronics-based chemical and biological sensors as well as in biotechnology. The R&D on (bio)chemical sensors is constantly growing in due to existing and surfacing needs for such sensors for a vast number of analytes in areas encompassing every aspect of life (e.g., medicine, environment, food and beverage, chemical industry, homeland security) as well as the need, which remains a challenge, for compact, field-deployable, inexpensive, versatile, and user-friendly sensors and sensor arrays. The organic electronics-based sensors—whether monitoring the effect of analytes on the photoluminescence of, e.g., analyte-sensitive dyes, where the excitation source is a thin OLED pixel array, or monitoring the effect of analytes on the attributes of the OFETs—are very promising in alleviating existing sensor-related issues such as the limited portability and high cost and maintenance. Importantly, such sensors are not as sensitive to the limiting issues in organic semiconductors (e.g., long-term stability), in particular when considering inexpensive disposable devices. Indeed, attributes such as small, potentially miniaturized size, compatibility with microfluidic architectures, and high sensitivity have been demonstrated in organic electronics-based sensors. The efficacy of such sensors for simultaneous detection of multiple analytes using small-size sensor arrays has also been shown. Such sensors build on the ability to fabricate (micro)arrays of multiple OLED pixels and OFETs. As an example, tens of OLED pixels, ranging in size from millimeters down to nanometers, can be fabricated combinatorially on compact substrates. Each pixel (or a small group of pixels) can be associated with a different analyte. Such pixels can be of single or multiple colors. Moreover, OLED-based sensors can be further integrated with organic-based or other thin-film photodetectors to generate very thin, portable sensors. In OFETs, where charge mobility is low in comparison to crystalline Si, the promise is in their potential lower cost and design flexibility. For example, for biomedical applications the advantage is in the possibility to fabricate devices on large areas on unusual substrates such as paper, plastic, or fabrics.

The use of organic semiconductors in other biotechnological applications is drawing significant interest as well. As an example, in cell biology where the interface between an aqueous fluid and a solid surface is of great importance, electric biasing is a promising approach for dynamic control of surface properties and thus for advancing research in this field. Demonstrated solid-state ion pumps based on conducting polymers are also promising for such studies.

This volume covers various aspects of ongoing R&D in organic electronics for sensors and biotechnology. Chapter 1 describes scaling effects in organic transistors and on the sensing response to organic compounds. Chapter 2 describes sensing of inorganic compounds using OFETs, including gold nanoparticle-modified FET sensors. Chapter 3 describes organic semiconductor-based strain and pressure sensors. The chapter presents the state-of-the-art technologies and applications, including sensors on conformable, large surfaces. Chapter 4 deals with the characterization of the electronic properties of organic materials by impedance spectroscopy and the integration of organic thin-film transistors and capacitive pyroelectric polymer sensors to form integrated flexible pyroelectric sensors. Chapter 5 summarizes recent advances and challenges in the OLED-based chemical and biological sensing platform, including simultaneous sensing of multiple analytes, monitoring of foodborne pathogens, and integration with thin-film photodetectors. Chapter 6 is an introduction to organic photodetectors, whose performance characteristics are compared to those of conventional photomultiplier tubes and solid-state detectors. Chapter 7 discusses the fabrication and the properties of organic semiconductor lasers with some focus on the aspects of low-cost replication. It also addresses the integration of organic lasers into optical sensor chips and the choice of sensing principles. Chapter 8 discusses organic electronics in memory elements and sensing applications for light, temperature, and pressure monitoring. Chapter 9 discusses the use of luminescent conjugated polymers as optical sensors for biological events, especially protein aggregation diseases. Chapter 10 describes applications of electrophoretically deposited polymers for organic electronics, and Chapter 11 discusses the use of conjugated polymer–based electrochemical surface switches and ion pumps for cell biology studies.

With the broad aspects covered, which are at different stages of research and development, it is hoped that the book will be useful as a reference guide for researchers established in the field, as well as an introduction for scientists entering the field.

Ruth Shinar
Joseph Shinar

Organic Electronics in Sensors and Biotechnology

CHAPTER 1

Scaling Effects in Organic Transistors and Transistor-Based Chemical Sensors

Liang Wang, Daniel H. Fine, and Ananth Dodabalapur
The University of Texas at Austin

1.1 Scaling Behavior in Organic Transistors

1.1.1 Charge Transport in Polycrystalline Organic Semiconductors (Intragrain and Intergrain)

In the past few years, much work has been done with small organic molecules and conjugated polymers[1] to manufacture devices that take advantage of the unique properties of their conjugated π-electron systems, such as organic light-emitting diodes (OLEDs), solar cells, smart cards, resistive chemical sensors, and field-effect transistors.[2–5] Among these applications, the structure of field-effect transistors, with an organic or polymer semiconductor as the electrically active layer, allows for the study of the two-dimensional transport properties of the semiconductor material modulated by the charge carrier density due to increased gate bias.[6–8] Organic electronics will not replace classic Metal Oxide Semiconductor Field-Effect Transistors (MOSFETs) in most applications due to the low mobility values in organic materials compared to those in single crystalline silicon. However, organic devices are still of considerable interest to the industry because of the low cost and mechanical flexibility of organic materials. To preserve this advantage, inexpensive methods have to be developed to fabricate organic devices. Organic field-effect transistors have been made with relatively high mobilities [of order 1 $cm^2/(V \cdot s)$] from small organic molecules, such as pentacene, deposited by vacuum-based sublimation[9,10] and soluble conjugated polymers, such as regio-regular polythiophene, deposited by spin-coating, solution-casting, or ink-jet printing processes.[11–13] Due to the relatively high cost in vacuum systems for subliming small molecules, conjugated polymers fabricated onto devices by solution-based processes are preferred for industrial applications. To obtain high performance of polymer electronic devices, improvements in synthesis techniques are essential for high-quality polymers with high purity, low residual doping, and air stability. On the other hand, a solution-based precursor technique for small molecules such as pentacene has also been developed to be used in low-cost processes such as ink-jet or spin-coat,[14–16] which shed some light on small molecules for their potential in industrial applications. We also note that 300 K mobilities of > 1 $cm^2/(V \cdot s)$ have been reported for a few single crystals and crystalline materials.

Figure 1.1 lists some of the small molecule and oligomeric organic materials which have been commonly used in organic field-effect transistors. Both α-4T and α-6T are oligomers of thiophene, tetracene and pentacene are polyacenes, while CuPc (copper phthalocyanine) has a coordinate structure. All the aforementioned are nominally p-type semiconductors with holes as the primary charge carriers. In these materials, the mobility of holes is generally much higher than that of electrons, due to the presence of electron traps, particularly at the interface with most dielectrics. We note that modifying the interface with the gate insulator can reduce the electron trap concentration. n-type organic semiconducting materials have also been drawing a lot of interest for applications such as OLED and Complementary Metal Oxide Semiconductor (CMOS) circuits requiring both types of charge carriers. C_{60}, copper hexadecafluorophthalocyanine ($F_{16}CuPc$), and perylene derivatives are commonly used n-type organic semiconductor materials which have been used as channel semiconductor materials in n-channel organic transistors. The semiconducting property of these organic materials originates from the delocalized π-electron (remaining p_z electron after sp^2 hybridization) in conjugated bonds (alternating single/double bond) between carbon atoms within one molecule or one polymer chain. In semiconducting organic molecules and polymers, there exist two type of coupling between neighboring carbon atoms: single bond (σ bond only) or double bond (σ bond plus π bond). The σ bond is the coupling of sp^2 hybridized orbitals from neighboring carbon atoms, while the π bond is the coupling of the

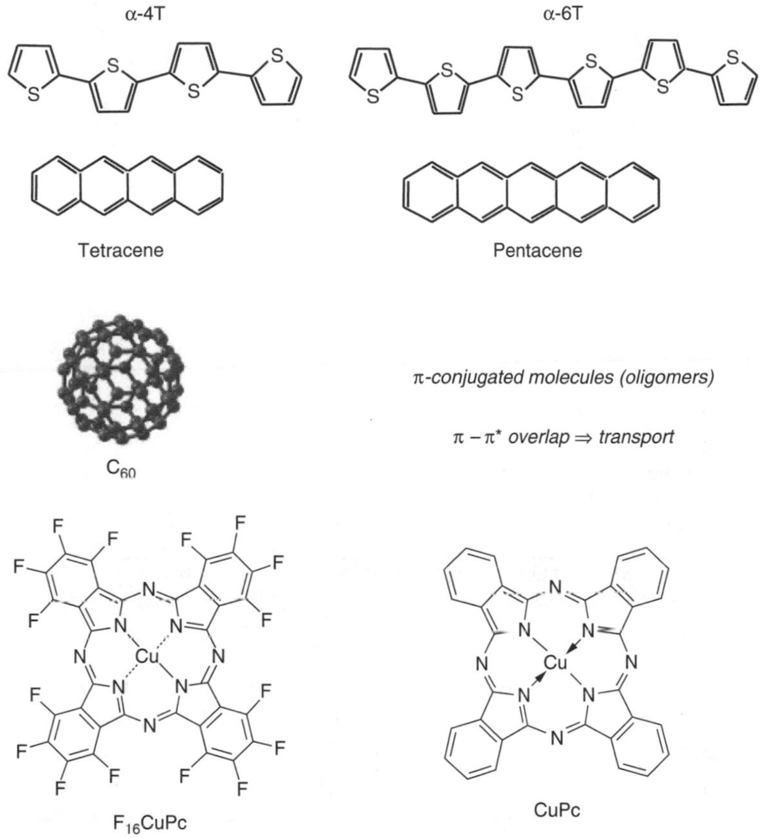

FIGURE 1.1 Commonly used organic molecules. Both α-4T and α-6T are a chain of thiophene rings, and tetracene and pentacene are polyacenes (fused benzene rings). C_{60} possesses a fullerene-type ball structure. CuPc and F-CuPc have a coordinate structure. (*Liang Wang, "Nanoscale Organic and Polymeric Field-Effect Transistors and Their Applications as Chemical Sensors," Ph.D. dissertation, The University of Texas at Austin.*)

remaining p_z orbitals. The σ bond forms the structure of the oligomer molecule or the backbone of a polymer chain, and the electrons on σ bonds are localized. Electrons associated with π bonds are delocalized, and the alternating configuration of single bond and double bond gives rise to the semiconducting properties.

In organic semiconductors, intramolecular electron transport is facile, as shown in Fig. 1.2*a*. The interaction between molecules is through van der Waals forces. Additionally, organic semiconductors are characterized by a strong carrier-phonon interaction that reduces the electronic bandwidth at high temperatures. Consequently the room-temperature mobility in organic molecular crystals is over two orders of magnitude lower than that in crystalline silicon. The intragrain transport mechanism between molecules within one grain of

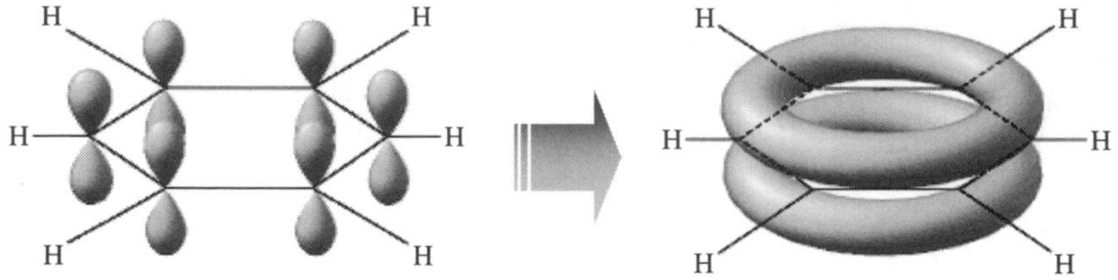

(*a*) Two-dimensional delocalized p bond forms in small organic molecules.

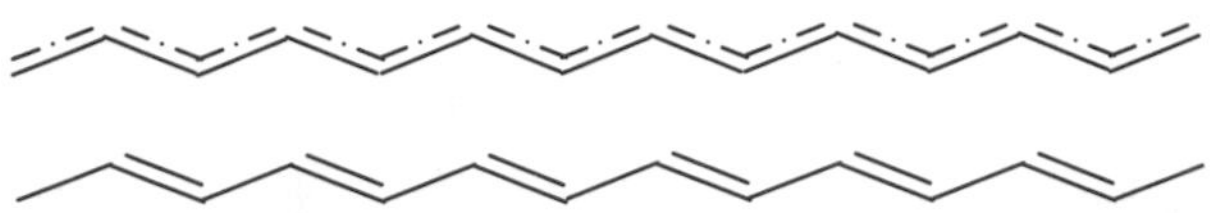

(*b*) Due to the resulting instability in one-dimensional system, delocalized p bond depicted as (I) does not exist in a polymer chain; instead, an alternating configuration of single/double bond forms as (II) shows.

FIGURE 1.2 Alternating configuration of single/double bonds. (*Liang Wang, "Nanoscale Organic and Polymeric Field-Effect Transistors and Their Applications as Chemical Sensors," Ph.D. dissertation, The University of Texas at Austin.*)

organic semiconductors could be bandlike transport,[17, 18] thermal-activated independent hopping,[19] correlated hopping,[20, 21] or tunneling.[22]

There exists a two-dimensional delocalized π bond inside a small organic molecule. However, in case of polymers which are one-dimensional chains, delocalized π bonds spanning the entire chain cannot be formed due to its higher energy and the consequent instability of the system. Instead, the system chooses a conjugated state with a lower energy, where a π bond forms between every other pair of nearest carbon atoms so that single bonds (σ bond) and double bonds (σ and π bonds) alternate along the entire chain, as shown in the phase II of Fig. 1.2(*b*). This is the ground state of a polymer chain. In an excited state, with thermal activation or electron/hole dopants, a polymer chain will form a polaron in most cases.

In conjugated polymers, the charge carrier contributing to electrical transport is a polaron (a polaron is a quasiparticle consisting of a charge accompanied by the associated lattice deformation). The intrachain transport mechanism is bandlike polaron transport along the chain with scattering by long-wavelength acoustic phonons.[23] The interchain transport mechanism is phonon-assisted polaron hopping (perpendicular to the chain direction).[24–25]

In semiconducting organic materials such as molecular crystals, the room temperature bandwidth is small so that we apply the terms *HOMO* (highest occupied molecular orbital) and *LUMO* (lowest unoccupied molecular orbital) instead of those used for crystalline silicon (bottom of conduction band and top of valence band). For transport within one grain, charge transport could occur through such mechanisms as bandlike transport, hopping, or tunneling. This is the basic picture for transport within a single grain (a domain of well-ordered molecules or well-aligned polymer chains). The situation for transport between grains is complicated. Generally the presence of domain or grain boundaries (the region between grains) limits the mobility of the polycrystalline organic semiconductors, because the disorder at grain boundaries leads to more localized states within the energy gap which trap the mobile charge carriers.[26] Based on this concept, the *density of states* (DOS) in polycrystalline systems can be described by bands with tails which extend into the bandgap. Inside the band the DOS varies slowly with energy while its tails decays exponentially into bandgap.[26] Around the edge of HOMO or LUMO, the band tail, composed of localized states, has a much wider distribution at grain boundaries than within each grain, due to the increased disorder at grain boundaries.[27] According to the multiple trapping and release model,[28] the gate-induced charges in an organic field-effect transistor are composed of two parts: the free (mobile) charges in the channel, which contribute to the channel conductance, and the trapped charges in trapping levels within the energy gap, which do not contribute to the channel conductance. The channel conductivity is a product of free charge density and free charge mobility. There is another way (which is the convention in this area) to view the channel conductivity; namely, it could also be considered as a product of total gate-induced charge density (including both free charges and trapped charges) and effective mobility. Therefore the value of the effective mobility is determined by the ratio of free charges to the sum of free and trapped charges.[28–30]

1.1.2 Characterization of Nanoscale Organic Transistors

In addition to the above-mentioned transport mechanisms within the active organic semiconductor layer, the performance of an organic field-effect transistor depends also on many extrinsic factors, especially at the interfaces. The dielectric/semiconductor interface property affects the molecular ordering within one/two monolayers (roughly equal to the channel depth) of the organic semiconductor. Interface traps at grain boundaries also play a role in device mobility. The contact or electrode/semiconductor interface property determines the injection barrier. In nanoscale transistors, the injection at contacts limits the transport of charge carriers at low gate bias and low source-drain bias, as well as leads to a sensing mechanism which is quite different from that in large-scale devices. A bottom-gate transistor structure, as shown in Fig. 1.3, has been commonly used in organic FETs comprised of a substrate, usually a highly doped silicon wafer or conductor-covered plastic sheet, followed by a dielectric layer atop the substrate, such as silicon dioxide or an insulating polymer. The organic or polymer semiconductor and metallic source and drain contacts are then deposited on top of the gate dielectric with different patterning technologies.[31–33] This kind of structure reduces fabrication

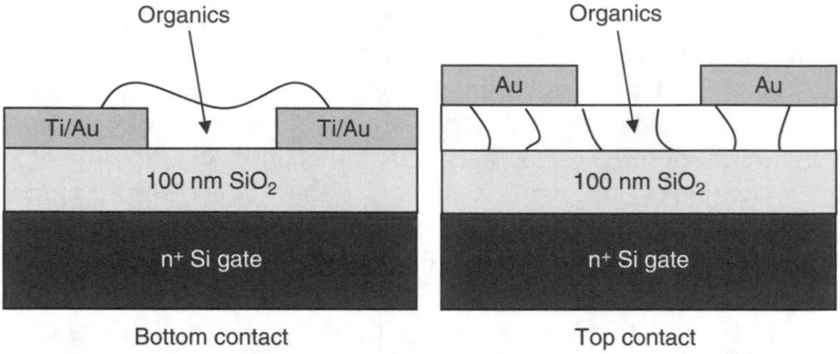

FIGURE 1.3 Two typical architectures (top contact and bottom contact) in upside-down structure for organic transistors. (*Liang Wang, "Nanoscale Organic and Polymeric Field-Effect Transistors and Their Applications as Chemical Sensors," Ph.D. dissertation, The University of Texas at Austin.*)

complexity and enables direct interaction between the active semiconducting layer and the ambient. We note that some of the organic semiconductors reported have a relatively high mobility.[10, 34] Such large mobilities are not really necessary for sensing applications, where the change in current or threshold voltage produced by an analyte is more important.

Nanoscale organic field-effect transistors have been investigated by a few groups. It is technically difficult to pattern the active semiconductor area of devices with such small channel lengths. For transistors with a channel length near and below 10 nm, a large width-to-length (W/L) ratio is not favorable due to a higher chance of shorted electrodes[35] and worse line edge roughness (LER) of the order of the channel length. Consequently, the spreading currents which travel outside the intended channel cannot be ignored for devices of small W/L ratios, and it becomes a concern if W/L is less than 10.[36] We have fabricated a large number of devices in which the channel length is less than 50 nm, with a small W/L ratio, and in which the active semiconductor layer and gate are not patterned. In such devices the spreading current which travels outside the defined channel will contribute significantly to the total current. To collect the spreading current, we designed a separated pair of guarding electrodes near the two sides of the channel, unconnected to and kept at the same potential as the drain. By this design, these guarding electrodes collect the spreading currents so that the drain current measured is the current from source to drain, excluding contributions from macroscale spreading currents, as shown in Fig. 1.4. Devices were fabricated with below 50 nm channel lengths, small W/L ratios, and non-patterned active semiconductor layer and gate. The distance between a channel and its side guards in the fabricated devices is in the range from 20 to 50 nm, almost comparable to the channel lengths. Measurements of the drain current were taken on the same device without biasing the

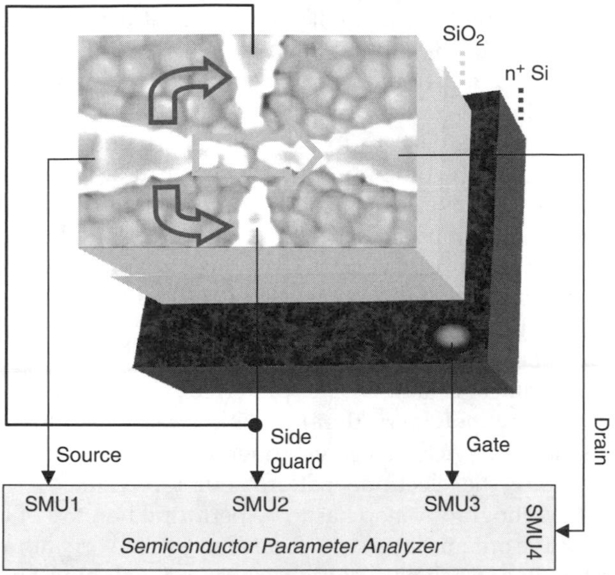

FIGURE 1.4 Three-dimensional device structure and circuit diagram for the function of side guards in a nanoscale transistor used as chemical sensor. To collect spreading currents traveling through the area outside the defined channel, two side guards were designed on the two sides of the channel, unconnected to and kept at the same potential as the drain. The three layers from top are bottom-contact pentacene and Au/Ti electrodes (surface shown by a SEM image of real device), SiO_2 as dielectric, and n^+ Si as gate. Each SMU (source measurement unit) of the Semiconductor Parameter Analyzer (Agilent 4155C) supplies voltage and measures current independently. SMU1, 2, 3, 4 serves as the source, side-guard, gate, and drain, respectively. SMU1 = ground; SMU3 = V_g; SMU2 and SMU4 were set at the same value V_{ds}. *(Reprinted with kind permission from Springer Scientific + Business Media.*[68])

side guards (I_{open}) and also with the side guards biased (I_{ds}). For nanoscale devices, I_{ds} and I_{open} manifested different behaviors. I_{open} behaved more as a long-channel FET, which indicates a substantial component of spreading current. For each of the measured devices, the maximum value of I_{ds} was significantly lower than that of I_{open} under the same voltage configuration, and the ratio I_{ds}/I_{open} was below 70%. This ratio was found to positively correlate to the W/L ratio. The distance from a channel to its side guards and the geometry of the electrodes may actually affect its I_{ds}/I_{open} ratio.

1.1.3 Channel Length and Temperature Dependence of Charge Transport in Organic Transistors

Although there have been reports of electrical characteristics at room temperature for OTFTs with submicron and nanoscale channel lengths,[37–40] no experimental study has been conducted systematically for these devices with such small dimensions at different temperatures down to 4.2 K. On the other hand, there have been many reports on the transport of various organic semiconductor materials with large channel length transistors at low temperatures with different measurement techniques.[41–45] To investigate the possible transport mechanisms which become dominant when scaling the device size from micron-scale down to nanoscale, we systematically fabricated thin-film field-effect transistors of a series of channel lengths from 5 μm down to 80 nm, with pentacene as the active organic semiconductor layer. All the investigated transistors of different channel lengths were fabricated by the same nominal geometry ratio $W/L = 10$, for the purpose of consistent scaling.

Bottom-contact structures are preferred in the fabrication of nanoscale organic/polymeric thin-film field-effect transistors. To fabricate submicron devices, e-beam lithography is a commonly used tool in defining the electrode pattern. For top-contact structures, however, the lithography step has to be performed on top of the predeposited active organic layer, which can degrade the organic material upon immersion in solvent solutions during developing and liftoff steps, or damage the organic material while evaporating metal atoms onto it. Hence, the bottom-contact configuration is chosen over the top-contact one. A heavily doped n-type silicon wafer serves as the mechanical substrate and the gate. The single-crystal silicon substrate was heavily doped by n-type dopants (phosphorous or arsenic) so that the depletion in the silicon as a gate was minimized. To serve as conductive gate, the doping concentration of the Si substrate needs to be higher than $3.74 \times 10^{18}/cm^3$ for phosphorous dopant or $8.5 \times 10^{18}/cm^3$ for arsenic dopant.[46] A SiO_2 layer that serves as the gate dielectric was then thermally grown on the substrate. Transistors of channel length greater than 1 μm utilize a 100 nm thermally grown SiO_2 layer as the gate dielectric, whereas for nanoscale transistors, a 5 nm SiO_2 layer as the gate dielectric was grown by rapid thermal annealing in dry oxygen. The electrode patterns were made by e-beam lithography. Metal electrodes with an adhesive layer (Ti was chosen) that improves sticking to SiO_2 were then deposited by e-beam evaporation in high vacuum, followed by a liftoff process. If the Ti layer is thinner than the accumulation channel depth, then carriers are injected directly from the high work function metal layer atop the Ti layer, which remarkably reduces the series resistance. Therefore the Ti layer was chosen to be 1 to 3 nm thick. Gold was chosen as the electrode material, due to its high work function aiding the injection of holes into organic material, its air stability, and its ability to form an ohmic contact with organic materials under certain optimal conditions.

Device fabrication was completed by subliming small organic molecules (e.g., pentacene) at different growth rates and different substrate temperatures for different grain sizes. Slower growth rates and higher substrate temperatures during sublimation tend to produce an organic layer with large grain size and planar morphology. The substrate temperature has a stronger influence than the sublimation rate. However, attempts to deposit large pentacene grains onto nanoscale channels at elevated substrate temperatures did not yield favorable results. This could be attributed to the repulsion of the Au electrodes from the pentacene molecules at elevated substrate temperatures. Usually 300 to 350 Å is the minimal thickness for the organic semiconductor to uniformly cover the channel region, whereas too thick an organic layer will not increase the gate-induced channel conduction (the channel is only one or two monolayers deep) but increase the non-gate-induced bulk conduction and therefore reduce the on/off current ratio of the field-effect transistor. To form an organic semiconductor layer with high mobility in field-effect transistors, the requirements for the quality and the morphology of the organic thin film are exacting. The source material has to be in high purity, usually post-purchase purified with cycling many times under a temperature gradient in vacuum. The grown film has to be uniform and flat and have an ordered molecular stacking. For this purpose, after UV ozone or gentle oxygen plasma cleaning, a surface treatment is desired to form a well-ordered self-assembled monolayer of alkyl groups, before depositing an organic semiconductor layer. Two commonly used materials for a self-assembled monolayer (SAM) on top of SiO_2 are HMDS (hexamethyldisilazane) and OTS [trichloro (octadecyl)silane], because these molecules have two functional groups: silane group on one end to bond with SiO_2 and alkyl group on the other end to provide a hydrophobic layer for the molecules of organic semiconductor to grow. This self-assembled monolayer will improve the molecular ordering of the pentacene layer and reduce the trapped charges at the dielectric-semiconductor interface, which will enhance the field-effect mobility and improve the subthreshold slope.[47,48] Within the other type of SAM agent molecule 4-nitrobenzenethiol to treat the interface between organic semiconductor and electrodes, its surface active head group (thiol) can anchor to the electrodes through a dative bond between the sulfur atom and the metals such as Au, Pd, Pt, and Ag. This treatment can significantly reduce the contact resistance by a factor of 10 at room temperature and even more at low temperatures,[48] which is important for the low-temperature measurements described here. Ideally, HMDS only forms a monolayer on SiO_2, without chemisorption onto electrodes, and 4-nitrobenzenethiol only adsorbs onto electrodes. This ideally gives an orthogonal self-assembly without interaction between dielectric treatment and contact treatment.[47]

For the device characterization, a Table Top Manipulated Probe System from Desert Cryogenics was employed, at varying temperatures and over two orders of magnitude of electric field (from 10^4 to 10^6 V/cm). The DC I–V characteristics for all transistors of different

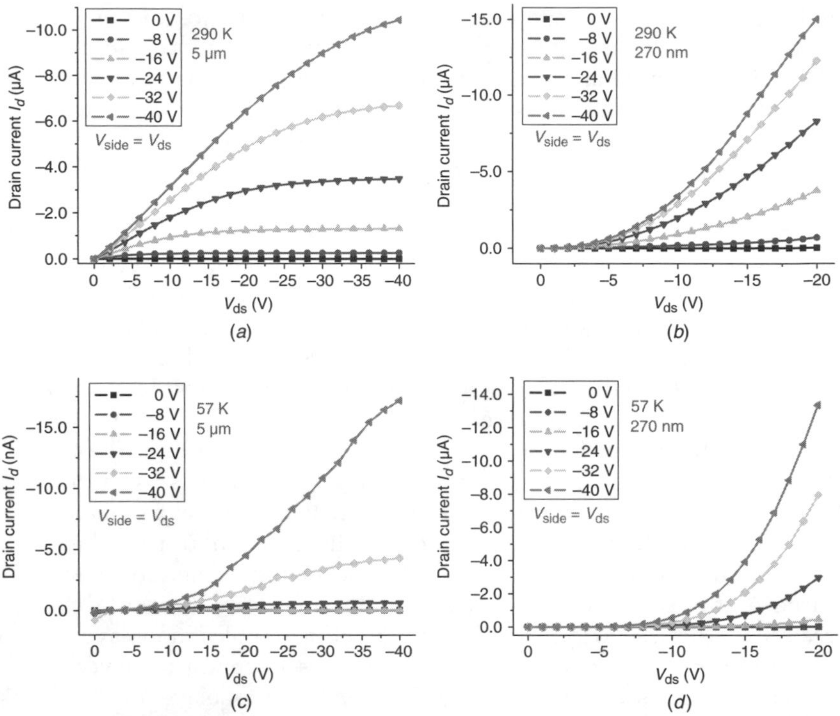

FIGURE 1.5 The DC characteristics of pentacene transistors, with channel lengths of 5 μm and 270 nm at temperatures of 290 and 57 K. Different symbols on each panel represent different gate biases. (*Reprinted with permission from Ref. 60. Copyright 2007, American Institute of Physics.*)

channel lengths were measured with a semiconductor parameter analyzer in mTorr vacuum from room temperature down to liquid helium temperatures, with great care taken to ensure good thermal contact and thermal equilibrium before performing the measurements. Figure 1.5*a* and *b* demonstrates the characteristics for transistors with channel length of 5 μm and 270 nm, respectively, at 290 K. For OTFTs with channel lengths longer than 1 μm, the DC characteristics show normal behavior; i.e., the drain current is in a linear relation to source-drain voltage at low V_{ds} bias while the drain current saturates at high V_{ds}. However, for OTFTs with channel lengths of 1 μm or shorter, the DC characteristics exhibit superlinear behavior over the entire range of longitudinal field. This scaling behavior of transition in characteristic can be seen clearly in Fig. 1.6. In Fig. 1.6, we show the drain current I_d as a function of source-to-drain voltage V_{ds} for transistors with different channel lengths at room temperature. Because the W/L ratios of all channels were kept at the same value of 10 in fabrication, this figure demonstrates charge transport scaling behavior in

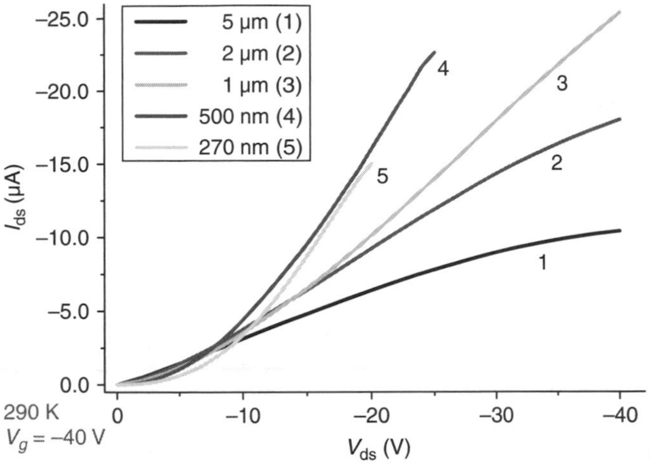

FIGURE 1.6 Drain current as a function of source-to-drain voltage for different channel lengths. The characterization was taken at room temperature and high density of charge carriers ($V_g = -40$ V, well beyond threshold voltages of each channel). For observation of the scaling behavior, the *W/L* ratios of all channels were kept at the same value of 10 in fabrication to exclude geometric factors. Clearly in the regime of $V_{ds} < V_g - V_{th}$, the current-voltage characteristic transitions from linear to superlinear upon scaling from micron to submicron channel length. For submicron channels there is an exponential dependence at very small V_{ds} due to the injection-limited transport through Schottky barrier at the metal-semiconductor contact. (*Reprinted with permission from Ref. 60. Copyright 2007, American Institute of Physics.*) (See also color insert.)

the organic semiconductor excluding the geometry factors of the devices. When temperature decreases, as Fig. 1.5*c* and *d* shows, the I_d–V_{ds} curve becomes more superlinear for both long and short channels. While the current level of a 5 μm channel markedly reduces at low temperatures, surprisingly the drain current of a 270 nm channel remains at the same level from 290 to 57 K.

In field-effect transistors using organic semiconductor as channel material, the charge transport occurs only within the first one or two monolayers of the organic semiconductor.[49] Furthermore, the gate-induced charges within the quasi-two-dimensional channel can be considered to be composed of the relatively free (mobile) charges in the channel, which contribute to the channel conductance, and the trapped charges on trapped levels within the energy gap, which do not contribute to the channel conductance and affect the mobility and threshold voltage of the transistor. The trap levels within the energy gap of organic semiconductors are the localized states that originate from disorder. For a polycrystalline thin-film layer of an organic semiconductor such as pentacene thermally sublimed in vacuum, most of the disorder is located at grain boundaries. Therefore for the

OTFTs with channel lengths of 1 μm and longer, which are much larger than the average grain size (100 to 150 nm) of pentacene deposited under the conditions described above, the electrical transport data could be explained by the multiple trapping and release model[28] with significant influence from the grain boundaries. However, when the channel length scales down to 500 nm and smaller, which is comparable to the grain sizes of the organic semiconductor layer, the average number of grains spanning the source-drain gap is reduced, as shown in Fig. 1.7, so that the dominant electrical transport mechanism becomes different from that in large-scale devices. For polycrystalline organic thin-film transistors, in addition to the traps at grain boundaries due to disorder which dominates the charge transport in a long channel, field-dependent mobility and limited injection through the Schottky barrier at metal-organic semiconductor contacts play important roles in channel conduction as the device geometry scales down to nanoscale dimensions. Injection-limited behavior is observed in small-channel-length devices even when the channel length is smaller than the average grain size of the organic semiconductor layer.

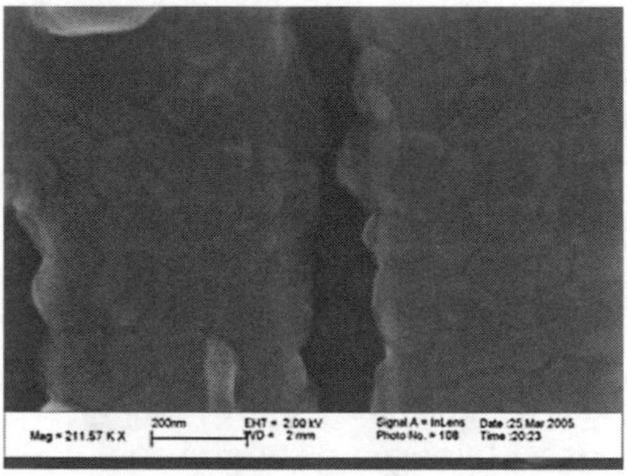

~1 grain on 180 nm channel

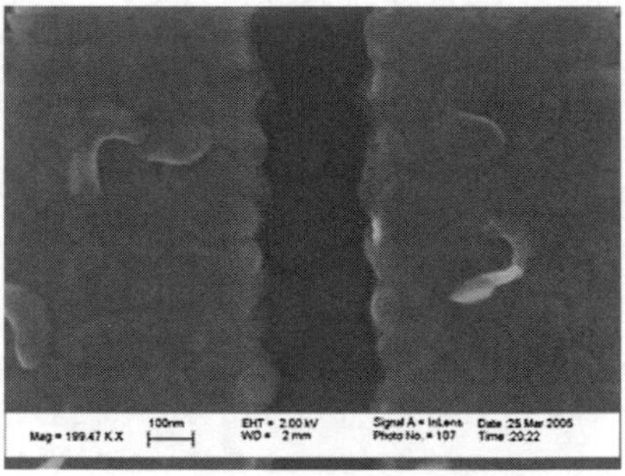

1–2 grains on 270 nm channel

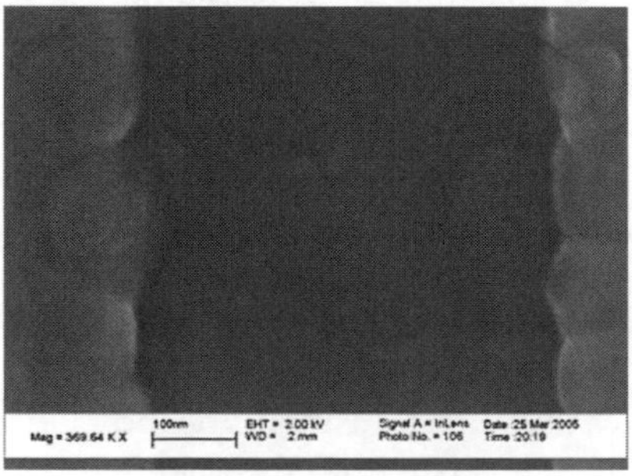

2–3 grains on 500 nm channel

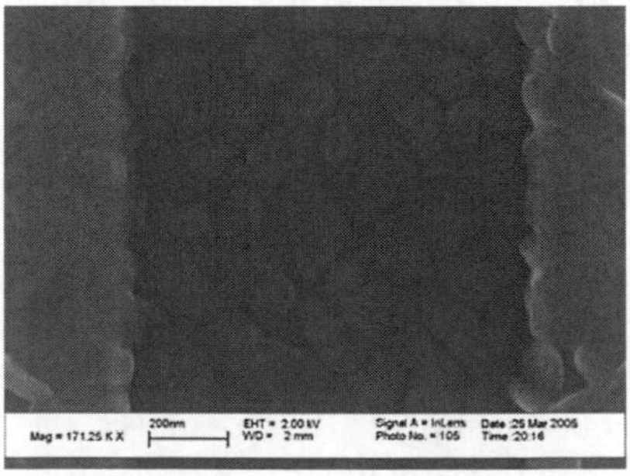

6–7 grains on 1 μm channel

FIGURE 1.7 The size of grains of the pentacene layer relative to channel geometry. There are roughly 6 to 7 grains within a 1 μm channel but only 1 to 2 grains within a 270 nm channel.

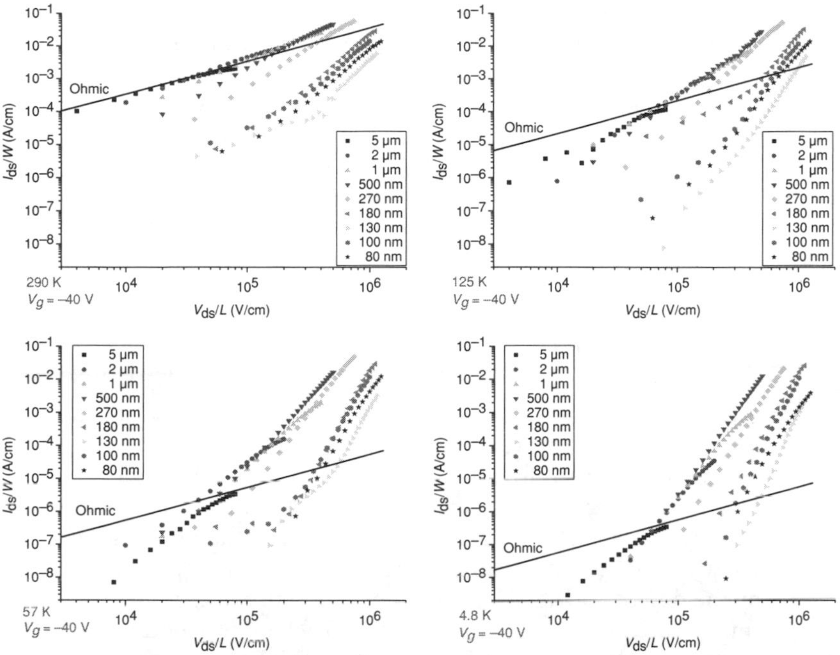

Figure 1.8 The current density vs. longitudinal field plots for various channel lengths at four different temperatures. The solid lines in these plots are the ohmic channel transport currents, calculated based on the mobility extracted from long-channel (5 μm) devices. These lines serve as the references to investigate the issues of contact injection-limited transport and field-dependent mobility. All the four figures are exactly on the same scale for the purpose of comparison. (See also color insert.)

Figure 1.8 shows the scaling behavior of charge transport in organic transistors, including field-dependent mobility, injection-limited transport, and temperature dependence. To study the scaling behavior, the apparent contribution of the device geometry was filtered out by taking the current density vs. longitudinal field and plotting that relation for various channel lengths at four different temperatures, under a certain high gate bias beyond the threshold voltage of the transistors (–40 V). The solid lines in these plots are the ohmic channel transport currents at V_g = –40 V calculated through the drain-current equation for linear-region operation of transistors, based on the mobility values extracted from the measurement data taken at the corresponding temperature on the long-channel (5 μm) devices where field dependence of the mobility is not significant at its operative longitudinal field and the channel resistance is much larger than contact resistance. At the same temperature, for all the devices with different channel lengths, their calculated ohmic channel transport behaviors, ignoring other factors, should fall onto the same solid line on the current density–longitudinal field plot. Therefore these solid lines of calculated ohmic channel currents at varying temperatures serve as a reference to investigate the effects of contact injection-limited transport and field-dependent mobility on the charge transport through the channel, both of which feature superlinear relation of current density-longitudinal field. Injection-limited transport occurs at low longitudinal field which results in lower currents and effective mobilities compared to the ideal gate-induced ohmic channel current, while field-dependent mobility takes place at high longitudinal field which is larger than the value expected when compared to the ideal gate-induced ohmic channel current.

Based on all the above understanding, in Fig. 1.8 the lower portion of the measured curves below the solid line represents the injection-limited transport, while the upper portion of the measured curves beyond the solid line represents the transport regime of field-dependent mobility. It can also be observed that the field dependence of mobility is stronger at lower temperatures. All the current-field curves in logarithmic scale measured for different channel lengths shift downward when the temperature decreases, as represented clearly by the downward shifts of the ohmic reference lines with lowering temperature. This is so because, for all channel lengths, the threshold voltage shifts toward higher gate bias at lower temperatures. Note that the data in the quadrant labeled with 4.8 K contain some self-heating effects which make the actual device temperature deviate from the readings. After careful analysis, it was determined that device's temperature was not significantly affected by the self-heating at temperatures beyond 40 K. Figure 1.9 shows that threshold voltages shift with temperature quasi-linearly, and this trend holds for all channels ranging from 5 μm to 270 nm. The temperature dependence of the threshold voltage in an organic field-effect transistor (OFET) is significantly larger than that in a Si-MOSFET,[50] and this is attributed to the presence of deep trap states in an organic semiconductor.[51] At low temperatures the thermal energy (kT) of charge carriers becomes very small, and thereby a large percentage of gate-induced charge carriers are falling into trap states whose levels are much deeper than kT and not being released and thus do not contribute to the channel transport. In this case much higher gate bias is required for the same amount of mobile charges to be present in the channel, and therefore at lower temperature the threshold voltage of an OFET significantly shifts toward higher gate bias.

1.1.4 Field-Dependent Mobility Model for the Scaling Behavior of Charge Transport

To understand the scaling behavior observed in organic transistors as shown in the previous section, a physical model is needed that explains the mechanisms of charge transport with regard to temperature, field,

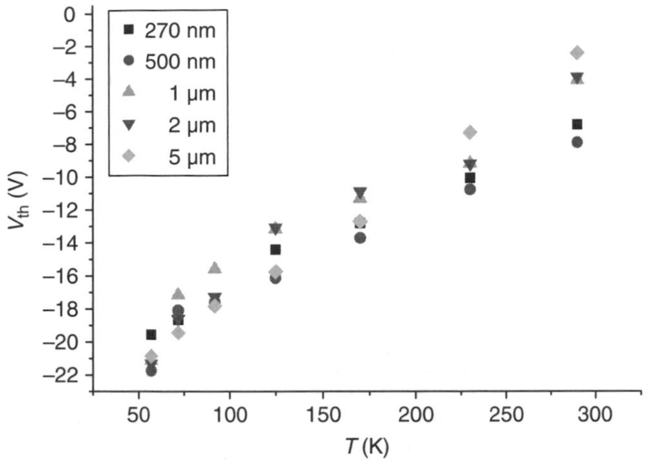

FIGURE 1.9 The temperature dependence of threshold voltage for scaled channel lengths. Threshold voltages V_{th} were extracted from the transconductance plots (I_d vs. V_g for $V_{ds} < V_g - V_{th}$; $(I_d)^{1/2}$ vs. V_g for $V_{ds} \geq V_g - V_{th}$) on high end of V_{ds} for each channel. V_{th} does not significantly change with longitudinal field. Channels of different lengths follow the same trend; namely, V_{th} shifts to high gate bias quasi-linearly with decreasing temperature. (*Reprinted with permission from Ref. 60. Copyright 2007, American Institute of Physics.*) (See also color insert.)

geometry, morphology, and interfaces. Most experimental studies performed so far have emphasized measurement and interpretation of temperature-dependent field-effect mobility.[52–55] In partially ordered organic semiconductors, however, it is the electric field dependence of the mobility at various temperatures that sheds the most light on the understanding of transport phenomena.[56–57] Temperature dependence of the mobility alone is insufficient to draw some of the important conclusions. Previous reports[58–59] contained evidence of a field-dependent mobility, but the measurements were not accurate enough for quantitative work. The work reported recently by Liang Wang et al.[60] represented the first quantitative measurement of the electric field dependence of the mobility in organic thin-film transistors when device geometries are scaled along the direction of charge transport and when accumulated carrier densities are at levels of practical importance. With systematic measurements, they reported temperature and electric field dependence of field-effect mobility in polycrystalline pentacene thin-film transistors of scaled channel geometries, and their results show a Frenkel-Poole type dependence of mobility on electric field.

In purified single crystals of polyacenes such as anthracene, band-like transport has been experimentally observed by Karl and coworkers and subsequently interpreted in terms of polaronic transport models.[61–62] In this picture, band transport can exist at low temperatures when the bandwidth is large compared to kT and the mean free path is large compared to intermolecular spacing. As the temperature is increased, both band and hopping contributions to transport can exist, leading eventually to hopping-dominated transport at high temperatures.[63] At the other extreme, in highly disordered semiconductors such as glasses and molecular doped polymers, transport has been explained in terms of the correlated disorder model which is characterized by a Frenkel-Poole type variation of mobility with electric field.[64] The nature of charge transport in an intermediate class (polycrystalline form) of semiconductors, which contains ordered crystallites separated by disordered regions (grain boundaries), can be expected to possess features of both crystalline organic semiconductors as well as disordered ones. For a polycrystalline thin-film layer of an organic semiconductor such as pentacene thermally sublimed in vacuum, most of the disorder is located at grain boundaries. Therefore, when the channel length scales down to 500 nm and smaller, which is comparable to the grain sizes of the organic semiconductor layer, the average number of grains within one channel reduces as shown in Fig. 1.7, so that the dominant electrical transport mechanism becomes different from that of large-scale devices where channel lengths are much larger than the grain sizes. For polycrystalline organic thin-film transistors, field-dependent mobility and limited injection through Schottky barrier at metal-organic semiconductor contacts play important roles as the device geometry scales down to these dimensions. To address this scaling behavior, the mobility vs. longitudinal field (nominally taken as V_{ds}/L) was plotted in logarithmic vs. square root scales, respectively, for a series of channel lengths at different temperatures, as shown by the scattered data in Fig. 1.10. The mobility turns out to be a function of the longitudinal field (increasing with the field) consistently throughout a wide range of scaled channel lengths for the same morphology of organic semiconductor layer deposited at a given fabrication batch.

It is very important that the contacts do not influence the measured mobility. For each channel length, as the source-drain voltage magnitude is increased from zero, the devices are initially injection-limited. In this regime, the drain current increases exponentially with voltage, as is expected when charge is injected over a Schottky barrier. In Fig. 1.6 it can be clearly seen that at small V_{ds}, there is an exponential dependence of I_d on V_{ds} (injection-limited regime). As the source-drain voltage is increased, the devices are operated away from the injection-limited regime. For this reason, in Fig. 1.10 we consider data only at high V_{ds} for each channel length when the device operation has moved away from the injection-limited regime to a regime where I_d increases slightly superlinearly with V_{ds} as is expected in the linear regime of operation of an organic transistor with some field-dependent mobility. For each scattered curve in Fig. 1.10 measured at a certain channel length, the "bottom part" of the curve (at relatively lower longitudinal fields,

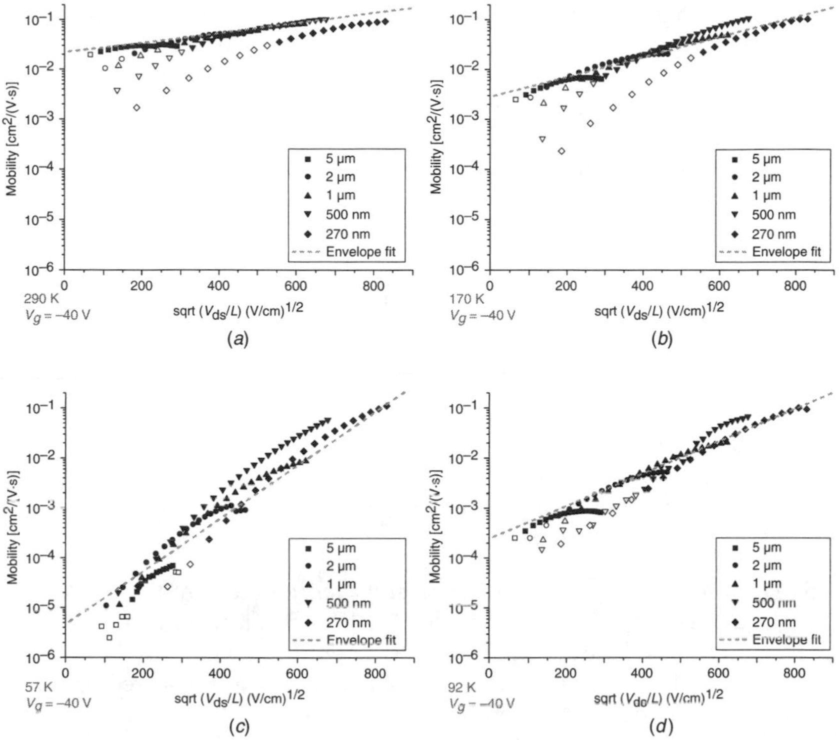

FIGURE 1.10 The plots of mobility at varying longitudinal field. They are in logarithmic vs. square root scales for a series of channel lengths at four different temperatures. For each scattered curve measured at a certain channel length, its "bottom part" (at relatively lower longitudinal field, marked as hollow symbols) is subject to injection-limited transport. The straight dash line in each panel is the envelope combining the high field parts (solid symbols) of all the scattered curves measured at different channel lengths to filter out the injection limitation at low fields. All the four panels utilize the same data label and are on the same scale for comparison. The data fit Frenkel Poole's model:

$$\mu = \mu_0 \exp\left(\frac{\beta\sqrt{E}}{kT}\right)$$

(Reprinted with permission from Ref. 60. Copyright 2007, American Institute of Physics.)

marked as hollow symbols in Fig. 1.10) is subject to injection-limited transport, while its "top part" (at high fields, solid symbols) is less affected by injection-limited transport. Therefore the envelope combining the "top parts" of all curves at different channel lengths represents the true behavior of field-dependent mobility. These envelopes (as straight dashed lines in Fig. 1.10) were found to roughly fit the field-dependent mobility model outlined by Frenkel-Poole's law as described below.

In a disordered organic semiconductor system, charge transport occurs mainly by hopping between adjacent or nearby localized states which are induced by disorder. The physical effect of the longitudinal electrical field is then to effectively reduce the hopping barrier. From this basic concept and the reasonable assumption of Coulomb potential type for hopping barrier, the hopping probability, and therefore the mobility, will demonstrate a dependence on electrical field which follows a Frenkel-Poole relationship[64, 65]

$$\mu = \mu_0 \exp\left(\frac{\beta\sqrt{E}}{kT}\right) \tag{1.1}$$

where k = Boltzmann's constant
T = temperature
E = electrical field
μ_0 = mobility at zero field
β = field-dependent coefficient originally proposed by Gill in an empirical relation[66]

This law predicts the experimental results represented by the four straight dashed lines in the panels of Fig. 1.10. Field dependence of mobility becomes more severe at lower temperature, as Fig. 1.10 indicates and Frenkel-Poole's law predicts. When temperature decreases, the field dependence of the mobility becomes stronger as indicated in Eq. (1.1); namely, the slope of the logarithmic mobility-field curve in Fig. 1.10 is steeper at lower temperature for the same channel length under the same longitudinal field. The temperature-dependent behavior of the field-dependent mobility for charge transport in these organic field-effect transistors is well exhibited in Fig. 1.11, from 44 K to room temperature. Figure 1.11 is a collection of fitting lines each of which represents the behavior of field-dependent mobility at a certain temperature, obtained from experimental data in the same way as the four straight dashed lines in Fig. 1.10. It is interesting and important to notice the converging point in Fig. 1.11; i.e., at the field ~7.3×10^5 V/cm, mobilities at all temperatures fall onto the same value ~0.15 $cm^2/(V \cdot s)$, corresponding to a zero hopping barrier at such a high field. This is also predicted by Frenkel-Poole's model. Recalling that zero-field mobility μ_0 in Eq. (1.1) can be expressed as

$$\mu_0 = \mu_i \exp\left(-\frac{\Delta}{kT}\right) \tag{1.2}$$

based on the basic description for hopping transport,[65] the Frenkel-Poole's expression for field-dependent mobility can be comprehensively written as

$$\mu = \mu_i \exp\left(\frac{\beta\sqrt{E} - \Delta}{kT}\right) \tag{1.3}$$

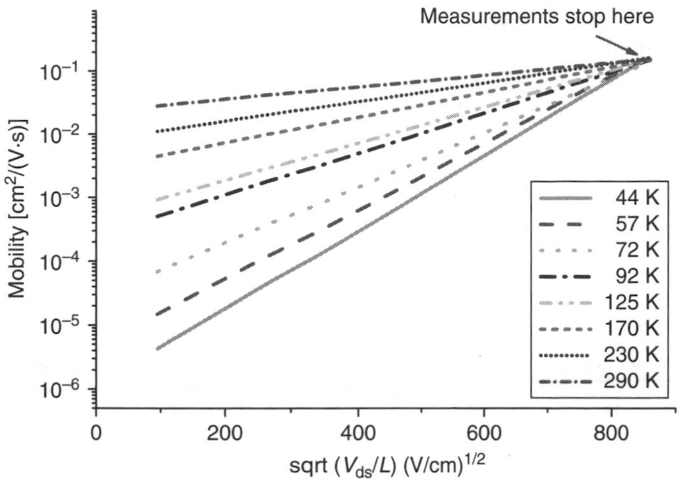

FIGURE 1.11 The temperature dependence of field-dependent mobility. Each line represents the field-dependent mobility at a certain temperature (the mobility increases with increasing temperature), obtained from experimental data in the same way as the four straight dashed lines in Fig. 1.10 and taken within the field range spanned by the experimental measurements. The converging point of straight lines at different temperatures is well predicted by Frenkel-Poole's expression for field-dependent mobility

$$\mu = \mu_i \exp\left(\frac{\beta\sqrt{E} - \Delta}{kT}\right)$$

(Reprinted with permission from Ref. 60. Copyright 2007, American Institute of Physics.)

where Δ is the zero-field hopping barrier or low-field activation energy and μ_i is the intrinsic mobility at zero hopping barrier. A charge traveling in an organic semiconductor will encounter a hopping barrier which is reduced by the field as $\Delta - \beta\sqrt{E}$. A critical field $E_0 = (\Delta/\beta)^2$ as high as 7.3×10^5 V/cm is enough to balance out the zero-field hopping barrier Δ in the investigated devices so that under such a high field the device mobility shows no dependence on temperature, as indicated by the converging point in Fig. 1.11. Fitting the data in Fig. 1.11 into Frenkel-Poole's model gives the values of μ_i, β, and Δ as 0.15 $cm^2/(V\cdot s)$, 5.8×10^{-5} $eV(V/cm)^{-1/2}$, and 50 meV, respectively. These parameters could vary with the charge density modulated by gate bias. The existence of the converging point in Fig. 1.11 also indicates that the hopping barrier is the same in all the devices of different channel lengths, which is consistent with the fact that the organic semiconductor layer for all channel lengths was fabricated within the same batch.

By combining field and temperature dependence studies, a physical picture of charge transport in polycrystalline organic field-effect transistors was drawn, which was consistent over a range of scaled channel lengths. In these partially ordered systems, experimental data suggest that the charge transport mechanism is thermally activated and field-assisted hopping transport, and the hopping transport between disorder-induced localized states dominates over the intrinsic polaronic hopping transport seen in organic single crystals. Figure 1.12 shows that the field-dependent mobility in the hopping regime predicted by the polaron model[67]

$$\mu = \mu_0 \frac{2kT}{eEa} \sin\text{h}\left(\frac{eEa}{2kT}\right) \tag{1.4}$$

deviates significantly from the experimental data of field-dependent mobility in polycrystalline pentacene thin-film transistors, whereas the hopping transport between disorder-induced localized states with Frenkel-Poole type field dependence fits the data quite well. The low-field activation energy for the localized states induced by disorder is 130 meV in polycrystalline pentacene thin-flim field-effect transistors,[68] which is much larger than the intrinsic polaron binding energy (21 to 35 meV) in polyacenes.[69] Therefore charges are more appropriately thought of as being trapped in localized states than being dressed by intrinsic lattice distortions, which leads to the preference of the transport model in polycrystalline organic transistors as demonstrated in Fig. 1.12.

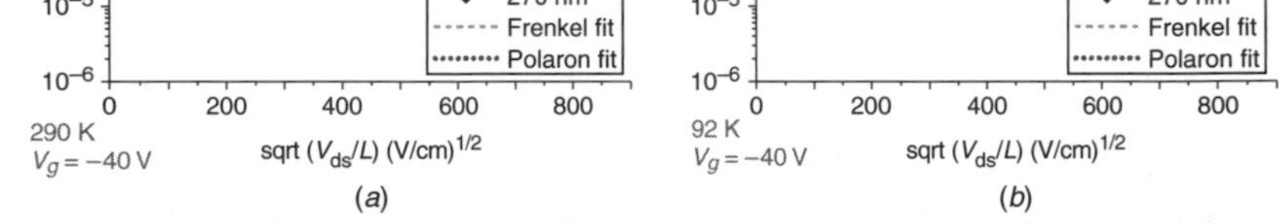

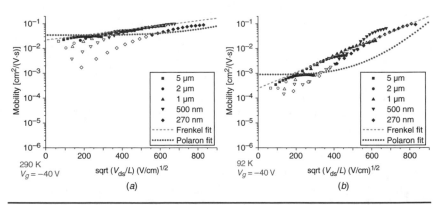

FIGURE 1.12 Comparison of Frenkel-Poole's model vs. polaronic model. Shown are representative plots at two temperatures to fit the experimental data with Frenkel-Poole's model (gray dashed line) and polaronic model. Frenkel-Poole's model and the polaronic model follow different laws

$$\mu = \mu_0 \exp\left(\frac{\beta\sqrt{E}}{kT}\right) \quad \text{and} \quad \mu = \mu_0 \frac{2kT}{eEa} \sin\text{h}\left(\frac{eEa}{2kT}\right)$$

respectively. Frenkel-Poole's model fits the data quite well whereas the polaronic model deviates obviously. (*Reprinted with permission from Ref. 60. Copyright 2007, American Institute of Physics.*)

1.1.5 Charge Transport in sub-10-nm Organic Transistors

Organic thin-film transistors have received great interest from the scientific community due to their use in potentially low-cost, large-area circuits.[10, 70] Characteristics of top contact organic thin-film transistors have been reported for channel lengths down to 30 nm fabricated with e-beam lithography.[71] Recently several groups reported their work on the performance of bottom contact organic thin-film transistors down to 50 nm[72] and 30 nm[73] channel lengths, also defined by e-beam lithography. Compared to the expensive e-beam lithography process, emerging techniques such as nanoimprint lithography combined with dry etching process have captured the attention of industry due to their potential as fast and inexpensive candidates to fabricate nanoscale devices. There have been several reports for utilizing these techniques to fabricate organic/polymeric transistors with channel lengths of 500 nm[74] and 70 nm.[75] The transport mechanisms would be different when the channel length of a transistor shrinks to sub 10 nm, since tunneling effects become important at these dimensions and it is also likely that charge transport between source and drain takes place through a single grain. Charge injection from the contact would play a very important role in such a small geometry device as well. Indeed, a situation could exist in which the local morphology of organic semiconductor in the vicinity of the channel dominates the device behavior, leading to huge variations in individual transistor responses. At the high longitudinal electric fields present in devices with very small channel lengths, the carrier velocity rather than mobility becomes important. This velocity may saturate at high fields, leading to a reduction in the effective mobility at high fields.

The field-effect transistors of channel length below 10 nm were fabricated on a heavily doped silicon substrate serving as the gate and a thermally grown SiO_2 layer as the gate dielectric. The electrode patterns were defined by e-beam lithography, metallization, and lift-off process. Electrode patterns of channel lengths ranging from 40 nm down to sub 10 nm were obtained. A pattern of a 5 nm channel with side guards as close as 20 nm away is shown in Fig. 1.13*a*. Bottom-contact devices were completed by subliming pentacene molecules, resulting in pentacene layer with an average grain size of about 100 nm. The dc electrical characteristics of the transistors were measured in air at room temperature. After the measurements, all the devices were examined by SEM. Figure 1.13*b* shows the post-measurement SEM image of a bottom-contact pentacene transistor of 9 nm channel length. By applying the same potential as the drain on the two side guards, the true source-to-drain current ($-I_{ds}$) was observed as shown in Fig. 1.14*a*. The behavior of I_{ds} exhibited reasonable gate modulation. Due to the thickness of the gate dielectric layer (100 nm SiO_2), the operating gate voltage (V_g, up to –30 V) in most cases is much

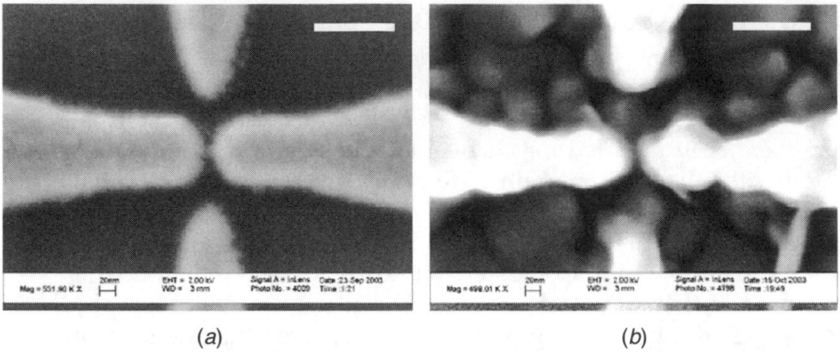

Figure 1.13 SEM images of electrodes and semiconductor layer. (*a*) SEM image of a 5 nm channel just before pentacene evaporation. (*b*) SEM image of a 9 nm channel after *I–V* measurement. The white scale bars are 100 nm. (*Reprinted with permission from Ref. 40. Copyright 2004, American Institute of Physics.*)

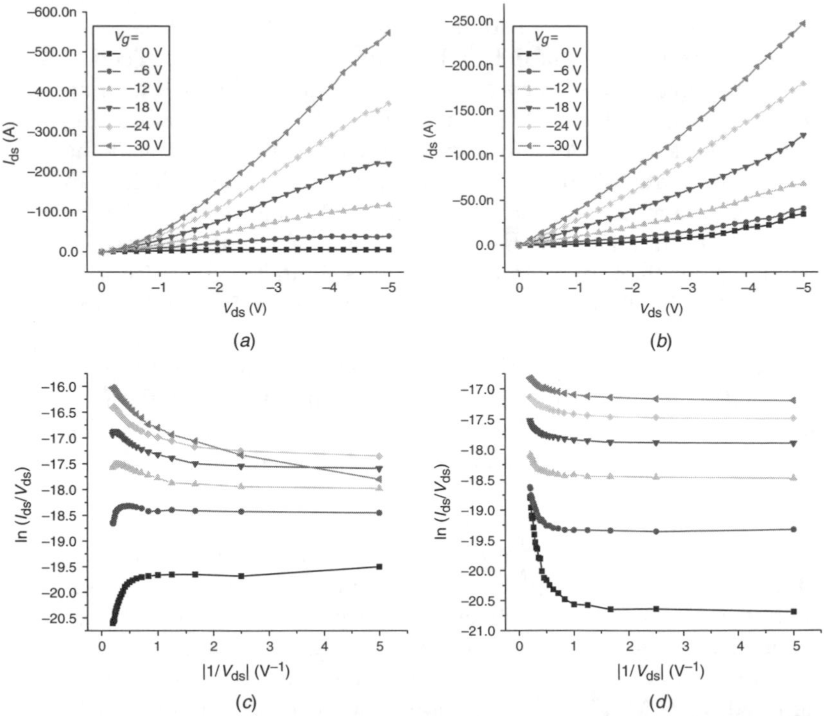

Figure 1.14 The DC characteristics of sub-10-nm pentacene FETs. (*a*) The DC *I–V* measurement of the device in Fig. 1.13*b*, with the side guards biased at the same potential as the drain. (*b*) The DC *I–V* measurement of a 19 nm channel device with the side guards biased; I_{ds}/I increases with increasing $|V_g|$. (*c*) $\ln(I_{ds}/V_{ds})$ vs. $1/|V_{ds}|$ plot of the data in (*a*). (*d*) $\ln(I_{ds}/V_{ds})$ vs. $1/|V_{ds}|$ plot of the data in (*b*). (*Reprinted with permission from Ref. 40. Copyright 2004, American Institute of Physics.*) (See also color insert.)

higher than the operating drain voltage (V_{ds}, up to −5 V), which causes the FET to operate in the linear region. The calculation of field-effect mobility derived from the transconductance of I_{ds} in the linear region is 0.046 $cm^2/(V \cdot s)$, and the on/off ratio (at $V_{ds} = -5$ V) is 97 for this device. The superlinear characteristics at the high V_g end are probably due to the possibility that channel resistance becomes small enough at high gate biases that contact resistance plays a role. The measured devices can be broadly classified into two groups. One group typically shows the behavior similar to Fig. 1.14*a*, with fairly good gate modulation, manifesting linear and saturation behavior upon increasing longitudinal field. The other group exhibits a different kind of characteristic, comprising much less gate modulation and always superlinear behavior of the drain current vs. V_{ds}, as shown in Fig. 1.14*b* for a 19 nm channel device. In this group of devices there is an additional barrier to overcome before charges are injected into the channel from the source.

The above-mentioned injection barrier at the metal-organic semiconductor contacts influences the shape of the current–voltage curves. These results are similar to those by Collet et al., who reported superlinear currents in α-6T transistors with 30 nm channel length which they attributed to Fowler-Nordheim tunneling through the metal/organic semiconductor interfacial barrier.[71] When the injection barrier is high, the current I_{ds} is not determined by the conduction in the semiconductor channel, but is limited by the injection properties of the contact. The hole injection barrier has been reported to be 0.5 to 0.85 eV in the presence of surface-dipole contribution (determined from photoemission spectroscopy experiments for large-area pentacene thin-film deposited atop Au at room temperature).[76–78] Tunneling through the source/pentacene barrier is preferable only when the longitudinal field is high enough to reduce the width of the barrier. To investigate the relation between the conductivity and the longitudinal field, we rescaled Fig. 1.14*a* and *b* into plots of $\ln(I_{ds}/V_{ds})$ vs. $1/|V_{ds}|$ as shown in Fig. 1.14*c* and *d*, respectively. Both parts figures show nearly constant conductivities at small V_{ds}. In Fig. 1.14*d*, at large V_{ds}, $\ln(I_{ds}/V_{ds})$ vs. $1/|V_{ds}|$ follows the Fowler-Nordheim tunneling model where its slope decreases with V_g, corresponding to the gate-lowered tunnel barrier. Assuming a triangular potential barrier as a first-order approximation, we have[79]

$$I_{ds} \propto F \cdot e^{-F_0/F} \qquad F = \left|V_{ds}/L\right| \quad F_0 = \frac{4\sqrt{2m^*}}{3e\hbar}(\phi_B)^{3/2} \qquad (1.5)$$

where the tunnel barrier height ϕ_B (derived from the slope) is 0.13 to 0.034 eV at $V_g = 0$ to −30 V (taking effective mass m^* as 1.7 times free-electron mass[80, 81]). Here the prefactor in the above equation is F instead of F^2, the usual expression for Fowler-Nordheim tunneling, because the tunneling occurs through the pentacene channel which is semiconducting with charges induced from gate control. In usual cases, Fowler-Nordheim tunneling happens through a barrier which is made of insulator, and therefore the charges contributing to current are solely injected from electrodes, which is similar to space-charge limited current (SCLC) and follows the same square law vs. source-drain voltage. In the devices shown as Fig. 1.14*d*, the charges constituting the drain current are induced in the channel by applying gate bias, and thereby the prefactor of this tunneling current is a linear term. The onset field (V_{ds}/L) of tunneling slightly increases with V_g, probably due to the fact that the ability of the drain bias to make the barrier thinner is screened by the gate-induced charge, and the slope at high field decreases with V_g, corresponding to the gate-lowered tunnel barrier. On the other hand, in Fig. 1.14*c*, at large V_{ds} the slope increases with V_g, which indicates that in this case the transport mechanism is not tunneling through the contact barrier, but rather is determined by the intrinsic properties of pentacene, which suggests that in the devices of the first group the metal electrode and the organic semiconductor form good contacts.

For nanoscale pentacene transistors with non-ideal contacts, the overall transport is limited by the gate-modulated metal/organic semiconductor barrier, as described in Fig. 1.14*b* and *d*. If the contacts are good, transport is not injection-limited (Fig. 1.14*a* and *c*). In long-channel polycrystalline organic transistors, the overall transport is limited by the transport through grain boundaries rather than that through the bulk of the grains.[82, 83] On the other hand, when channel lengths are less than the average grain size, transport through a single grain may become important; however, the injection properties of the contact (which is possibly affected by the local morphology of the pentacene grain) are the most important factor that influences the electrical response. The grain sizes of the pentacene layer were found to be from 40 up to 200 nm, much larger than the channel lengths of the measured devices (from 8 to 35 nm). Therefore for these devices the number of grain boundaries inside a channel is significantly reduced; i.e., a channel is composed of only one grain or has only a single grain boundary inside. For these nanoscale transistors (channel length from 8 to 35 nm) with small W/L ratios, field-effect mobility calculated for each of the measured devices was found to have no significant correlation to the channel length. The histogram in Fig. 1.15 gives a distribution of mobility by statistical counts of the number of devices in every mobility interval. Nearly 50% of the measured devices performed with a mobility higher than $0.01 \mathrm{cm}^2/(\mathrm{V \cdot s})$. The wide-range distribution of mobility is attributed to the variability of the semiconductor-electrode interface nature and the local morphology of pentacene coverage on the channel.

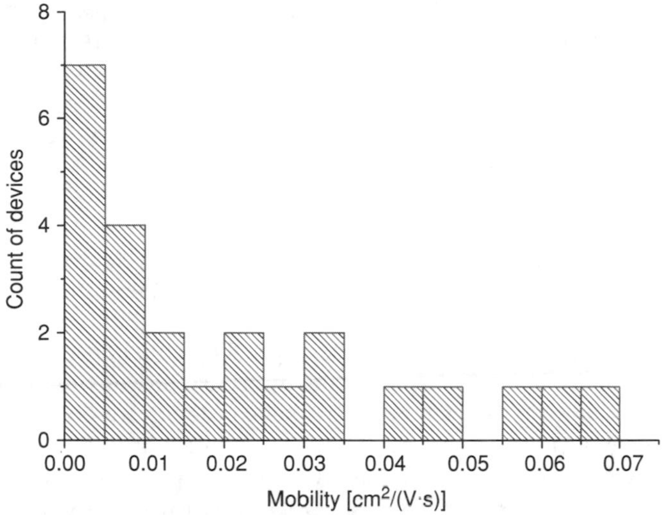

FIGURE 1.15 Statistics on the mobility of the measured devices (L = 8–35 nm). (*Reprinted with permission from Ref. 40. Copyright 2004, American Institute of Physics.*)

1.2 Scaling Behavior of Chemical Sensing with Organic Transistors

1.2.1 General Introduction to Organic Transistors for Sensing Applications

In recent years chemical sensing has become increasingly important not only from an industrial standpoint but also from a homeland security perspective. Sensors with the ability to detect chemicals, such as volatile organic compounds (VOCs), and biological species, including DNA and proteins, have been given more and more attention by the scientific and industrial communities. This is due to their great potential in an array of applications comprising manufacturing, transportation, environmental monitoring, process control, health care, homeland security, and national defense.[84–88] However, most of the commercial products available at present are bulky and costly and require long times for sampling and analyzing data. A novel scheme that employs portable and inexpensive devices with the advantage of fast data retrieval is desired. Miniaturization is demanded for all types of sensors because of the needs of better portability, higher sensitivity, lower power dissipation, and better device integration.[89, 90] There have been many types of sensors evaluated for this purpose, utilizing various techniques of transduction such as acoustic wave devices,[91, 92] electrochemical resistive sensors,[93, 94] and field-effect transistors with a chemically sensing gate (CHEMFETs).[95, 96] Among these sensing schemes, field-effect transistors have attracted more and more interest due to their ability to amplify in-situ, gate-modulate channel conductance, and allow for compatibility with well-developed microelectronic fabrication techniques that enable miniaturization. A simple resistive sensor probes only the change in bulk conduction, and simple capacitive sensors probe only the change in permittivity. Since the drain current in a FET reflects the transport through the two-dimensional electron gas (2DEG) at the semiconductor-dielectric interface, instead of the conduction through the bulk, a FET sensor would directly detect the effects on 2DEG transport caused by the analyte through the change in the drain current. The organic thin-film transistor (OTFT) is a promising sensor device for an electronic olfaction platform that possesses all the required features (sensitivity, reliability, and reproducibility) at low cost. Compared to CHEMFETs or chemiresistor sensors, an OTFT sensor can provide more information from changes in multiple parameters upon exposure to analyte, namely, the bulk conductivity of the organic thin film, the field-induced conductivity, the transistor threshold voltage, and the field effect mobility.[97]

In the second portion of this chapter we will first introduce polycrystalline organic and polymeric thin-film field-effect transistors and then cover such topics as the proper detection of sensing signals truly from nanoscale active area, the geometry (for device and material) dependence of the sensing behavior, and discussions for the sensing mechanisms in these sensors. We will also address several aspects of the interactions which produce sensing effects in electronic devices. The chemical sensors made of organic or conjugated polymeric transistors are operated at room temperature, which gives an advantage compared to inorganic oxide semiconductor sensors. Upside-down (see Fig. 1.3) OTFT sensors use the organic semiconductor active layer as the transducer, which interacts with airborne chemical species, referred to as analytes. This kind of structure provides analytes a direct access to the active semiconducting layer and enables the investigation of how the sensing behaviors depend on its morphology and interface properties. The interaction given by analytes directly affects the conductive channel of an OTFT sensor, unlike the sensors made of inorganic MOSFETs[98–101] or the insulated gate FETs (IGFETs use the polymer layer as the gate for a silicon FET[89]) where the sensing events occur at the gate or gate/insulator boundary and indirectly modulate the drain current through capacitive coupling. This means conductivity in the upside-down structure can be affected by changes in mobility (as well as changes in charge density/threshold) which is not possible in the other sensor configurations. These upside-down organic and polymer sensors can be refreshed by reverse-biasing the gate (a high positive voltage for p-channel, a high negative voltage for n-channel) to remove the trapped charges which result from the semiconductor/analyte interaction. Additionally, compared to inorganic semiconductor sensors, organic/polymeric transistors possess the advantages of being able to add specific functional groups on the semiconductor molecules/backbones able to selectively bind analytes, and the compatibility to incorporate small receptor molecules for better sensitivity and selectivity. Organic and polymeric field-effect transistors employing different active layers are able to detect a variety of analyte molecules with good stability and significant sensitivity.[102] Chemical detection is possible through direct semiconductor-analyte interactions, specific receptor molecules percolated in the semiconductor layer for selective analytes, varying the end/side groups of the semiconductor material, and controlling the thin-film morphology of the semiconductor layer. These advantages of OTFTs offer a basis to construct combinatorial arrays of sensors with different responses to the components of an odor mixture. Furthermore, gas-sensing complementary circuits and logic gates with OTFTs have been demonstrated.[103] These advances, in addition to circuitry for pattern recognition, could lead to an electronic nose.

1.2.2 Vapor Sensing in Micron-Sized Organic Transistors and Trapping at Grain Boundaries

The vapor-sensing behavior of an organic transistor depends on the morphological structure and interface properties of the device because analyte molecules are able to act at different interaction sites of the device and correspondingly modulate the overall conductance of the device. In previous work on micro-sized OTFTs, fabricated with a variety of active semiconductor layers,[102, 104] Crone et al. and Torsi et al. investigated the relation of vapor sensing to thin-film morphology upon exposure to different analyte molecules. A correlation of the vapor response characteristics to the length of end groups (flexibility at the molecular level) and grain size (porosity at the morphological level) of the semiconductor was demonstrated by observing the transient source-drain current under vapor flow and performing transmission electron microscopy (TEM) for morphological characterization.

These experiments have demonstrated that the sensing response of an OTFT channel to analytes of moderate dipole moments (e.g., alcohols) is enhanced with decreased grain size and looser molecular packing of the organic semiconductor layer. Smaller grains yield more grain boundaries which provide more interaction sites for sensing events. The response also becomes stronger with increasing film thickness, again due to the increased number of grain boundaries (the surface morphology becomes more structured as the films grow thicker from the flat and featureless ultrathin film). Analytes binding to the disordered and thinner grain boundaries are closer to the channel than those on the top surfaces of the grains and thus exert greater influence on the charge transport in channel. It was also found that the degree of sensing response increased as the length of the organic semiconductor's hydrocarbon end group increased. This is due to the elongated lamellar morphology and looser molecular packing which allow greater access of analyte vapor and increased surface area, as well as change the electronic or spatial barriers between grains. Therefore the alkyl chains facilitate the adsorption of the analyte molecules by the sensing film. This adsorption mechanism could be a combination of hydrophobic interactions, intercalation to fill defect vacancies, and simple surface binding. All these processes are favored at grain boundaries.

The interaction between the alcohol and the organic semiconductor film in OTFT (dHα6T) does not involve the bulk of the crystalline grains of the film since no change in refractive index and no swelling or thickness change of the film were observed with a single wavelength ellipsometer during exposure of the film to the analyte vapor.[104] It was suggested that the sensing interaction is a surface-type interaction involving grain boundaries. This result also helped to rule out a chemical reaction in the organic semiconductor upon exposure to the analyte. This is believed to be similar for CuPc and pentacene.[105] It is now beneficial to determine where physically on the device the current modulation occurs. The study performed by Torsi et al. also showed that there is very little penetration of the organic analytes into the highly ordered crystalline grains evidenced by the fact that no appreciable swelling was measured,[104] most likely due to the tight crystalline packing of the molecules in the grains[105] (in the case of conducting polymers, swelling is a known result of exposing them to organic vapors[104]). This would indicate that there is very likely little interaction, chemical or otherwise, between the molecules in the ordered grains and the organic vapors. In the same study, dHα6T was deposited on a quartz crystal microbalance, and a change in the mass of the semiconductor was measured when exposed to varying concentrations of 1-pentanol.[104] This demonstrated that although the 1-pentanol did not penetrate into the film, it nonetheless added to its mass. This means that the analyte adsorbed to the surface of the organic layer.[104] This leads to a conclusion that a decrease or increase in source-to-drain current must then be attributed to something that occurs outside of the crystalline grain. Another study by Someya et al. exploring the interaction between the semiconductor α,ω-dihexylquarterthiophene and 1-pentanol demonstrated that the magnitude of the sensing response was directly related to the number of grain boundaries in the channel when the interaction resulted in a decrease in I_{ds}.[106] Given that almost all the charge within the channel of an organic semiconductor lies close to the semiconductor/dielectric interface, it must be the case that the analytes percolate down to this dielectric interface through the grain boundaries.

The interaction between organic semiconductor and analyte molecules is still not completely understood. The adsorbed polar analyte could influence the charge transport in the channel by trapping some charges and consequently decrease the mobility of other mobile charges in channel. However, the major effect of the trapped or otherwise immobile charges at the dielectric interface is to induce a shift of threshold voltage in the transistor characteristics. For alcohol sensing, a constant threshold shift was observed at low gate voltages, which became a gate voltage-dependent mobility change at high gate voltages. It was also observed that a reverse gate bias can facilitate the restoration of drain current to near its original value. In many experiments, exposure of p-channel organic FETs to alcohols produced an initial increase in channel current (presumably the effect of dipoles) followed by a decrease (due to trapping). The initial increase may be due to dipoles in the polar analyte, inducing more charge in the channel.

1.2.3 Transition of Sensing Response by Organic Transistors from Micron-Scale to Nanoscale

There have been reports exploring the chemical sensing effects of organic and conjugated polymer transistors.[102, 103, 107–109] It has been shown that on large-scale devices ($L > 1$ μm), alkyl chain alcohol analytes interact with polythiophene thin films at grain boundaries and at the dielectric interface rather than the bulk of the films.[110] Recently, there have been reports on the role of the side chain in chemical sensing of polymer/oligomer based field-effect transistors.[109, 111] Other articles have delved into the dependence of the sensing effects on the channel length–grain size relationship in large-scale organic transistors.[112, 113] There is also an increasing need to develop sensors with very small active areas, not only for lower power requirement, but also for the possibility of higher sensitivity.[114] However, according to our experimental findings and analysis, scaling down the geometry of an OFET device is not a simple way, as expected, to enhance the sensitivity. Liang Wang et al.[115] first systematically investigated the scaling behavior of chemical sensing in OFETs with channel lengths ranging from microns to tens of nanometers and found that the sensing mechanism of where and how analyte molecules affect the electrical transport in an organic transistor becomes quite different when devices scaled from micron-scale to nanoscale dimensions. The main reason is that unlike the classic MOSFET with crystalline silicon channel, the electrical transport and chemical sensing behaviors of a polycrystalline OFET heavily depend on the morphology structure of the channel material and the properties of interfaces. For large-scale OFET devices, grain boundaries play the dominant role for both electrical transport and chemical sensing behaviors. For smaller channel lengths, the number of grain boundaries within a channel decreases so that the weight of influence of grain boundaries on electrical transport and chemical sensing reduces and other factors become more important. At smaller channel dimensions, especially when the channel length is comparable to or smaller than the grain size of polycrystalline organic molecules or conjugated polymers, we might possibly observe the electrical transport and chemical sensing behaviors within the body of grains which may exhibit a mechanism different from that in large-scale devices where grain boundaries dominate. In addition, contact barrier at the interface between electrode and semiconductor will play an important role in scaling since the resistance through the semiconductor channel becomes smaller. We believe it is the injection of current at the source/drain contacts that gets modulated by the analyte molecules.[116] Thus the behavior of nanoscale OFET sensors is markedly different from that of larger-channel-length devices.

In their study on the scaling behavior of chemical sensing in organic transistors,[115] Liang Wang et al. employed pentacene as the active channel responsible for both charge transport and chemical sensing, and 1-pentanol was employed as the analyte, because pentacene is a typical organic semiconductor due to its relatively high mobility and wide use in organic electronics and sensors and 1-pentanol is a prototypical alcohol analyte to represent the sensing behaviors of the alcohol group. The channel length of the device and grain size were both varied to investigate the role of scale in organic transistor sensing behaviors. The device structure (bottom-contact devices) and experimental setting are shown in Fig. 1.16. This configuration allows the organic semiconductor to be operated simultaneously as both the

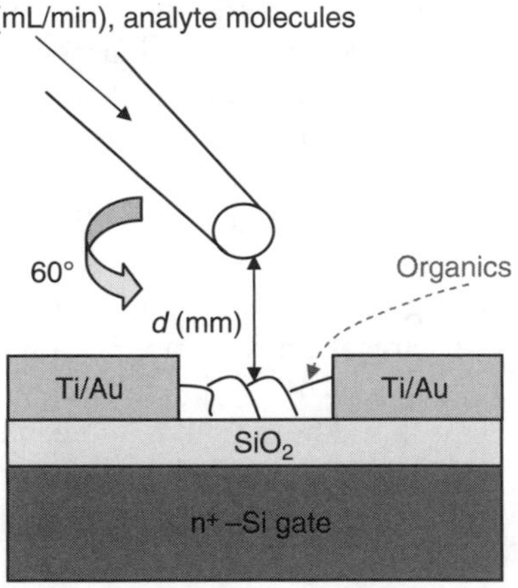

FIGURE 1.16 The schematic structure of a bottom-contact organic thin-film transistor used as chemical sensor. Its organic semiconductor channel, which serves as the sensing layer, is exposed to the analyte vapor delivered with a controlled flux through a carefully positioned syringe. (*Reprinted with kind permission from Springer Science and Business Media from Ref. 68*)

transistor channel and the sensing material which the analyte molecules can directly access. The transient variation of drain current I_{ds} in chemical sensing mode was examined under fixed gate (V_g) and source-drain voltage (V_{ds}) in air at room temperature, with saturation vapor of analyte molecules delivered to the proximity of organic-semiconductor channel surface. Currents were measured throughout the entire measurement period while the analyte was only delivered for a certain period of duration, to facilitate a comparison of the drain current behavior before, during, and after exposure to the analyte. Before every sensing measurement, each device was tested with air stream (the carrier to deliver various analytes), and no sensing or kinetic effect on drain current was detected. The currents in the absence of analyte continuously decrease over time due to bias stress effect. This is a well-known effect in OFETs in which the current slowly decreases with time at a fixed bias because the field-induced carriers (holes) fall into deep trap states where they are less mobile, which results in a reduction of drain current.[117,118] The sensor response is superposed on this background. After each sensing measurement, the device characteristics were recovered by using a reverse-bias configuration (high positive gate voltage with a small drain current flowing for 1 min).

With the design of side-guard electrodes as described in Sec. 1.1.2, we can investigate the transport behavior of OTFTs in truly nanoscale dimensions excluding the spreading currents in large area, without the difficulty to pattern the semiconductor layer into such a small size as the channel length. More importantly, with these devices as chemical sensors, the side-guarding function ensures that the active sensing area is truly nansocale. This novel four-terminal geometric design provides a powerful capability to investigate the sensing behavior within the nanoscale active area through the defined channel without any undesirable background from large-scale parasitic conduction pathways around the channel. To investigate the sensing response from the sensor active area in truly nanoscale geometries and the role of the side-guarding electrodes in the sensing measurement, the drain current transients upon exposure to 1-pentanol without and with applying side-guarding function, as shown in Fig. 1.17*a* and *b*, respectively, were recorded on a transistor with a 7 nm channel made by trapping bridge and break junction technique.[119, 120] Surprisingly, Fig. 1.17*a* and *b* manifested opposite sensing response directions. Without the side-guarding function, the transient shown in Fig. 1.17*a* represents the sensing behavior of the large area around the channel, which is a decrease in drain current by 20 to 25%. With the side-guarding function to eliminate most of the spreading current (two side-guarding electrodes biased at the same potential as the drain), the transient in the drain current (Fig. 1.17*b*) represents the true sensing behavior of the nanoscale active area, which is an increase in drain current by approximately a factor of 2. Figure 1.17*c* exhibits the

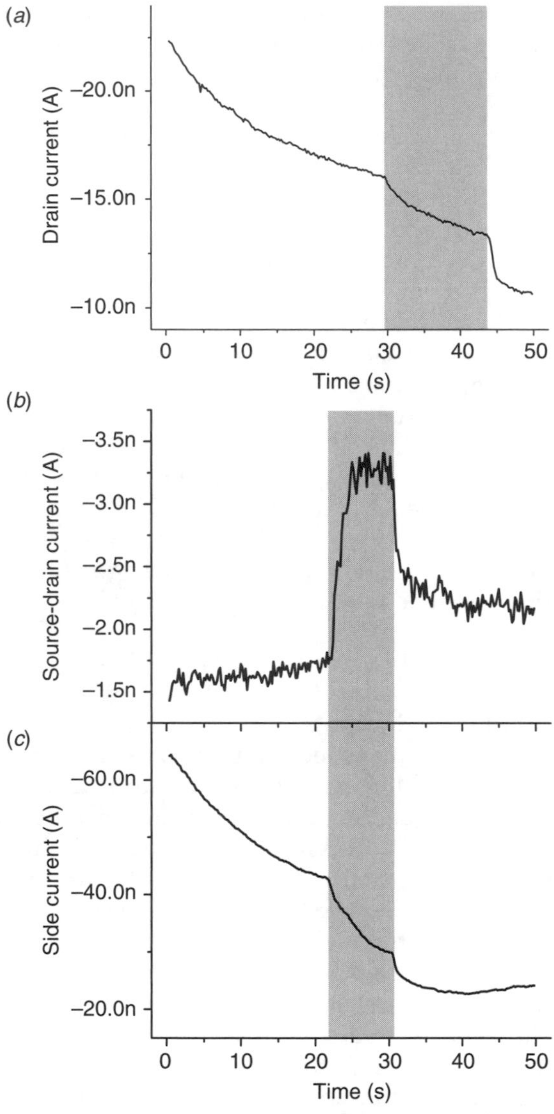

FIGURE 1.17 The sensing effects of a 7 nm channel P3HT transistors upon exposure to the saturated 1-pentanol vapor. (*a*) The sensing response at $V_g = -20$ V and $V_{ds} = -1.5$ V, without applying side-guarding electrodes. (*b*) and (*c*) Sensing responses of currents simultaneously collected at the drain and side electrodes, respectively, with side-guarding electrodes at the same potential as the drain to collect spreading current, at $V_g = -20$ V and $V_{ds} = V_{side} = -1.5$ V. (*Reprinted from Ref. 130. Copyright 2005, with permission from Elsevier*.)

variation of the side current (collected at side-guarding electrodes) corresponding to that of the drain current (Fig. 1.17*b*), which were measured simultaneously. The side current response to the analyte is markedly different compared to the drain current in Fig. 1.17*b*; namely, the side current shows a decrease in response to the analyte whereas the drain current simultaneously shows an increase upon exposure to the analyte. Since Fig. 1.17*b* and *c* was recorded simultaneously and showed different kinds of sensing behavior, it is evident that the side current is the spreading current traveling outside the defined channel and the current collected simultaneously at the drain is the direct current through the nanoscale channel. Therefore, utilizing the side-guarding function, we can eliminate most of the contribution from the spreading current and detect the true sensing response of a nanoscale channel. The experimental findings obtained with this unique side-guard function are presented below for the chemical sensing responses dependent on the scaling geometry, analyte delivery, as well as different type of analytes.

We investigated the response of I_{ds} (operated in saturation region) upon exposure to the saturated vapor of 1-pentanol, with a series of channel length and varied grain sizes of pentacene under the same experimental conditions. As shown in Fig. 1.18*a*, while the long-channel-length devices all exhibited a decrease in drain current upon delivery of the analyte, the small-channel-length devices showed an increase. There are two mechanisms influencing sensor behavior: one causes a decrease in current (dominant in large *L* devices) and the other causes an increase (dominant in small *L* devices). There is a crossover between these two types of response behavior which depends on grain size, occurring in the interval of channel length between 150 and 450 nm for ~80 nm grain size. Under the same condition, when the average grain size of pentacene is increased to 250 nm, the sensors exhibits the crossover behaviors at larger channel lengths (from 450 nm to 1 μm), as shown in Fig. 1.18*b*.

Figure 1.19 shows the SEM image (taken after all measurements) for geometric relation of the same channel length (150 nm) with different grain sizes of the pentacene layer. Figure 1.20*a* and *b* is the sensing responses of long-channel devices with pentacene grain sizes of 140 nm and 1 μm, respectively. For all devices with channel lengths of 2 μm or greater, I_{ds} manifested decreasing responses upon analyte delivery. The amplitudes of decreasing signal for 2 μm channels were smaller than those of longer channels. This effect is stronger with larger pentacene grains (Fig. 1.20*b*). These results are consistent with the reported work for sensing effects dependent on organic grain sizes and channel lengths in large scale.[104, 106] The sensing responses shown in Figs. 1.18 and 1.20 are reproducible for different devices with the same channel lengths and grain sizes, leading one to conclude that

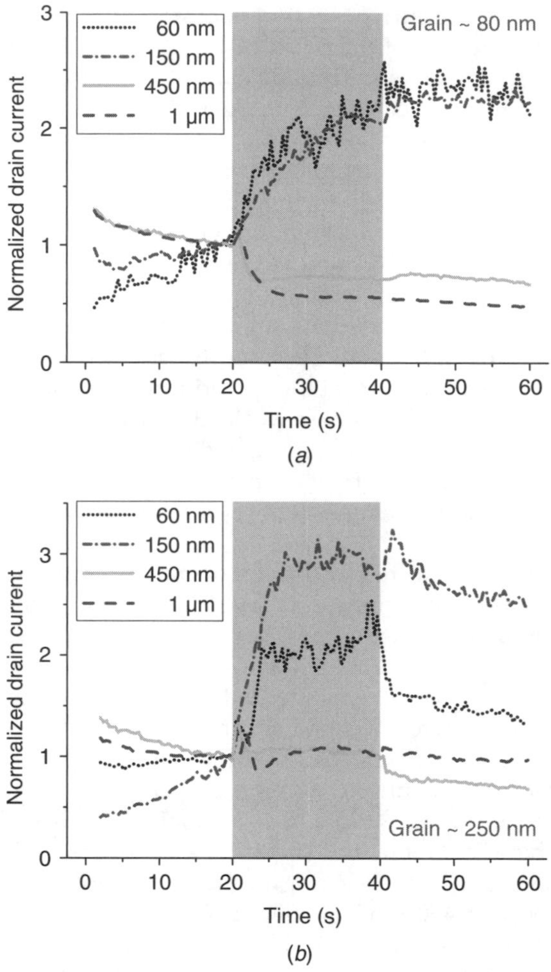

FIGURE 1.18 The sensing effects of nanoscale pentacene transistors upon exposure to 1-pentanol. (*a*) Sensing data of I_{ds} (normalized to that measured just before the analyte took effect) for 80 nm pentacene grain size and different nanoscale channel lengths (same $W/L = 10$), measured at $V_g = V_{ds} = V_{side} = -2.5$ V (two side guards were kept at the same potential as the drain), v (analyte flux) = 45 mL/min, d (distance from syringe nozzle to device) = 2 mm. (*b*) Sensing data of normalized I_{ds} for 250 nm pentacene grain size, measured at the same conditions as (*a*). (*Reprinted with permission from Ref. 115. Copyright 2004, American Institute of Physics.*)

the response of pentacene transistors to the 1-pentanol vapor changes from decreasing I_{ds} to increasing I_{ds}, when the channel length shrinks from micron to 100 nm scale, with a crossover happening in a transition interval of channel length which is related to the grain sizes of pentacene.

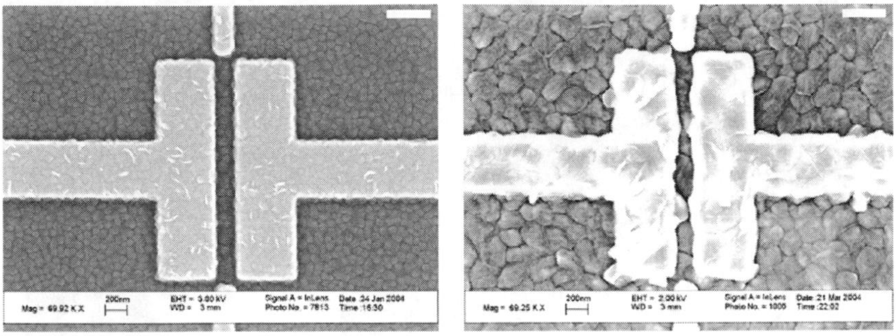

(*a*) (*b*)

FIGURE 1.19 SEM image taken after sensing measurements of a 150 nm channel with different average pentacene grain sizes of (*a*) 80 nm and (*b*) 250 nm, scale bar = 400 nm. The grains appearing in the figure are pentacene. (*Reprinted with permission from Ref. 115. Copyright 2004, American Institute of Physics.*)

The analyte flux (v) and the syringe nozzle-device distance (d) were varied to examine the influence of analyte delivery on the sensing responses. It turns out that for all the channel lengths and grain sizes, increasing v and decreasing d have similar influences, i.e., to increase the amplitude of the sensing signal. Figure 1.21 gives an example of this sensing test on a 22 nm channel with an average grain size of 80 nm, measured under operation in the linear region.

1.2.4 Discussions on the Scaling Behavior of Sensing Response: Role of Grain Boundaries and Contact

We discovered that by scaling down the device geometry to nanoscale dimensions, the sensing behavior is remarkably different from that of larger devices composed of the same materials for the same analyte. The direction and amplitude of sensing responses were found to be correlated to the channel length and the grain sizes of the organic semiconductor as sensing layer. These results follow the same trend as the reported work for sensing effects dependent on channel lengths relative to grain sizes in large-scale organic transistors.[104, 106] These organic and conjugated polymer thin-film field-effect transistors have some features of similarity with polycrystalline oxide semiconductor sensors. In both, grain boundaries play a key role in large-scale devices, and the analyte influences the electrical transport through grain boundaries and thereby modulates the channel conductivity. For large-scale transistors, in which a number of grain boundaries are located within a channel, the analyte molecules at grain boundaries play a dominant role in the sensing response, where they trap the mobile charge carriers in active channel and mainly result in a threshold voltage shift of transistor, which leads to a decrease in drain current.[102] For devices with smaller dimensions, there are fewer grain

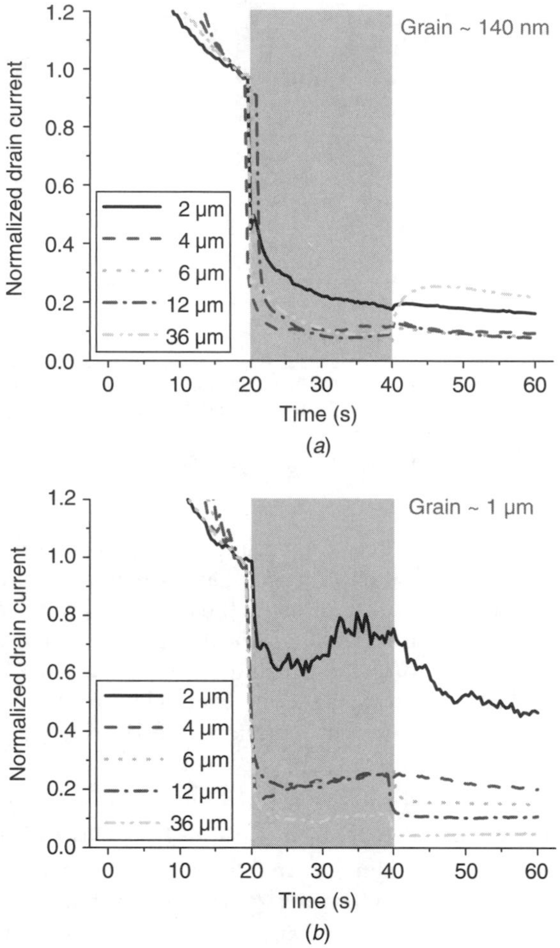

FIGURE 1.20 Sensing data of large-scale pentacene transistors upon exposure to 1-pentanol: normalized I_{ds} under the condition of $V_g = V_{ds} = -25$ V, $v = 45$ mL/min, and $d = 2$ mm for different microscale channel lengths, with average pentacene grain size of (*a*) 140 nm and (*b*) 1 μm, respectively. (*Reprinted with permission from Ref. 115. Copyright 2004, American Institute of Physics.*) (See also color insert.)

boundaries per channel so that the effect of decrease in drain current reduces. When the channel length is close to or smaller than the average grain size, the dominant factor is then a reduction of the contact injection barrier by the interaction from analyte, which leads to an increase in drain current. This is supported by the sensing experiments conducted in our group for the same channel material with contact metals of different work functions.[116] The overall sensing response is the result of a combination of these competing effects from different mechanisms: one causes a decrease in drain current and the other causes an increase, dominating at different length

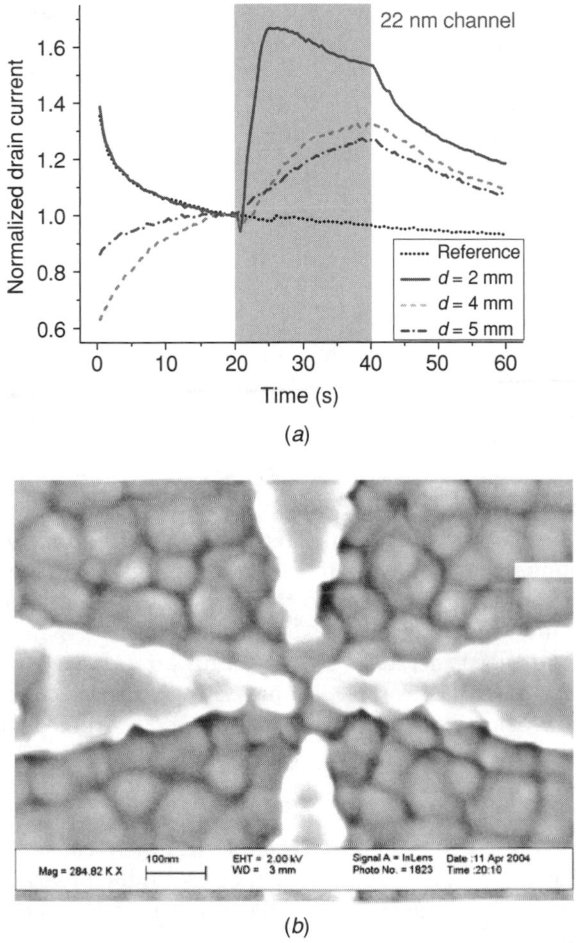

FIGURE 1.21 The effect of analyte delivery on nanoscale OTFT sensor. (*a*) Sensing data of a pentacene transistor of 22 nm channel in response to 1-pentanol, with $V_g = -2$ V, $V_{ds} = V_{side} = -0.4$ V and $v = 45$ mL/min for different *d* (nozzle-device distance), reference = absence of analyte. (*b*) SEM image of the device in (*a*) taken after measurements, grain ~80 nm, scale bar = 100 nm. The appearing grains are pentacene. (*Reprinted with permission from Ref. 115. Copyright 2004, American Institute of Physics*.)

scales. It is indicative of scale being a key element in the sensing process with organic transistor sensors.

A structural explanation of where the current modulation occurs therefore becomes clear. The decrease in current must be due to a phenomenon which occurs in the grain boundary, and an increase in current must be the result of a phenomenon which occurs at the contact (and also possibly an increase in carrier density in the channel due to dipole effects). Then it is necessary to understand whether the channel material is actually chemically reacting with the analyte or interacting with the analyte in another non-chemical way, such as an electrostatic interaction. There are examples of analytes which do chemically dope the semiconductor, such as electrophilic gases like NO_2, which remove electrons from CuPc and dope the material with holes.[121] The fact that the responses of many of the microscale organic semiconductor devices exposed to many of the organic vapors are reversible and reproducible reductions in current (not increases) would likely exclude any type of chemical reaction, including doping, as the predominant current modulation mechanism for most of these combinations. The result of no change in refractive index and no swelling or thickness change of the organic semiconductor that were observed with an ellipsometer upon exposure to the analyte[104] also suggested that the interaction between the organic semiconductor and the organic vapor is not a product of a chemical reaction.

In a unified picture, the chemical sensing effects at grain boundaries and metal-organic semiconductor contacts both arise from the dipole nature of analyte molecules. Due to its polaron nature, the charge transport in organic semiconductors is fairly sensitive to the local polar environment. Changes in the local crystal structure nearby charge carriers and thus changes to the polarizability of the lattice could drastically affect the local distribution of energy states. This problem was further exacerbated in the grain boundaries due to a large amount of disorder. Most of the analyte organic molecules used in this study have one thing in common: they all have dipoles (hexane does not, but it did not produce any appreciable response). The predominant mechanism that leads to a decrease in the magnitude of the current is increased trapping of carriers in the grain boundaries due to a modulation of the local electronic environment caused by the presence of the polar organic vapors (an increase in the polarizability of the semiconductor in the grain boundaries). An increased number of traps in the grain boundary would lead to an increase in the activation energy for hopping through the grain boundaries which was demonstrated by Sharma et al. using top-contact pentacene transistors exposed to ethanol.[116] The measured activation energy changed from 77 to 92 meV when the analyte concentration was changed from pure nitrogen to 100 ppm of ethanol.[102]

It has been experimentally shown that the analyte molecules of stronger dipole moments trigger stronger responses from the same OTFT chemical sensor.[122] Also results reported by Torsi et al. demonstrated the importance of the analyte's alkyl chain length in terms of its interaction with the organic transistor.[110] The longer the carbon chain length, the greater the interaction of that analyte molecule with the semiconductor and the higher the mass uptake. The same group also showed that increased mass uptake occurred when the side chain of the polythiophene derivative was made to be polar (by putting an ester in the side chain).[110] This enhancement of the mass uptake was even more pronounced than the increase produced by longer analyte carbon chains.[110] Therefore, in addition to the polar nature of analyte molecules, sensing response could be adjusted with the use of polar vs. nonpolar side chain moieties of the organic semiconductor.[110] For instance, ethanol showed no electrical response with the nonpolar side chain but did show a response with the polar side chain. This was most likely due to a poor interaction of the short ethanol alkyl chain to the nonpolar side chain and a much better interaction of the polar ethanol to the polar side chain.

In case of nanoscale OTFT sensors where the injection at contacts dominates charge transport, the increases in magnitude of the source-to-drain current upon exposure to the analyte is due to changes in the nature of how charge is injected into the channel. When carriers are injected into the semiconductor, they accumulate close to the contact interface and induce the image charge in the metal electrode and thus form an interface dipole.[123, 124] The polarity of the analytes can work to shield carriers in the semiconductor from the reverse electric field which results from the interface dipole. The larger the dipole moment in the analyte, the greater the shielding effect and thus the stronger the response in drain-current increase. Also, the smaller the channel length, the stronger this effect owing to greater domination of the injection-limited charge transport. For example, the injection barrier at contacts between Au and pentacene is attributed to the existence of interface dipoles. As Fig. 1.22*a* shows, due to the work function of Au (5.1 eV) and the electron affinity (2.6 eV)/energy gap (2.5 eV) of pentacene, the Fermi level of Au meets with the HOMO level of pentacene.[125, 126]

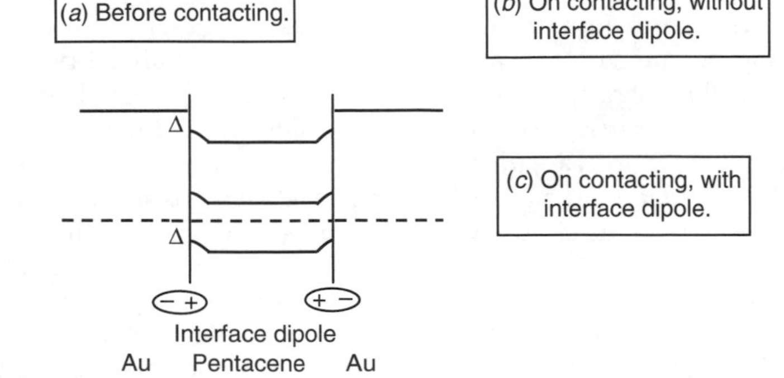

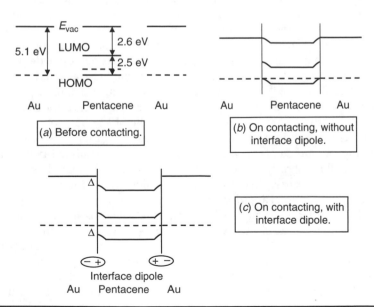

FIGURE 1.22 The diagram for the formation of injection barrier at contacts between Au and pentacene, due to the existence of interface dipoles.

Therefore on contacting each other, no injection barrier is expected between Au and pentacene, as shown in Fig. 1.22*b*. However, due to the accumulation of holes at the pentacene side of the contact, these positive charges and their image charges (negative) at Au side of contact form a type of interface dipole, and the interface dipoles shift the vacuum level at the Au side of the contact by an amount Δ.[123, 124] This shift in the vacuum level changes the band bending at the contact and forms a barrier to hole injection from Au into pentacene, as shown in Fig. 1.22*c*.

1.2.5 Sensor Response to Different Analytes and the Function of Receptors

It would be interesting and meaningful to investigate the scaling behavior of chemical sensing with organic semiconductor-analyte combinations other than pentacene and 1-pentanol. P3HT is a soluble conjugated polymer with relatively high mobility and thus has a great potential to be manufactured into chemical sensors with low-cost techniques such as ink-jet printing.[127–129] Similar to pentacene transistors, for micron scale or larger channel lengths, the drain currents of P3HT transistors decreased in response to the analyte 1-pentanol, whereas an increase in current was observed for nanoscale channel lengths.[130] Various types of analytes have been applied to investigate the chemical sensing responses of OTFTs. Among these analytes, vanillin is widely used in pharmaceuticals, perfumes, and flavors. P3HT transistors with a series of channel lengths in the submicron range were employed to examine the sensing responses of the conjugated polymer upon exposure to another type of analyte—vanillin. As shown in Fig. 1.23*a* for submicron channels, the 215 nm or larger channels exhibited a decreasing response in drain current upon delivery of vanillin, whereas the 125 nm or smaller channels behaved in the opposite direction, i.e., an increase in current as the sensing response. The crossover of response behavior exists in the interval of channel length 125 to 215 nm. Figure 1.23*b* shows the SEM image of a 75 nm channel taken before depositing P3HT. Based on the results for the sensing measurements of various channel lengths to analyte 1-pentanol and vanillin, there are two mechanisms influencing sensor behavior: one causes a decrease in current (dominant in large L devices) and the other causes an increase (dominant in small L devices).

By incorporating small receptor molecules, it is possible to enhance the sensitivity and selectivity of chemical sensing without the necessity of chemically editing the molecule of the organic semiconductor. The small receptor molecules can be incorporated by drop-casting onto the organic semiconductor layer from a solution of receptor molecules in chloroform after the fabrication of an OTFT, or directly mixing the solution with the solution containing the polymer chains and then depositing the mixture onto the prefabricated electrode pattern and

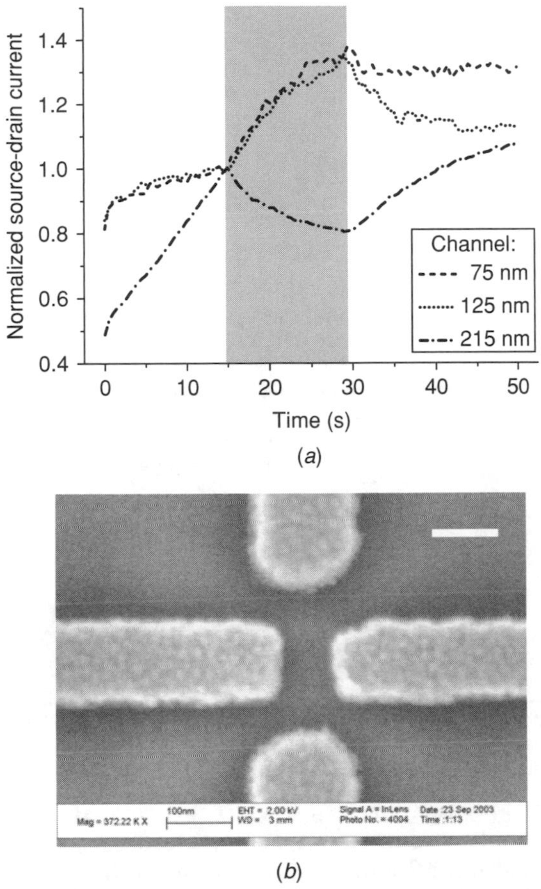

FIGURE 1.23 The sensing effects of P3HT transistors upon exposure to vanillin. (*a*) Sensing data with $V_g = -25$ V, $V_{ds} = V_{side} = -10$ V, and -15 V for $L < 100$ nm and $L > 100$ nm, respectively (two side guards were kept at the same potential as the drain), with different channel lengths and the same W/L of 3. (*b*) SEM image of a 75 nm channel taken before depositing P3HT, scale bar = 100 nm. (*Reprinted from Ref. 130. Copyright 2005, with permission from Elsevier.*)

forming the channel. This will enable the OTFT to select the specific analyte molecules among several candidates with a much more pronounced response. In an example of demonstration,[122] the sensing response of macro-scale P3HT transistors to the analyte vanillin was studied with and without incorporating the receptor (which we designate as Circle K) by means of depositing the solution mixture. As shown in Fig. 1.24, the response of neat P3HT transistor to vanillin vapor was ~30% decrease in drain current, while upon incorporation of receptor Circle K, the response was enhanced to ~72%.

The receptor Circle K was designed to form hydrogen bonds with the analytes. By investigating the chemical sensing behavior with

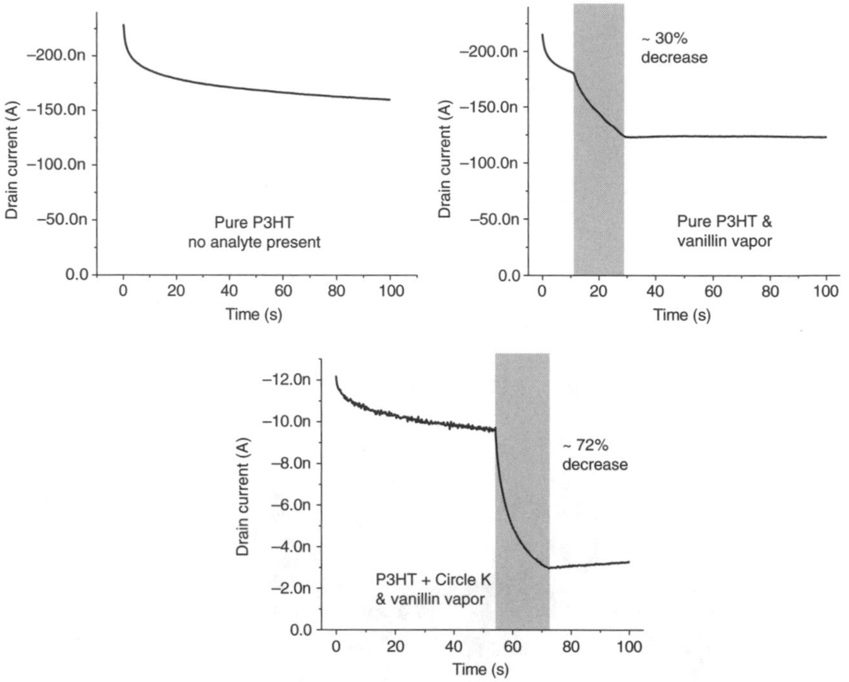

FIGURE 1.24 Another degree of freedom in chemical sensing: incorporation of receptor molecules. Mixing receptor with pure P3HT makes sensing response stronger. Channel length = 300 μm, channel width = 5 mm, gate dielectric is polyimide.

various analyte molecules and the receptor molecules of different type of hydrogen-bond roles (hydrogen donor or acceptor), incorporated with the organic semiconductor, the basic interaction between these receptors and the analyte molecules was determined to be most likely hydrogen bonding[122] instead of dipole-dipole interaction. When the very electronegative atoms of nitrogen, oxygen, or fluorine are covalently bound to a hydrogen atom, they draw a large amount of the electron density off the hydrogen, forming a very strong dipole.[131] Since hydrogen is a small atom with most of its electron density drawn off by its more electronegative covalently bound neighbor, it can get very close to the lone pair of electrons of the nitrogen, oxygen, or fluorine of the neighboring molecule and form a strong electrostatic interaction.[131] These bonds can also be slightly covalent since electrons can be shared between both of the electron withdrawing neighbors (O, N, and F) and the hydrogen in between them.[131] These bonds are not quite as strong as a full covalent bond but are much stronger than other interatomic forces such as van der Walls forces. The receptors have alkane side chains which make them soluble in organic solvents and allow them to permeate into the grain boundary. The amine (hydrogen donor in this case) groups designed into Circle K molecules could then form hydrogen bonds with the oxygen containing analytes (hydrogen acceptor) and significantly increase their interaction with (including the time that the analyte is held in the grain boundary) and their percolation depth into the grain boundaries of P3HT.

In an unpublished work,[122] it was demonstrated that the incorporation of receptor molecules not only can improve the sensitivity of chemical sensing, but also can show different magnitude and rate of response to the analyte molecules with different alkyl chain lengths or different polar strengths. This work also indicated that different receptor molecules differentiate the same group of analyte molecules in different manners. For example, for some analyte-semiconductor combinations, the receptor contributing three hydrogen bonds appears better in sensitivity enhancement than that contributing two hydrogen bonds. Furthermore this work showed that the OTFTs modified with a molecular-cage receptor can make distinctions between the size of the analyte molecule as well as the molecular position of the functional group (based on rate of change of the response and the extent of the response change). Devices modified with small molecule receptors showed the similar ability to be refreshed as unmodified OTFT sensors. The drain current of a receptor-modified OTFT sensor, however, did not fully recover to the original level, probably due to the strong and prolonged receptor/analyte binding. These devices also had lower mobilities than the neat OTFTs, due to the presence of the relatively unconductive receptors in the grain boundaries. A better understanding of the relationship between sensing response and mobility adjustment to receptor quantity, which has not been studied to date, will alleviate these issues to a large extent. Furthermore, the added flexibility which receptors offer in tuning the responses of organic semiconductors to various analytes allows for the selection of the semiconductors that possess the highest mobility as the most common host material making up a channel in an array of OTFT sensors. In spite of these challenges, it is still our prediction that fabricating an array of OTFTs with specific receptor incorporated, in combination with the circuitry for pattern recognition, could lead to an electronic nose. Small molecule receptors seem to be a very promising direction to pursue when attempting to further enhance the selectivity and sensitivity of the analyte/semiconductor interaction. These enhancements would reduce the need for full fingerprint pattern recognition and could do so without greatly increasing device fabrication complexity.

1.3 The Unified Picture of Scaling Behavior of Charge Transport and Chemical Sensor

The vapor sensing behavior of nanoscale organic transistors is different from that of large-scale devices due to the fact that electrical transport in an OTFT depends on its morphological structure and interface properties, and thereby analyte molecules are able to interact with

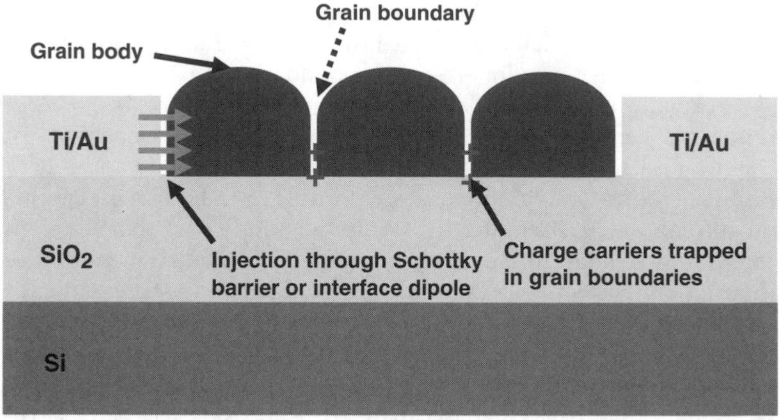

FIGURE 1.25 A graphic summary of the origins of the sensing response. The response arises because of trapping in the grain boundaries and/or the mediation of carrier injection at the contacts.

different parts of a device and in combination modulate the overall conductance of the device which gives rise to chemical sensing effect. Figure 1.25 outlined the sites in an OTFT device playing important roles in charge transport and thus chemical sensing, which become dominant at different scales. In a large-scale device (channel length much larger than the grain size of active semiconductor layer), the charge transport is dominated by the localized charges in tail states at grain boundaries due to high disorder there, and therefore the sensing mechanism is analyte dipole-induced charge trapping at grain boundaries, which generally leads to a decrease in device current upon exposure to a polar analyte. In some cases, the analyte dipoles result in a transient increase in channel conductivity, presumable because of more charges being induced in the channel. For smaller channel lengths, the number of grain boundaries within a channel decreases so that the weight of influence of grain boundaries on electrical transport and chemical sensing gives its place to organic semiconductor-metal interfaces. At smaller channel dimensions, especially when the channel length is comparable to or smaller than the grain size of polycrystalline organic molecules or conjugated polymers, the charge transport is dominated by the injection through the Schottky barrier and interface dipole at the metal-organic semiconductor contact as well as influenced by the local coverage of organic semiconductor material on the nanoscale channel, both of which overwhelm the intrinsic transport within the body of grains. Correspondingly during chemical sensing events in nanoscale OTFTs, the analyte molecules diffuse near the source/drain contacts through the porous semiconductor layer and modulate the charge carrier injection at the source and drain contacts. The smaller the channel length, the stronger this effect is. It is in this fashion that nanoscale organic transistors exhibit a remarkably different behavior in charge transport and chemical sensing from the microscale counterparts.

According to the mobility study under low temperatures down to 77 K,[68] the microscale transistors presented in the charge transport and chemical sensing work exhibited a transport mechanism of thermally activated hopping, which is mainly attributed to hopping at grain boundaries, following the relation as $\mu \sim \exp[-E_a/(kT)]$. As measured in an Arrhenius plot of temperature-dependent mobility under different gate bias for a pentacene transistor with 2 μm channel length,[68] the activation energy E_a (energy barrier at a grain boundary) is 130 meV under –5 V gate bias, and reduces to 52 meV under –30 V gate bias. The mobility in the OTFTs presented in this work increases with increasing gate bias, which is attributed to filling of the tail states of the density of states (DOS). Also the gate dependence is stronger at lower temperatures. These phenomena are well known for disordered organic field-effect transistors where the band tail of localized states has a much wider distribution at grain boundaries than within each grain body, due to the increased disorder at grain boundaries.[27–28] It would be meaningful to investigate the influence of the sensing event on the distribution of tail states through the combination of sensing experiments and temperature-dependent charge transport measurements. However, this requires much experimental care. Most VOCs possess a high vapor pressure so that the analytes will not remain at the sensing sites of the device for long enough time. Therefore the comparison of temperature dependence of transistors before and after sensing might not give an accurate view for what happened during the sensing action. Furthermore, performing temperature dependence experiments in the presence of analyte molecules might be difficult due to the condensation effect of VOCs at low temperatures.

References

1. H. Shirakawa, E. Louis, A. G. MacDiarmid, C. K. Chiang, and A. J. Heeger, *J. Chem. Soc. Chem. Commun.*, 16:578 (1977).
2. D. B. Mitzi, K. Chondroudis, and C. R. Kagan, *IBM J. Res. & Dev.*, 45(1):29–45 (Jan. 2001).
3. C. D. Dimitrakopoulos and D. J. Mascaro, *J. Res. & Dev.*, 45(1):11 (2001).
4. James R. Sheats, *J. Mater. Res.*, 19(7):1974–1989 (July 2004).
5. Tommie W. Kelley, Paul F. Baude, Chris Gerlach, David E. Ender, Dawn Muyres, Michael A. Haase, Dennis E. Vogel, et al., *Chem. Mater.*, 16:4413 (2004).
6. H. Koezuka, A. Tsumura, and T. Ando, *Synth. Met.*, 18:699 (1987).
7. J. L. Bredas, J. P. Calbert, D. A. da Silva Filho, and J. Cornil, *PNAS*, 99(9): 5804–5809 (Apr. 2002).
8. M. Kiguchi, M. Nakayama, T. Shimada, and K. Saiki, *Phys. Rev. B*, 71(3), Art. no. 035332 (Jan. 2005).
9. Hagen Klauk, Marcus Halik, Ute Zschieschang, Gunter Schmid, Wolfgang Radlik, and Werner Weber, *J. Appl. Phys.*, 92(9):5259 (2002).

10. D. J. Gundlach, Y. Y. Lin, T. N. Jackson, S. F. Nelson, and D. G. Schlom, *IEEE Electron Dev. Lett.*, 18(3):87–88 (1997).
11. Zhenan Bao, Ananth Dodabalapur, and Andrew J. Lovinger, *Appl. Phys. Lett.*, 69(26):4108 (1996).
12. H. Sirringhaus, N. Tessler, and Richard H. Friend, *Science*, 280:1741 (1998).
13. Jürgen Krumm, Elke Eckert, Wolfram H. Glauert, Andreas Ullmann, Walter Fix, and Wolfgang Clemens, *IEEE Electron Dev. Lett.*, 25(6):399 (June 2004).
14. A. Brown, R. A. Pomp, C. M. Hart, and D. M. de Leeuw, *Science*, 270:972 (1995); A. R. Brown, A. Pomp, D. M. de Leew, D. B. M. Klaassen, E. E. Havinga, P. T. Herwig, and K. Muellen, *J. Appl. Phys.*, 79:2136 (1996).
15. A. Afzali, C. D. Dimitrakopoulos, and T. L. Breen, *JACS*, 124(30):8812–8813 (2002).
16. Steven K. Volkman, Steven Molesa, Brian Mattis, Paul C. Chang, and Vivek Subramanian, *Mat. Res. Soc. Symp. Proc.*, 769, H11.7.1/L12.7.1, 2003.
17. N. Karl, J. Marktanner, R. Stehle, and W. Warta, *Synth. Met.* 42:2473 (1991).
18. W. Warta, R. Stehle, and N. Karl, *Appl. Phys. A: Solids Surf.* A36:163 (1985).
19. T. Holstein, *Ann. Phys. (N.Y.)*, 8:325 (1959).
20. D. Emin, *Phys. Rev. B*, 4:3639 (1971).
21. D. Emin, *Phys. Rev. Lett.*, 25:1751 (1970).
22. R. M. Glaser and R. S. Berry, *J. Chem. Phys.*, 44:3797 (1966).
23. S. Jeyadev and E. M. Conwell, *Phys. Rev. B*, 35(12):6253 (1987).
24. E. M. Conwell, H.-Y. Choi, and S. Jeyadev. *Synth. Met.*, 49–50:359–365 (1992).
25. S. Jeyadev and J. R. Schrieffer. *Phys. Rev. B*, 30:3620 (1984).
26. A. Salleo, T. W. Chen, A. R. Volkel, Y. Wu, P. Liu, B. S. Ong, and R. A. Street, *Phys. Rev. B*, 70:115311 (2004).
27. R. A. Street, J. E. Northrup, and A. Salleo, *Phys. Rev. B*, 71:165202 (2005).
28. Gilles Horowitz, Riadh Hajlaoui, and Philippe Delannoy, *J. Phys. III Fance*, 5:355–371 (1995).
29. Christopher R. Newman, Reid J. Chesterfield, Jeffrey A. Merlo, and C. Daniel Frisbie, *Appl. Phys. Lett.*, 85(3):422 (2004).
30. D. Knipp, R. A. Street, and A. R. Volkel, *Appl. Phys. Lett.*, 82(22):3907 (2003).
31. A. Dodabalapur, L. Torsi, and H. E. Katz, *Science*, 268(5208):270–271 (1995).
32. C. D. Dimitrakopoulos and P. R. L. Malenfant, *Adv. Mater.*, 14:99 (2002).
33. Y. M. Sun, Y. Q. Liu, and D. B. Zhu, *J. Mater. Chem.*, 15(1):53–65 (2005).
34. Hagen Klauk, Marcus Halik, Ute Zschieschang, Gunter Schmid, Wolfgang Radlik, and Werner Weber, *J. Appl. Phys.*, 92(9):5259 (2002).
35. Yuanjua Zhang, Jason R. Patta, Santha Ambily, Yulong Shen, Daniel C. Ralph, and George G. Malliaras, *Adv. Mater.*, 15(19):1632 (2003).
36. C. D. Dimitrakopoulos, S. Purushothaman, J. Kymissis, A. Callegari, and J. M. Shaw, *Science*, 283:822 (1999).
37. J. Collet, O. Tharaud, A. Chapoton, and D. Vuillaume, *Appl. Phys. Lett.*, 76(14):1941 (2000).
38. Yuanjua Zhang, Jason R. Patta, Santha Ambily, Yulong Shen, Daniel C. Ralph, and George G. Malliaras, *Adv. Mater.*, 15(19):1632 (2003); M. Leufgen, U. Bass, T. Muck, T. Borzenko, G. Schmidt, J. Geurts, V. Wagner, et al., *Synth. Met.*, 146:341–345 (2004).
39. Ch. Pannemann, T. Diekmann, and U. Hilleringmann, *Microelectronic Engg.*, 67–68:845–852 (2003); Michael D. Ausin and Stephen Y. Chou, *Appl. Phys. Lett.*, 81(23):4431–4433 (2002).
40. Liang Wang, Daniel Fine, Taeho Jung, Debarshi Basu, Heinz von Seggern, and Ananth Dodabalapur, *Appl. Phys. Lett.*, 85(10):1772 (2004); Josephine B. Lee, Paul Chang, J. Alex Liddle, and Vivek Subramanian, *IEEE Trans. Electron Dev.*, 52(8):1874–1879 (Aug. 2005).
41. Gilles Horowitz, Riadh Hajlaoui, and Philippe Delannoy, *J. Phys. III Fance*, 5:355–371 (1995); Gilles Horowitz, Mohsen E. Hajlaoui, and Riadh Hajlaoui, *J. Appl. Phys.*, 87(9):4456–4463 (2000).
42. S. F. Nelson, Y.-Y. Lin, D. J. Gundlach, and T. N. Jackson, *Appl. Phys. Lett.*, 72(15):1854–1856 (1998).
43. Eric L. Granstrom and C. Daniel Frisbie, *J. Phys. Chem. B*, 103:8842–8849 (1999); Anna B. Chwang and C. Daniel Frisbie, *J. Phys. Chem. B*, 104:12202 (2000).

44. J. Takeya, C. Goldmann, S. Haas, K. P. Pernstich, B. Ketterer, and B. Batlogg, *J. Appl. Phys.*, 94(9):5800–5804 (2003); Christopher R. Newman, Reid J. Chesterfield, Jeffrey A. Merlo, and C. Daniel Frisbie, *Appl. Phys. Lett.*, 85(3): 422–424 (2004).
45. V. Y. Butko, X. Chi, D. V. Lang, and A. P. Ramirez, *Appl. Phys. Lett.*, 83(23): 4773–4775 (2003).
46. P. F. Newman and D. F. Holcomb, *Phys. Rev. Lett.*, 51(23):2144 (1983).
47. David J. Gundlach, LiLi Jia, and Thomas N. Jackson, *IEEE Electron Dev. Lett.*, 22(12):571–573 (2001).
48. J. Takeya, C. Goldmann, S. Haas, K. P. Pernstich, B. Ketterer, and B. Batlogg, *J. Appl. Phys.*, 94(9):5800–5804 (2003).
49. A. Dodabalapur, L. Torsi, and H. E. Katz, *Science*, 268(5208):270–271 (1995).
50. S. M. Sze, *Physics of Semiconductor Devices*, 2d ed., New York: Wiley, 1982, pp. 451–453.
51. V. Podzorov, E. Menard, A. Borissov, V. Kiryukhin, J. A. Rogers, and M. E. Gershenson, *Phys. Rev. Lett.*, 93(8):086602 (Aug. 2004).
52. Gilles Horowitz, Riadh Hajlaoui, and Philippe Delannoy, *J. Phys. III Fance*, 5:355–371 (1995).
53. N. Karl, *Synth. Met.*, 133–134:649–657 (2003).
54. J. Takeya, C. Goldmann, S. Haas, K. P. Pernstich, B. Ketterer, and B. Batlogg, *J. Appl. Phys.*, 94(9):5800–5804 (2003).
55. Christopher R. Newman, Reid J. Chesterfield, Jeffrey A. Merlo, and C. Daniel Frisbie, *Appl. Phys. Lett.*, 85(3):422–424 (2004).
56. P. M. Borsenberger and E. H. Magin, *J. Sci, Physica B*, 217:212–220 (1996).
57. E. J. Meijer, D. B. A. Rep, D. M. de Leeuw, M. Matters, P. T. Herwig, and T. M. Klapwijk, *Synth Met.*, 121:1351–1352 (2001).
58. L. Torsi, A. Dodabalapur, and H. E. Katz, *J. Appl. Phys.*, 78(2):1088 (1995).
59. Anna B. Chwang and C. Daniel Frisbie, *J. Phys. Chem. B*, 104:12202–12209 (2000).
60. Liang Wang, Daniel Fine, Debarshi Basu, and Ananth Dodabalapur, *J. Appl. Phys.*, 101:054515 (2007).
61. N. Karl, J. Marktanner, R. Stehle, and W. Warta, *Synth. Met.*, 41–43:2473 (1991).
62. W. Warta, R. Stehle, and N. Karl, *Appl. Phys. A: Solids Surf.*, A36:163 (1985).
63. N. Karl, *Synth. Met.*, 133–134:649–657 (2003).
64. L. B. Schein, A. Peled, and D. Glatz, *J. Appl. Phys.*, 66(2):686–692 (1989).
65. J. Frenkel, *Phys. Rev.*, 54:647–648 (1938).
66. W. D. Gill, *J. Appl. Phys.*, 43(12):5033–5040 (1972).
67. Paul J. Freud, *Phys. Rev. Lett.*, 29(17):1156–1159 (1972).
68. Liang Wang, Daniel Fine, Deepak Sharma, Luisa Torsi, and Ananth Dodabalapur, *Anal. Bioanal. Chem.*, 384(2):310–321 (Dec. 2005).
69. Martin Pope and Charles E. Swenberg (eds.), *Electronic Processes in Organic Crystals and Polymers*, Oxford University Press, Oxford, 1999, pp. 343–369.
70. Hagen Klauk, Marcus Halik, Ute Zschieschang, Gunter Schmid, Wolfgang Radlik, and Werner Weber, *J. Appl. Phys.*, 92(9):5259 (2002).
71. J. Collet, O. Tharaud, A. Chapoton, and D. Vuillaume, *Appl. Phys. Lett.*, 76(14):1941 (2000).
72. M. Leufgen, U. Bass, T. Muck, T. Borzenko, G. Schmidt, J. Geurts, V. Wagner, et al., *Synth. Met.*, 146:341–345 (2004).
73. Yuanjua Zhang, Jason R. Patta, Santha Ambily, Yulong Shen, Daniel C. Ralph, and George G. Malliaras, *Adv. Mater.*, 15(19):1632 (2003).
74. Ch. Pannemann, T. Diekmann, and U. Hilleringmann, *Microelectronic Engg.*, 67–68:845–852 (2003).
75. Michael D. Ausin and Stephen Y. Chou, *Appl. Phys. Lett.*, 81(23):4431–4433 (2002).
76. N. Koch, A. Kahn, J. Ghijsen, J.-J. Pireaux, J. Schwartz, R. L. Johnson, and A. Elschner, *Appl. Phys. Lett.*, 82(1):70 (2003).
77. P. G. Schroeder, C. B. France, J. B. Park, and B. A. Parkinson, *J. Appl. Phys.*, 91(5):3010 (2002).
78. N. J. Watkins, Li Yan, and Yongli Gao, *Appl. Phys. Lett.*, 80(23):4384 (2002).
79. S. M. Sze, *Physics of Semiconductor Devices*, 2d ed., New York: Wiley, 1982, pp. 520–527, 469–486.

80. J. Takeya, C. Goldmann, S. Haas, K. P. Pernstich, B. Ketterer, and B. Batlogg, *J. Appl. Phys.*, 94(9):5800 (2003).
81. Gilles A. de Wijs, Christine C. Mattheus, Robert A. de Groot, and Thomas T. M. Palstra, *Synth. Met.*, 139:109 (2003).
82. H. Sirringhaus, N. Tessler, and Richard H. Friend, *Science*, 280:1741 (1998).
83. G. Horwitz, M. E. Hajlaoui, and R. Hajlaoui, *J. Appl. Phys.*, 87(9):4456 (2000).
84. J. W. Gardner, and P. N. Bartlett, *Electronic Noses: Principles and Applications*, Oxford University Press, Oxford, 1999.
85. T. C. Pearce, S. S. Schiffman, H. T. Nagle, and J. W. Gardner (eds.), *Handbook of Machine Olfaction*, Wiley-VCH, Weinheim, 2003.
86. S. Ampuero, and J. O. Bosset, *Sensors and Actuators B*, 94:1 (2003).
87. E. R. Thaler, D. W. Kennedy, and C. W. Hanson, *Am. J. Rhinol.*, 15:291 (2001).
88. K. J. Albert, N. S. Lewis, C. L. Schauer, G. A. Sotzing, S. E. Stitzel, T. P. Vaid, and D. R. Walt, *Chem. Rev.*, 100:2595 (2000).
89. J. Janata, *Proc. IEEE*, 91(6):864–869 (June 2003).
90. D. Vincenzi, M. A. Butturi, V. Guidi, M. C. Carotta, G. Martinelli, V. Guarnieri, S. Brida, et al., *Sensors and Actuators B*, 77:95 (2001).
91. D. S. Ballantine, R. M. White, S. J. Martin, A. J. Ricco, E. T. Zellers, G. Frye, and H. Wohltjen, *Acoustic Wave Sensors: Theory, Design, and Physicochemical Applications*, San Diego: Academic Press, 1997.
92. N. Cioffi, I. Farella, L. Torsi, A. Valentini, L. Sabbatini, and P. G. Zambonin, *Sensors and Actuators B*, 93:181–186 (2003).
93. D. Tyler McQuade, Anthony E. Pullen, and Timothy M. Swager, *Chem. Rev.*, 100:2537–2574 (2000).
94. J. Wollenstein, J. A. Plaza, C. Cane, Y. Min, H. Bottner, and H. L. Tuller, *Sensors and Actuators B*, 93:350 (2003).
95. G. Haranyi, *Polymer Films in Sensor Applications*, Technomic Publishing Co., Lancaster, Penn., 1995, p. 58; B. R. Eggins, *Biosensors: An Introduction*, Wiley, 1996.
96. M. Burgmair, M. Zimmer, and I. Eisele, *Sensors and Actuators B*, 93:271–275 (2003).
97. L. Torsi, A. Dodabalapur, L. Sabbatini, and P. G. Zambonin, *Sensors and Actuators B*, 67:312–316 (2000).
98. H. Wingbrant, I. Lundstrom, and A. Lloyd Spetz, *Sensors and Actuators B*, 93:286–294 (2003).
99. Danick Briand, Helena Wingbrant, Hans Sundgren, Bart van der Schoot, Lars-Gunnar Ekedahl, Ingemar Lundstrom, and Nicolaas F. de Rooij, *Sensors and Actuators B*, 93:276–285 (2003).
100. H. L. Tuller and R. Mlcak, *Curr. Opin. Solid State Mat. Sci.*, 3:501 (1998).
101. N. J. P. Paulsson and F. Winquist, *Forensic Sci. Int.*, 105:95 (1999).
102. B. Crone, A. Dodabalapur, A. Gelperin, L. Torsi, H. E. Katz, A. J. Lovinger, and Z. Bao, *Appl. Phys. Lett.*, 78(15):2229 (2001).
103. B. K. Crone, A. Dodabalapur, R. Sarpeshkar, A. Gelperin, H. E. Katz, and Z. Bao, *J. Appl. Phys.*, 91(12):10140 (June 2002).
104. L. Torsi, A. J. Lovinger, B. Crone, T. Someya, A. Dodabalapur, H. E. Katz, and A. Gelperin, *J. Phys. Chem. B*, 106:12563 (2002).
105. Luisa Torsi and Ananth Dodabalapur, "Organic thin-film transistors as plastic analytical sensors," *Anal. Chem.*, 77(19): 380A–387A, (Oct. 1, 2005).
106. Takao Someya, Howard E. Katz, Alan Gelperin, Andrew J. Lovinger, and Ananth Dodabalapur, "Vapor Sensing with α,ω-Dihexylquarterthiophene Field-Effect Transistors: The Role of Grain Boundaries," *Appl. Phys. Lett.*, 81(16):3079–3081 (Oct. 14, 2002).
107. Jiri Janata and Mira Josowicz, *Nature Mater.*, 2:19 (Jan. 2003).
108. L. Torsi, N. Cioffi, C. Di Franco, L. Sabbatini, P. G. Zambonin, and T. Bleve-Zacheo, *Solid-State Electronics*, 45:1479 (2001).
109. Luisa Torsi, M. Cristina Tanese, Nicola Cioffi, Maria C. Gallazzi, Luigia Sabbatini, and P. Giorgio Zambonin, *Sensors and Actuators B*, 98:204 (2004).
110. L. Torsi, A. Taifuri, N. Cioffi, M. C. Gallazzi, A. Sassella, L. Sabbatini, and P. G. Zambonin, *Sensors and Actuators B*, 93:257 (2003).

111. L. Torsi, M. C. Tanese, N. Cioffi, M. C. Gallazzi, L. Sabbatini, P. G. Zambonin, G. Raos, et al., *J. Phys. Chem. B*, 107:7589 (2003).
112. Takao Someya, Howard E. Katz, Alan Gelperin, Andrew J. Lovinger, and Ananth Dodabalapur, *Appl. Phys. Lett.*, 81(16):3079 (2002).
113. L. Torsi, A. J. Lovinger, B. Crone, T. Someya, A. Dodabalapur, H. E. Katz, and A. Gelperin, *J. Phys. Chem. B*, 106:12563 (2002).
114. Marc J. Madou and Roger Cubicciotti, *Proc. IEEE*, 91(6):830 (June 2003).
115. Liang Wang, Daniel Fine, and Ananth Dodabalapur, *Appl. Phys. Lett.*, 85:6386 (2004). Copyright 2004, American Institute of Physics.
116. Deepak Sharma and Ananth Dodabalapur, unpublished result.
117. M. Matters, D. M. de Leeuw, P. T. Herwig, and A. R. Brown, *Synth. Met.*, 102:998 (1999).
118. H. L. Gomes, P. Stallinga, F. Dinelli, M. Murgia, F. Biscarini, D. M. de Leeuw, T. Muck, et al., *Appl. Phys. Lett.*, 84(16):3184 (2004).
119. Saiful I. Khondaker and Zhen Yao, *Appl. Phys. Lett.*, 81(24):4613 (2002).
120. Liang Wang, Taeho Jung, Daniel Fine, Saiful I Khondaker, Zhen Yao, Heinz von Seggern, and Ananth Dodabalapur, *Proc. 3d IEEE Conf. Nanotechnol.*, 2:577–580 (2003).
121. Marcel Bouvet, "Phthalocyanine-Based Field-Effect Transistors as Gas Sensors," *Anal. Bioanal. Chem.*, 384:366–373 (2006).
122. Daniel Fine, Liang Wang, Debarshi Basu, and Ananth Dodabalapur, unpublished material.
123. Hisao Ishii and Kazuhiko Seki, *IEEE Trans. Electron Dev.*, 44(8):1295 (1997).
124. Hisao Ishii, Kiyoshi Sugiyama, Eisuke Ito, and Kazuhiko Seki, *Adv. Mater.*, 11(8):605 (1999).
125. T. Li, J. W. Balk, P. P. Ruden, I. H. Campbell, and D. L. Smith, *JAP*, 91(7):4312 (2002).
126. E. A. Silinsh and V. Capek, *Organic Molecular Crystals*, AIP, New York, 1994.
127. H. Sirringhaus, P. J. Brown, R. H. Friend, M. M. Nielsen, K. Bechgaard, B. M. W. Langeveld-Voss, A. J. H. Spiering, et al., *Nature*, 401(6754):685–688 (1999).
128. H. Sirringhaus, T. Kawase, R. H. Friend, T. Shimoda, M. Inbasekaran, W. Wu, and E. P. Woo, *Science*, 290(5499):2123–2126 (2000).
129. Beng S. Ong, Yiliang Wu, Ping Liu, and Sandra Gardner, *J. Am. Chem. Soc.*, 126(11):3378 (2004).
130. Liang Wang, Daniel Fine, Saiful I. Khondaker, Taeho Jung, and Ananth Dodabalapur, *Sensors and Actuators B*, 113(1):539–544 (2005).
131. David W. Oxtoby, H. P. Gillis, and Norman H. Nachtrieb, *Principles of Modern Chemistry*, 4th ed., Fort Worth: Saunders College Publishing, 1999, pp. 147–148.

CHAPTER 2

Organic Thin-Film Transistors for Inorganic Substance Monitoring

L. Torsi,[*,†] N. Cioffi,[*,†] F. Marinelli,[*] E. Ieva,[*] L. Colaianni,[*] A. Dell'Aquila,[‡] G. P. Suranna,[‡] P. Mastrorilli,[‡] S. G. Mhaisalkar,[§] K. Buchholt,[¶] A. Lloyd Spetz,[¶] and L. Sabbatini[*,†]

2.1 Inorganic Substance Monitoring for Early Diagnosis

The relevance in monitoring toxic volatile analytes arises from the increasing human activities that, especially in the last decades, caused the emission into the atmosphere of a wide range of pollutants at high concentration levels. Among the inorganic gaseous pollutants, nitrogen oxides (NO, NO_2), carbon monoxide (CO), and hydrogen sulfide (H_2S) are enumerated as the most toxic gases with very harmful effects on human health. In addition to their toxic effect on the human body, such gases are produced by specific cells at low concentrations, and their antiviral and immunodefensive

[*]Dipartimento di Chimica, Università degli Studi di Bari, Bari (Italy).
[†]Centro di Eccellenza TIRES, Università degli Studi di Bari, Bari (Italy).
[‡]Department of Water Engineering and of Chemistry, Polytechnic of Bari, Bari (Italy).
[§]School of Materials Science and Engineering, Nanyang Technological University, Singapore.
[¶]Department of Physics, Chemistry and Biology, Linköping University, Linköping, Sweden.

role in a wide range of inflammatory illnesses has been recently highlighted.[1–3] In the following, a brief overview of both the toxic and physiological role of the above-mentioned inorganic gases is reported.

Carbon monoxide is a colorless and odorless gas produced during the incomplete combustion of organic fuels (natural gas, wood, petrol, and charcoal). In urban context, motor vehicle emissions are the primary source of CO, and about 90% of the CO released in such condition is due to road traffic. Average carbon monoxide concentration in the outdoor air is about 100 ppb, but it may reach up to 200 ppm in the urban areas during rush hours.

The toxicity of carbon monoxide resides in its greater capability to bond hemoglobin with respect to the oxygen, eventually reducing the amount of oxygen that human cells need for mitochondrial respiration. Progressive exposures to CO can result in fatigue, angina, reduced visual perception, reduced dexterity, and finally death. The elderly, children, and people with preexisting respiratory diseases are particularly susceptible to carbon monoxide pollution. The recommended exposure limit set by the Occupational Safety and Health Administration (OSHA) is 50 ppm during a typical 8 h day with a ceiling level of 200 ppm.

Despite the toxic effects, CO has a potential protective role against oxidative stress.[4, 5] CO is produced endogenously in the human body by a class of enzymes known as heme oxygenase (HO-1 and HO-2).[6] HO enzymes are activated by various stimulant factors including proinflammatory cytokines and nitric oxide.[7] HO-1 catalyzes the initial and rate-determining step in the oxidative degradation of heme to the antioxidant bilirubin. Employing NADPH (reduced nicotinamide adenine dinucleotide phosphate) and molecular oxygen, the HO-1 enzyme cleaves a meso carbon of the heme molecule, producing biliverdin, free iron, and CO. Biliverdin is subsequently converted to bilirubin by the bilirubin reductase. CO produced in the bilirubin synthesis has several biological activities including stimulation of the guanylate cyclase, which eventually activates the immunodefensive mechanisms against inflammatory diseases. CO is normally present in the exhaled air of healthy subjects at detectable concentrations of 1 to 3 ppm,[8] while it increases up to 3 to 7 ppm in patients with inflammatory pulmonary diseases such as bronchial asthma, bronchiectasis, upper respiratory tract infections, and seasonal allergic rhinitis.[3, 8] Treatments with inhaled and oral corticosteroids, which have been shown to reduce airway inflammation, have been associated with a reduction in the exhaled levels of CO in asthma patients.[9] Based on these findings, it has been proposed that the measurements of CO concentration in exhaled air may serve as an indirect marker of airway inflammation. Moreover, recent reports demonstrated an increase of exhaled CO concentrations also in other diseases such as cystic fibrosis and diabetes mellitus.[10–11]

Nitrogen oxide (NO) and nitrogen dioxide (NO_2), usually termed NO_x, enter the atmosphere from polluting sources as well as from natural sources such as lightning and biological trials. NO_x gases of anthropic origin mainly arise from the combustion of fossil fuels usually used to feed vehicle engine and from house heating. NO_x cause several detrimental effects on the environment such as the photochemical smog [a mixture of nitric acid, NO_x, inorganic and organic nitrates, peroxyacetylnitrate (PAN), ozone, and other reactive oxygen species], and acid rains. NO_x inhalation may result in several pulmonary inflammations, and a constant exposure to 500 ppm of such gases may cause death within 2 to 10 days. The allowed exposure limit given by OSHA for NO is 25 ppm (averaged throughout an 8 h exposure), while for NO_2 a 5 ppm limit is set as a ceiling level. For both gases, however, the alarm threshold is lower than 1 ppm, and Italian legislation set their concentration limit at 100 and 200 ppb for NO and NO_2, respectively.

As previously discussed for CO, nitrogen oxide is also produced in the human body (lung), and it is involved in several metabolic and inflammatory processes. NO is synthesized endogenously by NO synthases (NOS).[12] Three isoforms of NOS, which are products of three separate genes, have been identified including neuronal (nNOS or NOS I), endothelial (eNOS or NOS III), and iNOS (or NOS II). These enzymes convert L-arginine to NO and L-citrulline in a reaction requiring oxygen, NADPH, and many cofactors.[13]

NO produced mainly by NOS II plays a key role in immunodefense and antiviral mechanisms of airway epithelium cells, and alteration of its concentration in exhaled breath is a signal of a pathological status. Indeed, a concentration increase of exhaled NO has been demonstrated to be related to chronic inflammatory airway diseases such as asthma, bronchiectasis, and other respiratory infections including bacterial and viral illnesses.[14–16] Typically, NO concentration in exhaled breath of healthy subjects ranges from 5 to 10 ppb while for asthmatic patients it falls in the range of 50 to 100 ppb.[17–18] On the other hand, a much too low NO concentration has been related to the human immunodeficiency virus (HIV-1) and to cystic fibrosis, thus suggesting the existence of viral mechanisms that suppress the NO-related host defense.[14, 19, 20] More recently, the potential therapeutic role of NO gas in cancer treatment has also been investigated.[21]

Hence, NO can be considered as a signaling molecule of the inflammatory response to viruses. Such an immunodefense action is due to the NO activation of the guanylate cyclase[22] to produce guanosine 3′,5′-cyclic monophosphate (cGMP) that acts as a second messenger and relaxes the airway smooth muscles. Moreover, NO has an antiviral effect[23, 24] since it is able to activate specific defensive enzymes through protein modification processes such as nitration of tyrosine and nitrosylation of thiols.[25, 26] Unlike NO, NO_2 is not involved in any physiological mechanism in the human body.

Hydrogen sulfide (H_2S) is one of the most dangerous chemicals; it is a colorless gas with a strong rotten egg smell at concentrations lower than 30 ppm and a sickeningly sweet odor at concentrations up to 100 ppm. H_2S is an asphyxiant gas since it causes paralysis of the nerve centers responsible for the brain-controlled breathing. Exposure to 1000 ppm H_2S produces rapid paralysis of the respiratory system, cardiac arrest, and death within few minutes. In addition to the effects on the respiratory system, at lower concentrations (in the range of 20 to 150 ppm) it causes irritation of the eyes. Slightly higher concentrations may cause irritation of the upper respiratory tract and pulmonary edema for prolonged exposure time. The concentration limit (OSHA) for such toxic gas is 20 ppm, averaged on 10 min of exposure time.

In spite of its dramatic effects on human health, H_2S is well-known to be naturally synthesized in mammalian tissue from L-cysteine by two different enzymes: cystathionine-g-lyase (CSE) and cystathionine-b-synthetase (CBS).[27] Recent studies have shown the important role of H_2S in several inflammatory states such as acute pancreatitis,[28] diabetes mellitus,[29, 30] chronic obstructive pulmonary diseases (COPD),[31] and many other lung injuries. Unlike NO and CO, H_2S relaxes the vascular tissues without the activation of the cGMP pathway.[32] Indeed it exhibits a potent vasodilator activity via the activation of the K_{ATP} channels in the vascular smooth muscles.[33, 34] Nevertheless, the mechanism of the K_{ATP} channel activation as well as the H_2S antiviral and antiinflammatory properties are still unclear, and both require further investigations.

The discovery of biological functions related to gases believed until the last decades to be "poisons" has attracted increasing interest in worldwide research, so that the development of suitable analytical methods for their recognition and quantification has become a stringent demand. Selective detection of inorganic biomarkers in exhaled breath is a non-invasive analysis method that can offer essential information for determining the typology and/or the evolution of a specific illness. These analytical methods could be implemented in new diagnostic tools to be used in many different contexts such as hospital medical equipment, home diagnostic system for elderly and handicapped people, and safety guard sensor for both civil and military purposes. The main drawback of diagnostic breath analysis is related to the difficulties of detecting simultaneously and selectively the thousands of compounds contained in the human exhaled air. Moreover, the chemical composition of human breath depends on many other factors such as people's habit, age, and location. All these reasons make breath analysis very difficult to perform.

The most common analytical methods used to detect inorganic gases are chemiluminescence and spectroscopic techniques. Although these techniques are very sensitive (down to a few ppb), they are often expensive, non-portable, and time consuming. Several research efforts have been devoted to the development of chemical sensor technologies with the aim of overcoming the limitations of the "non-real-time" analytical methods. Ease of implementing in portable systems and the online monitoring of the inorganic volatile analytes are the main claimed advantages of the different sensor categories.

Nowadays, only few chemical sensor-based breath detectors (electrochemical transducers) are used to monitor the concentration of NO and CO while no portable systems for H_2S in exhaled breath are known (H_2S is usually measured via blood plasma analysis in the laboratory). Unfortunately, these detectors do not allow the simultaneous monitoring of NO, CO, and H_2S. The detection of such gases at the same time in human breath could give a more accurate assessment of inflammation and oxidative stress status of the patients.

The monitoring of inorganic substances (NO_x, CO, and H_2S) in the exhaled and inhaled air is therefore a very interdisciplinary field of research, and the development of stable, low-cost, and portable detectors is still a challenge. A wide range of sensors have been investigated in order to fulfill these needs, and although some of them are sensitive and reliable, they generally suffer from low selectivity and stability, or often require high working temperature. Examples of such sensors are potentiometric[35–37] and amperometric sensors,[38, 39] metal-oxide sensors (TiO_2, SnO_2, ZnO, WO_3),[40–47] and MOSFET-type sensors.[48, 49]

In this area, organic thin-film transistors (OTFTs) are a promising alternative since they can offer reliable and reversible "analytical signal" upon exposure to the target analyte, at room temperature. Besides, OTFTs can be miniaturized and give quite selective responses at ppm-ppb concentration level through the suitable choice of the sensitive layer, allowing the detection of a wide range of organic and inorganic compounds.

2.2 OTFT-Based Sensors: A Bird's-Eye View

Despite an impressive list of available analytical technologies, designing an inexpensive handheld or household sensor-based instrument is still an open challenge for the worldwide scientific community. In addition to sensitivity, selectivity, and robustness, low power consumption and compact size are stringent requirements that sensor devices have yet to fulfil.[50]

Recognition of *complex odors* is being addressed, so far, by sensor array-based systems called *electronic noses* (*e-noses*) that attempt to mimic the mammalian olfactory system.[51, 52] The e-noses commercially available today, however, remain well outside of the feasible range for consumer products. The introduction of e-noses into the consumer market could enable a *plethora* of applications ranging from medical self-diagnosis kits to indoor air quality monitoring. Moreover, at the industrial level, sensor arrays could be integrated into devices for the quality control of beverages, olive oil, explosives, pathogenic bacteria, and many other products.

Currently, the commercially available sensors are metal-oxide or conducting polymer-based chemiresistors and inorganic field-effect sensors. The most widely used MOS (Metal Oxide Semiconductor) devices are based on tin oxide resistors doped with different catalytic metal additives such as platinum or palladium in order to modulate its sensor response. p-type MOS devices are also available. Such devices are based on copper oxide, nickel oxide, and cobalt oxide, which respond to oxidizing odorants like oxygen, nitrous oxide, and chlorine. The gas analyte is detected by means of its effect on the electrical resistance of a metal-oxide semiconductor active layer.

Modern MOS devices are typically produced through sputtering processes and are patterned with microfabrication techniques to minimize the device size. However, integration of multiple metal oxides, or even the same metal oxide containing different dopants, onto the same substrate is difficult and expensive using traditional microfabrication techniques, because an extensive subtractive processing is required, and the production yield quickly drops with each additional layer of material used.

Generally MOS devices offer the advantages of low cost, a good level of insensitivity to humidity, and an output signal that is easy to read and process. Disadvantages, however, include high operating temperatures, signal drift over time, limited selectivity, high power consumption, and only modest sensitivity. For these reasons, the use of organic active layer, such as conductive polymers (CPs) instead of metal-oxide, is being widely investigated in chemiresistors. The sensor response is produced when ambient vapors absorb into the polymer, inducing physical or chemical interactions that change the conductivity of the film.[53, 54] The wide range of organic material that can be synthesized enables the fabrication of CP chemiresistors with sensitivities over a broad range of organic compounds. CPs are cost-effective and easily synthesized, and they present fast responses to a large number of volatile analytes with low power consumption.

Nevertheless, reliability in CP-based chemiresistors is still an issue. The biggest disadvantage of CP sensors is their sensitivity to humidity and their susceptibility to poisoning due to irreversible binding of vapor molecules to the sensing material.[55]

Organic thin-film transistors, as discrete elements or implemented in plastic circuits, are currently the most fascinating technology for chemical and biological sensing. OTFTs are field-effect devices with organic or polymer semiconductor thin film as channel materials. They can act as multiparametric sensors,[56] with remarkable response repeatability and as semi-CP-based sensing circuits.[57, 58] As a matter of fact, transistors based on organic semiconductors are known to exhibit gas sensor responses, and they may be able to provide a low-cost replacement for the gas sensor arrays currently used in e-noses.

OTFTs are three-terminal devices in which charge conduction between two terminals, the source and the drain, is controlled by modulating the electrical potential of a third terminal, the gate. A typical OTFT structure is shown in Fig. 2.1. The gate can be used to switch the transistor "on" (high source-drain current) and "off" (negligible source-drain current). The organic active layer is generally a few tens of nanometers made of a conducting polymer or oligomer. The semiconducting film is typically deposited by solution processing (such as solution casting, spin coating, Langmuir-Shäfer or Langmuir-Blodgett techniques) or by thermal evaporation. The deposited thin films are generally polycrystalline, to be more precise, films composed of contiguous grains having size of few hundreds of nanometers (see the magnification in Fig. 2.1).

In a simplified view, for a p-type semiconductor, the application of a negative gate voltage leads to an accumulation of positive charge carriers, and the thin film conductance increases (accumulation mode), while the application of a positive gate voltage depletes the major carriers and reduces the film conductance (depletion mode). When enough charge carriers have been accumulated, the conductivity increases dramatically, providing a conductive channel between the source and drain electrodes. The typical *I–V* curves reported in

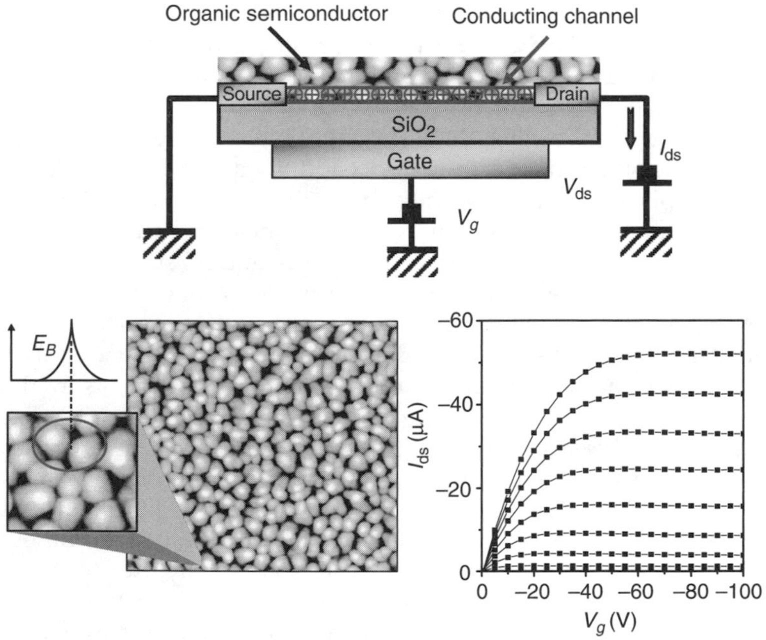

Figure 2.1 Schematic structure of an organic thin-film transistor (OTFT) with a polycrystalline active layer and a typical I_{ds}–V_{ds} characteristic curves.

Fig. 2.1 show a linear region for $V_{ds} \leq V_g - V_t$ and a saturation region for $V_{ds} > V_g - V_t$. Here V_t is the device threshold voltage, that in TFTs is equal to qNd/C_i,[59] where N is the film doping density, C_i is the dielectric capacitance per unit area, and d is the gate dielectric thickness. The threshold voltage V_t marks the passage from the *off* to the *on* conductivity regime.[60] In transistors with undoped active semiconductor materials, the threshold voltage extracted from the *I–V* curves depends on the density of the low-mobility electronic levels (trap states) as it corresponds to the voltage required to fill such trap states,[61, 62] located primarily at the interface with the gate dielectric.

Generally speaking, the device performance is described by two figures of merit: the field-effect mobility μ_{FET} and the on/off ratio. The μ_{FET} is the field-effect mobility of the charge carriers moving in the organic layer, which should be as large as possible. It is often extracted from the *I–V* curves by using the standard analytical equations developed for inorganic semiconductor devices[59, 63]

$$\sqrt{I_{ds}^{SAT}} = \sqrt{\frac{W}{2L} C_i \mu} \cdot (V_g - V_t) \tag{2.1}$$

Strictly speaking, these equations apply only to TFTs exhibiting constant charge carrier mobility, which is not the case for polycrystalline OTFTs; thus μ_{FET} values extracted should be considered as estimates. Typical μ_{FET} extracted values are in the 10^{-3} to 10^{-1} $cm^2/(V \cdot s)$ range, but values as high as 1 to 10 $cm^2/(V \cdot s)$ [a-Si:H TFTs have μ_{FET} of 0.1 to 1 $cm^2/(V \cdot s)$] can be reached.[63] The *on/off ratio*, defined as the ratio of the I_{ds} current in the on ($V_g > V_t$) and off ($V_g = 0$ V) states, is indicative of the switching performance of the device.

Using a three-terminal OTFT structure instead of the two-terminal chemiresistor construction introduces additional process complexity and complicates signal processing, but judicious use of the gate bias can enhance sensitivity, discrimination, and measurement repeatability.[64–68] Moreover, the extraction of many parameters such as μ_{FET}, V_t, and bulk resistivity allows the partial deconvolution of various analyte-sensor interaction related parameters, which is not possible in the single-parameter chemiresistor devices. As a matter of fact, OTFT devices have been successfully employed as chemical sensors for the detection and monitoring of organic volatile analytes and less frequently for inorganic ones. OTFTs based on various oligothiophenes and phthalocyanines, when exposed to a variety of volatile analytes such as alcohols, ketones, thiols, nitriles, and esters, have shown a typical pattern for each single analyte, demonstrating that the variety of organic semiconductor materials can be successfully used for fabricating a selective array of gas sensors.[58] More recently phthalocyanine-based transistors have been used to measure the concentration of strong oxidizing species, such as ozone and NO_2, demonstrating that the use a field-effect transistor rather than conventional chemiresistors offers several advantages, one of which is the improvement of the reversibility of the adsorption process thanks to the reverse bias applied.[69] Very recently the potentialities of the OTFT gas sensors have been extended to the chiral discrimination of enantiomeric compounds. A chiral bilayer OTFT gas sensor—comprising an outermost layer with built-in enantioselective properties—has been demonstrated to exhibit field-effect amplified sensitivity that allows differential detection of optical isomers in the tens of ppm concentration range.[70] Moreover, although response enhancement has been already reported,[50, 67, 71–74] this is the first direct evidence of gate field-induced sensitivity enhancement. This opens interesting perspectives for the use of microscopic OTFTs as ultrasensitive electronic transducers that could also operate as sensing switches.

It can be concluded that organic materials are useful for e-nose applications for several reasons:

- The carbon backbones of the organic semiconductors make the sensitive film more chemically active than most of their inorganic counterparts, thus amplifying the electrical responses and the sensitivity.
- Organic electronic materials can be readily modified using synthetic chemistry, allowing their chemical sensitivities to be controlled by careful design of the organic semiconductors or by the introduction of selected functional groups.
- Organic molecules are commonly soluble at room temperature in common solvents. This is especially important for applications such as the e-noses, where the construction of an integrated array of different chemical sensors can be made by printing techniques on plastic substrates.
- Detection limits and sensitivity also benefit from the signal amplification that is inherent in transistor devices, allowing transistor-based sensors to outperform chemiresistors even better than amperometric and potential sensors.

2.3 Anthracene-Based Organic Thin-Film Transistors as Inorganic Analyte Sensors

In this section, the field-effect and gas sensing properties of the OTFT devices based on functionalized 9,10-ter-anthrylene-ethynylene oligomers (D3ANT) are presented. The oligomer was characterized by ^{1}H-NMR, mass spectrometry, IR, UV–vis, photoluminescence (solution and solid state), and cyclic voltammetry (CV). Moreover, the morphological and structural characterizations of the molecule have been investigated by means of AFM, STM, and grazing incidence X-ray diffraction (GIXRD) techniques. D3ANT oligomers have shown good semiconducting properties with an average mobility in top-contact configuration of 1.2×10^{-2} cm^2/(V·s) with on/off ratios higher than 10^4, while the highest mobility obtained was 0.055 cm^2/(V·s). Gas sensing measurements performed on D3ANT-based OTFT showed high selectivity toward the NO_x gases with detection limit at sub-ppm levels.

2.3.1 Introduction

Until now, polycyclic aromatic hydrocarbons (PAHs) have been among the most studied organic materials for OTFT applications. Particular attention has been paid to linear PAHs composed of laterally fused aromatic rings and called simply *acenes*.[75] Among all the studied acenes, with no doubt *pentacene* is known to have the highest hole mobility [>1.0 cm^2/(V·s)] for high vacuum deposited thin films,[76–78] and it is used as reference standard for all newly synthesized p-type organic semiconductors. Thermally evaporated pentacene OTFTs have been used in sensor applications as well. Pentacene OTFTs were tested as humidity sensors[79] by Zhu et al. reporting a saturation current reduction up to 80% after a relative humidity (RH) change from 0 to 30%. The gas sensing response was attributed to a decrease of the hole mobility due to the water molecule. Moreover, the sensitivity depended on the thin-film thickness. Other experimental work was conducted to shed light on the sensing mechanisms of pentacene.[80, 81] The reduction of both hole mobility and source-drain current has been related to the interaction at grain boundary of the charge carriers with the polar water molecules.

The gas sensing properties of pentacene OTFT sensors have been explored also toward 1-pentanol[82] by employing devices with channel lengths ranging from 36 μm to 20 nm. The source-drain current was found to decrease upon analyte exposure for the larger-scale devices whereas the opposite happened surprisingly for the smaller ones. These results were explained by considering the effects of two competing mechanisms: a reduction in mobility due to charge carrier trapping at grain boundaries and an increase in charge density associated with the interactions between the analyte molecules and pentacene grains. However, the role of the contact resistance was not ruled out. Moreover, it was observed that channel length corresponding to gas sensing crossover point was dependent on the pentacene grain size.

To fully realize the advantages of organic materials for sensor devices, all the components of organic circuits and in particular the semiconductor layer should be constructed by solution or printing methods at ambient temperature and pressure. Unfortunately, even though pentacene has shown the highest charge mobility, its limited solubility prevents the use of solution deposition techniques. For this reason many research groups have set out to synthesize a processable version of this molecule. The first method for solubilizing pentacene was proposed by Brown et al.[83] The adopted strategy is based on preparing a soluble pentacene precursor that is not a semiconductor but can be converted to the active form upon heating at 140 to 220°C, obtaining a mobility of 0.2 $cm^2/(V \cdot s)$.[84] Similar synthetic approaches have been proposed for obtaining soluble pentacene derivates,[85–87] and OTFT based on such molecules showed field-effect mobilities ranging from 8.8×10^{-3} to 0.89 $cm^2/(V \cdot s)$.

The preparation of soluble precursors is, however, synthetically challenging and often a costly annealing process is required to obtain high-order and semiconducting thin films. Moreover, even though the thermal conversion processes have been claimed to be quantitative, undetectable amounts of precursors or by-products can introduce trap states in the films, reducing the mobilities.[88]

Alternatively, pentacene can be structurally modified to achieve better solubility through wet techniques, to obtain desirable supramolecular order in the solid state and to improve stability toward oxygen and light. The seminal work of Anthony and coworkers has shown that all these objectives can be simply achieved in a one-pot reaction from 6,13-pentacenequinone introducing bulky silyl groups separated from the acene by alkyne spacers at the 6,13-pentacene position.[89–92]

These novel bis-silylethynylated pentacenes exhibit several remarkable features compared to the respective parent pentacene molecule. First, the bis-silylethynylation, apart from increasing the solubility, lowers the triplet and LUMO (Lowest Unoccupied Molecular Orbital) energy of pentacene and thus enhances the oxidative stability.[93] Moreover, the sterically demanding substituents prevent the dimerization through Diels-Alder reaction. The improved stability of the TIPS pentacene (triisopropylsilylethynyl pentacene) was demonstrated by its 50 times slower degradation in air saturated tetrahydrofurane solution compared to the unsubstituted compound. In the dark, even oxygen-saturated solutions of TIPS pentacene did not show considerable decomposition after 24 h.[93]

Of all functionalized pentacene only the TIPS pentacene that exhibits 2D π-stacking interactions yields high-performance thin-film devices: in OTFT prepared by vacuum deposition a hole mobility of 0.4 $cm^2/(V \cdot s)$ was observed, while in a spin-coated device a mobility of 0.17 $cm^2/(V \cdot s)$ and on/off current ratios of 10^6 were measured.[91] Notably the electrical measurements were performed in air at room temperature, proving that the introduction of these substituents improves the device stability.

The high mobilities observed in the TIPS pentacene were explained by invoking the two-dimensional self-assembly of the aromatic moieties into π-stacked arrays that enhance the intermolecular overlap. Indeed, TIPS pentacene does not adopt the typical pentacene herringbone pattern, but stacks in two-dimensional columnar arrays with significant overlap of the pentacene rings (the interplanar spacing of the aromatic rings is 3.47 Å for TIPS pentacene, compared with 6.27 Å for pentacene). Recently better results were obtained with drop-casted films. Hole mobilities as high as 1.8 $cm^2/(V \cdot s)$ and on/off current ratios of 10^8 have been achieved by drop-casting a $1\%_w$ toluene solution of TIPS pentacene on top of a bottom-contact FET device substrate.[94] The approach proposed by Anthony has also been extended to other acene derivatives such as pentacene ethers,[95] acenedithiophenes,[96] and longer hexa- and heptacenes.[97] Remarkably, drop-casted films of a bis-triethylsilylethynyl anthradithiophene yielded devices with mobilities as high as 1.0 $cm^2/(V \cdot s)$ and on/off current ratios of 10^7.[92]

Although acenes result in appealing materials for OTFT application, only few sensing application examples are found in literature. In addition to pentacene thin film employed in alcohols and humidity OTFT sensors, other experimental works deal with the use of naphtalene tetracarboxylic derivates film as sensitive materials for water vapor,[56, 98] ketones, thiols, and nitriles.[58] Differently from acenes, oligo- and polythiophene have been widely investigated for the sensing of alcohols,[58, 72, 99] esters,[58] thiols,[58] ammonia,[100] chloroform,[101] lactid acid, and glucose.[102] Moreover, in the case of the regio-regular polythiophenes the fundamental role of the alkyl- and alkoxy-side chain in conferring selectivity[72] to an OTFT has been fully demonstrated, pointing out the importance of the chemical functionalization for the control of the sensing element selectivity.

In conclusion, the ability to functionalize organic semiconductors in order to control their chemical-physical properties, the type of interaction occurring in solid state, and the gas sensing properties of the thin film are a critical aspect in the perspective to develop and commercialize a new sensing platform with high performance and low power consumption.

2.3.2 New Materials for OTFT Sensing Applications

As seen, more stable and solution processable pentacene derivatives have been synthesized, some of which showed high field-effect mobilities comparable to or even greater than the parent pentacene molecule. An alternative approach to obtain semiconductive materials consists of the construction of suitable oligomeric structures with smaller polycyclic benzenoid systems.[103–107]

The adoption of this strategy is justified by the following reasons.

- Smaller acenes are less conjugated and have a greater oxidative stability; this characteristic is fundamental in applications where durability and reproducibility are the deciding factor.

- The several available synthetic methods permit the construction of a plethora of extended π-systems. This allows the fine tuning of the electronic properties.
- The synthesis of extended π-systems may enhance the π–π interactions which are crucial for efficient interchain charge transport.

Due to the higher oxidation potential and the more accessible synthetic protocols, anthracene is fast becoming a very deeply investigated moiety for use as active layer in OTFT devices. Historically, anthracene has been widely studied as an organic photoconductor and for OLED devices[108–112] while little attention has been paid to its application in field-effect transistors. Hole mobilities in anthracene single crystal were measured by the time of flight technique and were found to reach up to 3 $cm^2/(V \cdot s)$[113] at 300 K. While in single crystal OTFT devices the higher mobility was 0.02 $cm^2/(V \cdot s)$ measured at 170 K.[114]

In the last few years a considerable number of anthracene derivatives were reported. The best results for thermal evaporated thin films were obtained by Meng and coworkers reporting for a 2,6-bis [2-(4-pentylphenyl)vinyl] anthracene a hole mobility of 1.28 $cm^2/(V \cdot s)$ and on/off ratio greater than 10^7.[115] The observed high charge mobilities have been attributed to the densely packed crystal structure that in principle would ensure a high transfer integral between the adjacent molecules.

For solution deposited thin films, the best mobility was reported recently by Park et al. for a 9,10-bis(triisopropylsilylethynyl) substituted anthracene-thiophene oligomers.[116] A hole mobility of 4.0×10^{-3} $cm^2/(V \cdot s)$ and on/off ratio of 10^6 were observed for a chloroform spin-coated OTFT. Remarkably the introduction of the bulky TIPS substituents led to a π-stacked crystal structure with an interplanar distance of 3.49 Å, similar to the value observed for functionalized pentacene.

Here we report on synthesis, characterization,[131] and NO_x gas sensing properties of a new p-type semiconductor based on a 9,10-*ter*-anthryleneethynylene bearing decyl alkyl chains as side groups, namely, D3ANT. To the best of our knowledge, no example has been provided so far of 9,10 substituted anthracenes exhibiting field-effect modulation properties and high selectivity toward NO_x gases.

In the *ter*-anthrylene ethynylene structure, three neighboring anthracenes are chemically linked by two ethynylene bonds, hence facilitating the effective charge transport to be extended beyond a single anthracene unit. With respect to 9,10-anthrylenes, the introduction of ethynylene moieties between the anthracene units was conceived to build a planar π-framework potentially endowed with good intermolecular interactions also in a disordered film.

9,10-bis[(10-decylanthracen-9-yl)ethynyl]anthracene was prepared following the synthetic approach depicted in Fig. 2.2. 9,10-Dibromoanthracene (1) was mono-lithiated with phenyl lithium and subsequently reacted with decyl bromide to yield 9-bromo-10-decylanthracene (2). A Negishi coupling of 2 with (trimethylsilyl)ethynyl zinc chloride yielded the trimethylsilyl-ethynyl derivative 3. Deprotection of 3 with tetrabutylammonium fluoride gave 9-ethynyl-10-decyl-anthracene (4) that was readily reacted with 1 in a Sonogashira coupling to obtain D3ANT. The final product was purified from soluble by-products by subsequent Soxhlet washings with methanol, acetone, and *n*-hexane. Finally the product was extracted using chloroform and reprecipitated twice from chloroform-methanol solution. D3ANT was obtained in good yield (60%) as a purple-red powder slightly soluble in chlorinated solvents at room temperature. The proposed structure is fully supported by its molecular mass (APCI-MS), as well as by the ^{1}H-NMR spectrum recorded in toluene at 80°C.

To investigate the thermal degradation of D3ANT, thermogravimetric analyses (TGA) and differential scanning calorimetry (DSC) were carried out. A good thermal stability was revealed with a decomposition temperature of 373.7°C at 5% weight loss. The thermal behavior below

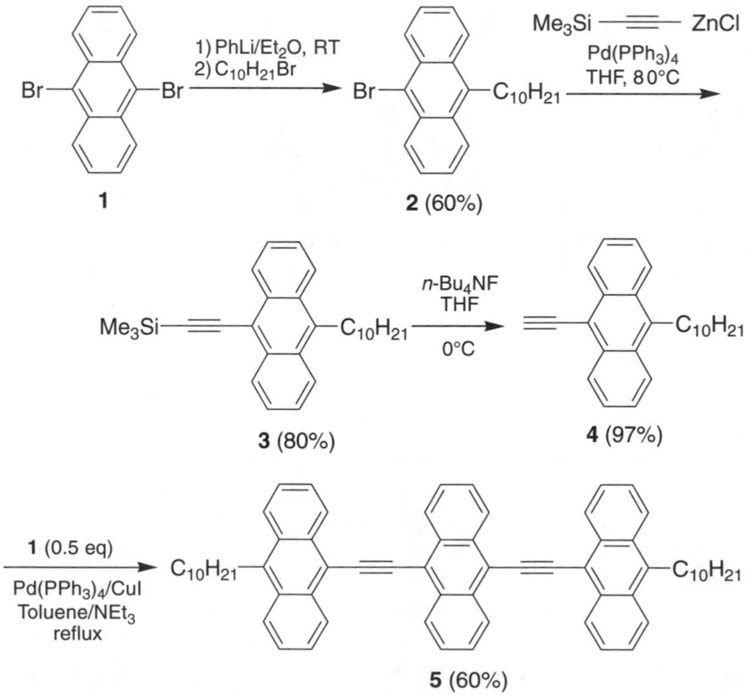

Figure 2.2 Scheme 1. Synthesis of D3ANT. (*Reproduced by permission of The Royal Society of Chemistry, Ref. 131.*)

the decomposition temperature was investigated by DSC analysis, evidencing only fusion occurs at 247.1°C.

The optical properties of D3ANT were investigated by UV-vis absorption and fluorescence spectroscopy. The UV-vis spectrum of D3ANT measured in $CHCl_3$ solution (Fig. 2.3*a*) showed a structureless absorption band with a maximum at 504 nm ascribed to the π–π* transition of the conjugated backbone. From the onset of the absorption spectra the optical HOMO/LUMO gap is 2.19 eV, almost the same observed for pentacene in toluene solution (2.12 eV).[117] The introduction of ethynylene spacers between the three anthracene units effectively extends the conjugation beyond the single anthracene chromophore (λ_{max} = 376 nm in cyclohexane solution). Moreover, the absence of well-resolved vibronic replica, typical for anthracene-based systems, can be ascribed to the low rigidity of the *ter*-anthrylene ethynylene architecture caused by a low torsional energy barrier between the anthracene moieties. In fact it has been shown that the energy barrier for the rotation about the alkyne-aryl bond is relatively low[118, 119] because the energy difference between the fully coplanar structure and the perpendicularly twisted structure has been estimated to be less than k_BT at room temperature.[120] Therefore, the rotation around the ethynyl group between the coplanar form and the twisted form, in absence of steric constraint, is allowed. Moreover, in most aryleneethynylene derivatives the coplanar structure over the neighboring aryl groups is the most stable both in the ground and in the excited state, which affords the well-extended π-conjugated system.

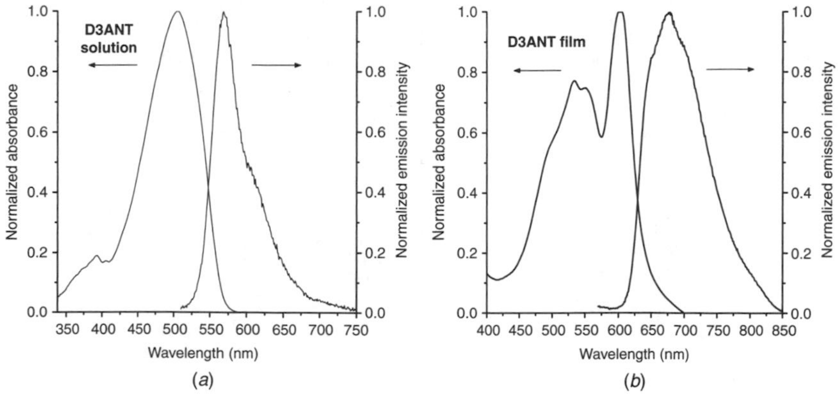

FIGURE 2.3 (*a*) Normalized UV-vis absorption and photoluminescence emission spectra for D3ANT in $CHCl_3$ solution and (*b*) as thin films drop-cast from 0.05 wt % chloroform solution. For solution and solid states, respectively, λ_{ex} was 504 and 532 nm. (*Reproduced by permission of The Royal Society of Chemistry, Ref. 131.*)

The solution emission spectrum is characterized by a maximum 568 nm localized in the green-yellow region accompanied by a shoulder at longer wavelength, ascribable to a poorly resolved vibronic replica. The measured Stokes shift is equal to 64 nm. This value suggests that appreciable nonradiative processes take place before the absorbed photons are reemitted. The nonradiative processes can originate from either molecular rearrangements or solvent-molecule interactions upon photoexcitation.

The optical properties of the D3ANT were also investigated in the solid state (Fig. 2.3*b*). The absorption spectrum is characterized by a marked red-shifted band at 535 nm accompanied by a new more intense red-shifted band at 603 nm. The latter clearly indicates strong electronic interaction between molecules in the films and is generally recognized as the aggregation band. Actually for oligo and poly arylene ethynylene systems, the situation is somewhat complicated by the conformational freedom of the conjugated backbone; in fact the red-shifted band might originate from the cooperative effect of both molecular planarization of conjugated backbone and electronic orbital interaction between different conjugated backbones.[121–124] The emission spectrum at the solid state is characterized by a consistent red shift of the maxima (677 nm) with respect to those in solution, independently of the exciting wavelength. We suppose that in the solid state the emission arises only from the intermolecular aggregates.

To further analyze the optical properties of D3ANT, *solvent-induced aggregation* studies were performed. We expected that the addition of nonsolvents to a D3ANT chloroform solution would possibly induce molecular aggregation. As shown in Fig. 2.4*a*, D3ANT in $CHCl_3$ exhibits a strong absorption band at about 504 nm. As the methanol (nonsolvents) component increases to about 40 Vol %, a new absorption band

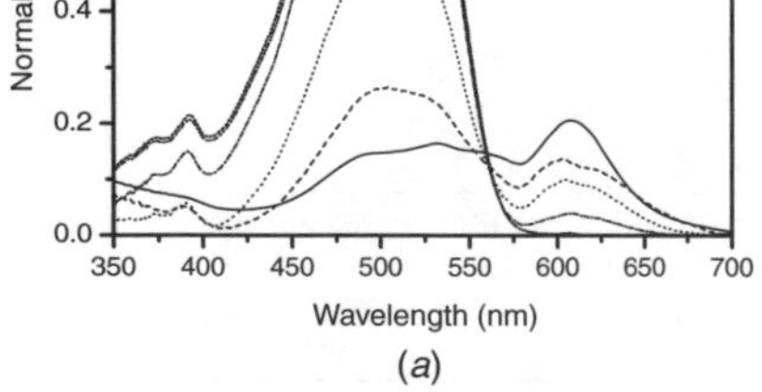

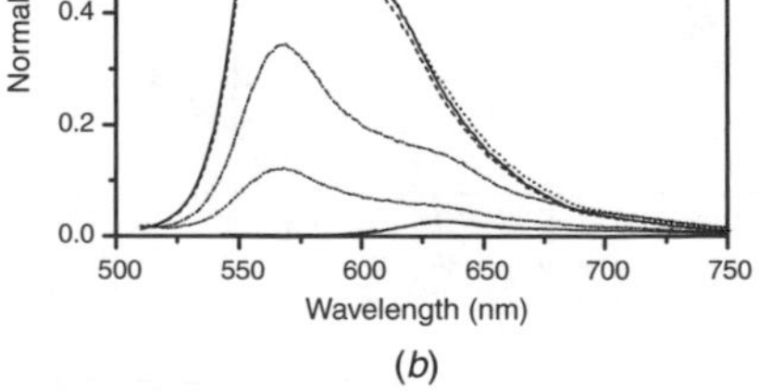

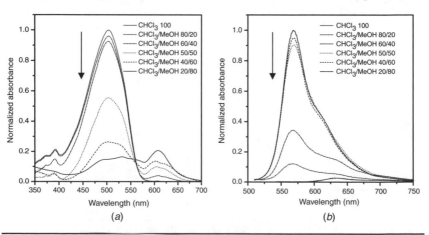

FIGURE 2.4 (*a*) Absorption spectra and (*b*) fluorescence spectra of D3ANT measured in $CHCl_3$/MeOH with various compositions. D3ANT was excited at 500 nm, $C = 7.5 \times 10^{-6}$ *M*.

at 607 nm is developed along with decrease of the absorption band at 504 nm. The wavelength of this new absorption band is comparable to that of thin-film D3ANT (603 nm), indicating formation of molecular aggregation. Fluorescence of D3ANT in the solvent/nonsolvent mixture reveals a similar trend (Fig. 2.4*b*). As the methanol concentration increases, the fluorescence intensity around 568 nm tended to decrease. Simultaneously, the relative band intensity around 635 nm increased. Moreover, the fluorescence intensity of both bands dramatically decreases with increasing the methanol concentration.

The appearance of the red-shifted fluorescence band and the strong decrease of the fluorescence intensity can be interpreted by the intermolecular aggregation as in the film state. The formation of strong intermolecular interaction in the solid state is one of the essential requisites for efficient charge transport. The observation of aggregate formation for the synthesized oligomers is a good clue for the application of the oligomers as channel materials in OTFTs.

The electrochemical behavior of D3ANT was studied in CH_2Cl_2 using Ag/Ag^+ as reference electrode and employing ferrocene as internal standard for the potential calibration. The cyclic voltammogram showed two quasi-reversible oxidation peaks (1.05 and 1.20 V) and an irreversible oxidation peak at 1.49 V. This behavior is likely due to the sequential abstraction of three electrons from the three anthracene moieties. Two irreversible reduction peaks at –1.26 and –1.53 V were also observed. Moreover, taking –4.8 eV as the HOMO level of the ferrocene system,[125] the HOMO and LUMO energy levels were estimated from the onset of the first oxidation and reduction as –5.23 and –3.15 eV, respectively, resulting in an energy gap of 2.1 eV. The obtained electrochemical HOMO/LUMO gap is in reasonable agreement with the optical gap in $CHCl_3$ (2.2 eV). A HOMO level of 5.23 eV aligns well with the work function of gold (4.8 to 5.1 eV) and in principle should lead to a good injection of charge carriers from the Au electrodes.

2.3.3 Device Performance

OTFT devices were fabricated by using both bottom- and top-contact geometry. In the bottom-contact configuration, drain and source electrodes (Ti 10 nm/Au 50 nm) at various channel widths and lengths were fabricated on the Si/SiO_2 surface by photolithography. The silicon substrate serves as gate electrode. The substrates were cleaned as follows: the wafers were first immersed in a *piranha* solution for 8 min to remove traces of adsorbed organic materials, rinsed with deionized water (DI), and eventually dried with a nitrogen flow. They were then washed according to the standard cleaning (SC) procedures.[126, 127] Thin films of D3ANT were deposited onto the SiO_2 surface by spin coating from 0.05 % (w/w) solutions in anhydrous $CHCl_3$ at 2000 rpm for 30 s. Some of the films were subjected to an annealing procedure consisting of a treatment at 100°C for 30 min under a nitrogen atmosphere.

The devices were tested as p-channel materials with an Agilent 4155 C semiconductor parameter analyzer kept in a glove box at room temperature. The field-effect mobilities in saturation regimes were extracted using the well-known equation[59]

$$I_{ds} = C_i \mu (W/2L)(V_g - V_t)^2 \text{ at } V_{ds} > V_g \tag{2.2}$$

where I_{ds} = drain-source current
C_i = capacitance per unit area of the gate dielectric layer
V_g = gate voltage
V_t = threshold voltage

Voltage V_t was extrapolated from the $(I_{ds})^{1/2}$ vs. V_g plot.

Channel mobilities as high as 1.6×10^{-3} cm^2/(V·s) with an on/off ratio of 2×10^4 were reached with a bottom-contact geometry. These figures were slightly improved by annealing the substrate film for 30 min at 100°C, resulting in a mobility of 2.8×10^{-3} cm^2/(V·s) and an on/off ratio of 6×10^4. Postthermal annealing treatments have been known to improve molecular ordering and grain sizes of the thin film and frequently result in better device performance. Moreover, annealing may reduce also the concentration of adsorbed impurity dopants (moisture and oxygen), increasing the OTFT properties.[128–130]

Field-effect mobility greater than one order of magnitude was achieved for spin-coated and annealed top-contact OTFT. Top-contact devices were fabricated using a highly n-doped silicon wafer (resistivity 20 Ω·cm) as gate contact on which 100 nm of dielectric (SiO_2) was thermally grown. Gold was used as the source and drain electrodes, and it was deposited on organic active layer through a shadow mask with a channel width (*W*) of 500, 1000, 2000, and 4000 μm and a channel length of 150, 100, 100, and 200 μm, respectively.

Figure 2.5*a* shows the current–voltage characteristics (I_{ds} vs. V_{ds}) at different gate bias for top-contact devices of an active channel having $W/L = 500/150$ (μm/μm). Figure 2.5*b* shows I_{ds} and $I_{ds}^{1/2}$ vs. V_g transfer

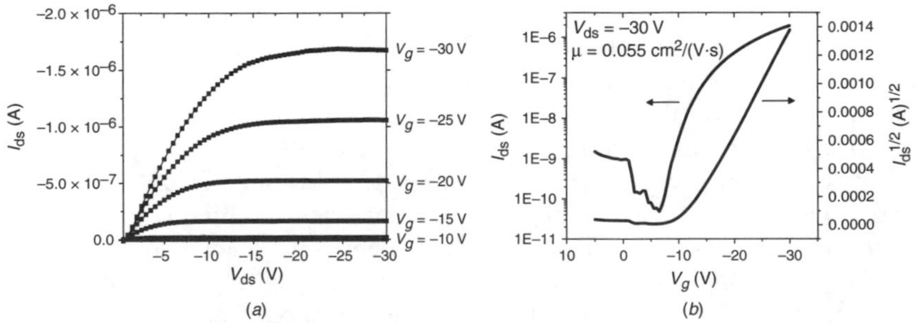

FIGURE 2.5 (*a*) I_{ds}–V_{ds} output characteristics at different gate voltage for top-contact D3ANT OTFT. (*b*) I_{ds}–V_g transfer characteristic curves and plot of $I_{ds}^{1/2}$ vs. V_g at constant V_{ds} = −30 V. (*Reproduced by permission of the Royal Society of Chemistry, Ref. 131.*)

characteristics at source-drain voltage fixed at –30 V. Such plots show that D3ANT-based OTFT exhibits a well shaped p-channel response, with defined linear and saturation regions. An averaged mobility of 1.3×10^{-2} cm^2/(V · s) was observed with on/off ratio varying from 10^3 to 10^5. The highest mobility reached was 0.055 cm^2/(V · s) while optimized on/off ratios reached the value of 4×10^5.[131] Data reported in Fig. 2.5 are relevant to the best mobility device (spin and annealed film). The observed mobilities of ~10^{-2} cm^2/(V · s) are the highest reported so far for solution processed anthracene oligomers.[132–133]

Thin-film morphology of D3ANT deposited onto a Si/SiO_2 substrate was studied by means of atomic force microscopy (AFM). Spin-coated thin films showed a good surface coverage, as can be noted from the AFM topographical image reported in Fig. 2.6*a*. Moreover, a grainlike structure with continuous grains having a size of about 0.04 μm can be noted. The film thickness of D3ANT was also evaluated to be 10 nm.

The structural characterization of D3ANT thin films was carried out by means of out-of-plane GIXRD analysis. Thin films of D3ANT were deposited by spin coating on SiO_2/Si substrates under the same conditions used for the construction of the OTFTs. All the GIXRD diffractograms are featureless, indicating the absence of long-range structural order in the bulk active layer and in the out-of-plane direction. Similar GIXRD patterns were observed for thin films of molecules of the same family, i.e., bearing two (trimethoxyphenyl) ethynyl substituents at the 9,10 positions of anthracene, which are also inherently disordered though they do not show any field-effect transistor behavior.[134]

The most plausible reason for the improved charge mobility is related to the introduction of two 10-decylanthr-9-yl-ethynyl groups at the 9,10 positions of anthracene that extends the interaction between

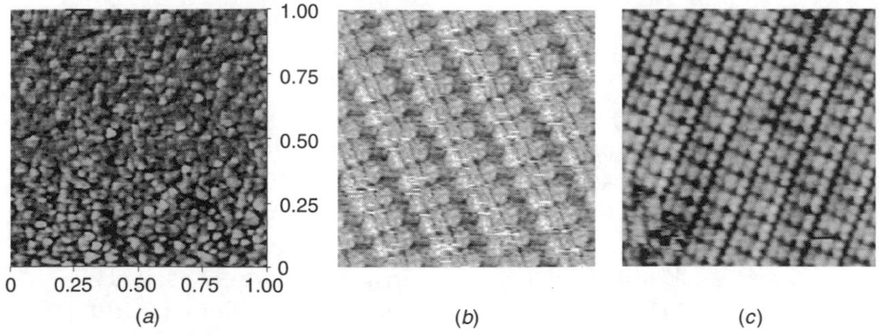

FIGURE 2.6 (*a*) AFM topographical image of D3ANT deposited on Si/SiO_2 (1 × 1 μm) and typical constant-current STM images of self-organized monolayers of D3ANT adsorbed (*b*) at the *n*-tetradecane–HOPG interface (16 × 16 nm^2; V_t = –333 mV; I_t = 27 pA) and (*c*) at the *n*-tetradecane–Au(111) interface (16 × 16 nm^2; V_t = –62 mV; I_t = 83 pA). (*Reproduced by permission of The Royal Society of Chemistry, Ref. 131.*) (See also color insert.)

π-systems, independently from the structural packing of the D3ANT molecules in the solid active layer. However, as a bulk technique, GIXRD is unable to provide structural information on molecular arrangement of the first monolayer at ultrathin organic semiconductor/ SiO_2 interface where field effect takes place. XRD studies only allow the interpretation of oligomer chain packing up to the resolution of several monolayers.

To shed more light on the two-dimensional self-organization of D3ANT, we investigated monolayers on highly organized pyrolitic graphite (HOPG) and Au(111) by scanning tunneling microscopy (STM) in a liquid environment of *n*-tetradecane.[135–137] The STM study reveals 2D organization of the molecular architectures induced by the substrate-molecule and molecule-molecule interactions on conducting and atomically flat substrates.

Two-dimensional layer consisting of highly ordered monodomains that extend over distances of 100 to 400 nm on highly oriented pyrolytic graphite (HOPG) was formed, as seen in Fig. 2.6*b*. Each domain consists of a closed-packed arrangement of linear and parallel rows of molecules. The distance between adjacent rows is 1.3 nm (in dark contrast) corresponding to the decyl chains in a slightly tilted position. The bright contrast areas correspond to the main π-conjugated backbone of D3ANT with a length of ~2.0 nm, in good agreement with the calculated core length (1.9 nm) and the profile of frontier wave functions, which is mainly defined by three anthracene subunits. The 2D unit cell parameters were measured to be $a = 3.1$ nm, $b = 1.2$ nm, $\alpha = 94°$. Moreover, from the high-resolution STM figure, the three anthracene subunits of each individual molecule are clearly visible on HOPG surface. Such evidence proves a nearly planar adsorption of the entire molecule on HOPG with no twist angle and full conjugation between the anthracene subunits.

D3ANT molecules adsorbed on reconstructed Au(111) (Fig. 2.6*c*) also form self-organized domains with linear and parallel row 2D structure very similar to that on HOPG. However, noticeable differences in the organizing behavior of the D3ANT can be observed by comparing both images. The self-organized domains on Au(111) surface do not extend over distances as large as on HOPG, and the distance between adjacent D3ANT rows on Au(111) (~0.3 nm) is markedly smaller than that on HOPG. On Au surface, D3ANT cannot accommodate decyl chains which are probably folded below or above the anthracene subunits, and this results in a more densely packed 2D structure on Au(111) as compared to HOPG. Again, the individual anthracene subunits of each molecule are clearly visible.

As seen, for spin-coated and annealed thin films of D3ANT, the GIXRD analysis indicated the lack of bulk ordered, but at the same time the STM revealed the strong tendency to form high-order 2D self-assembled monolayer on both HOPG and Au(111) surface.

Indeed, *solvent-induced aggregation* studies point out that the formation of ordered structures should be an intrinsic property of D3ANT.

Generally for amorphous thin films, field-effect mobilities in the range of 10^{-6} to 10^{-4} $cm^2/(V \cdot s)$ have been observed.[138] In the case of D3ANT the quite high mobilities of ~10^{-2} $cm^2/(V \cdot s)$ obtained for spin coating suggest that a totally amorphous organization of the thin films is unlikely. This observation is further supported by the STM studies as well as by the solvent-induced aggregation studies. Finally, we believe that during the spin coating process few ordered monolayers are formed on the dielectric surface wherein charge transport takes place while the subsequent layer nucleates in a disordered manner, affording the observed GIXRD spectra.

2.3.4 Gas Sensing Measurements

Gas sensing measurements were performed by using bottom-contact OTFT with channel width and length of 11,230 and 30 μm, respectively. D3ANT was deposited as thin film by spin coating as previously discussed. The devices, after an annealing treatment, were tested in a clean-room environment exhibiting good field-effect properties with a mobility of 1.2×10^{-3} $cm^2/(V \cdot s)$ and an I_{on}/I_{off} ratio of 4×10^4 similar to the performance observed in a nitrogen atmosphere.

Nitrogen oxides, carbon monoxide, and hydrogen sulfide were supplied by certificated cylinders (5 ppm NO, CO; 2 ppm H_2S and NO_2) connected to a gas delivery station by means of four different fed systems to prevent cross contamination. Nitrogen was used as carrier gas and to dilute the test gases. The gas mixing was carried out with computerized flowmeters. All the sensor testing was performed at room temperature and at a pressure of 760 Torr by using a constant flow rate of 200 sccm. The ambient humidity condition was also kept constant at about 40% RH. The thin films were exposed to each gas at different concentration levels through a nozzle located about 3 mm above the device.

The I_{ds} transient curves, at fixed source-drain and gate voltage ($V_g = -40$ V and $V_{ds} = -40$ V) were measured while exposing the D3ANT OTFT to each gas for 60 s. Afterward the device was exposed for 120 s of pure N_2. The results are reported in Fig. 2.7*a*. It is apparent that the I_{ds} current systematically increases at each NO_2 exposure cycle (dotted line), and the responses in the working range (0.2 to 2 ppm) are also linearly dependent on the analyte concentration. A similar behavior can be seen (solid line in Fig. 2.7*a*) for the interaction with NO molecules. The interaction of strong electron-accepter molecules, such as NO_2, with a p-type organic semiconductor, such as D3ANT, leads to a change of the transport properties, ascribable to a doping process, as already reported in previous studies. Indeed, similar current increases in metal and metal-free phthalocyanines,[139–141] porphyrine,[142] and tailored phenylene-thienylene copolymer[143] thin film

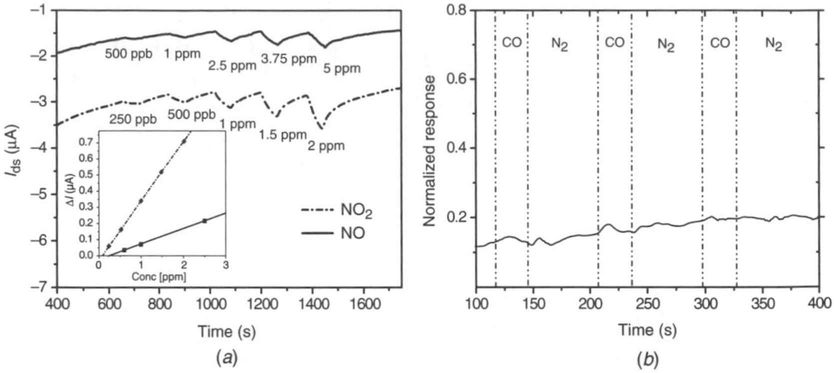

FIGURE 2.7 (*a*) I_{ds} transient response curves upon exposure of D3ANT OTFT to different concentration flows of NO and NO_2. The inset (ΔI vs. conc.) shows that D3ANT exhibits differential sensibility toward NO and NO_2. (*b*) Normalized time–response curve for three exposures of CO at concentration of 5 ppm.

exposed to oxidant analytes have been already reported and the occurrence of charge transfer processes is postulated.[144] Although phthalocyanines are sensitive to very low concentration levels (0.1 to 10 ppm levels) of NO_2, the working temperatures required are relatively high (70 to 100°C).[140, 145] At room temperature the conductivity change needs several minutes to reach equilibrium; similarly the relatively high operation temperature is needed to ensure the device full recovery. This is not the case for the D3ANT OTFT that shows a quite fast response already at room temperature, and no heating steps are required to reach a full recovery.

D3ANT thin film exposed to different concentrations of NO gas, ranging from 0.6 to 5 ppm, gave sensing responses (continuous line) similar to those observed for NO_2, as shown in Fig. 2.7*a*. It is important to outline that all the measurements performed in this study have been carried out at room temperature. It is clear that the average current change for unit concentration was significantly smaller for NO compared to NO_2. This can be explained by considering that NO_2 has a strong electron acceptor nature while NO is a weaker electron-withdrawing molecule. Differently from NO and NO_2, a closed electronic shell molecule, such as CO, and a reducing gas, such as H_2S, were almost undetectable in the 0.6 to 5 ppm and the 0.2 to 2 ppm concentration ranges, respectively. This can be readily seen in Fig. 2.7*b* where three replicates of CO normalized responses (I_{dsCO}–I_{dsN2}/I_{dsN2}) are shown. Such data demonstrate very low cross-sensitivities to hydrogen sulfide and carbon monoxide while showing differential sensibility toward NO and NO_2. Indeed, the OTFT sensor sensibility, calculated as the slope of the regression calibration line, is 0.37 μA/ppm for NO_2, about 4 times greater than for NO (0.096 μA/ppm).

The I_{ds}–V_g transfer characteristics in N_2 and at different concentrations of NO_2 were measured as well. The transfer characteristics are the I_{ds} curves measured as a function of V_g, keeping V_{ds} constant at −40 V. Such an operating regime, already used in previous works,[146, 56] allows one to extract multiparametric information to fully investigate the gas sensing properties of the active layer. Besides, since the device is driven in the depletion regimes before each measurement run, response repeatability is seen to improve.[70] The I_{ds}–V_g curve transfer characteristics were taken alternatively in N_2 atmosphere and in streams of different concentrations of NO_2. The scan rate allowed measurement of each curve in 25 s. The dotted line in Fig. 2.8*a* is the transfer characteristic measured in the presence of 1 ppm NO_2, and the solid one is relevant to the device in pure nitrogen. The extent of the response was estimated as $\Delta I_{ds} = I_{ds}(NO_2) - I_{ds}(N_2)$ from the relevant curves and is apparently a function of the gate bias. The maximum drain-current change after NO_2 exposure was 3.92 μA. For NO molecules, similar responses were detected, but the average change was less pronounced than for NO_2, as previously seen for I_{ds} transient responses. Again, CO and H_2S sensing responses were almost negligible. Such experimental evidence suggests that the sensing response can be the result of an electrophilic interaction of NO_2 and NO molecules to the anthracene units π-orbital system. Molecules adsorbed at the active layer produce therefore an increase in the hole concentration that leads to a source-drain current increase. By comparing the figure of merit extracted from the transfer characteristics in N_2 and in the analytes atmosphere, we noted that the change in current is mainly due to an increase in field-effect mobility while no change of

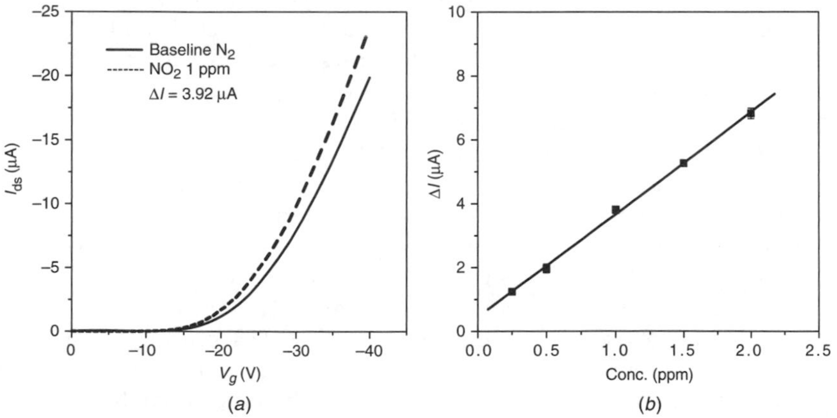

FIGURE 2.8 (*a*) I_{ds}–V_g transfer characteristics at V_{ds} = −30 V in nitrogen flow (solid line) and in a stream of 1 ppm of NO_2 (dotted line). (*b*) Gas sensing responses ΔI at V_g = −40 V and V_{ds} = −40 V relative to NO_2 exposures in 0.25 to 2 ppm. The calibration curve fitting of the data points (with the error bars), averaged over three replicates, is shown.

the threshold voltage was observed. Besides, the reversible charge transfer between analyte and sensitive active layer creates a charge dipole that could influence the transport too. The responses of D3ANT OTFT exposed to various NO_2 concentrations in the 0.25 to 2 ppm are shown in Fig. 2.8*b*. Three replicates for each concentration were gathered and used to perform the linear regression analysis. Response repeatability (expressed as relative standard deviation, RSD) was in the 2 to 12% range. Similar results have been already reported for dihexyl sexithiophene (DH-α6T) OTFTs used as alcohol sensors.[50, 51] Linear regression of the data (averaged over three replicates) has a linearity coefficient *R* of 0.999. The slope of the calibration curve representing the sensor sensibility was 3.21 (μA/ppm), and a detection limit c_d of 100 ppb (referred to as a signal-to-noise ratio = 3, noise taken as the standard error of the fit) was estimated from the equation

$$c_d = \frac{r_d S_{y/x}}{b} \tag{2.3}$$

where c_d = detection limit
r_d = signal/noise ratio (3 for LOD)
$S_{y/x}$ = standard deviation of fit
b = slope of regression curve

When an organic semiconductor is used as a sensitive membrane for inorganic gas monitoring, the detection limits are generally below 1 ppm. However, cross sensitivities are not always fully investigated. This was not the case in the study carried out for a poly(phenylene-thienylene) bearing alkoxy groups (POPT) as side chains.[143] POPT chemiresistor sensor showed large responses upon exposure to NO_2, and concentrations as low as 50 ppb could be detected, and no responses were seen for potential interfering gases such as carbon monoxide, sulfur dioxide, and ammonia. However, a quite high working temperature (60 to 100°C) was required.

The amorphous bulk structure of the D3ANT active layer could be also responsible for the good sensitivity and the fast recovery. It has been already suggested that NO_2 sensing with carbon nanotubes and phthalocyanines proceeds via absorption and interaction occurring preferentially at defect sites.[147,148] Recently, a nitrogen dioxide sensor based on amorphous poly(triarylamine) (PTAA) sensitive layers has been proposed as a room temperature OTFT sensor.[149] The lowest NO_2 concentration detectable with PTAA FET devices integrated in a pulsing oscillator circuit was 10 ppb. Unfortunately, no data are reported on the linearity range.

As already demonstrated for DH-α6T based OTFTs and for differently substituted thiophene oligomers exposed to organic and inorganic species, for the D3ANT OTFT a sensibility that increased with the gate bias can also be seen. In Fig. 2.9, the slopes of the calibration

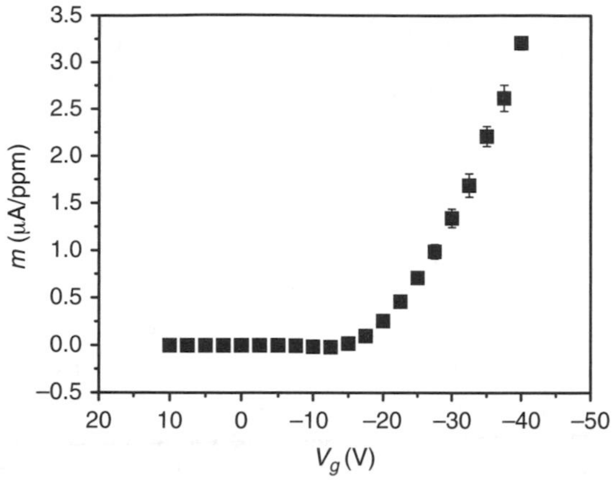

Figure 2.9 The slopes of the linear calibration curves of D3ANT OTFT exposed to NO_2 at gate voltage ranging from +10 to −40 V with associated error bars.

curves for NO_2 reported as a function of the gate voltage are shown with the associated error bars. OTFT sensor exhibits a sensitivity enhancement of 4 orders of magnitude while the device switches from the off to the on regime. The sensitivity enhancement induced by the gate bias is a general property of an OTFT that makes it particularly promising for sensing applications.

In conclusion, in this chapter we described a new p-type organic semiconductor with high stability in air, processable from solution and with an amorphous structure, exhibiting differential sensibility to NO and NO_2 down to the ppb level and very low cross-sensitivity to potential interfering gases such as carbon monoxide and hydrogen sulfide. The future aim of this research project is to develop low-cost and reliable NO_x sensors easily implementable in portable detectors for a fast environmental pollutant monitoring as well as for diagnostic applications. Such goals could allow the control of the physiological state of unhealthy subjects exposed to unsafe atmosphere, as well as the pollution level of the area surrounding them.

2.4 Gold Nanoparticle-Modified FET Sensors for NO_x Detection

2.4.1 Introduction

Nanoparticle (NP) based gas sensors have been strongly investigated in the last decade since they can offer several advantages, such as the increased surface area–volume ratio, and new reactivity properties resulting in brand new or improved sensing features, in terms of sensitivity, selectivity, as well as response and recovery time.[150] Moreover, the size dependence of NP properties may be exploited to tune the sensor performance level, through a proper choice of NP morphology and structure.[151] Frequently, metal NPs are synthesized in the presence of organic capping agents, thus giving rise to *core-shell* structures in which an inorganic core is surrounded and stabilized by an organic shell;[152] the latter may play a crucial role in driving the final NP properties, including its sensing performance. The present section deals with FET devices based on the use of *gold* core–*quaternary ammonium* shell NPs. The use of this transition metal was due to its peculiar reactivity toward nitrogen oxide target analytes,[153] and will be discussed in detail in the next sections. In fact, the active layer of such FET devices is not purely organic: it is a hybrid nanocomposite film. The analytical results reported in the following show that the organic capping agent plays a key role in preserving the catalytic activity of the Au nanophases and significantly influences the properties of the active layer.

Gold nanoparticles (Au-NPs) have attracted a great interest in the last decades because of their unique chemical and physical properties, but also for the wide range of potential applications (optics,[154] sensing,[155] electronics,[156] catalysis,[157] etc.). As catalyst active material, Au-NPs have not fulfilled the initial expectations because of their substantially low catalytic activity due to their completely filled d-band. Nevertheless, recently, it has been found that a relatively high number of chemical reactions can be catalyzed by gold structures in the nanometer range.[158] In a recent report, the catalytic activity of Au-NPs of different size was investigated in the reduction of aromatic nitro compounds. The authors showed the strict relationship between nanoparticle size and the reaction rate for a wide range of particle diameters: in particular, the bigger the particles, the slower the reaction.[159] The first study on Au-NPs based gas sensor was reported by Wohltjen and Snow. They reported that, using a catalytic thin film composed by metallic NPs stabilized by an organic thiol, a good sensibility in the detection of toluene, tetrachloroethene, 1-propanol, and water can be reached. The electrical conductivity of the particle film showed a strong dependence on the core size and the thickness of the organic capping layer.[153] A great number of applications of Au-NPs as sensing layer have since been published, particularly based on the use of core-shell NPs stabilized by thiols[160–161] and/or with several different functionalities,[162] self-assembled coatings from Au-NPs and dendrimers,[163] gold nanotriangles,[164] etc. A promising feature of Au-NPs based sensing devices was their ability to perform a selective detection of target analytes, the sensitivity of such systems being affected by the particle size.[165] Subsequently, other studies were published on further developments of these concepts, focusing on the introduction of functional groups in the organic layer in order to tune the sensor toward various organic vapours.[160,162,165–166]

In the last years, several types of NO_x sensors have been developed, most based on changes of conductance of different metal-oxide

(SnO_2, TiO_2, WO_3) film or organic materials during adsorption on NO_x.[167] In the intense research for the design of innovative gold-based sensors, important efforts were aimed to the detection of NO_x, due to a particularly strong affinity of NO_x to a gold surface underlined by Lu and coworkers[168] as a support of the idea of Au sensing layers for NO_x detection. More recently, Langmuir-Schaeffer layers of thiol-stabilized Au-NPs have been used by Hanwell and coworkers for NO_2 monitoring at room temperature.[169] The authors pointed out the strong influence on the sensor performance of both the particle size and the composition of outer functional groups.

In 2001, an innovative field-effect gas sensor based on a thermally evaporated nanostructured Au film was proposed for the detection of NO_2.[170] The sensor showed a very low sensitivity to interferents such as H_2 and CO and a preferential detection of NO_2 with respect to NO; better results have been shown to correspond to thinner gate layers with smaller Au grain sizes. The high sensitivity of this type of sensor could be explained by the large surface area arising from the adsorption on nanometer-size particles. Two studies addressed the effects of particle size on sensor features such as sensitivity.[165,171] Baratto and coworkers have proposed the use of Au-doped microporous silicon layers for selective and sensitive sensing of NO_x. As a result, they obtained a device response to NO that was comparable to that to NO_2.[172] Steffes and coworkers have pointed out the improvement in the sensing properties of In_2O_3 toward NO_2 by adding finely distributed gold nanoparticles.[173] Langmuir-Schaeffer layers of thiol-stabilized Au-NPs have proved to be sensitive to NO_2 at concentration levels of 0.5 ppm, sensor performance being influenced by the composition of outer functional groups bound to the thiol stabilizer and the particle size.[169] Noteworthy, recent studies on the NO_x/Au system demonstrated that a slow response and/or recovery affected the performance of sensing devices.[169,170] This evidence can be interpreted in terms of the peculiar interaction between the NO_2 molecules and the Au-NPs surface, leading to residual polarization phenomena in the active gates.[170] Finally, a device was proposed by assembling a gold nanostructured film on the top of zinc oxide nanowires. This sensor was sensitive to both reducing (methanol) and oxidizing (nitrogen dioxide) gases at high temperature.[174] These examples show that such inorganic-organic hybrid systems enable highly selective detection of different compounds.

In our laboratory, electrochemically synthesized Au-NPs were used as active gate materials in FET sensor for the NO_x monitoring.[175–176]

2.4.2 New Materials

As mentioned, Au-NPs are of great academic and industrial interest, due to their possible application in several fields, such as optics,[154] catalysis,[157] medicine,[177] microelectronics[156] and sensor technology.[155]

Because of their size- and shape-dependent properties[178–182] NPs in the size range of 1 to 100 nm may show an intermediate behavior between the atomic and the bulk material, although frequently they are demonstrated to possess brand-new properties. A deep understanding of how the electron energy distribution changes from discrete levels (for atoms and very small particles) to continuous bands (for bulk metals) is at the basis of many useful applications. The high relative abundance of surface atoms (and electronic states) that is typical of nano-sized materials dramatically increases the importance of surface chemistry in tuning the macroscopic properties of the nanomaterial. Au-NPs show surprising properties: while in the bulk form, gold is commonly considered an inert material; when it is finely dispersed, at the nanometer scale, it shows promising catalytic properties[158] that can be tuned by controlling the particle size and structure.[183–185]

Tens of different preparation strategies of Au-NPs have been proposed, and their detailed discussion is beyond the scope of this chapter. Many reviews on this topic and useful information can be found in Ref. 158 and Refs. 185 to 197.

In the work described here, the synthesis of gold nanoparticles was carried out according to an electrochemical process called the *sacrificial anode electrolysis* (SAE), which was reported for the first time by Reetz and Helbig in 1994.[198]

The SAE synthesis was carried out in a three-electrode cell, filled with an electrolytic solution composed of a quaternary ammonium salt (tetra-octyl-ammonium chloride, TOAC) dissolved in tetrahydrofuran/acetonitrile mixed solution (see reference for experimental details).[198] In SAE processes, the ammonium salt acts both as supporting electrolyte and as NP stabilizer, thus leading to a stabilized colloidal suspension of core-shell NPs. The shell of such nanostructures has been demonstrated to be composed of a monolayer of quaternary ammonium moieties, and its thickness approximately corresponds to the length of the alkyl chains.[199]

SAE electrochemical route offers several advantages in terms of reduction of the overall cost, high morphological and chemical stability, as well as the possibility of easily tuning the nanoparticle size, the NP diameter being correlated to the process parameters and particularly to the applied current density.

2.4.3 Key Features of the Nanostructured Active Layers

Since particle size is expected to have a substantial impact on sensor performance, a great deal of work was devoted to investigate the Au core modulation by means of the Reetz and Helbig's SAE approach. An inverse correlation between the applied current density and the NP core diameter is generally expected in galvanostatic SAE.[198] In the case of gold, however, we found that this results in just a slight size modulation of the Au-NPs cores.

A set of electrosyntheses were carried out by applying different current density values J_{app} to the cell, ranging from –0.5 mA/cm^{-2} to –15.0 mA/cm^{-2}. At the end of the process, the J_{app} values were corrected by taking into account the effective electroconversion yield, and the resulting effective current density J_{eff} was used for further quantitative considerations.[200] Both UV-vis spectroscopy and Transmission Electron Microscopy (TEM) were used to study the electrosynthesized NPs, and the correlation between the results and the J_{eff} values is shown in Fig. 2.10.[201]

In particular, UV-vis spectroscopy was used to study the surface plasmon resonance (SPR) peak position of Au-NPs. This is a well-known size-dependent spectral feature and can be used to achieve information on the NP size modulation.[202–203] In Fig. 2.10*c*, the change in the SPR position vs. the current density is reported. An even more evident size-dependent correlation can be achieved by comparing the Full Width at Half Maximum (FWHM) of NP samples synthesized at different current densities (Fig. 2.10*b*). When the current density decreased, a corresponding decrease in the FWHM was observed. According to Link and coworkers, this evidence is correlated to NP size changes, and, in particular, a FHWM decrease corresponds to an increase of the nanophase mean diameter.[204] The differences in the NP-core size resulted in slight changes of the SPR peak position and much more evident changes of its FWHM value. These results provide useful diagnostic tools for spectroscopic investigation of Au-NPs. Furthermore, they are in excellent agreement with the theoretical predictions and with the TEM investigations reported in the Fig. 2.10*a*. The latter panel shows that milder electrosynthesis conditions led to greater core diameter (7.9 nm) with a narrower size dispersion (±1.0 nm), while higher current densities resulted in smaller core diameters (5.5 nm) and broader diameter distribution (±2.2 nm).

Before use as a gate material in field-effect sensors, Au-NPs were subjected to a mild thermal treatment at 200°C for 1 h (higher temperatures or longer annealing times resulted in excessive change of the Au-NP morphology). The effect of the annealing treatment on the Au-NP morphology was studied using Scanning Electron Microscopy (SEM) and TEM (data not shown). Both techniques clearly show that the material is still nanostructured, and there is a homogeneous in-plane distribution of Au-NPs. Gold clusters show a spherical morphology, but an appreciable increase in their size (up to 50 nm) can be observed with respect to pristine particles.[175]

X-ray photoelectron spectroscopy (XPS) was used to study the surface chemical composition of both pristine and thermally annealed nanomaterials. Independent of the electrolysis conditions, carbon, nitrogen, and chlorine are the most abundant elements on the surface

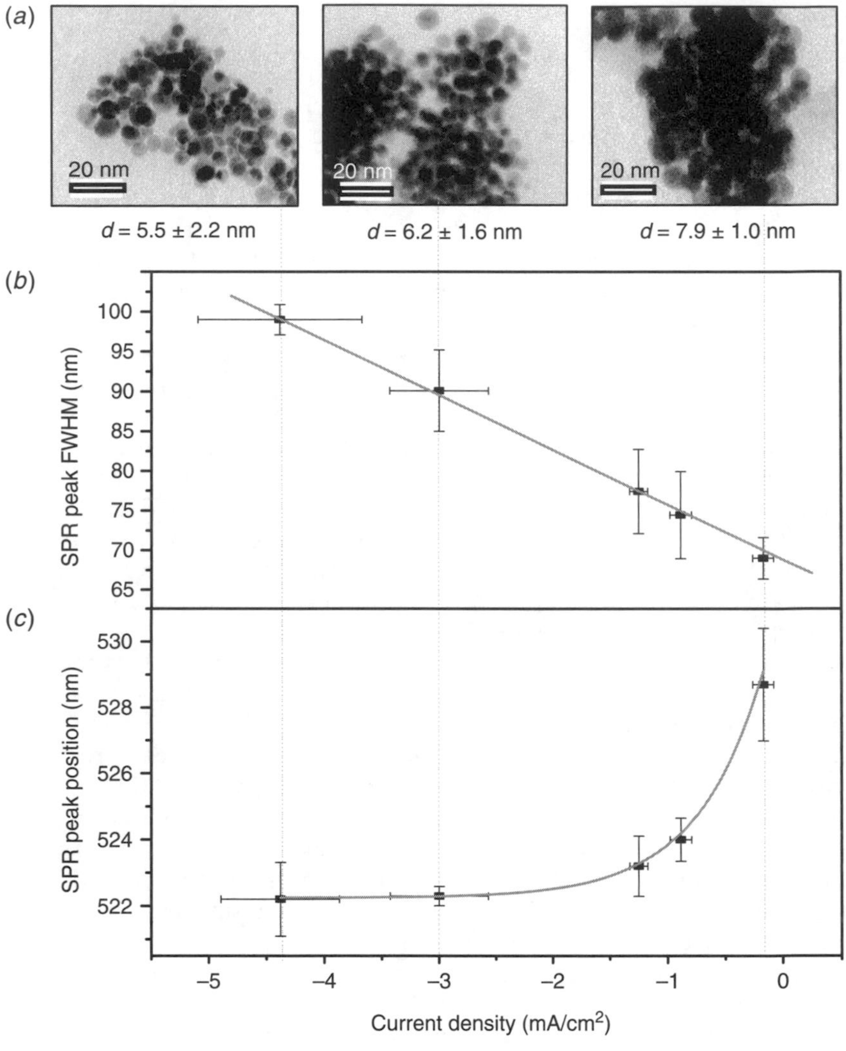

FIGURE 2.10 (*a*) TEM micrographs of Au-NPs electrosynthesized at different current densities in the presence of TOAC; the average core diameter was calculated over more than 500 particles; (*b*) dependence of surface plasmon resonance (SPR) band FWHM (Full Width at Half Maximum); and (*c*) position on the applied current density. Current densities have been corrected by the process yield, namely, by the percentage of the electrolysis charge effectively spent for the Au_{bulk} to Au-NPs conversion. Data have been obtained as the mean value over at least three replicate experiments. (*Reprinted with permission from Ref. 201: E. Ieva and N. Cioffi, in Nanomaterials: New Research Developments, Egor I. Pertsov (ed.), Nova Science Publishers, New York, 2008.*)

Element	C	Au	N	Cl	O	Si	C/Au Ratio
Pristine Au-NPs	92.3%	0.1%	3.9%	2.4%	1.3%	—	923
Annealed Au-NPs	34.6%	1.3%	3.2%	—	36.2%	24.7%	27

TABLE 2.1 Surface Atomic Concentrations Recorded by XPS on Pristine and Annealed Au-NPs. The error in the atomic percentages is ±0.1% for gold and ±0.3% for the other elements.

of pristine Au-NP films. This is due to the high concentration of the surfactant present as base electrolyte in the colloidal solution. Gold and oxygen are detected as well, but they are present at lower concentration levels. For pristine Au-NP films, the carbon/gold surface atomic ratio is close to 10^3. Thermal annealing causes an appreciable removal of the organic matter from the sample surface: carbon and nitrogen surface concentrations are noticeably decreased, while chlorine is completely removed. Gold surface concentration is increased significantly; consequently, the C/Au ratio is lowered by more than one order of magnitude. Finally, oxygen and silicon signals increase, and this occurs since a part of the $Si/SiO_2/SiO_x$ substrate becomes exposed upon heating (see Table 2.1 for details).

Interestingly, the heating treatment allowed a larger catalytically metal area to be exposed to the gas molecules, increased the electrical conductivity of nanostructured gold, and greatly improved its thermal stability.

High-resolution XP spectra of pristine and annealed Au-NP are reported in Fig. 2.11, where they can be compared to a reference spectrum recorded on bulk metallic gold.

The Au4f region of pristine NPs (Fig. 2.11*a*) is composed by two doublets, relevant to two chemical states: nanostructured elemental Au (binding energy $BE_{Au4f7/2} = 83.0 \pm 0.1$ eV) and Au(I) chlorides, most probably as $(NR_4)AuCl_2$ species ($BE_{Au4f7/2} = 84.5 \pm 0.1$ eV).[175]

The Au4f region of annealed NPs (Fig. 2.11*b*) is composed only by one doublet, ascribed to nano-$Au^{(0)}$ ($BE_{Au4f7/2} = 83.7 \pm 0.1$ eV). The higher BE value of the latter peak, as compared to what detected in case of pristine NPs, is ascribed to size-dependent chemical shifts[205] and is in perfect agreement with the size increase observed by electron microscopies.

The thermal annealing procedure is critical since, on the one hand, it is necessary to increase the chemical stability and the conductivity of the nanostructured film while, on the other hand, it caused the increase of the particle dimensions and partial degradation of the protective organic shell. A sketch outlining the changes induced on Au-NPs by thermal annealing is shown in Fig. 2.12.

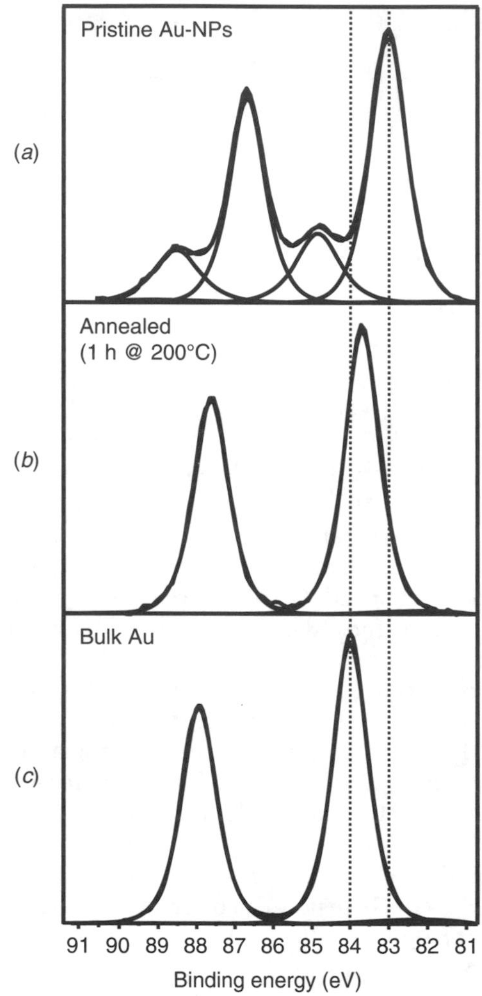

FIGURE 2.11 High-resolution Au4f XPS of (*a*) pristine and (*b*) annealed Au-NPs, stabilized by TOAC. (*c*) The same region, recorded in case of a bulk gold sample, is reported for comparison.

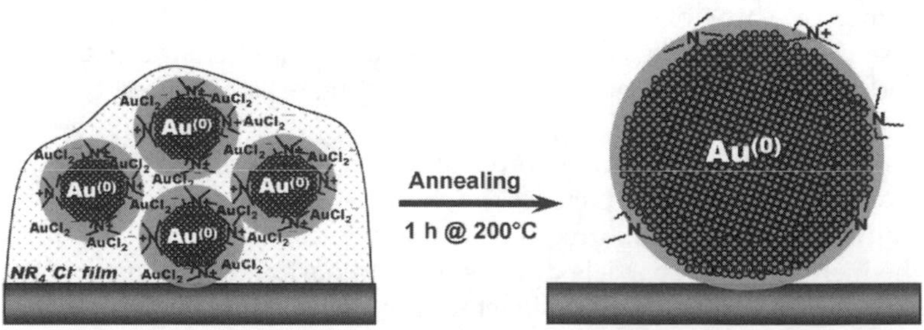

FIGURE 2.12 Sketch of the changes induced on Au-NPs by thermal annealing.

2.4.4 Gas Sensing Results and Perspectives of the Study

Annealed Au-NP layers were used as active material in capacitive FET sensing devices consisting of p-doped Si semiconductor with a thermally grown SiO_2 insulating layer. The ohmic backside contact consisted of evaporated, annealed Al. Bonding pads of evaporated Cr/Au were then deposited on the insulator. The sensor chip, a ceramic heater, and a Pt-100 element for temperature control were mounted on a 16-pin holder and then bonded.

A fixed volume (0.5 μL) of the colloidal gold solution was drop-cast on the SiO_2 surface of the capacitor, partially overlapping the bonding pad and subjected to the thermal heating, prior to gas sensing measurements.

The first set of experiments aimed to investigate the effects of the operating temperature on the sensing performance. Measurements were performed at 150 and 175°C under a N_2/O_2 (90 to 10%) carrier gas flow. Typical calibration curves of FET capacitive sensors are reported in Fig. 2.13; note that a couple of preliminary injections (200 ppm of NO_2) were used to condition the sensor, and then two

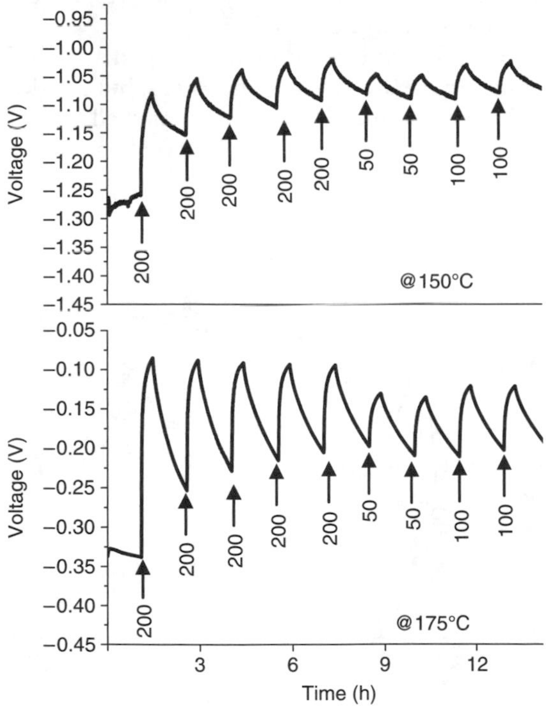

FIGURE 2.13 Response curves of a sensor based on TOAC-stabilized Au-NPs and exposed to NO_2 in a nitrogen/oxygen carrier flow at 150 and 175°C. (*Reprinted with permission from Ref. 175: E. Ieva, K. Buchholt, L. Colaianni, N. Cioffi, L. Sabbatini, G. C. Capitani, A. Lloyd Spetz, P. O. Käll, and L. Torsi, Sensor Letters, 6:577–584, 2008.*)

replicate injections were carried out at any investigated concentration level. At both the working temperatures (150 and 175°C) Au-NP-FET sensors were able to detect NO_2 in a concentration range ranging from 50 to 200 ppm, with no significant difference when NO was employed. Furthermore, the signal-to-noise (*S/N*) ratio was almost doubled upon increasing the operating temperature from 150 to 175°C.

Note, however, that after the NO_x pulse, the sensor did not recover back to the initial baseline, thus indicating that some irreversible interactions take place between NO_x and the Au-NP film. Such evidence has already been reported in Refs. 169 and 170, relevant to similar Au-NPs, capped by other stabilizers. Data of Fig. 2.13 confirm that slow sensor features are intrinsically related to the Au-NO_x system, regardless of the capping agent and particle size.

The selectivity was evaluated by exposing the sensor to different gases: CO, H_2, NH_3, and C_3H_6. Typical sensor response to interferent and target species (at 175°C) is reported in Fig. 2.14.

Exposing the sensor to NO_x and NH_3 caused quite similar peak intensities, although changes of the output voltage occurred in opposite directions, due to the different oxidizing/reducing character of the analytes.[170] The sensor showed also a small response to H_2, while it did not respond at all to CO and C_3H_6.

As the operation temperature was changed to 150°C (data not shown), the selectivity toward NO_x increased, but a reduction in the response and recovery time was observed as well, with the sensing response strongly influenced by the operation temperature.

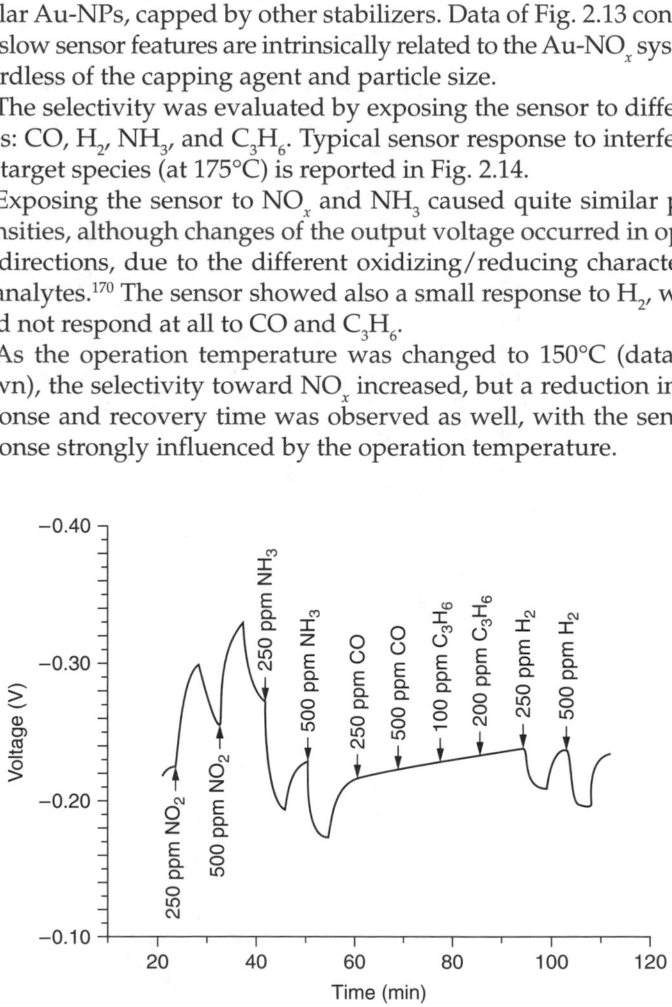

FIGURE 2.14 Responses of the Au-NP sensor to NO_2 and interfering species, measured at 175°C. (*Reprinted with permission from Ref. 175: E. Ieva, K. Buchholt, L. Colaianni, N. Cioffi, L. Sabbatini, G. C. Capitani, A. Lloyd Spetz, P. O. Käll, and L. Torsi, Sensor Letters, 6:577–584, 2008.*)

The results confirm that Au-NP based FET sensors are able to detect NO_x at concentration levels ranging from 50 to 200 ppm, with high selectivity toward other species The increase of the operating temperature from 150 to 175°C doubled the S/N ratio and improved the sensitivity and recovery features, although a certain decrease in the selectivity was observed as well.

We recently obtained some promising results on In_2O_3-NPs doped with electrosynthesized Au-NPs as catalytic sensing layer in capacitive FET sensors for NO_x monitoring.[206] The Au-In_2O_3 mixed system was shown to exhibit enhanced sensing properties, at temperatures as high as 400°C.

For low-temperature applications, active layers exclusively composed of Au-NPs remain an interesting sensing material for NO_x detection, and further research is in progress to improve their stability at higher operating temperatures.

The possibility to work at temperatures higher than 200°C will also enable faster response and recovery times, and this will allow the sensor implementation in real systems, such as the manifold of cars, for the online monitoring of NO_x.

References

1. K. Alving, E. Weitzberg, and J. M. Lundberg, *Eur. Respir. J.*, 6:1368–1370 (1993).
2. Y. H. Chen, W Z. Yao, B. Geng, Y. L. Ding, M. Lu, M.W. Zhao, and C. S. Tang, *Chest*, 128:3205–3211 (2005).
3. I. Horvath, S. Loukides, and T. Wodehouse, *Thorax*, 53:867–870 (1998).
4. L. E. Otterbein, F. H. Bach, J. Alam, M. Soares, V. Lu, M. Wysk, R. J. Davis, et al., *Nat. Med.*, 6:422–428 (2000).
5. J. D. Antuni, S. A. Kharitonov, D. Hughes, M. E. Hodson, and P. J. Barnes, *Thorax*, 55:138–142 (2000).
6. M. D. Mines, *Annu. Rev. Pharmacol. Toxicol.*, 37:517–554 (1997).
7. H. J. Vreman, R. J. Wong, and D. K. Stevenson, "Carbon Monoxide in Breath, Blood, and Other Tissues," in *Carbon Monoxide Toxicity*, D. G. Penney (ed.), CRC Press, Boca Raton, Fa., 2000, pp. 19–59.
8. M. Yamaya, M. Hosoda, S. Ishizuka, M. Monma, T. Matsui, T. Suzuki, K. Sekizawa, et al., *Clin. Exp. Allergy*, 31:417–422 (2001).
9. S. Lim, D. Groneberg, A. Fischer, T. Oates, G. Caramori, W. Mattos, I. Adcock, et al., *Am. J. Respir. Crit. Care Med.*, 162:1912–1918 (2000).
10. P. Paredi, P. L. Shah, P. Montuschi, P. Sullivan, M. E. Hodson, S. A. Kharitonov, and P. J. Barnes, *Thorax*, 54:917–920 (1999).
11. M. Scharte, H. G. Bone, H. Van Aken, and J. Meyer, *Biochem. Biophys. Res. Commun.*, 267:423–426 (2000).
12. H. H. Schmidt, H. Hofmann, U. Schindler, Z. S. Shutenko, D. D. Cunningham, and M. Feelisch, *Proc. Nat. Acad. of Sci. USA*, 93(25):14492–14497 (1996).
13. S. Moncada, M. W. Radomski, and R. M. Palmer, *Biochem. Pharmacol.*, 37:2495–2501 (1988).
14. W. Xu, S. Zheng, R. A. Dweik, and S. C. Erzurum, *Free Radical Bio. Med.*, 41:19–28 (2006).
15. B. Gaston, J. M. Drazen, J. Loscalzo, and J. S. Stamler, *Am. J. Respir. Crit. Care Med.*, 149:538–551 (1994).
16. P. J. Barnes, and M. G. Belvisi, *Thorax*, 48:1034–1043 (1993).
17. S. Kharitonov, K. Alving, and P. J. Barns, *Eur. Resp. J.*, 10:1683–1693 (1997).

18. N. Kisson, L. Duckworth, K. Blake, S. Murphy, and P. E. Silkoff, *Pediatr. Pulmonol.*, 28:282–296 (1999).
19. M. O. Loveless, C. R. Phillips, G. D. Giraud, and W. E. Holden, *Thorax*, 52:185–186 (1997).
20. I. M. Balfour-Lynn, A. Laverty, and R. Dinwiddie, *Arch. Dis. Child.*, 75:319–322 (1996).
21. M. Bhatia, L. Li, and P. K. Moore, *Drug Discov. Today: Disease Mechanism*, 3:71–75 (2006).
22. D. J. Stuehr, *Biochim. Biophys. Acta*, 1411:217–230 (1999).
23. G. Folkerts, W. W. Busse, F. P. Nijkamp, R. Sorkness, and J. E. Gern, *Am. J. Respir. Crit. Care Med.*, 157:1708–1720 (1998).
24. S. Zheng, B. P. De, S. Choudhary, S. A. Comhair, T. Goggans, R. Slee, B. R. Williams, et al., *Immunity*, 18:619–630 (2003).
25. K. S. Aulak, T. Koeck, J. W. Crabb, and D. J. Stuehr, *Am. J. Physiol. Heart Circ. Physiol.*, 286:H30–H38 (2004).
26. A. J. Gow, C. R. Farkouh, D. A. Munson, M. A. Posencheg, and H. Ischiropoulos, *Am. J. Physiol. Lung Cell. Mol. Physiol.*, 287:L262–L268 (2004).
27. X. Chen, K. H. Jhee, and W. D. Kruger, *J Biol. Chem.*, 279:52082–52086 (2004).
28. M. Bhatia, F. L. Wong, D. Fu, H.Y. Lau, S. M. Moochhala, and P. K. Moore, *FASEB J.*, 19:623–625 (2005).
29. M. Yusuf, B. T. Kwong Huat, A. Hsu, M. Whiteman, M. Bhatia, and P. K. Moore, *Biochem. Biophys. Res. Commun.*, 333:1146–1152 (2005).
30. R. L. Jacobs, J. D. House, M. E. Brosnan, and J. T. Brosnan, *Diabetes*, 47:1967–1970 (1998).
31. Y. H. Chen, W. Z. Yao, B. Geng, Y. L. Ding, M. Lu, M. W. Zhao, and C. S. Tang, *Chest*, 128:3205–3211 (2005).
32. P. K. Moore, M. Bhatia, and S. Moochhala, *Trends Pharmacol. Sci.*, 24:609–611 (2003).
33. Y. Cheng, J. F. Ndisang, G. Tang, K. Cao, and R. Wang, *Am. J. Physiol. Heart Circ. Physiol.*, 287:H2316–H2323 (2004).
34. G. Tang, L. Wu, W. Liang, and R. Wang, *Mol. Pharmacol.*, 68:1757–1764 (2005).
35. W. Gopel, G. Reinhardt, and M. Rosch, *Solid State Ionics*, 136,:519–531 (2000).
36. N. Miura, G. Lu, and V. Yamazoe, *Solid State Ionics*, 136:533–542 (2000).
37. F. H. Garzon, R. Mukundan, and E. L. Brosha, *Solid State Ionics*, 136:633–638 (2000).
38. G. Reinhardt, R. Mayer, and M. Rosch, *Solid State Ionics*, 150:79–92 (2002); N. Miura, G. Lu, M. Ono, and N. Yamazoe, *Solid State Ionics*, 117:283–290 (1999).
39. P. Schmidt-Zhang, K. P. Sandow, F. Adolf, W. Gopel, and U. Guth, *Sens. Actuat. B*, 70:25–29 (2000).
40. A. Forleo, L. Francioso, M. Epifani, S. Capone, A. M. Taurino, and P. Siciliano, *Thin Solid Films*, 490:68–73 (2005).
41. S. T. Shishiyanu, T. S. Shishiyanu, and O. I. Lupan, *Sens. Actuat. B*, 107:379–386 (2005).
42. H. Meixner and U. Lampe, *Sens. Actuat. B*, 33 :198–202 (1996).
43. S. Nicoletti, L. Dori, G. C. Cardinali, and A. Parisini, *Sens. Actuat. B*, 60:90 (1999).
44. C. Pijolat, C. Pupier, M. Sauvan, G. Tournier, and R. Lalauze, *Sens. Actuat. B*, 59:195–202 (1999).
45. N. Koshizaki and T. Oyama, *Sens. Actuat. B*, 66:119–121 (2000).
46. E. Comini, L. Pandolfi, S. Kaciulis, G. Faglia, and G. Sberveglieri, *Sens. Actuat. B*, 127:22–28 (2007).
47. E. Comini, *Anal. Chim. Acta*, 568:28–40 (2006).
48. R. P. Gupta, Z. Gergintschew, V. Schipanski, and P. D. Vyas, *Sens. Actuat. B*, 56:65–72 (1999).
49. M. Law, V. Kind, V. Messer, V. Kim, and V. Yang, *Angew. Chem.*, 41:2405–2408 (2002).
50. L. Torsi and A. Dodabalapur, *Anal. Chem.*, 77:380A–387A (2005).
51. V. Persaud and G. H. Dodd, *Nature* (*London*), 299:352–355 (1982).

52. J. W. Gardner and V. Bartlett. *Electronic Noses—Principles and Applications*, Oxford Science Publication, Oxford, 1999.
53. K. C. Persaud, *Mater. Today*, 8:38–44 (2005).
54. D. M. Wilson, S. Hoyt, J. Janata, K. Booksh, and L. Obando, *IEEE Sens. J.*, 1:256–274 (2001).
55. J. Janata and M. Josowicz, *Nature*, 2:19–24 (2003).
56. L. Torsi, A. Dodabalapur, L. Sabbatini, and P. G. Zambonin, *Sens. Actuat. B*, 67:312–316 (2000).
57. B. K. Crone, A. Dodabalapur, R. Sarpeshkar, A. Gelperin, H. E. Katz, and Z. Bao, *J. Appl. Phys.*, 91:10140–10146 (2002).
58. B. Crone, A. Dodabalapur, A. Gelperin, L. Torsi, H. E. Katz, A. J. Lovinger, and Z. Bao, *Appl. Phys. Lett.*, 78:2229–2231 (2001).
59. G. Horowitz, *Adv. Mater.*, 10:365–376 (1998).
60. L. Torsi, A. Dodabalapur, and H. E. Katz, *J. Appl. Phys.*, 78:1088–1093 (1995).
61. G. Horowitz and P. Delannoy, *J. Appl. Phys.*, 70:469–475 (1991).
62. T. W. Kelley and C. D. Frisbie, *J. Phys. Chem. B*, 105:4440–4538 (2001).
63. C. D. Dimitrakopoulos and P. R. L. Malenfant, *Adv. Mater.*, 14:99–117 (2002).
64. F. Liao, C. Chen, and V. Subramanian, *Sens. Actuat. B*, 107:849–855 (2005).
65. M. C. Tanese, D. Fine, A. Dodabalapur, and L. Torsi, *Microelectron. J.*, 37:837–840 (2006).
66. B. Crone, A. Dodabalapur, A. Gelperin, L. Torsi, H. E. Katz, A. J. Lovinger, and Z. Bao, *Appl. Phys. Lett.*, 78:3965 (2001).
67. K. C. See, A. Becknell, J. Miragliotta, and H. E. Katz, *Adv. Mater.*, 19:3322–3327 (2007).
68. L. Torsi, M. C. Tanese, N. Cioffi, M. C. Gallazzi, L. Sabbatini, and V. Zambonin, *Sens. Actuat. B*, 98:204–207 (2004).
69. M. Bouvet, *Anal. Bioanal. Chem*, 384:366–373 (2006).
70. L. Torsi, G. M. Farinola, F. Marinelli, M. C. Tanese, O. H. Omar, L. Valli, F. Babudri, et al., *Nat. Mater.*, 7:412–417 (2008).
71. D. A. Bernards, D. J. Macaya, M. Nikolou, J. A. DeFranco, S. Takamatsu, and G. G. Malliaras, *J. Mater. Chem.*, 18:116–120 (2008).
72. L. Torsi, M. C. Tanese, N. Cioffi, M. C. Gallazzi, L. Sabbatini, PG. Zambonin, G. Raos, et al., *J. Phys. Chem. B*, 107:7589–7594 (2003).
73. M. C. Tanese, D. Fine, A Dodabalapur, and L. Torsi, *Biosens. Bioelectron*, 21:782–788 (2005).
74. T. Someya, H. E. Katz, A. Gelperin, A. J. Lovinger, and A. Dodabalapur, *Appl. Phys. Lett.*, 81:3079–3081 (2002).
75. J. E. Anthony, *Chem. Rev.*, 106:5028–5048 (2006).
76. M. Halik, H. Klauk, U. Zschieschang, T. Kriem, G. Schmid, W. Radlik, and K. Wussow, *Appl. Phys. Lett.*, 81:289–291 (2002).
77. T. W. Kelly, L. D. Boardman, T. D. Dunbar, D. V. Muyres, M. J. Pellerite, and T. P. Smith, *J. Phys. Chem. B*, 107:5877–5881 (2003).
78. T. W. Kelley, D. V. Muyres, P. F. Baude, T. P. Smith, and T. D. Jones, *Mater. Res. Soc. Symp. Proc.*, 771:169–179 (2003).
79. Z. T. Zhu, J. T. Mason, R. Dieckmann, and G. G. Malliaras, *Appl. Phys. Lett.*, 81:4643–4645 (2002).
80. Y. Qiu, Y. Hu, G. Dong, L. Wang, J. Xie, and V. Ma, *Appl. Phys. Lett.*, 83:1644–1646 (2003).
81. D. Li, E.-J. Borkent, R. Nortrup, H. Moon, H. Katz, and Z. Bao, *Appl. Phys. Lett.*, 86:042105 (2005).
82. L. Wang, D. Fine, and A. Dodabalapur, *Appl. Phys. Lett.*, 85:6386–6388 (2004).
83. A. R. Brown, A. Pomp, C. M. Hart, D. M. de Leeuw, D. B. M. Klassen, E. E. Havinga, P. Herwig, et al., *J. Appl. Phys.*, 79:2136–2138 (1996).
84. P. T. Herwig and K. Müllen, *Adv. Mater.*, 11:480–483 (1999).
85. A. Afzali, C. D. Dimitrakopoulos, and T. L. Breen, *J. Am. Chem. Soc.*, 124: 8812–8813 (2002).
86. A. Afzali, C. D. Dimitrakopoulos, and T. O. Graham, *Adv. Mater.*, 15:2066–2069 (2003).

87. K.-Y. Chen, H.-H. Hsieh, C.-C. Wu, V. Hwang, and T. J. Chow, *Chem. Commun.*, pp. 1065–1067 (2007).
88. M. P. Cava and D. R. Napier, *J. Am. Chem. Soc.*, 79:1701–1709 (1957).
89. J. E. Anthony, J. S. Brooks, D. L. Eaton, and S. R. Parkin, *J. Am. Chem. Soc.*, 123:9482–9483 (2001).
90. J. E. Anthony, D. L. Eaton, and S. R. Parkin, *Org. Lett.*, 4:15–18 (2002).
91. C. D. Sheraw, T. N. Jackson, D. L. Eaton, and J. E. Anthony, *Adv. Mater.*, 15:2009–2011 (2003).
92. M. M. Payne, S. R. Parkin, J. E. Anthony, C.-C. Kuo, and T. N. Jackson, *J. Am. Chem. Soc.*, 127:4986–4987 (2005).
93. A. Maliakal, K. Raghavachari, H. Katz, E. Chandross, and T. Siegrist, *Chem. Mater.*, 16:4980–4986 (2004).
94. S. J. Park, C.-C. Kuo, J. E. Anthony, and T. N. Jackson, *Tech. Dig. Int. Electron Dev. Meet.*, 113–116 (2006).
95. M. M. Payne, J. H. Delcamp, S. R. Parkin, and J. E. Anthony, *Org. Lett.*, 6:1609–1612 (2004).
96. M. M. Payne, S. A. Odom, S. R. Parkin, and J. E. Anthony, *Org. Lett.*, 6:3325–3328 (2004).
97. M. M. Payne, S. R. Parkin, and J. E. Anthony, *J. Am. Chem. Soc.*, 127:8028–8029 (2005).
98. L. Torsi, A. Dodabalapur, N. Cioffi, L. Sabbatini, and P. G. Zambonin, *Sens. Actuat. B*, 77:7–11 (2001).
99. L. Torsi, A. Tafuri, N. Cioffi, M. C. Gallazzi, A. Sassella, L. Sabbatini, and P. G. Zambonin, *Sens. Actuat. B*, 93:257–262 (2003).
100. A. Assadi, G. Gustafsson, M. Willander, C. Svensson, and O. Ingas, *Synth. Met.*, 37:123–130 (1990).
101. Y. Ohmori, H. Takahashi, K. Muro, M. Uchida, V. Kawai, and K. Yoshino, *Jpn. J. Appl. Phys.*, 30:L1247–L1249 (1991).
102. T. Someya, A. Dodabalapur, A. Gelperin, H. E. Katz, and Z. Bao, *Langmuir*, 18:5299–5302 (2002).
103. V. A. L. Roy, Y.-G. Zhi, Z.-X. Xu, S.-C. Yu, P. W. H. Chan, and V. Che, *Adv. Mater.*, 10:1258–1261 (2005).
104. T. Yasuda, K. Kashiwagi, Y. Morizawa, and T. Tsutsui, *J. Phys D: Appl. Phys.*, 40:4471–4475 (2007).
105. H. Tian, J. Wang, J. Shi, D. Yan, L. Wang, Y. Geng, and F. Wang, *J. Mater. Chem.*, 15:3026–3033 (2005).
106. H. Tian, J. Shi, S. Dong, D. Yan, L. Wang, Y. Geng, and F. Wang, *Chem. Commun.*, 33:3498–3500 (2006).
107. T. Yasuda, K. Fujita, T. Tsutsui, Y. Geng, S. W. Culligan, and S. H. Chen, *Chem. Mater.*, 17:264–268 (2005).
108. Y. H. Kim, D.-C. Shin, S.-H. Kim, C.-H. Ko, H.-S. Yu, Y.-S. Chae, and S. K. Kwon, *Adv. Mater.*, 13:1690–1693 (2001).
109. Z. L. Zhang, X. Y. Jiang, W. Q. Zhu, X. Y. Zheng, Y. Z. Wu, and S. H. Xu, *Synth. Met.*, 137:1141–1142 (2003).
110. X. H. Zhang, M. W. Liu, O. Y. Wong, C. S. Lee, H. L. Kwong, S. T. Lee, and S. K. Wu, *Chem. Phys. Lett.*, 369:478–482 (2003).
111. J. Shi and C. W. Tang, *Appl. Phys. Lett.*, 80:3201–3203 (2002).
112. T.-H. Liu, W.-J. Shen, C.-K. Yen, C.-Y. Iou, H. -H. Chen, V. Banumathy, and C. H. Chen, *Synth. Met.*, 137:1033–1034 (2003).
113. N. Karl and J. Marktanner, *Mol. Cryst. Liq. Cryst.*, 355:149–173 (2001).
114. A. N. Aleshin, J. Y. Lee, S. W. Chu, J. S. Kim, and J.-H. Park, *Appl. Phys. Lett.*, 84:5383–5385 (2004).
115. H. Meng, F. Sun, M. B. Goldfinger, F. Gao, D. J. Londono, W. J. Marshal, G. S. Blackman, et al., *Am. Chem. Soc.*, 128:9304–9305 (2006).
116. J.-H. Park, D. S. Chung, J.-W. Park, T. Ahn, H. Kong, Y. K. Jung, J. Lee, et al., *Org. Lett.*, 9:2573–2576 (2007).
117. H. Yamada, Y. Yamashita, M. Kikuchi, H. Watanabe, T. Okujima, H. Uno, T. Ogawa, et al., *Chem. Eur. J.*, 11:6212–6220 (2005).
118. K. Okuyama, T. Hasegawa, M. Ito, and N. Mikami, *J. Phys. Chem.*, 88:1711–1716 (1984).

119. J. M. Seminario, A. G. Zacarias, and J. M. Tour, *J. Am. Chem. Soc.*, 120:3970–3974 (1998).
120. M. Levitus, K. Schmieder, H. Ricks, K. D. Shimizu, U. H. F. Bunz, and M. A. Garcia-Garibay, *J. Am. Chem. Soc.*, 123:4259–4265 (2001).
121. T. Miteva, L. Palmer, L. Kloppenburg, D. Neher, and U. H. F. Bunz, *Macromolecules*, 33:652–654 (2000).
122. Q. Chu and Y. Pang, *Macromolecules*, 36:4614–4618 (2003).
123. U. H. F. Bunz, *Chem. Rev.*, 100:1605–1644 (2000).
124. U. H. F. Bunz, *Acc. Chem. Res.*, 34:998–1010 (2001).
125. D. M. de Leeuw, M. M. J. Simenon, A. R. Brown, and R. E. F. Einerhand, *Synth. Met.*, 87:53–59 (1997).
126. W. Kern (ed.), *Handbook of Semiconductor Wafer Cleaning Technology: Science, Technology, and Applications*, Noyes Publications, Park Ridge, N. J., 1993.
127. W. Kern and D. A. Puotinen, *RCA Rev.*, 31:187–206 (1970).
128. M. Stolka and M. A. Abkowitz, *Synth. Met.*, 54:417 (1993).
129. H. Sirringhaus, N. Tessler, and R. H. Friend, *Science*, 280:1741–1744 (1998).
130. L. Torsi, A. Dodabalapur, A. J. Lovinger, H. E. Katz, R. Ruel, D. D. Davis, and K. W. Baldwin, *Chem. Mater.*, 7:2247–2251 (1995).
131. A. Dell'Aquila, F. Marinelli, J. Tey, P. Keg, Y.-M. Lam, O. L. Kapitanchuk, P. Mastrorilli, et al., *J. Mater. Chem.*, 18:786–791 (2008).
132. J.-H. Park, D. S. Chung, J.-W. Park, T. Ahn, H. Kong, Y. K. Jung, J. Lee, et al., *Org. Lett.*, 9:2573–2576 (2007).
133. W. Cui, X. Zhang, X. Jiang, H. Tian, D. Yan, Y. Geng, X. Jing, et al., *Org. Lett.*, 8:785–788 (2006).
134. R. Schmidt, S. Gottling, D. Leusser, D. Stalke, A.-M. Krause, and F. J. Wurthner, *Mater. Chem.*, 16:3708–3714 (2006).
135. A. Marchenko, N. Katsonis, D. Fichou, C. Aubert, and M. Malacria, *J. Am. Chem. Soc.*, 124:9998–9999 (2002).
136. N. Katsonis, A. Marchenko, and D. Fichou, *J. Am. Chem. Soc.*, 125:13682–13683 (2003).
137. A. Nion, P. Jiang, A. Popoff, and D. Fichou, *J. Am. Chem. Soc.*, 129:2450–2451 (2007).
138. Y. Shirota and H. Kageyama, *Chem. Rev.*, 107:953–1010 (2007).
139. W. Hu, Y. Liu, Y. Xu, S. Liu, S. Zhou, D. Zhu, B. Xu, et al., *Thin Solid Films*, 360:256–260 (2000).
140. B. Wanga, X. Zuoa, Y. Wua, Z. Chena, and Z. Lia, *Mater. Lett.*, 59:3073–3077 (2005).
141. J. Bruneta, A. Paulya, L. Mazeta, J. P. Germaina, M. Bouvet, and B. Malezieux, *Thin Solid Films*, 490:28–35 (2005).
142. C. Gu, L. Sun, T. Zhang, and T. Li, *Thin Solid Films*, 490:863–865 (1996).
143. F. Naso, F. Babudri, D. Colangiuli, G. M. Farinola, F. Quaranta, R. Rella, R. Tafuro, et al., *J. Am. Chem. Soc.*, 125:9055–9061 (2003).
144. W. Qiu, W. Hu, Y. Liu, S. Zhou, Y. Xu, and D. Zhu, *Sens. Actuat. B*, 75:62–66 (2001).
145. Y.-L. Lee, C-Yi. Sheu, and R.-H. Hsiao, *Sens. Actuat. B*, 99:281–287 (2004).
146. A. Star, J.-C. P. Gabriel, K. Bradley, and G. Grüner, *Nano Lett.*, 3:459–463 (2003).
147. L. Valentini, F. Mercuri, I. Armentano, C. Cantaldini, S. Picozzi, L. Pozzi, S. Santucci, et al., *Chem. Phys. Lett.*, 387:356–361 (2004).
148. J. C. Hsieh, C. J. Liu, and Y. H. Ju, *Thin Solid Films*, 322:98–103 (1998).
149. A. Das, R. Dost, T. Richardson, M. Grell, J. J. Morrison, and M. L. Turner, *Adv. Mater.*, 19:4018–4023 (2007).
150. M. E. Franke, T. J. Koplin, and U. Simon, *Small*, 2:36–50 (2006).
151. G. Jimènez-Cadena, J. Riu, and F. X. Rius, *Analyst*, 132:1083–1099 (2007) and references therein.
152. M.-I. Baraton (ed.), *Synthesis, Functionalization and Surface Treatment of Nanoparticles*, American Scientific Publishers, Stevenson Ranch, Calif., 2003.
153. H. Wohltjen and A. W. Snow, *Anal. Chem.*, 70:2856–2859 (1998).
154. A. K. Sharma and B. D. Gupta, *Photonics Nanostr.*, 3:30–37 (2005).
155. C. A. Mirkin, R. L. Letsinger, R. C. Mucic, and J. J. Storhoff, *Nature*, 382:607–609 (1996).

156. T. L. Chang, Y. W. Lee, C. C. Chen, and F. H. Ko, *Microelectron. Eng.*, 84:1698–1701 (2007).
157. A. Abad, P. Concepción, and A. Corma, *Angew. Chem. Int. Ed. Engl.*, 44:4066–4069 (2005).
158. M. Haruta, *Catalysis Today*, 36:153–156 (1997).
159. S. Panigrahi, S. Basu, S. Praharaj, S. Pande, S. Jana, A. Pal, S. Kumar Ghosh, et al., *J. Phys. Chem. C*, 111:4596–4605 (2007).
160. D. S. Evans, S. R. Johnson, Y. L. Cheng, and T. Shen, *J. Mater. Chem.*, 10:183–187 (2000).
161. C. Grate, D. A. Nelson, and R. Skraggs, *Anal. Chem.*, 75:1868–1879 (2003).
162. H. L. Zhang, S. D. Evans, J. R. Henderson, R. E. Miles, and T. H. Shen, *Nanotechnology*, 13:439–444 (2002).
163. T. Vossmeyer, B. Guse, I. Besnard, R. E. Bauer, K. Muellen, and A. Yasuda, *Adv. Mater.*, 14:238–242 (2002).
164. A. Singh, M. Chaudhari, and M. Sastry, *Nanotechnology*, 17:2399–2405 (2006).
165. L. Han, D. R. Daniel, M. M. Mayer, and C. J. Zhang, *Anal. Chem.*, 73:4441–4449 (2001).
166. Y. Joseph, I. Besnard, M. Rosenberger, B. Guse, H.-G. Nothofer, J. R. Wessels, U. Wild, et al., *J. Phys. Chem. B*, 107:7406–7413 (2003).
167. S. Zhuiykov and N. Miura, *Sens. Actuat. B*, 121:639–643 (2007) and reference therein; T. H. Richardson, C. M. Dooling, L. T. Jones, and R. A. Brook, *Adv. Coll. Interf. Sci.*, 16:81–96 (2005).
168. X. Lu, X. Xu, N. Wang, and Q. Zhang, *J. Phys. Chem A*, 103:10969–10974 (1999) and references therein.
169. M. D. Hanwell, S. Y. Heriot, T. H. Richardson, N. Cowlam, and I. M. Ross, *Coll. Surf. A*, 379:284–285 (2006).
170. D. Filippini, T. Weiss, R. Aragon, and U. Weimar, *Sens. Actuat. B*, 78:195–201 (2001).
171. A. W. Snow, and H. Wohltjen, U.S. Patent, 6221673 B1, 2001.
172. C. Baratto, G. Sberveglieri, E. Comini, G. Faglia, G. Benussi, V. La Ferrara, L. Quercia, et al., *Sens. Actuat. B*, 68:74–80 (2000).
173. H. Steffes, C. Imawan, F. Solzbacher, and E. Obermeier, *Sens. Actuat. B*, 78:106–112 (2001).
174. P. M. Parthangal, R. E. Cavicchi, and M. R. Zachariah, *Nanotechnology*, 17:3786–3790 (2006).
175. E. Ieva, K. Buchholt, L. Colaianni, N. Cioffi, L. Sabbatini, G. C. Capitani, A. Lloyd Spetz, et al., *Sens. Lett.*, 6:577–584 (2008).
176. E. Ieva, K. Buchholt, L. Colaianni, N. Cioffi, I. D. van der Werf, A. Lloyd Spetz, P. O. Käll, et al., in *The IEEE International Workshop on Advances in Sensors and Interfaces*, June 26–27, 2007, Bari, Italy. IEEE catalog no. 07EX1794; ISBN 1-4244-1244-7, 2007.
177. D. Pissuwan, S. M. Valenzuela, and M. B. Cortie, *Trends Biotech.*, 24:62–67 (2006).
178. G. Schmid, *Cluster and Colloids: From Theory to Applications*, VCH, New York, 1994.
179. A. J. Henglein, *Phys. Chem.*, 97:5457–5471 (1993).
180. D. L. Feldein and C. D. Keating, *Chem. Soc. Rev.*, 27:1–12 (1998).
181. J. D. Aiken and R. G. Finke, *J. Mol. Catal. A*, 145:1–44 (1999).
182. M. A. El-Sayed, *Acc. Chem. Res.*, 34:257–264 (2001).
183. C. J. Zhong and M. M. Maye, *Adv. Mater.*, 13:1507–1511 (2001).
184. C. Mohr, H. Hofmeister, J. Radnik, and P. Claus, *J. Am. Chem. Soc.*, 125:1905–1911 (2003).
185. G. J. Hutchings, *Catalysis Today*, 100:55–61 (2005).
186. M. A. Hayat, *Colloidal Gold, Principles, Methods and Applications*, Academic Press, New York, 1989.
187. J. S. Bradley, *Clusters and Colloids*, G. Schmid (ed.), VCH, Weinheim, 1994, Chap. 6, pp. 459–544.
188. J. H. Fendler and F. C. Meldrum, *Adv. Mater.*, 7:607–631 (1995).
189. M. C. Daniel and D. Astruc, *Chem. Rev.*, 104:293–346 (2004).

190. M. Brust and C. Kiely, *J. Coll. Surf. A*, 202:175–186 (2002).
191. M. G. Warner and J. E. Hutchison, *Synthesis, Functionalization and Surface Treatment of Nanoparticles*, M. I. Baraton (ed.), American Scientific Publishers, Calif., 2003, Chap. 5, pp. 67–89.
192. F. Porta and L. Prati, *Rec. Res. Dev. Vacuum Sci. Techn.*, 4:99 (2003).
193. G. Schmid and B. Corain, *Europ. J. Inorg. Chem.*, 17:3081–3098 (2003).
194. L. Pasquato, P. Pengo, and P. Scrimin, *J. Mat. Chem.*, 14:3481–3487 (2004).
195. R. Kumar, A. Ghosh, C. R. Patra, P. Mukherjee, and M. Sastry, *Nanotechn. Cat.*, 1:111 (2004).
196. V. H. Perez- Luna, K. Aslan, P. Betala, and C. Pravin, *Encyclopedia of Nanoscience and Nanotechnology*, H. S. Nalwa (ed.), American Scientific Publishers Stevenson Ranch, Calif., 2004, Chap. 2, pp. 27–49.
197. J. Perez-Juste, I. Pastoriza-Santos, L. M. Liz-Marzan, and P. Mulvaney, *Coord. Chem. Rev.*, 249:1870–1901 (2005).
198. M. T. Reetz and W. Helbig, *J. Am. Chem. Soc.*, 116:7401–7402 (1994).
199. M. T. Reetz, W. Helbig, S. A. Quaiser, U. Stimming, N. Breuer, and R. Vogel, *Science*, 267:367–369 (1995).
200. A part of the electrolysis charge was not involved in the preparation of gold nanophases, but it was spent for side processes; at the end of the electrolysis, the weight loss of the metal reservoir, i.e., the gold electrode, allowed a direct quantification of the amount of gold species that had been converted to Au NPs. The process yield was then calculated as a ratio between the overall electrolysis charge and that effectively spent to produce Au NPs, and finally the effective current density J_{eff} was obtained as product of J_{app} and the yield value.
201. E. Ieva and N. Cioffi, "Electrochemical Synthesis of Colloidal Gold Nanoparticles," in *Nanomaterials: New Research Developments*, Egor I. Pertsov (ed.), Nova Science Publishers, N. Y., 2008, pp. 269–293.
202. G. Mie, *Ann. Phys.*, 25:377–445 (1908).
203. C. Burda, T. Green, C. Landes, S. Link, R. Little, J. Petroski, and M. A. El- Sayed, in *Characterization of Nanophase Materials*, Z. L. Wang (ed.), Wiley-VCH, Weinheim, DE, 2000, pp. 197–241.
204. S. Link and M. A. El Sayed, *J. Phys. Chem. B*, 103:4212–4217 (1999).
205. J. Radnik, C. Mohr, and P. Claus, *Phys. Chem., Chem. Pys.*, 5:172–177 (2003).
206. D. Lutic, M. Strand, A. Lloyd Spetz, K. Buchholt, E. Ieva, P. O. Käll, and M. Sanati, *Topics in Catalysis*, 45:105–109 (2007).

CHAPTER 3

Strain and Pressure Sensors Based on Organic Field-Effect Transistors

Ileana Manunza and Annalisa Bonfiglio

Department of Electrical and Electronic Engineering, University of Cagliari, Italy, and INFM-CNR S3 Centre for nanoStructures and bioSystems at Surfaces, Modena, Italy

3.1 Introduction

Research in biomedicine and engineering during the last years has led to a remarkable interest in sensor technologies. For a broad range of sensing applications there is a large demand for small, portable, and inexpensive sensors. Silicon technology is not suitable for manufacturing low-cost large-area sensor devices that are preferably light, flexible, and even disposable (for some biomedical applications). Its inherent high-temperature fabrication processes make it very difficult to use inexpensive flexible substrate materials, resulting in high fabrication costs. Organic semiconductors have been studied so far mainly for their exceptional combination of electrical conductivity, mechanical flexibility, and, last but not least, low cost of deposition and patterning techniques. On the other hand, several drawbacks still affect these materials: among them, low carrier mobility that severely limits the possibility of applications in electronic circuits. Therefore, to fully exploit the great potential of these materials, it is advisable to focus on those applications where high performances in terms of switching speed are not required. For these reasons, organic semiconductor-based devices offer very interesting opportunities for sensor applications due to the low cost and easy fabrication techniques, and the possibility of realizing devices on large and flexible areas on unusual substrates such as paper, plastic, or fabrics. In fact, being able to obtain large sensing areas is certainly a benefit for a wide set of applications, and using printing techniques for creating sensing devices on unusual substrates could certainly widen the set of possible applications where sensing is required. For instance, many body parameters can be measured by using non-invasive sensors, among them geometric and mechanically related parameters such as respiration rate and amplitude, heart rate, blood pressure, position, detection of falls, monitoring of various daily activities, etc.

In the following sections, we will focus our attention on organic semiconductor strain/pressure sensors. We will present the state of the art of technologies and applications and will show how to take advantage of a flexible, free-standing dielectric film for obtaining sensors on conformable, large surfaces.

3.2 Working Principles of Organic Field-Effect Transistor Sensors

In a field-effect device, the input signal is a voltage, applied through a capacitive structure to the device channel. This signal, named *gate voltage*, modulates the current flowing into a narrow portion of semiconductor (the channel) comprised between two ohmic contacts, the source and drain. A field-effect sensor is based on the idea that in the presence of the parameter to be sensed (whatever it is, for instance, a chemical compound), the current is reversibly affected (even by keeping constant the voltages applied to gate and drain) and this modification can be exploited to detect the parameter itself. An organic (semiconductor based) field-effect transistor—(OFET) or organic thin-film transistor (OTFT)—is usually realized in a thin film configuration which is a structure that was developed for the first time for amorphous silicon devices.[1] Several reviews have described how OTFTs operate, so we will present only the most salient aspects.[2]

An OTFT is composed of a multilayered structure where the gate capacitor is formed by a metal, an insulator, and a thin organic semiconductor layer. On the semiconductor side, two metal contacts, source and drain, are used for extracting a current that depends both on the drain-source voltage and on the gate-source voltage. The conductance of the organic semiconductor in the channel region is switched on and off by the gate electrode, which is capacitively coupled through a thin dielectric layer. The gate bias V_g controls the current I_d flowing between the source and drain electrodes under an imposed bias V_d.

The low thickness of the semiconductor is due to the fact that this device does not work in inversion mode; i.e., the carriers that accumulate in the channel are the same that normally flow in the bulk in the off state. Therefore, if the semiconductor were too thick, the off current would be too high and the switching ability of the device would be compromised. For the same reason, this structure is used only for low mobility semiconductors, as high mobility semiconductors would give rise, even in thin film devices, to high off currents. An individual I_d–V_d curve has a linear region at $V_d << V_g - V_t$ and a saturation region at $V_d > V_g - V_t$, where V_t is the device threshold voltage.

The basic device performances are described by the field-effect mobility μ and the on/off ratio. The equations developed for inorganic semiconductor devices are often used to extract the value of the field-effect mobility from the *I*–*V* curves.[3] Strictly speaking, these equations apply only to TFTs that exhibit constant charge carrier mobility; this is not the case for polycrystalline OTFTs. Thus, extracted mobility values should be considered only as estimations of the real value. Typical extracted values are in the 10^{-3} to 10^{-1} cm^2/(V·s) range, but they can be as high as 1 to 10 cm^2/(V·s) with pentacene.[4]

The on/off ratio, defined as the ratio of I_d in the on ($V_g > V_t$) and off ($V_g << V_t$) states, indicates the switching performance of the device. A low off current is desirable to ensure a true switching of the transistor to the off state; this is achieved by keeping the doping level of the organic semiconductor as low as possible. Because of non-intentional doping, this is not easily accomplished with polycrystalline organic active layers. However, on/off values as high as 10^6, suitable for most applications, have been reached with several different organic materials. Another important parameter is the threshold voltage V_t, which marks the passage from the off to the on conductivity regime.[5] As organic semiconductors can in principle support both hole and electron conduction, in principle V_t should be zero.[3] But, as a matter of fact, trap states for electrons or holes heavily affect the transistor performance and, in particular, the value of V_t. Therefore, the threshold voltage V_t corresponds to the voltage required to fill the trap states in the organic material or at the interface with the gate dielectric.[6] To obtain a sensing device from an OTFT, it is necessary to have a specific sensitivity to the parameter to be sensed. In chemical sensors, typically, the sensing mechanism is due to a chemical interaction between the chemical compound to sense and the semiconductor itself, normally affecting its mobility.[2] In detectors for physical parameters (for instance, deformation or pressure sensors), the external stimulus must reversibly affect one of the different layers of the device and result in a variation of one or more of the electronic parameters (mobility, threshold voltage, etc.) that can be extracted from the output curves. In the next paragraph, we describe pressure/strain sensors that can be obtained starting from an organic field-effect transistor.

3.3 Strain and Pressure Sensors

The effect of strain on the mechanical and electronic properties of organic semiconductors is an emerging research topic in fundamental physics and applications. Although mechanical flexibility is one of the main advantages of organic materials, relatively little progress has been made in the field of pressure or bending recognition mainly because mechanical sensing requires attributes of conformability and flexibility and three-dimensional large-area shaping that in many cases are difficult to achieve even for organic devices.

On the other hand, in the application domain, artificial sense of touch is considered an essential feature of future generations of robots, and wearable electronics has become one of the hottest themes in electronics, aiming at the design and production of a new generation of garments with distributed sensors and electronic functions.

3.3.1 State of the Art in Strain and Pressure Sensors Based on Organic Materials

The first example of a large-area pressure network fabricated on a plastic sheet by means of integration of organic transistors and rubbery pressure-sensitive elements was reported by Someya and Sakurai[7] in 2003. The device structure is shown in Fig. 3.1.

To fabricate the organic transistor, at first glass resin was spin-coated and cured on a 50 μm thick polyimide film with an 8 μm thick copper film which acts as the gate electrode. Then a pentacene semiconductor layer of nominal thickness of 30 nm was vacuum-sublimed on these films at ambient substrate temperature. Finally, source and drain gold electrodes were deposited by means of vacuum evaporation through shadow masks. On the other hand, pressure sensors are made by using pressure-sensitive conductive rubbery sheets sandwiched between two 100 μm wide metal lines that cross at a right angle. One of the metal lines is connected to an organic transistor while the other line is connected to the ground. The pressure-sensitive sheet is 0.5 mm thick silicone rubber containing graphite. In this chapter

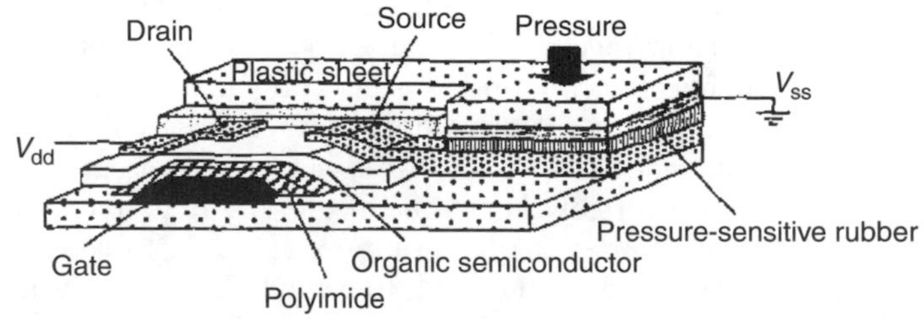

FIGURE 3.1 Schematic of the device structure reported in Ref. 7. (*Reprinted with permission from Ref. 7. Copyright 2003, IEEE.*)

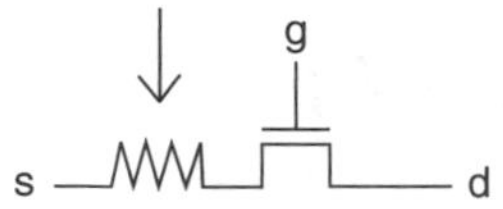

FIGURE 3.2 Equivalent circuit diagram of the device. (*Reprinted with permission from Ref. 7. Copyright 2003 IEEE.*)

organic transistors are used to address the rubber pressure-sensitive elements in a sensor array. The equivalent circuit diagram is shown in Fig. 3.2.

In 2004, Someya et al.[8] improved the fabrication technique and realized an electronic artificial "skin." In this work, once again, organic transistors are not used as sensors in themselves but as an addressing element of a flexible matrix which is used to read out pressure maps from pressure-sensitive rubber elements containing graphite. The obtained electronic artificial "skin" is shown in Fig. 3.3.

The mobility of organic transistors at –100 V is comparable to that of amorphous silicon, but this operating voltage is not realistic for artificial skin applications. At –20 V the mobility is still large [0.3 cm^2/(V·s)] and the device is still functioning. In the active driving method presented, only one transistor needs to be in the on state for each cell where pressure are applied, so this design is suitable for low-power applications where a high number of cells are required over large areas, such as electronic skin.

The device can detect a few tens of kilopascals, which is comparable to the sensitivity of discrete pressure sensors, and the time response of the pressure-sensitive rubber is typically of the order of hundreds of milliseconds.

In a paper dated December 2006,[9] the low-cost manufacturing processes have been further optimized in order to realize the flexible active matrix using ink-jet printed electrodes and gate dielectric layers. This work demonstrates the feasibility of a printed organic FET active

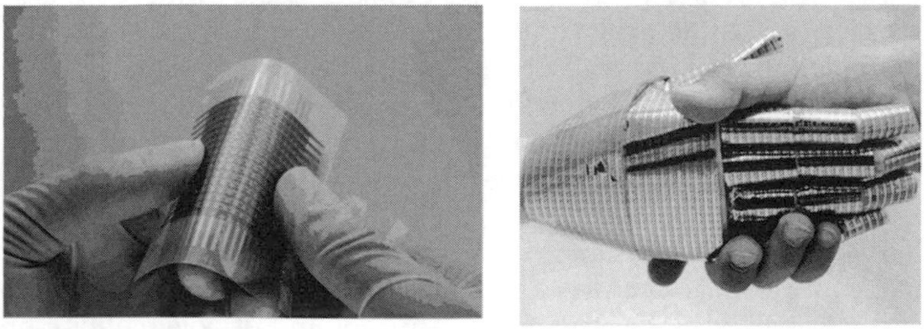

FIGURE 3.3 Electronic artificial skin. (*Reprinted with permission from Ref. 8. Copyright 2004, National Academy of Sciences, USA.*)

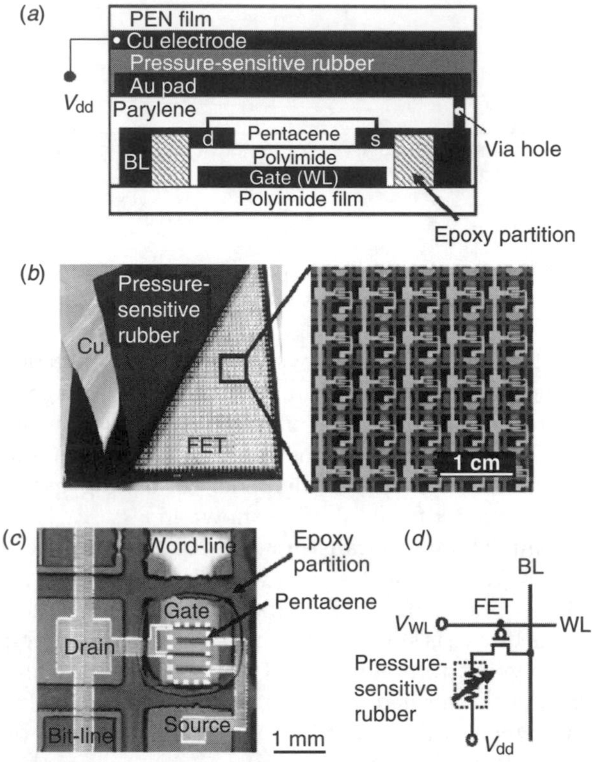

FIGURE 3.4 (*a*) Cross section of a pressure sensor. (*b*) An image of a pressure sensor comprising an organic FET active matrix, a pressure-sensitive rubber, and a PEN [poly(ethylene naphthalate)] film with a Cu electrode. A magnified image of the active matrix is also shown. (c) Micrograph of stand-alone pentacene FETs. (*d*) Circuit diagram of a stand-alone pressure sensor cell. (*Reprinted with permission from Ref. 9. Copyright 2006, American Institute of Physics.*)

matrix as a readout circuit for sensor application. Figure 3.4 shows the cross section of the device structure, an image of the large-area pressure sensor, an image of the stand-alone organic transistor, and the circuit diagram.

Other attempts concerned the direct use of organic semiconductors as sensing elements. For instance, Rang et al.[10] have investigated the hydrostatic pressure dependence of *I*–*V* curves in organic transistors. The device was realized on a heavily doped silicon substrate and measured in a hydrostatic pressure apparatus (a hydraulic press made by the Polish Academy of Sciences). The authors found a large and reversible dependence of drain current and of hole mobility on hydrostatic pressure and suggest that this kind of device could be suitable for sensor applications. However, the proposed device was

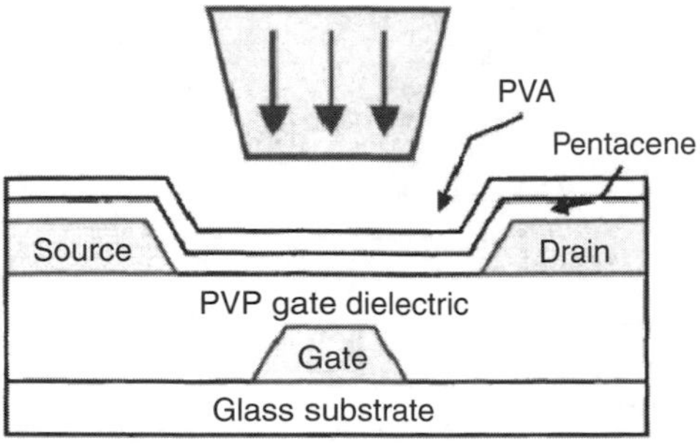

FIGURE 3.5 Schematic of the device structure reported in Ref. 11; PVA is poly(vinyl alcohol). (*Reprinted with permission from Ref. 11. Copyright 2005, American Institute of Physics.*)

not flexible, therefore not suitable for applications such as robot skin, e-textiles, etc.

Darlinski and coworkers[11] studied the possibility of realizing pressure sensors based solely on organic transistors, without the need of any additional sensing element. In this way the organic device itself acts as a sensing element. To study the pressure dependence of the electrical performance of these devices, the authors applied mechanical force directly on the transistors using a tungsten microneedle moved by a step motor, as shown in Fig. 3.5. During measurements, the device substrate is placed on a balance to measure the applied pressure.

The authors explained the force-induced change in the drain current in terms of the variation in the distribution and activity of trap states at or near the semiconductor/dielectric interface. A major drawback of the reported device is that, being realized on a stiff glass substrate, it is not flexible.

Strain sensors using an organic semiconductor as the sensitive (resistive) element of a strain gauge have been also reported by Jung and Jackson.[12] In conventional strain gauges, the large stiffness mismatch generated between the inorganic semiconductor element and the flexible (polymeric) substrate may lead to irreversible plastic substrate deformations, and this can be problematic. The stress in the sensitive element (i.e., the inorganic semiconductor or the metal, with a high Young modulus) is not representative of the stress present in the substrate (i.e., the polymeric material with the low Young modulus). In this way, as a result, the sensor's performance is reduced in terms of reliability and reproducibility. In contrast, it is expected that the use of organic semiconductors with low Young modulus (on the order of 5 GPa) as the sensing element would minimize the induced stress concentration. The sensor cross section is shown in Fig. 3.6.

For these sensors, 2 nm thick Ti and 20 nm thick Au were deposited on 50 μm thick polyimide substrates by thermal evaporation.

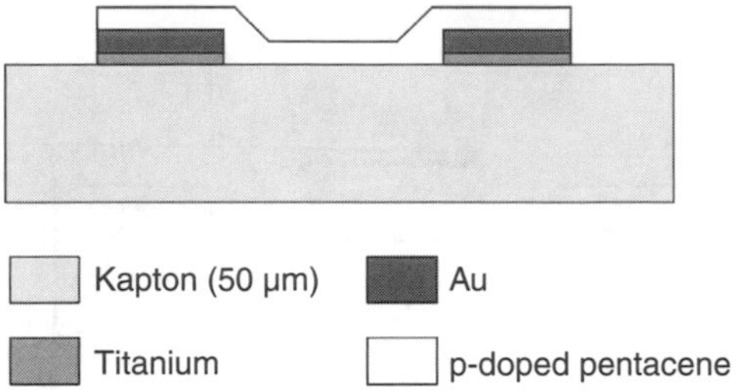

Figure 3.6 Cross section of the organic semiconductor strain sensor. *(Reprinted with permission from Ref. 12. Copyright 2003, IEEE.)*

Next, a 50 nm thick pentacene layer was deposited, again by thermal evaporation. The pentacene layer was then doped p-type by exposure to a 1% solution of ferric chloride in water. The maximum process temperature used to fabricate the organic strain sensors is 110°C.

The devices were tested using a Wheatstone bridge configuration, and the results indicate that it is possible to fabricate at low temperature a strain sensor with mechanical characteristics matched to low-Young-modulus substrates using organic semiconductors.

Jung et al. have also demonstrated[13] the possibility of combining these sensors with pentacene-based thin-film transistors as temperature sensors. The strain sensor consists of a Wheatstone bridge structure where the pentacene film acts as sensing layer of a strain gauge, while the temperature sensors adopt a bottom-contact pentacene transistor configuration in which the variations of the drain currents in the subthreshold regime are measured vs. temperature.

The effects of strain on pentacene transistor characteristics while changing the bending radius of the structure have been investigated by Sekitani and coworkers.[14] A cross section of their device structure is shown in Fig. 3.7.

First, a gate electrode consisting of 5 nm Cr and 100 nm Au was vacuum-evaporated on a 125 μm thick poly-ethylenenaphthalate (PEN) film. Polyimide precursors were then spin-coated and cured at 180°C to form 900 nm thick gate dielectric layers. A 50 nm thick

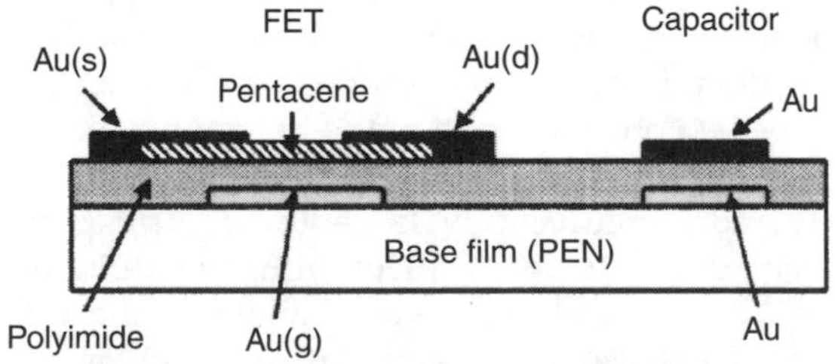

Figure 3.7 Cross section of the organic FET and capacitor reported in Ref. 14. *(Reprinted with permission from Ref. 14. Copyright 2005, American Institute of Physics.)*

pentacene film was vacuum-sublimed, and finally 60 nm thick Au drain and source electrodes were evaporated through a shadow mask. For comparison a capacitor was also manufactured simultaneously in the same base film. The authors observed large changes in the drain current that cannot be explained only by the deformation of the device structure. Then in the analysis of the electrical characteristics, changes in the structural parameters of the transistors (namely, channel width W and length L and the thickness of the gate insulator) were taken into account to evaluate possible variations in the mobility of the transistor. The transfer characteristics of the realized transistor and the mobility variations observed are shown in Fig. 3.8 as a function of strain or bending radius. The mobility increases monotonically when the strain is changed from tension to compression passing through the flat state.

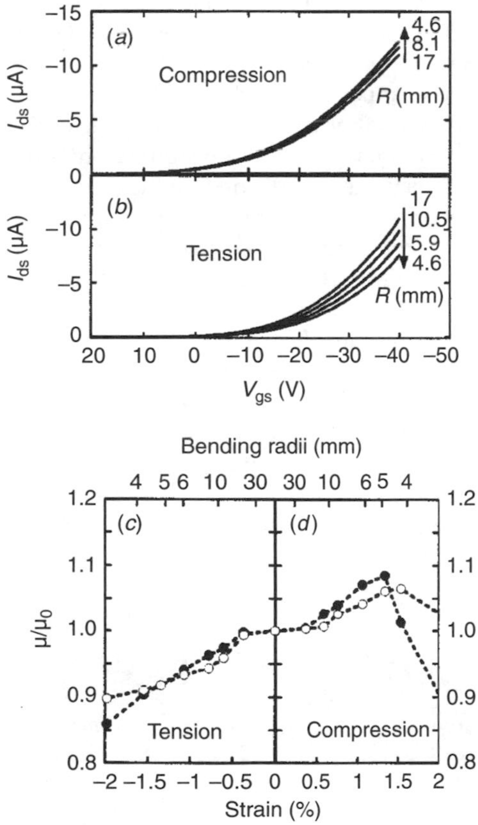

FIGURE 3.8 Transfer curves on (*a*) compressive and (*b*) tensile strains. Mobility as function of strain or bending radius in (*c*) tensile strain and (*d*) compressive strain.[14] Solid and open circles refer to transistors whose source-drain current path is, respectively, parallel and perpendicular to the strain direction. (*Reprinted with permission from Ref. 14. Copyright 2005, American Institute of Physics.*)

3.3.2 Substrate-Free Organic Thin-Film Strain and Pressure Sensors

In this section we describe organic semiconductor field-effect mechanical sensors based on a substrate-free OTFT structure. Papers regarding this kind of sensor have already been published by us.[15–17] The proposed device is completely flexible and combines both switching and sensing functions. These are very interesting hallmarks since there are only a few examples of organic mechanical sensors reported in the literature, as shown in the previous section, and none of them exploits all the advantages of organic devices.

The basic device structure is shown in Fig. 3.9. The device consists of a pentacene substrate-free structure with gold bottom-contact source and drain electrodes. A 1.6 μm thick PET (polyethylene terephtalate, Mylar) foil is used as gate insulator and also as mechanical support of the whole device. First, the Mylar foil is clamped to a cylindrical plastic frame (2.5 cm in diameter) to obtain a suspended membrane with both sides available for processing. Then bottom-contact gold electrodes (nominal thickness 100 nm) usually with $W/L = 100$ ($W = 5$ mm and $L = 50$ μm are the channel width and length, respectively) are thermally evaporated and patterned on the upper side of the flexible dielectric foil, using a standard photolithographic technique.

The gold gate electrode is patterned on the opposite side of Mylar film. Since the Mylar is transparent to UV light, source and drain may be used as shadow mask for the gate patterning. Therefore, a thin photoresist layer is spin-coated on the lower side of the Mylar layer, and then it is exposed to UV light projected through the Mylar itself. In this way, source and drain electrodes act as a mask for the UV light, and a perfect alignment between source and drain and the impressed photoresist is obtained. After the development process, a gold layer (nominal thickness 100 nm) is vacuum-sublimed and patterned by means of liftoff etching with acetone.

In this way, the channel area of the device included between source and drain contacts is precisely gate-covered on the opposite

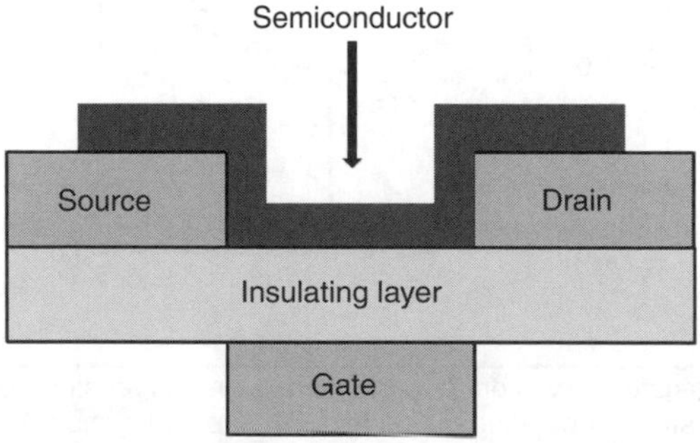

Figure 3.9 Flexible mechanical sensor structure. (*Reprinted with permission from Ref. 15. Copyright 2006, American Institute of Physics.*)

part of the insulator (see Fig. 3.9), thus limiting all the parasitic capacitance effects due to source-drain and gate metal overlapping.[18] Residual hysteresis can be interpreted both in terms of border effects and with trapping charge effects in the semiconductor.[19–21]

The marked sensitivity of the drain current to an elastic deformation induced by a mechanical stimulus, and the fact that the device is so thin and flexible that it can be applied to whatever surface, can be exploited to detect through the variation of the channel current any mechanical deformation of the surface itself. In particular, we investigated the sensitivity of the device to a pressurized airflow applied on the gate side of the suspended membrane and the sensitivity to a strain or a bending imposed by a deformation of the device.

Figure 3.10 shows the drain current I_d versus the drain voltage V_d at different values of V_g for different applied pressures. As seen, while the global shape of each curve is unvaried with respect to the curve taken without applying a pressure, there is a decrease of the current when a pressure is applied. Figure 3.11 shows the time variations of I_d while different increasing values of pressure are applied to the device.

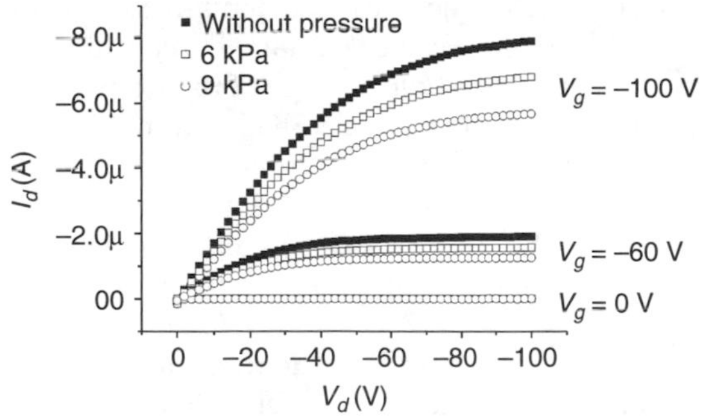

FIGURE 3.10 Output curves for different applied pressures. (*Reprinted with permission from Ref. 15. Copyright 2006, American Institute of Physics.*)

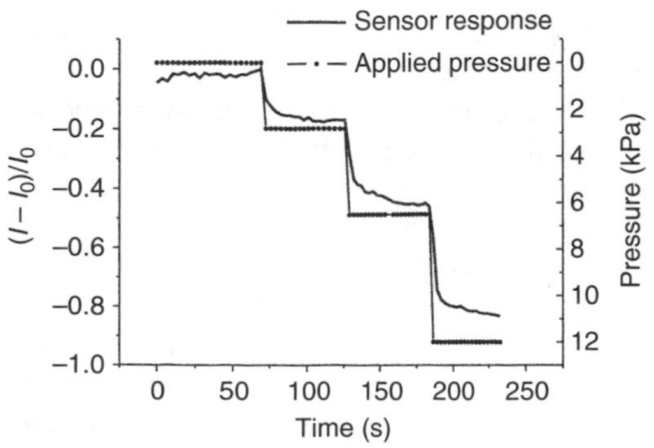

FIGURE 3.11 I_d vs. time for different applied pressures.

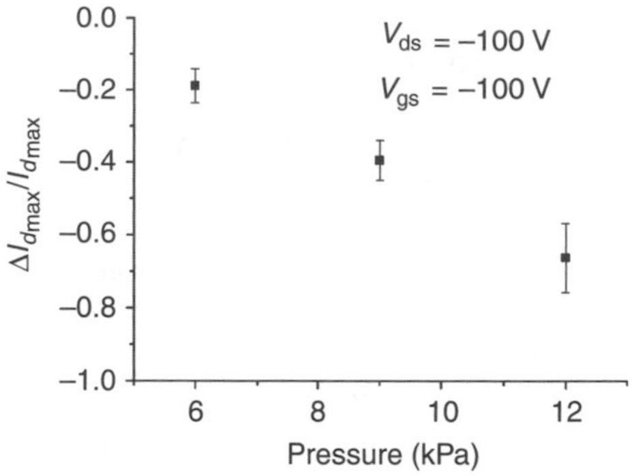

FIGURE 3.12 $\Delta I_{d\max}/I_{d\max}$ vs. pressure (for three samples). (*Reprinted from Ref. 16. Copyright 2007, with permission from Elsevier.*)

In Fig. 3.12 the relative variation of the maximum current recorded on a set of three samples (V_d = −100 V, V_g = −100 V) has been plotted against pressure. Notice that there is an uncontroversial linear dependence of this parameter on the pressure.

A careful analysis of the pressure dependence of the current[16] shows that this dependence can be explained in terms of variations in the mobility and in the threshold voltage of the transistor. Figure 3.13 shows (*a*) mobility and (*b*) threshold voltage plotted against pressure. The extracted values of mobility and threshold voltage show a similar linear dependence, but the standard deviation is higher. This can be partially attributed to the fact that mobility and threshold voltage result from an extrapolation that is affected by possible failures of the fitting model.

Despite the fact that the underlying mechanism of the observed pressure sensitivity is not completely clarified yet, pressure sensitivity seems to result from a combination of mobility variations in the channel and interface effects in the source/drain surrounding areas, likely due to morphological modifications of pentacene layer under stress.

To clarify the influence of structural effects (in particular of the contact/semiconductor interface) on the pressure sensitivity, we have also realized, on the same insulating layer, a couple of bottom-contact and top-contact devices with the same active layer as reported in Ref. 16. As a matter of fact, the different metal/semiconductor interface is expected to affect the behavior of the electrical characteristics of the transistors even if no pressure is applied. In fact, the morphology of the pentacene film in the device channel region close to the electrode edges is different in top-contact and bottom-contact devices. In the bottom-contact structure near the edge of the electrodes there is an area with a large number of grain boundaries that can act as charge carrier traps and are believed to be responsible for

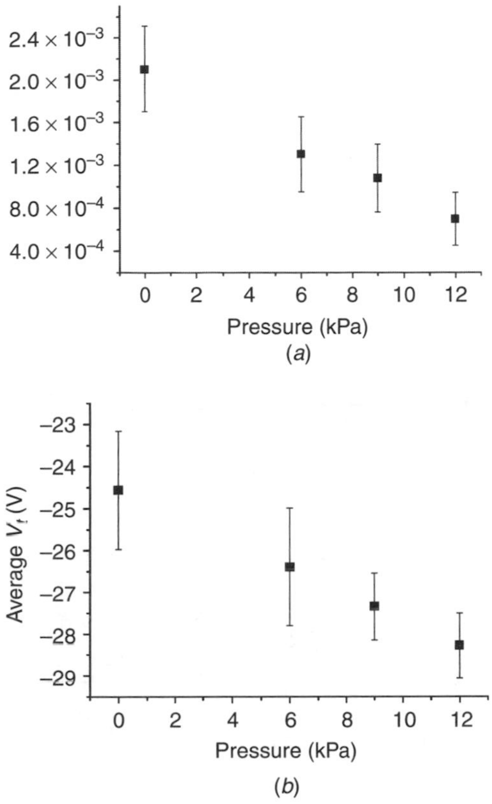

FIGURE 3.13 (*a*) Average mobility vs. pressure (three samples); (*b*) average threshold voltage vs. pressure (three samples). (*Reprinted from Ref. 16. Copyright 2007, with permission from Elsevier.*)

the reduced performance of bottom-contact devices when compared to top-contact devices. An example of the obtained results is shown in Figs. 3.14 and 3.15.

The threshold voltage decreases with pressure in bottom-contact devices while in top-contact devices it is pressure-insensitive. This observation confirms the role of insulator/semiconductor and metal/semiconductor interfaces in determining the pressure sensitivity of the device. On the other hand, mobility has a very similar behavior in top-contact and in bottom-contact devices, indicating a direct contribution of the semiconductor mobility in the channel to the observed sensitivity. Even if a complete explanation of the observed sensitivity needs additional investigation, it seems likely that the effect is quite reasonable since the transport properties of organic molecular systems follow the hopping model rather than the band model.[14] Under compressive strain, the energy barrier for hole hopping decreases due to the smaller spacing between molecules, resulting in an increase

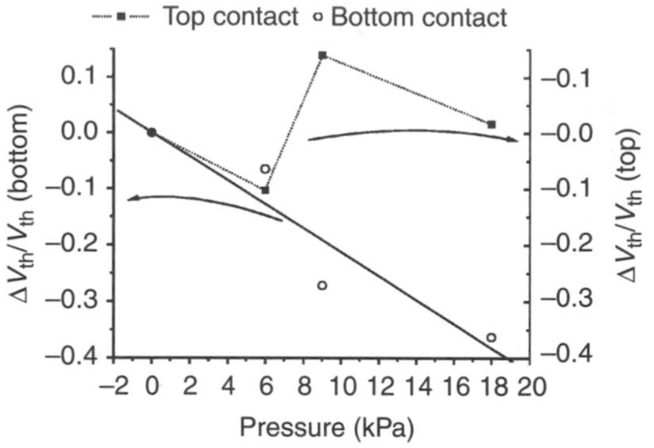

FIGURE 3.14 $\Delta V_t/V_{t_0}$ vs. pressure for a bottom-contact device (left axis) and for a top-contact device (right axis). (*Reprinted from Ref. 16. Copyright 2007, with permission from Elsevier.*)

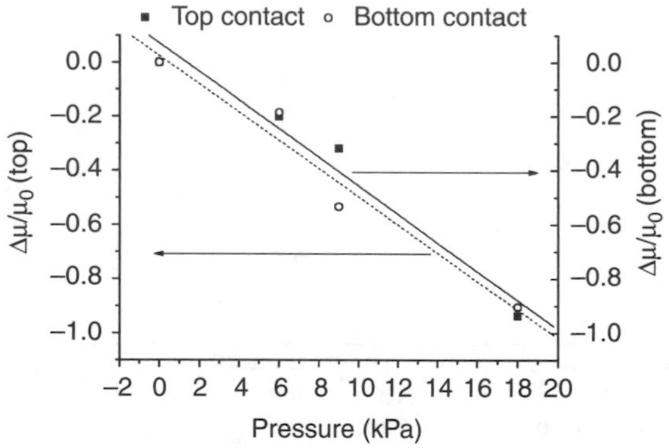

FIGURE 3.15 $\Delta\mu/\mu_0$ vs. pressure for a bottom-contact device (left axis) and for a top-contact device (right axis). (*Reprinted from Ref. 16. Copyright 2007, with permission from Elsevier.*)

of the current of the FETs. In contrast, tensile strain causes a larger spacing, resulting in a decrease of the current. This effect can be observed in Fig. 3.16 in which positive (tensile strain) and negative (compressive strain) pressures were applied to the device. As seen, the tensile stimulus results in a decrease of the drain current while a compressive stimulus induces an increase in the drain current as expected. This feature cannot be achieved using commercial (metallic) strain gauges and can be very useful in a wide range of applications (e.g., body parameters monitoring).

The main drawbacks of these devices seem to be the high operating voltage and a limited time stability that could be overcome by using a proper flexible encapsulation layer to protect the semiconductor layer

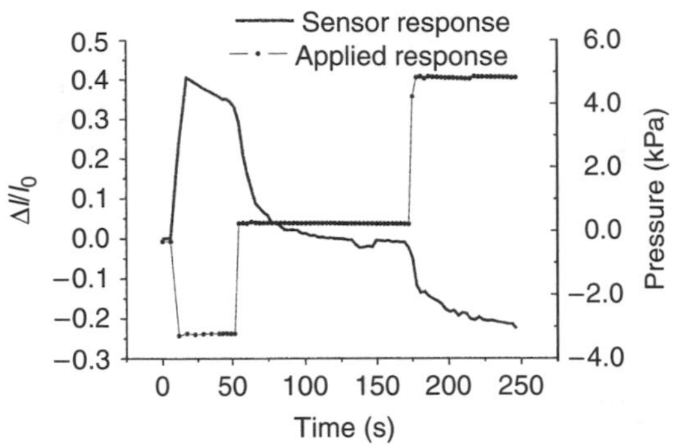

FIGURE 3.16 I_d vs. time for compressive and tensile applied pressures.

from exposure to external agents (i.e. humidity, light, etc) as these factors are known to cause drift and in general to negatively affect the device performance.

3.4 Applications for Organic Field-Effect Transistor Sensors

The possibility of obtaining low-cost physical sensors is attractive for a number of applicative fields. Organic field-effect sensors are still in their embryonic phase. As far as we know, no commercial application has yet been developed. Despite this, many groups are working toward innovative solutions that make use of this technology. Here we will focus on two cases that seem particularly interesting: the first is the realization of an artificial skin for robots, while the second concerns e-textiles. Both deal with a common need, i.e., the conformability of the final product to a 3D shape, a robot in the first case and the human body in the second. This requirement is fully satisfied by devices realized on substrates that are able to adapt their shape to the substrate, i.e., are flexible. Furthermore, large area is another desirable characteristic. In the following, an overview of these applicative fields is given with a special focus on requirements for future devices.

3.4.1 Artificial Sense of Touch

The skin is the largest organ of the human body. For the average adult human, the skin has a surface area of 1.5 to 2.0 m^2, most of it is 2 to 3 mm thick. The average square inch of skin holds 650 sweat glands, 20 blood vessels, 60,000 melanocytes, and more than a thousand nerve endings.

Skin is composed of three primary layers: the epidermis, which provides waterproofing and serves as a barrier to infection; the dermis, which serves as a location for the appendages of skin; and the hypodermis, which is called the basement membrane (see Fig. 3.17).

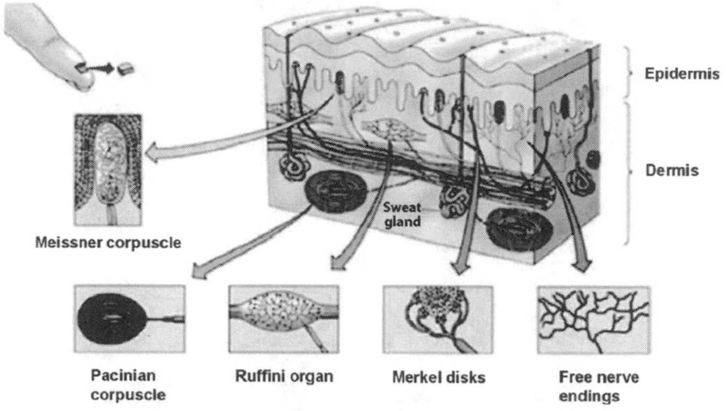

FIGURE 3.17 Cross section of human skin.[22]

The dermis is tightly connected to the epidermis by a basement membrane. It also contains many nerve endings that provide the sense of touch and heat.

Somatic sensation consists of the various sensory receptors that trigger the experiences labeled as touch or pressure, temperature (warm or cold), pain (including itch and tickle), and the sensations of muscle movement and joint position including posture, movement, and facial expression.

In human beings, touch is in fact a combination of different feelings, for instance, perception of pressure (hence shape, softness, texture, vibration, etc.), relative temperature, and sometimes pain. In addition, complex actions based on touch (such as grasping) are the result of a powerful sensory-motor integration which fully exploits the wealth of information provided by the cutaneous and kinesthetic neural afferent systems.[22] A very accurate description of tactile units is available in Ref. 23, where a classification of these units according to receptive fields and response time is given.

Obviously, reproducing the human sense of touch with an artificial system is a very challenging task, first, because the term *touch* is actually the combined term for several senses.

The tentative specifications for tactile sensors have been defined in Ref. 23 as follows:

1. The sensor surface or its covering should combine compliance with robustness and durability.
2. The sensor should provide stable and repeatable output signals. Loading and unloading hysteresis should be minimal.
3. Linearity is important, although only monotonic response is absolutely necessary. Some degree of nonlinearity can be corrected through signal processing.

4. The sensor transduction bandwidth should not be less than 100 Hz, intended as tactile image frame frequency. Individual sensing units should accordingly possess a faster response, related to their number, when multiplexing is performed.
5. Spatial resolution should be at least of the order of 1 to 2 mm, as a reasonable compromise between gross grasping and fine manipulation tasks.

The development of tactile sensors is one of the most difficult aspects of robotics. Many technologies have been explored, including a carbon-loaded elastomer, piezoelectric materials, and microelectromechanical systems. Artificial skin examples, able to detect pressure, already exist; but it is difficult to manufacture artificial skin in large enough quantities to cover a robot body, and it does not stretch. The most promising examples of "electronic skin-like" systems with large areas are based on organic semiconductors and have been reported by Takao Someya's group at University of Tokyo.[24] They have developed conformable, flexible, wide-area networks of thermal and pressure sensors in which measurements of temperature and pressure mapping were performed simultaneously. The device structure is shown in Fig. 3.18.

Someya has developed a skin that is stretchable and that remains as sensitive to pressure and temperature when it is at full stretch as when it is relaxed. In the presented design, both sensor networks contain their own organic transistor active matrices for data readout. This arrangement means that each network is self-contained and

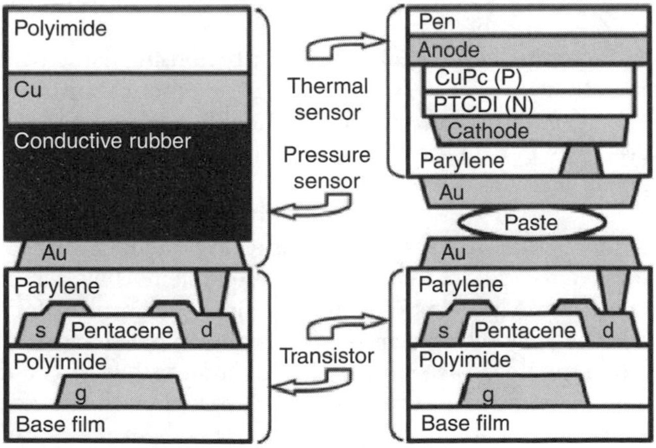

Figure 3.18 Schematic of the device structure reported in Ref. 24. A cross-sectional illustration of the pressure (left) and the thermal (right) sensor cells with organic transistors is shown. (*Reprinted with permission from Ref. 24. Copyright 2005, National Academy of Sciences, USA.*)

electrically independent and that organic transistors are only used to address sensitive (pressure or thermal) elements. The arrays are less sensitive than human skin, but already mark an improvement over previous efforts, while sensing temperatures in the range of 30 to 80°C. Moreover, the structure is flexible enough to be rolled or bent around a 2 mm bending cylinder. Someya estimates that his e-skin will be commercially available within a few years, and in the near future it will be possible to make an electronic skin that has a function that human skin lacks by integrating various sensors not only for pressure and temperature, but also for light, humidity, strain, sound, or ultrasonic.

Moreover, it could also be possible in the next years to develop electronic skin completely made of organic transistors. In particular, the possibility of realizing strain and pressure sensors that can act at the same time as switch and as sensor without the need of any further sensing element will be interesting. Moreover, flexible chemosensitive transistors, biosensors, and temperature sensors could be obtained with the same technologies, allowing new challenging and smart features for this application.

3.4.2 E-Textiles

There is an increasing interest in the emerging area of e-textiles, meaning with this term the idea of endowing garments and fabrics with new electronic functions, in particular aimed at monitoring physiological parameters in patients[25] and in subjects exposed to particular risks or external harsh conditions.[26]

Strain and pressure sensors for measuring body characteristics are particularly interesting for this kind of application because they could enable one to measure a wide set of parameters such as posture, breathing activity, etc., in a totally non-intrusive way. This characteristic is in fact very interesting for practical applications. For instance, it allows doctors to monitor the patient status in real time, 24 hours per day; additionally, it allows a better quality of life for patients who do not perceive them as invasive monitoring systems.

Basic specifications for this application are rather similar to those listed for electronic skin. In addition, these systems, being in contact (or close) with the human body, must comply with strict safety standards. Organic field-effect sensors developed on plastic flexible films are good candidates to accomplish this function as they can be assembled in arrays on flexible substrates to be applied on the fabric itself.

At present, first attempts of strain sensors on garments are made with piezoresistive stripes deposited on the garments.[27–32] In this case, the detection is made through piezoresistive tracks running on the fabric along, for example, a sleeve or parallel to the chest in a T-shirt (Fig. 3.19). In this way, the movement results in a deformation of the

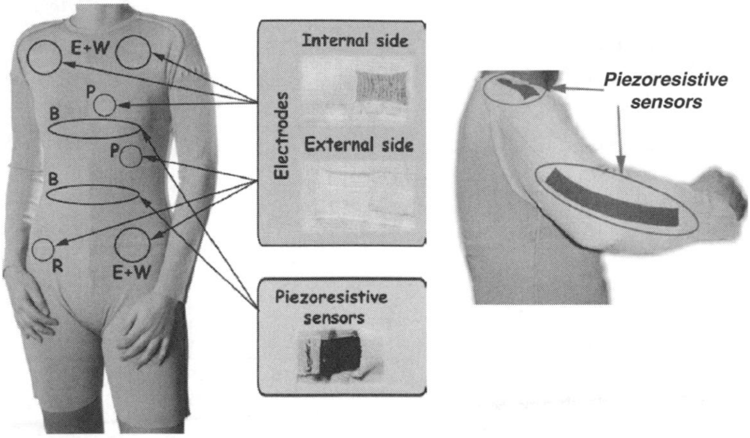

Figure 3.19 Examples of sensorized garments for recording body signals. (*Reprinted with permission from Ref. 36. Copyright 2005, IEEE.*)

track and a variation of its resistance. No spatial resolution is achievable with such a strategy that is based on the measurement of the resistance of the whole track.

Totally flexible organic field-effect sensors described in Sec. 3.3 can be integrated in different substrates to detect physiological body parameters such as the joint movements, the breathing signal, or the posture of a person.[33] In Fig. 3.20 the sensor is glued on a latex glove and positioned upon the joint between the first and the middle phalanx of the index finger and used to detect the finger movements.

As can be seen in Fig. 3.21, when the finger is clenched (tensile stress on the transistor channel), the current decreases whereas when the finger is forced to raise (compressive stress on the transistor channel), the current increases as expected.

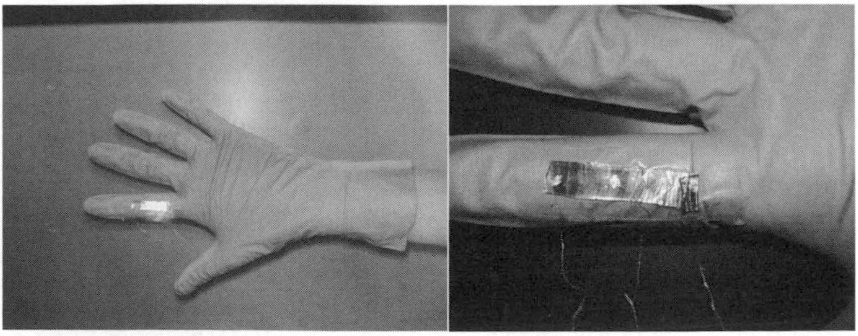

Figure 3.20 Organic field-effect sensor glued on a latex glove and used to detect finger bending movements.

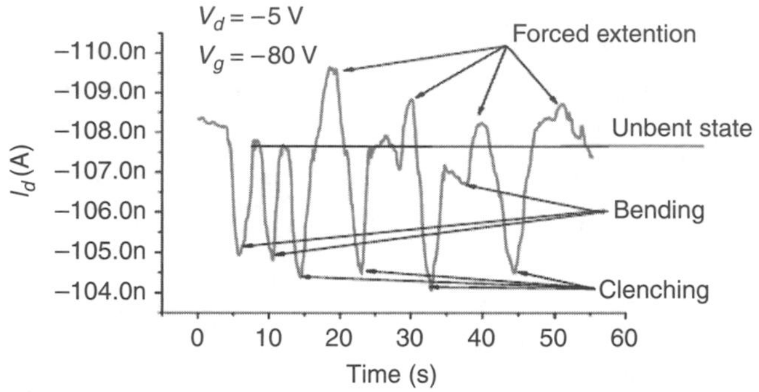

FIGURE 3.21 I_d vs. time monitored at $V_d = -5$ V and $V_g = -80$ V for the finger joint bending detection experiment.

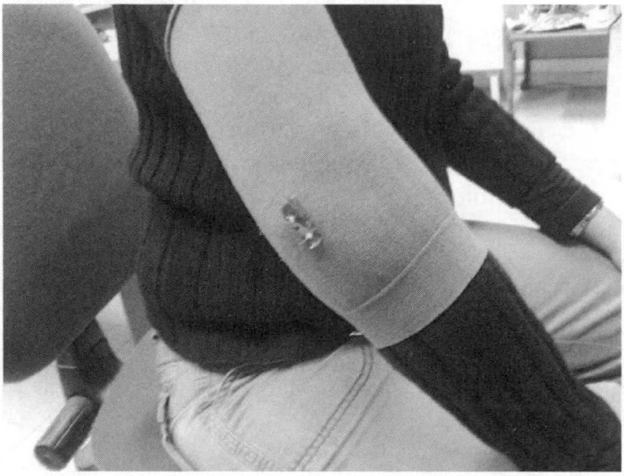

FIGURE 3.22 Application of a field-effect strain sensor to the elbow joint bending detection.

Changes were reversible even after numerous measurements, and the device response is reproducible, reversible, and prompt enough to be used to monitor relatively fast movements. In Fig. 3.22 a totally flexible mechanical sensor was used to detect the elbow joint movements. In this case, a device was glued to an elastic band and used for monitoring elbow bending. In Fig. 3.23 an example of the obtained results is shown.

We also tested our devices to detect the breathing activity. The sensor was mounted on a wearable elastic band placed around the diaphragm area of a volunteer, as shown in Fig. 3.24.

The sensor was able reveal variation in the breathing activity due to apnea, cough, laugh, hiccups, etc. In Figs. 3.25 and 3.26 examples of the obtained results for different breathing movements are shown.

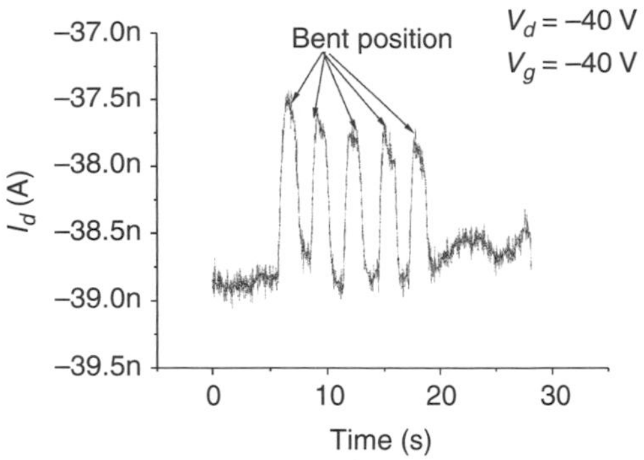

FIGURE 3.23 Experimental results for the elbow joint bending detection experiment.

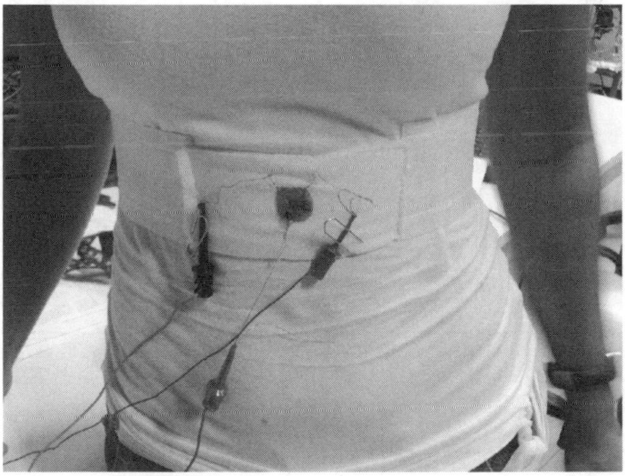

FIGURE 3.24 Wearable elastic band provided with a sensor for human breathing rhythm monitoring.

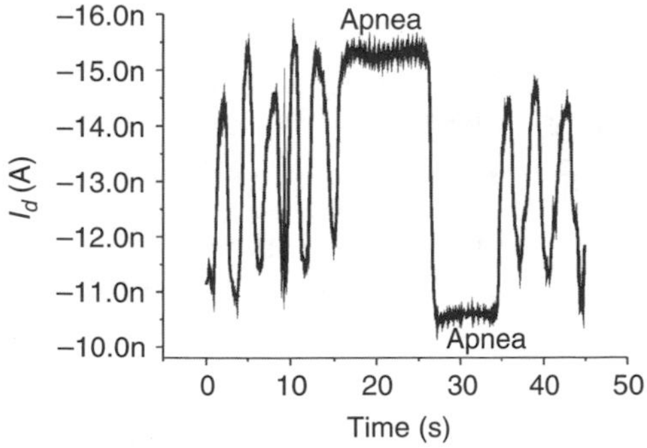

FIGURE 3.25 Sensor output for normal breathing and apnea.

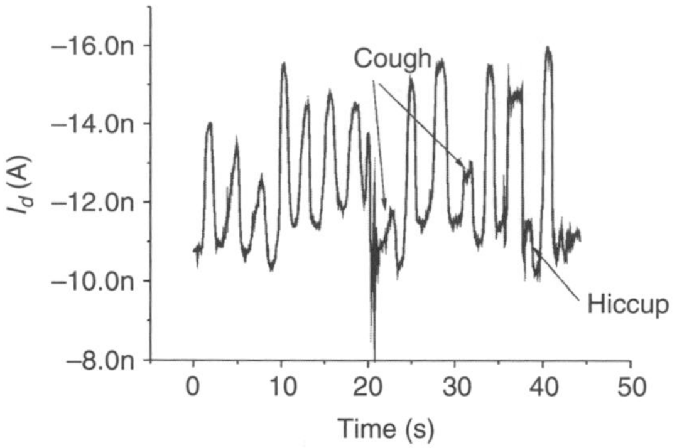

FIGURE 3.26 Sensor output for cough and hiccup.

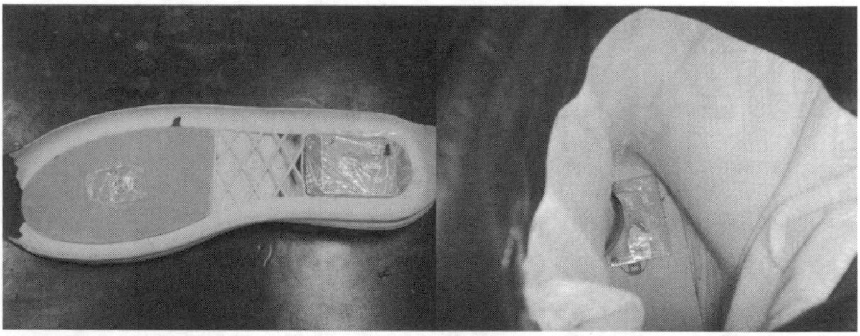

FIGURE 3.27 Experimental setup for detection of falls and gait analysis.

As can be seen the sensor response is reproducible and rather fast (hundreds of milliseconds).

This technology can be helpful also for detection of falls or for gait analysis through the detection of pressure exerted on shoes. In particular we performed experiments inserting sensors under the insole of a shoe to detect the pressure exerted by the foot, as shown in Fig. 3.27.

In the experiment shown in Fig. 3.28, two sensors were positioned under the heel and under the sole. As can be seen, it is possible to clearly distinguish when the subject is on tiptoes or on heels or is normally standing.

The employment of OFETs could enable building arrays and matrices able to give a full spatial resolution to the measurement. This feature is particularly useful for application as the gait analysis or the detection of posture in which a proper spatial resolution is needed.

Another possible, particularly interesting development is aimed at the realization of the functions described above directly on yarns.[34–36]

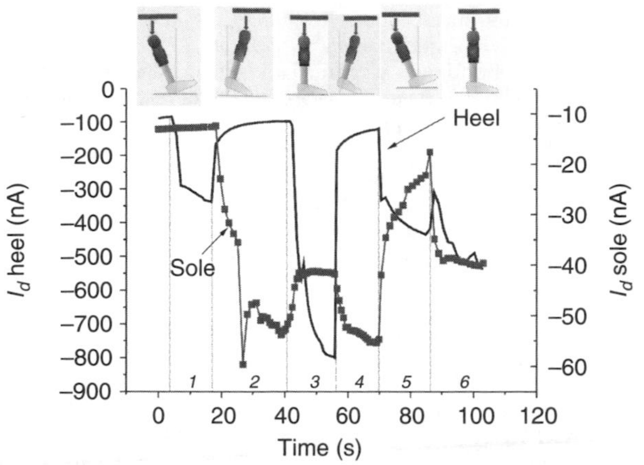

Figure 3.28 Heel sensor response (solid line) and sole sensor response (dotted line).

In this way, it could be possible to build a whole fabric made with sensitive elements that will be distributed over a large area. Not only is the distributed function in the fabric a real novel element, but also the possibility of leveraging from an existing industrial technology such as textiles for creating a new technological approach, though challenging, could really have a huge impact. Nevertheless, high operating voltages of devices and chemical safety of materials are still open questions and must be carefully considered in designing such an application.

References

1. P. K. Weimer, *Proc. Institute of Radio Engineering*, 50:1462 (1962).
2. L. Torsi and A. Dodabalapur, *Anal. Chem.*, 77:380 (2005).
3. G. Horowitz, *Adv. Mater.*, 10:365 (1998).
4. C. D. Dimitrakopoulos and P. R. L. Malenfant, *Adv. Mater.*, 14:99 (2002).
5. L. Torsi, A. Dodabalapur, and H. E. Katz, *J. Appl. Phys.*, 78:1088 (1995).
6. T. W. Kelley and C. D. Frisbie, *J. Phys. Chem. B*, 105:4538 (2001).
7. T. Someya and T. Sakurai, "Integration of organic field-effect transistors and rubbery pressure sensors for artificial skin applications," *Int. Electron. Dev. Meeting IEEE*, Washington D.C., 8(4):203–206 (2003).
8. T. Someya, T. Sekitani, S. Iba, Y. Kato, H. Kawaguchi, and T. Sakurai, *Proc. Natl. Acad. Sci. USA*, 101:27 (2004).
9. Y. Noguchi, T. Sekitani, and T. Someya, *Appl. Phys. Lett.*, 89:253507 (2006).
10. Z. Rang, M. I. Nathan, P. P. Ruden, R. Chesterfield, and C. D. Frisbie, *Appl. Phys. Lett.*, 85:23, 5760 (2004).
11. G. Darlinski, U. Böttger, R. Waser, H. Klauk, M. Halik, U. Zschieschang, Günter Schmid, et al., *J. Appl. Phys.*, 97:93708 (2005).
12. S. Jung and T. Jackson, *Dev. Res. Conf. Digest, IEEE*, 1:149 (2005).
13. S. Jung, T. Ji, and V. K. Varadan, *Smart Mater. Struct.*, 15:1872 (2006).
14. T. Sekitani, Y. Kato, S. Iba, H. Shinaoka, T. Someya, T. Sakurai, and S. Takagi, *Appl. Phys. Lett.*, 86:073511 (2005).

15. I. Manunza, A. Sulis, and A. Bonfiglio, *Appl. Phys. Lett.*, 89:143502 (2006).
16. I. Manunza and A. Bonfiglio, *Biosens. Bioelectron.*, 22:2775 (2007).
17. I. Manunza and A. Bonfiglio, "Organic Field-Effect Based Devices for Pressure Detection," in *Organic Electronics—Materials, Devices, and Applications*, F. So, G. B. Blanchet, and Y. Ohmori (eds.), (*Mater. Res. Soc. Symp. Proc.* 965E, Warrendale, PA, 2007), 0965–S10–03.
18. G. Horowitz, *J. Mater. Res.*, 19:1946 (2004).
19. A. Bonfiglio, F. Mameli, and O. Sanna, *Appl. Phys. Lett.*, 82(20):3550 (2003).
20. F. Garnier et al., *Science* (Washington, D.C., USA), 265:1684 (1994).
21. M. Halik, H. Klauk, U. Zschieschang, T. Kriem, G. Schmid, W. Radlik, and K. Wussow, et al., *Appl. Phys. Lett.*, 81:289 (2002).
22. D. De Rossi and E. P. Scilingo, "Skin-like Sensor Arrays," in *Encyclopedia of Sensors*, C. A. Grimes, E. C. Dickey, and M. V. Pishko (eds.), vol. 10, American Scientific Publishers, University Park, Pennsylvania, 2006, pp. 535–556.
23. D. De Rossi, *Meas. Sci. Tech.*, 2:1003 (1991).
24. T. Someya, Y. Kato, T. Sekitani, S. Iba, Y. Noguchi, Y. Murase, H. Kawaguchi, et al., *Proc. Natl. Acad. Sci. USA*, 102(35):12321 (2005).
25. E. P. Scilingo, A. Gemignani, R. Paradiso, N. Taccini, B. Ghelarducci, and D. De Rossi, *IEEE Trans. Inf. Tech. Biom.*, 9(3):345 (2005).
26. A. Bonfiglio, N. Carbonaro, C. Chuzel, D. Curone, G. Dudnik, F. Germagnoli, D. Hatherall, et al. "Managing catastrophic events by wearable mobile systems," *Proc. Mobile Response*, Springer-Verlag, Berlin, 4458:95–105 (2007).
27. F. Lorussi, W. Rocchia, E. P. Scilingo, A. Tognetti, and D. De Rossi, *IEEE Sens. J.*, 4(6):807 (2004).
28. F. Lorussi, E. Scilingo, M. Tesconi, A. Tognetti, and D. De Rossi, *IEEE Trans. Inf. Tech. Biom.*, 9(3):372 (2005).
29. A. Tognetti, N. Carbonaro, G. Zupone, and D. De Rossi, *28th Annual International Conference of the IEEE Engineering in Medicine and Biology Society*, New York, September 2006.
30. L. E. Dunne, S. Brady, B. Smyth, and D. Diamond, *J. Neuroeng. Rehab.*, 2:8 (2005).
31. P. T. Gibbs and H. H. Asada, *J. Neuroeng. Rehab.*, 2:7 (2005).
32. R. Paradiso, G. Loriga, and N. Taccini, *IEEE Trans. Inf. Tech. Biom.*, 9:337 (2005).
33. A. Bonfiglio, I. Manunza, A. Caboni, W. Cambarau, and M. Barbaro, in "Organic-based chemical and biological senors," R. Shinar and G. G. Malliaras (eds.), (*Proc. SPIE* 6659, SPIE, Bellington, WA, 2007) Article 665904.
34. J. B. Lee and V. Subramanian, *IEEE Trans. El. Dev.*, 52:269 (2005).
35. A. Bonfiglio, D. De Rossi, T. Kirstein, I. R. Locher, F. Mameli, R. Paradiso, and G. Vozzi, *IEEE Trans. Inf. Tech. Biom.*, 9:319 (2005).
36. R. Paradiso, G. Loriga, and N. Taccini, *IEEE Trans. Inf. Tech. Biomed.*, 9:337 (2005).

CHAPTER 4

Integrated Pyroelectric Sensors

Barbara Stadlober, Helmut Schön, Jonas Groten, Martin Zirkl, and Georg Jakopic

Institute of Nanostructured Materials and Photonics
Joanneum Research Forchungsgesellchaft
Weiz, Austria

4.1 Electrical Semiconductor and Dielectric Analysis

4.1.1 Impedance Spectroscopy (Basics, Impedance Elements, Ideal and Nonideal MIS Structures)

This chapter deals with the characterization of the electronic properties of organic materials by impedance spectroscopy. The motivation for this is novel organic devices, whose electronic properties are not yet fully known, such as OFETs, OLEDs, and OPDs, which are described in greater detail in other chapters. To understand an electronic device, it is essential to know its *equivalent circuit* (EC). If the EC contains capacitive or inductive elements, its impedance will show frequency dependence. Hence measuring the *impedance spectrum* (IS) of an electronic device over a large frequency range enables determination of the elements of the underlying EC.

In practice, the determination of EC can be complicated by several factors. First, the EC of the *device under test* (DUT) is often unknown. In this case the problem is not only quantitative but also qualitative. This requires an algorithm for the interpretation of the IS that cannot be given for the general case. Second, the IS of an EC is not necessarily unique—different ECs can have identical impedance spectra. Third, the DUT should not alter its electronic properties during the measurement, which takes ~1 min. Especially when the DUT is biased during measurement, it might experience electric stress leading to an ambiguous IS. When biasing a device, the measurement of the impedance has to be quasi-static; i.e., the EC at each frequency must have sufficient time to complete all charging and discharging processes. It will be shown that for certain ECs, which are typical of organic electronic devices, the related time constants can be high enough to limit the measurement speed.

The impedance is a measure of the electric resistance of AC circuits. It is a complex value with a magnitude and phase which are both measurable. An ohmic resistor does not alter the phase whereas an ideal capacitor has a phase ϕ of $-90°$ and an ideal inductor has a phase ϕ of $+90°$. Combinations of these basic elements have a phase value in between depending on the dominant element. At low frequencies capacitive elements dominate because their impedance Z_C is inversely proportional to the frequency f. For inductive elements it is the opposite since their impedance Z_L is proportional to the frequency. The impedance of ohmic elements Z_R is frequency-independent and identical to the ohmic resistance R.

Although each of these three basic impedance elements have individually a constant phase over the whole frequency range, combinations of these will show a characteristic frequency–dependent total phase. Thus it is surprising that the impedance spectrum of some organic devices shows constant phase between −90 and 0° over a large frequency range. For that reason we will provide a closer look at a class of impedance elements known from electrochemistry, the *constant-phase elements* (CPEs).

CPEs are impedance elements with a frequency-independent constant phase between −90 and 0°. One example is the Young element[1] whose characteristics suit well the properties of an organic semiconductor. A Young element is a parallel circuit of an ideal capacitance and a resistor with an exponential spatial dependence. According to Rammelt and Schiller,[2] its impedance Z_y is described by

$$Z_y = \frac{p}{i\omega C_y} \ln \frac{1 + i\omega\tau \exp(p^{-1})}{1 + i\omega\tau} \tag{4.1}$$

with $\tau = R(0)C_y$ $\quad R(0) = \rho(0)\delta$ $\quad$ and $\quad p = \delta/d$

where δ = characteristic penetration depth
p = relative penetration depth
d = thickness of capacitor
τ = Young time constant

The relative penetration depth p describes into which depth of the capacitor the conductive layer extends. A value of $p = 0.05$ indicates that at 5% of the capacitor thickness d the conductivity has decreased to $1/e$. The distribution of the local resistivity of the capacitive layer is described by

$$\rho(x) = \rho(0)e^{x/\delta} \tag{4.2}$$

where $\rho(0)$ = resistivity at interface (depth $x = 0$).

The phase of the Young element is frequency-dependent. However, for small values of p it can be approximated by the following expressions, of which the second is a frequency-independent value.

$$\phi \approx -90^\circ\left(1 - \frac{1}{\ln(\omega\tau) + p^{-1}}\right) \approx -90^\circ(1-p) \tag{4.3}$$

In Fig. 4.1 the phase spectrum of a Young element is plotted for different values of p. For large values the spectrum resembles the one of a parallel circuit of a capacitor whose phase is zero at low frequencies and –90° at high frequencies. However, at lower values of p, the phase levels at a value obtained by Eq. 4.3 in a certain frequency range. For very low values of p this level extends over a large frequency range in which the Young element is a CPE.

The total resistance R_{tot} of the conduction layer of the Young element is obtained by integration over the thickness:

$$R_{tot} = \int_0^d dR(x) = R(0)(e^{1/p} - 1) \overset{p<<1}{\approx} R(0)e^{1/p} \tag{4.4}$$

The complex impedance can be given either by its real and complex part or by its absolute value and phase. When measuring over a frequency range, the problem arises to plot this three-dimensional plot in two dimensions. One possibility is to plot the frequency-dependent impedance in the complex plane (Nyquist plot). However, this does

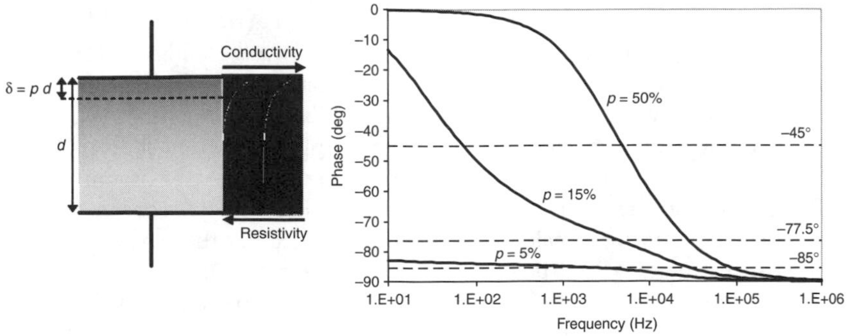

FIGURE 4.1 Left: specific resistance ρ of a Young element; right: phase spectrum of a Young element for different relative penetration depths p. The corresponding approximated constant phase values are indicated by horizontal lines (parameters: $R(0) = 10\ \text{k}\Omega$, $C_y = 1$ nF).

not enable identification of the frequency at which every single point was measured. This can be avoided by plotting both the real and imaginary parts of the impedance over frequency. The two plots can be combined into one if two different y axes are used. As an alternative to the real and imaginary parts, the absolute value and the phase of the impedance can also be plotted (Bode plot).

In this chapter the Bode plot is modified by multiplying the absolute value of the impedance by the frequency f. This is done to eliminate the frequency dependence of the capacitive impedance $Z_C = 1/(2\pi f C)$. By this the capacitive behavior of the impedance can be investigated more closely.

4.1.2 The IS of an Organic MIS Structure

The *Field-effect transistor* (FET) is probably the most important device in electronics due to its amplifying and switching properties. For this reason it is of great interest to understand the *metal-insulator-semiconductor* (MIS) structure which constitutes a FET by adding a source, drain, and gate contact. The transistor is switched on by accumulating charge carriers at the semiconductor-insulator interface, creating a conducting channel between the source and the drain contact. It is switched off by depleting this channel of charge carriers. This is done by applying an appropriate gate voltage. Assuming a p-type semiconductor, a negative voltage switches on the transistor and a small positive voltage switches it off. However, a sufficiently large positive voltage will accumulate minority charge carriers in the channel (inversion).

An ideal MIS structure behaves as a capacitor whose dielectric layer thickness varies with voltage. At accumulation and inversion the thickness corresponds to the thickness of the insulating layer whereas at depletion the thickness of the depletion zone must be added. The characteristics of the corresponding $C(V)$ curve give information not only about the layer thicknesses but also about several semiconductor parameters such as the charge carrier density.[3,4]

This method of device characterization, which is well established for silicon devices, can, however, not be applied on organic MIS structures. Because of the lower charge carrier density, mobility inversion does not occur. Moreover, the semiconductor layer has complicated contact properties compared to silicon. Finally, also the dielectrics used in organic FETs often do not have the ideal properties of silicon oxide regarding their leakage behavior, mobile ion density, and so on. For these reasons the EC of an organic MIS structure is much more complex, complicating the measurement of $C(V)$ characteristics. To determine quantitatively the EC of the organic device, its IS must be measured.

An organic MIS under bias basically consists of three capacitive layers. The first layer corresponds to the insulator, and the other two correspond to the depletion zone and the (remaining) bulk of the organic semiconductor. Since all three layers can show leakage to a

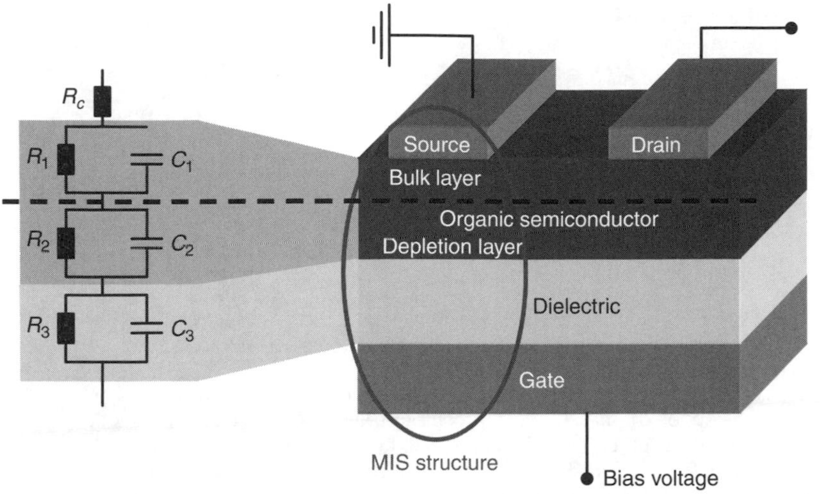

FIGURE 4.2 Three-layer EC of an organic MIS structure in an OFET.

certain extent, they are each modeled in a first approximation by a capacitor in parallel to a resistor. See Fig. 4.2.

If a layer shows no leakage, the parallel resistance is infinite and can be omitted. This is the case for ideal dielectrics and the depletion zone. Such layers cannot be distinguished and appear as a single layer. The remaining organic semiconductor bulk layer is characterized by a comparably low leak resistance. The involved capacitances frequently appear as Young elements mostly due to complex interfaces at the metal contact. In the following the IS of two organic MIS devices is discussed, showing some of the aforementioned effects (see Schoen[5]).

The first device consists of aluminum as top and bottom contacts, PVP (polyvinylphenol) as a dielectric, and MPP as the organic semiconductor (both spin-coated). The IS and $C(V)$ are shown in Fig. 4.3. The IS is taken at depletion and modeled by a two-layer EC. The first layer corresponds to the combination of dielectric layer and semiconductor depletion zone, which both have a high resistance (>GΩ). However, this layer shows the characteristics of a Young element with the relative penetration depth of 1% and a Young resistance of 10 kΩ. The Young element causes at low frequencies a constant phase of −89° and a linearly increasing $|Z| * f$ product. The peak at higher frequencies is caused by a thin second layer with a low leakage resistance corresponding to the remaining semiconductor bulk layer.

The $C(V)$ curve of this device shows a distinct time effect. At a normal measurement rate the $C(V)$ curve seems arbitrary. Only when measured extremely slowly (1 day for a 40 V sweep) do the typical $C(V)$ characteristics emerge. The fact that the fast measured

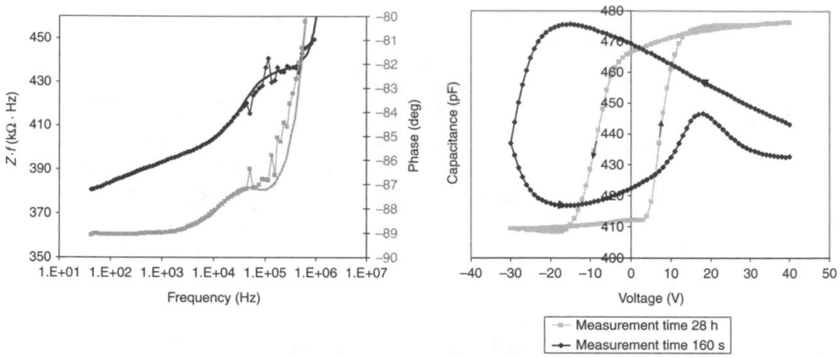

FIGURE 4.3 Left: IS of an MPP MIS structure at depletion (–40 V) (symbols–measured, lines–modeled) (fit parameters: contact resistance $R_c = 99\ \Omega$, inductivity $L = 10$ nH; capacitive layer 1: $C_1 = 391$ pF, $R_1 = 10\ \text{G}\Omega$; $p_1 = 0.98\%$, $R(0)_1 = 10.1$ kΩ; capacitive layer 2: $C_2 = 5.7$ nF, $R_2 = 704\ \Omega$, see text). Right: $C(V)$ curve of the same structure at two different measurement speeds (black curve: 875 mV/s, gray curve: 1.4 mV/s)

curve is mirrored at the reverse point indicates that slow voltages dependent charging effects play a role. This is discussed in greater detail in Sec. 4.2.

The second device is a pentacene MIS structure with BCB (benzocyclobutene) as dielectric. An analysis with impedance spectroscopy reveals the layer structure of the device. The IS is measured from accumulation (–6 V) to depletion (5 V). They can be fit by four layers: two layers have a constant capacitance over the entire bias voltage range and hence are attributed to the dielectric. However, to fit the IS one of the two layers is modeled by a Young element. The Young characteristic of this layer is due to the semiconductor-insulator interface with the contact.

The capacitances of the remaining two layers have an opposite dependence on the bias voltage. However, the sum of the two serial capacitances is constant. This indicates that the two layers correspond to the depletion and bulk layers of the semiconductor. The depletion layer is also modeled by a Young element, which accounts for the interface to the semiconductive bulk layer.

From the capacitances of the four layers, the device area, and the dielectric constants of pentacene (~3) and BCB (3.78), the layer thicknesses can be derived. As can be seen in Fig. 4.4, the bulk and depletion thicknesses add up to the thickness of the thermally grown pentacene layer (~50 nm); the thicknesses of the two BCB layers are in total 86 nm. This fits well to the thickness values obtained from direct measurements (oscillating crystal: ~45 nm pentacene, profilometer: ~85 nm BCB).

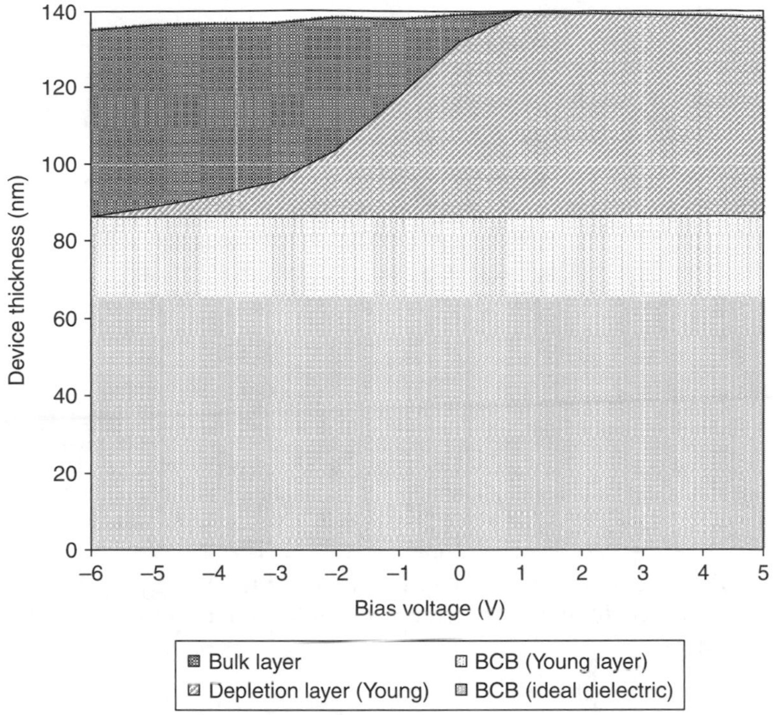

FIGURE 4.4 Thickness of the four capacitive layers of a BCB-pentacene MIS structure at different bias voltages obtained by impedance spectroscopy.

4.1.3 Charge-Time Behavior of Capacitive Multilayers

As can be seen from the example in Fig. 4.3 organic capacitive multilayer structures can show a significant temporal dependence when biased. There can be various reasons for this, e.g., mobile ions in the dielectric, sensitivity to light, heat, oxygen, or charging effects. One might think the latter reason would play a negligible role since the contact resistance is of the order of 1 kΩ, the device capacitance around 1 nF, and hence the charging time constant $\tau = RC = 1$ μs. However, it will be shown here that for multilayer structures several charging time constants apply that can be individually significantly higher.[5] To understand the temporal charging behavior of the three-layer structure, first the capacitive double-layer structure is described, as it is a special case of the three-layer structure.

The Capacitive Double-Layer Structure

The equivalent circuit of a double-layer structure corresponds to a contact resistor R_c in series with two capacitors $C_{1,2}$ which respectively have a leakage resistor $R_{1,2}$ in parallel (see Fig. 4.5).

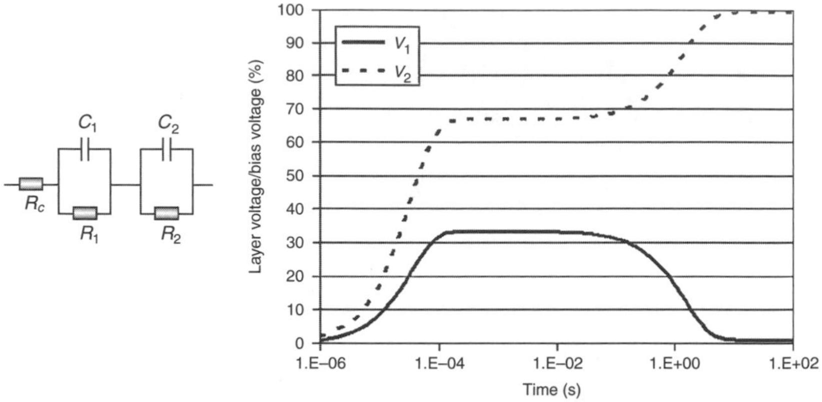

FIGURE 4.5 Evolution of layer voltages of a double-layer structure (V_1 is voltage at layer 1, V_2 is voltage at layer 2; parameters: $R_c = 1\ \text{k}\Omega$, $C_1 = 100$ nF, $R_1 = 10\ \text{M}\Omega$, $C_2 = 50$ nF, $R_2 = 1\ \text{G}\Omega$; hence $\tau_s = 33\ \mu\text{s}$, $\tau_l = 1.5$ s, $V_{1,s,0}/V = 33\%$, $V_{2,s,0}/V = 67\%$, $V_{1,l,0}/V = 99\%$, $V_{2,l,0}/V = 1\%$).

The evolution of the voltages at the layer is given by the following system of differential equations using Kirchhoff's current law for each layer.

$$\frac{V_1(t)}{R_1} + \frac{V - V_1(t) - V_2(t)}{R_c} + C_1 \frac{dV_1(t)}{dt} = 0 \tag{4.5}$$

$$\frac{V_2(t)}{R_2} + \frac{V - V_1(t) - V_2(t)}{R_c} + C_2 \frac{dV_2(t)}{dt} = 0 \tag{4.6}$$

The solutions were simplified assuming that the contact resistance is much smaller than the layer resistances ($R_c << R_1, R_2$). The solutions V_1 and V_2 of the system show the following behavior: When a voltage V (bias voltage) is applied on this circuit, the capacitors will charge with a common time constant τ_s, where τ_s is the product of the contact resistance R_c and the series capacitance C_{tot} (see Table 4.1). The

	Short time regime	Long time regime
Time constant	$\tau_s = C_{\text{tot}} R_c$	$\tau_l = C_1 + C_2 / 1/R_1 + 1/R_2$
Layer 1	$V_{1,s,0} = \frac{C_{\text{tot}}}{C_1} V$	$V_{1,l,0} = \frac{R_1}{R_{\text{tot}}} V$
Layer 2	$V_{2,s,0} = \frac{C_{\text{tot}}}{C_2} V$	$V_{2,l,0} = \frac{R_2}{R_{\text{tot}}} V$

TABLE 4.1 Time Constants and Layer Voltages of a Double-Layer Structure

relative voltage at the layers, i.e., the percentage $V_{1,s}/V$ with respect to $V_{2,s}/V$ of the applied voltage, exponentially approaches $V_{1,s,0}/V$ with respect to $V_{2,s,0}/V$ corresponding to the ratio between C_{tot} and layer capacitance (see Fig. 4.5).

$$\frac{1}{C_{\text{tot}}} = \frac{1}{C_1} + \frac{1}{C_2} \tag{4.7}$$

$$R_{\text{tot}} = R_1 + R_2 \ (+R_c) \tag{4.8}$$

However, simultaneously but with a much larger time constant τ_l, the relative layer voltages $V_{1,l}/V$ and $V_{2,l}/V$ exponentially approach $V_{1,l,0}/V$ and $V_{2,l,0}/V$ which is the ratio between layer resistance and total resistance R_{tot} (see Table 4.1).

In contrast to τ_s, the time constant τ_l is the product of *parallel* layer resistances and the *parallel* layer capacitances. Since the layer resistances are much larger than the contact resistance, τ_l is much larger than τ_s. In the ideal case of infinite layer resistances, τ_l is infinite and the layer voltage will stay at its first reached level V_s.

A very good approximation for the evolution of the *i*th layer voltage is given by the following expressions:

$$\begin{aligned} V_{i,s}(t) &= V_{i,s,0}(1 - e^{-t/\tau_s}) \\ V_{i,l}(t) &= V_{i,l,0}(1 - e^{-t/\tau_l}) \end{aligned} \tag{4.9}$$

$$V_i(t) = V_{i,s}(t)e^{-t/\tau_l} + V_{i,l}(t) \tag{4.10}$$

The Capacitive Three-Layer Structure

The behavior of the capacitive three-layer structure is largely analogous to that of the double-layer structure. However, the analytical solution of the underlying differential equation is more sophisticated. The equivalent circuit of the capacitive three-layer structure is shown in Fig. 4.6.

The system of differential equations for the evolution of the voltages at the three layers is analogous to that of the double-layer structure:

$$\frac{V_1(t)}{R_1} + \frac{V - V_1(t) - V_2(t)}{R_c} + C_1 \frac{dV_1(t)}{dt} = 0 \tag{4.11}$$

$$\frac{V_2(t)}{R_2} + \frac{V - V_1(t) - V_2(t)}{R_c} + C_2 \frac{dV_2(t)}{dt} = 0 \tag{4.12}$$

$$\frac{V_3(t)}{R_3} + \frac{V - V_1(t) - V_2(t)}{R_c} + C_3 \frac{dV_3(t)}{dt} = 0 \tag{4.13}$$

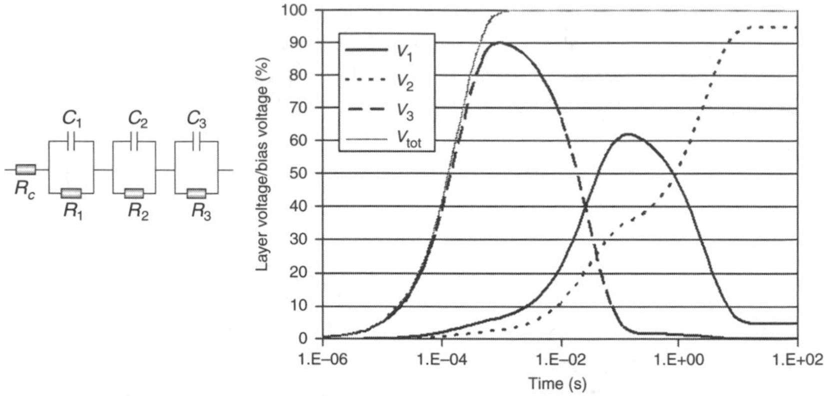

FIGURE 4.6 Evolution of layer voltages of a three-layer structure (V_1 is voltage at layer 1, V_2 is voltage at layer 2, V_3 is voltage at layer 3, V_{tot} is total voltage at all three layers; parameters: $R_c = 10\ \text{k}\Omega$, $C_1 = 100$ nF, $R_1 = 10\ \text{M}\Omega$, $C_2 = 50$ nF, $R_2 = 1\ \text{G}\Omega$; $C_3 = 2$ nF, $R_3 = 100\ \text{k}\Omega$; hence $\tau_s = 19\ \mu\text{s}$, $\tau_m = 3.5$ ms, $\tau_l = 1.5$ s, $V_{1,s,0}/V = 1.9\%$, $V_{2,s,0} = 3.8\%$, $V_{3,s,0}/V = 94.3\%$, $V_{1,m,0}/V = 32.4\%$, $V_{2,m,0} = 67.5\%$, $V_{3,m,0}/V = 0.1\%$, $V_{1,l,0}/V = 0.99\%$, $V_{2,l,0}/V = 99\%$, $V_{3,l,0}/V = 0.01\%$).

The solutions were simplified assuming that the contact resistance is much smaller than the layer resistances ($R_c \ll R_1, R_2, R_3$). As in the double-layer structure, the layers first charge with a time constant $\tau_s = C_{tot}R_c$ to a voltage determined by the layer capacitance (see Table 4.2). Moreover, the final layer voltage is determined by the respective layer resistances (see Fig. 4.6).

	Short time regime	Intermediate regime	Long time regime
Time constant	$\tau_s = C_{tot}R_c$	$\tau_m = \frac{A}{R_{tot}}\left(1-\sqrt{1-\frac{B}{A^2}}\right)$	$\tau_l = \frac{A}{R_{tot}}\left(1+\sqrt{1-\frac{B}{A^2}}\right)$
Layer 1	$V_{1,s} = \frac{C_{tot}}{C_1}V$	$V_{1,m} = V_{1,l}f(1,2,3)$	$V_{1,l} = \frac{R_1}{R_{tot}}V$
Layer 2	$V_{2,s} = \frac{C_{tot}}{C_2}V$	$V_{2,m} = V_{2,l}f(2,3,1)$	$V_{2,l} = \frac{R_2}{R_{tot}}V$
Layer 3	$V_{3,s} = \frac{C_{tot}}{C_3}V$	$V_{3,m} = V_{3,l}f(3,1,2)$	$V_{3,l} = \frac{R_3}{R_{tot}}V$

TABLE 4.2 Time Constants and Layer Voltages of a Three-Layer Structure

However, the final time constant τ_l is not given by the product of the parallel capacitances and resistances. In addition, an intermediate time constant with an associated voltage level for each layer arises.

$$\frac{1}{C_{\text{tot}}} = \frac{1}{C_1} + \frac{1}{C_2} + \frac{1}{C_3}$$
$$R_{\text{tot}} = R_1 + R_2 + R_3(+R_c) \tag{4.14}$$

$$A = \frac{C_1R_1(R_2+R_3) + C_2R_2(R_1+R_3) + C_3R_3(R_1+R_2)}{2} \tag{4.15}$$

$$B = C_1C_2C_3C_{\text{tot}}R_1R_2R_3R_{\text{tot}} \tag{4.16}$$

$$f(a,b,c) = \frac{-C_a(R_b+R_c)R_a\tau_l + C_bR_b[-C_c(R_b+R_c)R_c + C_aR_cR_a + R_b\tau_l] + C_cR_c(C_aR_bR_a + R_c\tau_l)}{C_a(R_b+R_c)R_a\tau_l + C_cR_c[-2C_aR_bR_a + (R_b+R_a)\tau_l] + C_bR_b[-2C_cR_cR_a - 2C_aR_cR_a + (R_c+R_a)\tau_l]} \tag{4.17}$$

In case one of the three layer resistances is much larger than the other two, say, $R_1 >> R_{2,3}$, one can find a simplified approximate expression for τ_l:

$$\tau_l = C_1(R_2+R_3) + C_2R_2 + C_3R_3 \tag{4.18}$$

If the time constants of the respective regimes are too close to each other, the first voltage levels of the respective time domain (regime) will not clearly show because the layer already charges up to the voltage level of the second time domain.

A very good approximation for the evolution of the ith layer voltage is given by the following expressions:

$$V_{i,s}(t) = V_{i,s,0}(1 - e^{-t/\tau_s})$$
$$V_{i,m}(t) = V_{i,m,0}(1 - e^{-t/\tau_m}) \tag{4.19}$$
$$V_{i,l}(t) = V_{i,l,0}(1 - e^{-t/\tau_l})$$

$$V_i(t) = [V_{i,s}(t)e^{-t/\tau_m} + V_{i,m}(t)]e^{-t/\tau_l} + V_{i,l}(t) \tag{4.20}$$

In case R_v is not much smaller than the layer resistances, a better approximation for the short time regime is obtained by adding a resistive prefactor:

$$V_i(t) = \left[\left(1 - \frac{R_v}{R_{\text{tot}}}\right)V_{i,s}(t)e^{-t/\tau_m} + V_{i,m}(t)\right]e^{-t/\tau_l} + V_{i,l}(t) \tag{4.21}$$

4.2 Integrated Pyroelectric Sensors

4.2.1 Introduction

Infrared sensors are used to detect thermal radiation in the mid- to far-infrared wavelengths. The wavelength region around 10 μm is of particular interest for it is there that the thermal radiation of living species reaches maximum intensity (at room temperature). Thermal radiation can be converted to electric signals by two groups of infrared detectors. The first group is formed by the photon detectors, which are wavelength-selective. They can be based on the photovoltaic, photoconductive, or photoelectric effect. They are made of semiconductor materials with a narrow energy gap, such as indium antimonide, and are extremely fast and sensitive. However, these quantum detectors require a minimal energy per photon for their operation and therefore often are cooled to cryogenic temperatures to obtain sufficient performance. Thermal detectors form the second group. They indicate the temperature rise of the sensor material by a change in resistance or thermoelectric power and are characterized by a slower response (and hence a low-frequency bandwidth) than that of the quantum detectors. They are sensitive to the entire absorbed radiation, regardless of its spectral composition, and are therefore particularly well suited for the detection of IR radiation. Their performance is limited solely by the spectral transmittance of the entrance window and of the optical imaging elements. However, thermal detectors are inferior to quantum detectors especially in sensitivity by several orders of magnitude. Bolometers, thermocouples and thermopiles, and pyroelectric detectors belong to this group.

The pyroelectric detector is the fastest of the thermal detectors since temperature changes at the molecular level are directly responsible for the detection process. Pyroelectricity is the electrical response of a material to a change in temperature. It is found in any dielectric material containing spontaneous or frozen polarizations resulting from oriented dipoles and occurs in 10 crystal classes, certain ceramics, and polymers that have been submitted to a special treatment. As discussed by S. B. Lang,[6] the pyroelectric effect has been known for 24 centuries when the Greek philosopher Theophrastus probably gave the earliest known description of the pyroelectric effect in his treatise "On Stones." Although pyroelectricity of polymers was already discovered in the 1940s,[7] it was not before 1971, when strong pyroelectricity was discovered in polyvinylidene fluoride (PVDF) by Bergmann et al.,[8] that the polymers received any serious attention due to the initially weak effects. Early applications then emerged very soon—Glass et al.[9] and Yamaka[10] reported on polymeric pyroelectric infrared sensors, while Bergmann and Crane[11] demonstrated a pyroelectricity-based xerography process. Nowadays the nature of pyroelectricity in polymers is reasonably well understood (for a review see Bauer and Lang[12]), and a large variety of amorphous, semicrystalline, single crystalline and liquid crystalline polymers are known to show significant pyroelectric response.

A complete pyroelectric polymer sensing pixel typically is thought of as a capacitive pyroelectric sensor with an input unit comprised of infrared absorption and focusing elements such as absorption layers and micro-optics and an output unit that accounts for signal readout, impedance transforming, and signal amplification.[13] In the angle-selective motion sensor developed by Siemens (marketed under the brand name PID-21), PVDF is used as a freestanding unit in an array configuration that is glued to an appropriate frame and connected to the electronics, thus forming a hybrid PCB construction.[14] Another possibility is to directly integrate polymer pyroelectric sensor arrays on silicon substrates, which provide the readout electronic circuits (impedance transformers, amplifiers). This, however, faces serious problems concerning thermomanagement.[14–18]

Contrary to that, it would be very advantageous to directly integrate polymer sensors and transistors, which means that in the case of silicon-based electronics, the sensor has to be produced directly on the silicon wafer, acting as the common substrate for transistor and capacitive sensing unit. Therefore classical integrated infrared detectors are confined to rigid substrates and planar surfaces and do not provide continuous panoramic (360°) views.[16–18]

With this in mind, organic thin-film transistors (OTFTs) and capacitive pyroelectric polymer sensors are easily pulled together to form integrated flexible pyroelectric sensors if one accounts for the possibility of both device classes being fabricated on flexible substrates and large areas in cost-effective production processes, thus opening completely new application areas. Such applications are found, e.g., in the context of pedestrian protection in the automotive industry, novel concepts for human/machine interfaces used in mobile electronics, large-area security features, low-cost home electronics, and artificial skin.[19] To date, there is no example of a large-area integrated organic pyroelectric sensor on the market, due to the high demands made on the performance of the OTFTs.

4.2.2 Theoretical Background—Pyroelectricity

Pyroelectrics and Ferroelectrics

Ferroelectrics have raised a lot of interest in the last decades, because of their wide field of applications especially in flash memories, due to their high dielectric constants (in the range from 10 to 1000), allowing high storage densities.

The important property for pyroelectrics and ferroelectrics is the existence of a spontaneous polarization. This polarization arises from a polarity in the unit cell of an electrical anisotropic crystal or portions of a semicrystalline polymer such as PVDF. The important property common to both materials is that this polarization can be switched between at least two metastable states by applying an external electric field. For the presence of the pyroelectric effect it is sufficient that the overall spontaneous polarization change with the temperature.

This is always the case, when a spontaneous polarization **P** exists, since the volume *V* changes with the temperature, and the total spontaneous polarization is defined by the sum of the molecular dipoles p_i in the unit cell of a crystal, or alternatively (as is in our case) in the repeat unit of a polymer, divided by its volume.[20]

$$\bar{P} = \frac{1}{V}\sum_i \bar{p}_i \tag{4.22}$$

The polarization has the units of coulombs per square meter (C/m²) since the dipole moment has the units of coulomb-meters (C·m). The ferroelectric materials can thus be regarded as a subgroup of the pyroelectric materials. Even if this chapter deals only with the pyroelectric effect, the term *ferroelectric* will often be present. This is due to the fact that most effects are termed with respect to the more popular ferroelectric effect. For example, the phase where a spontaneous polarization is present is called the *ferroelectric phase*, even if the term *pyroelectric* would be more general.

Pyroelectricity

In general the term *pyroelectricity* describes the ability of a material to change its spontaneous polarization vs. temperature. The magnitude of this change is described by the pyroelectric coefficient. The pyroelectric coefficient is defined as[21]

$$\vec{p}_{\text{pyro}} = \left(\frac{d\vec{P}}{dT}\right)_{M,\vec{E}} \tag{4.23}$$

Here dP is the change in the spontaneous polarization in response to the change in the temperature dT, normally measured under the condition of no or constant external electrical field **E** and mechanical stress **M**. The bulk polarization **P** is represented classically as a macroscopic manifestation of mean field effects due to residual electric-dipole moments.[22, 23] Discontinuities at the sample boundaries create a surface charge density $\sigma = \mathbf{nP}$, where **nP** is the component of the polarization normal to the boundary surface.

In the case of a parallel plate capacitor, the magnitude of the pyroelectric coefficient can be defined with respect to the charge generated at its electrodes in response to a temperature change:

$$p_{\text{pyro}} = \frac{1}{A}\left(\frac{dQ}{dT}\right)_{M,\vec{E}} \tag{4.24}$$

where A is the surface area of the capacitor and Q is the charge generated.

The spontaneous polarization creates an electric displacement **D** in the material. This generates a net positive or negative charge at the surface of the sample. In air these charges are more or less shielded by free charge carriers from the surrounding atmosphere. When electrodes are situated on the surface, the potential is compensated by free charges in the conductive material. For a stable temperature, no current can be observed since the electric displacement in the sensor material remains constant. As soon as the material undergoes a temperature variation, the spontaneous polarization changes due to changes in the dipole orientation and by a change in the density of the dipoles per volume. The change in the spontaneous polarization induces a change of the charge density at the surface of the pyroelectric layer. This is compensated by charge carriers in the electric circuit, thus producing a measurable current via the connecting electrodes.

4.2.3 Pyroelectric Polymer Materials

Structure of Poly(Vinylldene Fluoride)

The various properties of poly(vinylidene fluoride) or PVDF make it suitable for a wide range of applications. Depending on the way of fabrication the PVDF can crystallize in several different phases which are called α, β, γ, and δ phases. Vinylidene fluoride has a large molecular dipole moment of $\mu_v = 7 \times 10^{-30}$ C·m that occurs between the positively charged hydrogen and the negatively charged fluorine atoms in the (—CF_2CH_2—) section of the polymer. In the ferroelectric β-phase, the polymer chain is in the all-trans (TTTT) conformation, having all the dipoles oriented in the same direction, leading to a polar axis over the whole polymer chain. In the crystalline structure of the β-phase, the orientation of the polar axis is the same for all the polymer chains, leading to a macroscopic spontaneous polarization of 130 μC/m^2.[24]

Since the properties such as ferro-, piezo-, and pyroelectricity depend on the existence of a macroscopic spontaneous polarization, the generation of samples with a high amount of the ferroelectric β-phase is important. The crystallization as paraelectric α- or ferroelectric β-phase depends on the way in which the thin films are fabricated. To get the ferroelectric phase directly from the melt, some amount of the copolymer poly(trifluoro ethylene) (PTrFE) has to be added. This is so because a substitution of one hydrogen atom by the larger fluorine atom in TrFE (tri-fluoroethylene) favors the formation of the all-trans conformation. From a theoretical point of view, the copolymer reduces the possible maximum of the macroscopic polarization because it is a less polar molecule than the PVDF, but it increases the overall crystallinity of the semicrystalline polymer material.

The semicrystalline nature of polymer ferroelectrics covers three different types of polarization sources. There is frozen-in polarization from dipoles in the amorphous phase, ferroelectric (thermodynamic stable) polarization in the crystalline phases, as well as charge-induced polarization from compensation charges that accumulate at the interfaces between amorphous and crystalline regions, to compensate for the large depolarization fields and to stabilize the ferroelectric polarization.

Preparation of P(VDF-TrFE) Thin Films and of the Reference Capacitance Structures

It has been reported[25, 26] that lyophilized gels of PVDF using γ-butyrolactone as solvent lead to stable gels favoring the formation of TT conformation resulting in β-phase PVDF. Therefore it was decided to transfer the γ-butyrolactone-based sol-gel process for PVDF to the P(VDF-TrFE) copolymer, which seems to be the most promising way to obtain highly crystalline, sufficiently flat, ferroelectric thin films by using a low-temperature spin-on process.[27, 28]

In this process γ-butyrolactone is heated up to 180°C using a reflux condenser to avoid solvent evaporation. A known amount of P(VDF-TrFE) pellets with a VDF:TrFE composition of 55:45, 65:35, and 76:24 is added to the solvent and dissolved. The solution is kept at this temperature for approximately 2 h. After cooling down, the solution is poured into a glass bottle and kept at room temperature for gelation. Prior to spin-on, viscid gels are transferred to a low-viscous sol by slow heating up. Depending on the spin-speed and the solid phase content, layer thickness values between 300 nm and 3 μm can be achieved easily.[27, 28] The dielectric constants of materials with VDF:TrFE composition of 55:45, 65:35, and 76:24 typically are 15 to 17 for materials with 55% VDF content and 10 to 12 for the others.

For the electrical characterization of the ferroelectric properties of the polymer layers, capacitancelike structures were prepared by means of standard process steps. Aluminum layers with a thickness of 50 nm are sputtered on glass or PET substrates (Melinex) serving as the bottom electrode onto which the sol is spin-coated. Subsequently a calcination step is carried out at 110°C, which increases the crystallinity of the dielectric thin films. The samples are kept at this temperature for 5 h and then slowly cooled down to room temperature. Finally top electrodes are deposited by thermal evaporation of 50 to 80 nm silver via shadow masks. The hysteresis measurements for the determination of the remnant polarization were done in a Sawyer-Tower configuration.

Ferroelectric Characteristics of P(VDF-TrFE)

In Fig. 4.7 at left, a typical hysteresis loop measurement of a P(VDF-TrFE) capacitor is shown. The different curves correspond to loop measurements with increasing maximum electric field. The coercive

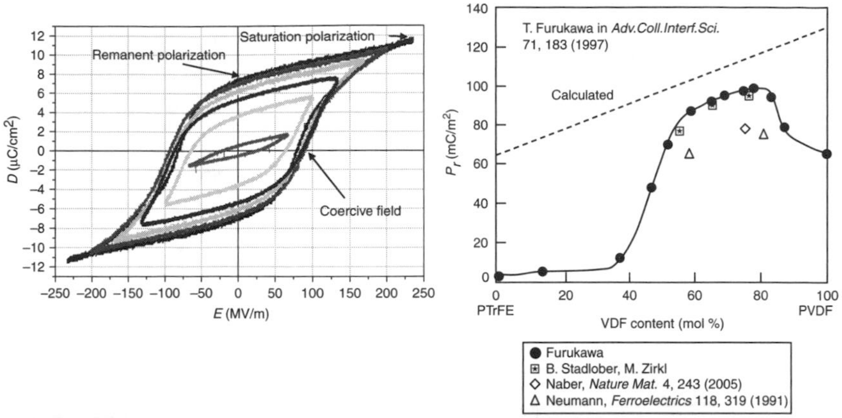

FIGURE 4.7 Left: Typical hysteresis loop of a flexible P(VDF-TrFE) sensor element (55:45) indicating the coercive field, the saturation polarization, and the remnant polarization. The different curves correspond to loop measurements with increasing maximum electric field. Right: Remnant polarization as a function of the VDF content derived by Furukawa[20] and other data from literature included as well as data from the sol-gel processed PVDF copolymer films. (*Reprinted from Ref. 20. Copyright 1997, with permission from Elsevier.*)

field strength, the field where all the dipoles in the material switch in the opposite orientation, is reached at about 80 MV/m. The maximum achievable polarization is called the *saturation polarization* and is about 200 MV/m.

In Fig. 4.7 at right, the theoretical and measured values of the remnant polarization P_r vs. the PVDF content (as derived by Furukawa[20]) are shown. It turns out that the highest polarizations are reached between 50 and 80% PVDF content. At very low PVDF concentrations, the few polar molecules prevent a high polarization and ferroelectricity is lost, while at high PVDF concentrations the creation of the α-phase is preferred and, moreover, the crystallinity is reduced to less than 50%. Between 80 and 50% there is a gradual decrease in the remnant polarization that is attributed to a decrease in the average dipole moment, because the dipole moment of a TrFE unit is one-half that of a VDF unit. The P_r values from the sol-gel processed P(VDF-TrFE) copolymers nicely agree with the Furukawa values (the highest reported so far), whereas other reported values are significantly smaller, thus indicating less ferroelectric activity.

All ferroelectric materials have a Curie temperature. At this temperature they undergo a transition from the ferroelectric to the paraelectric phase. Even if the polymer chains stay in the ferroelectric all-trans configuration, the macroscopic spontaneous polarization is lost under this transition because the alignment of the polymer chains, with the dipole moments pointing in a preferential direction, is lost.

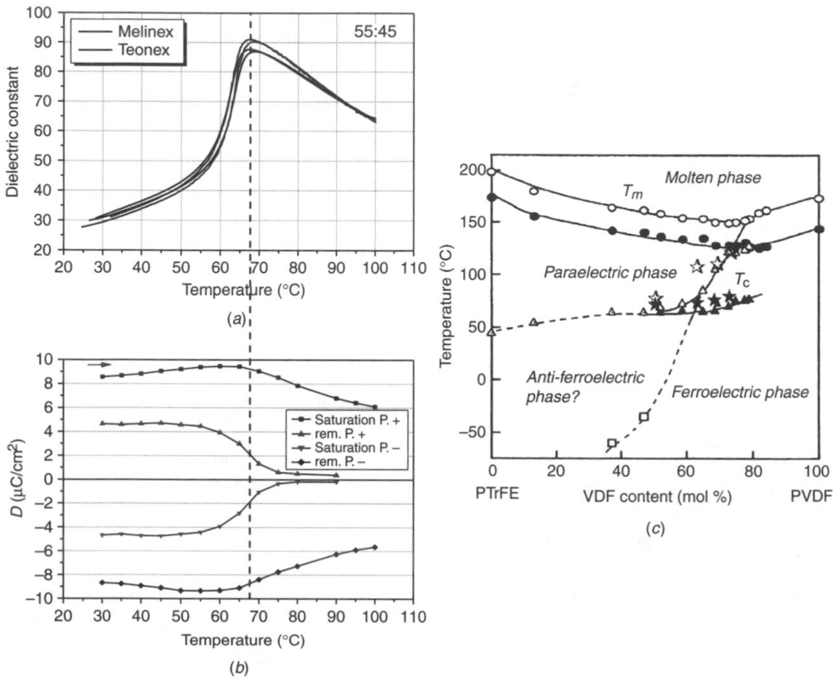

FIGURE 4.8 (*a*) Temperature dependence of the dielectric constant of a flexible P(VDF-TrFE) sensor element (55:45) indicating the phase transition temperature upon cooling and heating (dashed line). (*b*) Temperature dependence of the dielectric constant of the same flexible P(VDF-TrFE) sensor element indicating the phase transition temperature at the point where the remnant polarization vanishes. (*c*) Phase diagram as derived by Furukawa.[20] The stars are the data points for ferroelectric films based on the sol-gel procedure. All filled symbols correspond to transition temperatures upon cooling, whereas the empty symbols correspond to values obtained upon heating. (*Reprinted from Ref. 20. Copyright 1997, with permission from Elsevier.*)

This ferroelectric-paraelectric phase transition can be clearly seen in the temperature dependence of the polarization where the remnant polarization vanishes beyond the Curie temperature T_c (Fig. 4.8*b*). The temperature dependence of the dielectric constant (permittivity) also reveals the transition from the ferroelectric to the paraelectric phase at the Curie temperature, which, in the case of 55% VDF content, is the same upon heating and cooling, thus indicating a second-order phase transition (Fig. 4.8*a*). The behavior of the permittivity was also more deeply investigated by temperature- and frequency-dependent dielectric spectroscopy (Fig. 4.9), revealing the phase transition peak at T_C and the slight decrease of the dielectric constant with frequency that is most pronounced around the maximum of ε in the vicinity of T_C.

For materials with increased VDF content, a hysteresis develops between heating and cooling, thus corresponding to first-order phase

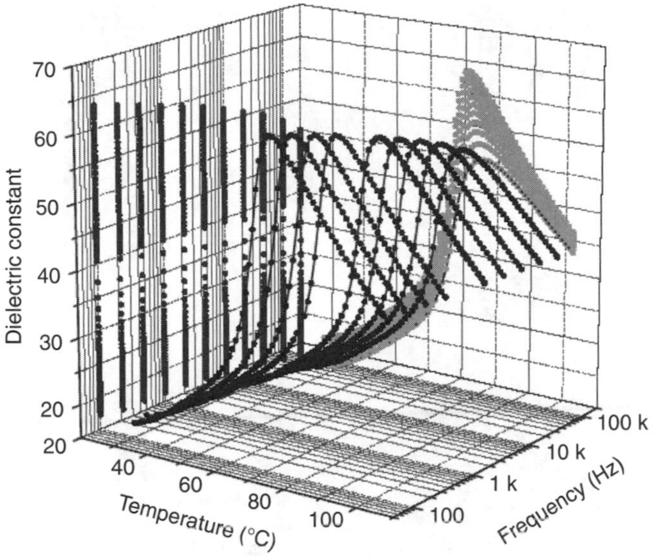

FIGURE 4.9 Temperature- and frequency-dependent dielectric spectroscopy of a P(VDF-TrFE) layer with a PVDF content of 55%. The phase transformation occurs at all frequencies at $T_c = 65°C$.

transitions. In the phase diagram derived by Furukawa (Fig. 4.8*c*), the phase transitions are plotted as a function of the VDF content. It is obvious that the transition temperatures decrease with the amount of PTrFE added, especially upon heating. The uppermost two parallel lines in this diagram correspond to the melting points upon heating and cooling of the material. For very high contents of PVDF, the Curie temperature is above the melting point, implying that the material is always ferroelectric in the solid state (assuming that the material is in the all-trans conformation).

Here again the values for phase transition temperatures of the sol-gel-based ferroelectric copolymer samples (Fig. 4.8*c*) nicely correspond to Furukawa's data. Therefore it can be concluded that the sol-gel processed P(VDF-TrFE) thin films are excellent ferroelectrics and therefore good candidates for materials with high pyroelectric activity.

4.2.4 Description of the Sensor Part

Setup for the Measurement of the Pyroelectric Response

To measure the macroscopic pyroelectric response of the sensor element, a laser diode (80 mW, 808 nm) is intensity-modulated with a sine or a square wave and is placed to illuminate the sample at the surface of the top electrode. The electrode is coated with a black graphite absorber to achieve maximal absorption. The induced temperature variations in

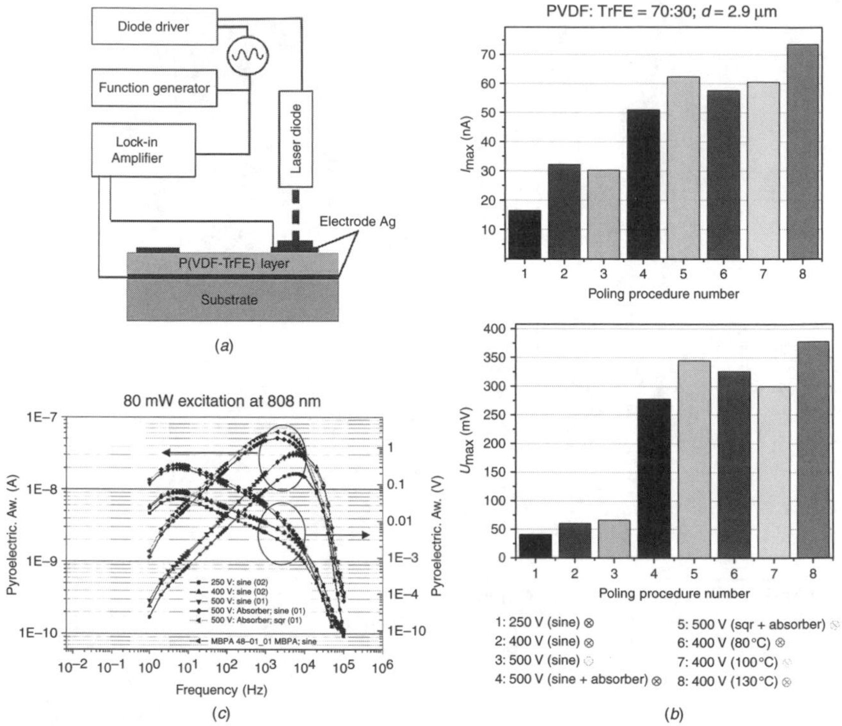

FIGURE 4.10 (*a*) Setup for the measurement of the pyroelectric response. (*b*) Influence of poling voltage, poling temperature, laser excitation waveform and absorber structure on the maximum of the pyroelectric current and voltage response (see part *c*), measured for a sample with 2.9 μm thick P(VDF:TrFE) layer and 70% VDF content on glass substrate at an excitation power of 80 mW. (*c*) Frequency dependence of the pyroelectric response of the same sample in voltage and current modes. The voltage response points at very low frequencies (below the cutoff frequency of the lock-in amplifier setup) were taken with a high-impedance parametric analyzer (MBPA) and correspond to poling at 500 V and sine excitation for a capacitance with absorber.

the sample generate an alternating current (compare to Sec. Pyroelectricity), which can be measured using a lock-in amplifier. The reference frequency is taken from the function generator which also drives the laser diode. The measurement setup for the detection of pyroelectric responses is shown as an inset in Fig. 4.10.

Poling

To measure a pyroelectric effect, it is important to align the dipoles in a preferential direction in order to achieve a macroscopic spontaneous polarization. This treatment is called *poling*. The poling can be achieved by applying an appropriate voltage to the electrodes, thus generating an external field, which should be higher than the coercive field of the sample.[29, 30] Since the coercive field of the P(VDF-TrFE) samples is about 80 MV/m, the applied coercive field should be taken well above this value (around 150 MV/m). A stepwise enhancement of the voltage and intermediate times with zero voltage applied to the sample was used as reported in the literature.[16] The effectiveness of the poling with respect to dipole alignment is strongly influenced not only by the poling voltage (the coercive field) but also by the temperature during poling as is depicted in Fig. 4.10*a* for a sample with 2.9 μm layer thickness and VDF content of 70%. According to this, the highest response (current and voltage) corresponding to the highest polarization can be achieved by increasing the poling field stepwise to about 140 MV/m at 130°C.

Phenomenology of the Pyroelectric Response

The pyroelectric response as obtained by the measurement setup shown in Fig. 4.10*a* can be detected in the voltage and current mode. The frequency dependence of the voltage as well as the current response for the copolymer sample with $d = 2.9$ μm is shown in Fig. 4.10*b*. The maximum voltage response values for the setup with the lock-in amplifier are achieved at 5 Hz, whereas the maximum current response is achieved between 2 and 6 kHz. According to the equivalent circuit and the relation

$$|V_{\text{pyro}}| = |I_{\text{pyro}}| \cdot \frac{R}{\sqrt{1+\omega^2 R^2 C^2}} \qquad C = C_p + C_i \qquad R = \left(\frac{1}{R_p} + \frac{1}{R_i}\right)^{-1} \tag{4.25}$$

the voltage response varies as IR below and $I/(\omega C)$ above the cutoff. The cutoff frequency ω_c is determined by the RC time constant of the whole equivalent circuit as $\omega_c = 1/RC$. Here R_p and C_p correspond to the resistance and capacitance of the pyroelectric element and R_i and C_i to the input resistance and capacitance of the measurement circuit, respectively.

From an inspection of Fig. 4.10 it becomes clear that, apart from influences of the dipole alignment (poling voltage and temperature), the absorption of the incident light (existence of absorber structure) and the waveform of the excitation have an influence on the magnitude of the response. With a graphite absorber layer, the voltage response is more than doubled, and the use of a square waveform of the laser excitation additionally yielded 30% signal (compare values for 500 V poling voltage in Fig. 4.10).

For the calculation of the voltage and current sensitivities of the pyroelectric sensor element, the magnitude of the pyroelectric response has to be divided by the power of the incident radiation. This is done in Fig. 4.11 for a thin pyroelectric sensor element fabricated on Melinex substrate. Here the inverse frequency dependence of the voltage response with respect to that of the current response [which is basically expected from Eq. (4.25) for $\omega > \omega_c$] is nicely observed.

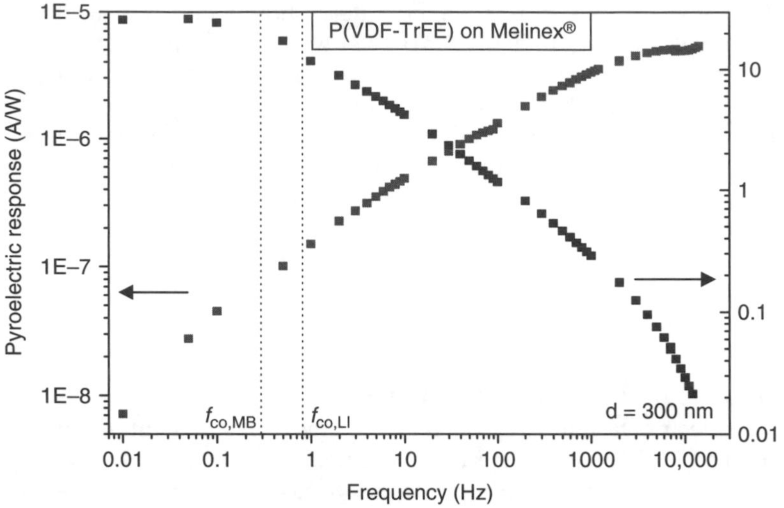

FIGURE 4.11 Pyroelectric voltage and current response of 300 nm samples with VDF:TrFE = 55:45 composition fabricated on PET substrate (Melinex). The pyroelectric sensitivities are obtained for an incident radiation of 45 mW. The points at frequencies below 1 Hz were obtained with a high input impedance parametric analyzer. The cutoff frequencies for measurement with lock-in amplifier and parametric analyzer are indicated, respectively.

Sol-gel Thin Films vs. Stretched Commercial Films To compare the pyroelectric sensitivities of sol-gel derived thin films with commercially available reference PVDF films with respect to sensitivity level and output voltage, a 25 μm thick uniaxially stretched PVDF foil (purchased from Piezotech SA) was provided with either aluminum or silver electrodes on both sides and characterized with the measuring setup described above. The results are shown in Fig. 4.12.

The voltage sensitivity R_V was determined in the low-frequency range by means of a high-impedance parametric analyzer from mb-technologies ($Z_{MBPA} = 1$ GΩ, if the MBPA is operated as a voltmeter) and for intermediate and high frequencies by a standard lock-in amplifier with a rather low input impedance of $Z_{LI} = 100$ MΩ. Between 5 and 5000 Hz the response voltage is inversely proportional to the modulation frequency of the input signal, as expected by pyroelectric theory (see Eq. 4.25). Moreover, the response is proportional to the infrared light intensity over three orders of magnitude. However, the response becomes constant below the cutoff frequency f_{co} that is decided by the $1/CR$ time constant of the measuring circuit. For the MBPA we calculated $f_{co} = 1.5$ Hz, whereas $f_{co} = 12$ Hz for the lock-in amplifier, resulting in a respective reduction of the voltage signal for $f < f_{co}$ (see Fig. 4.12 top).

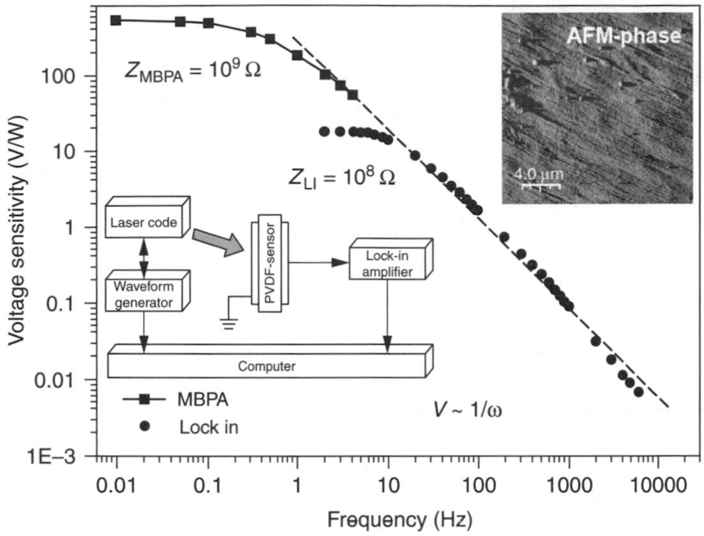

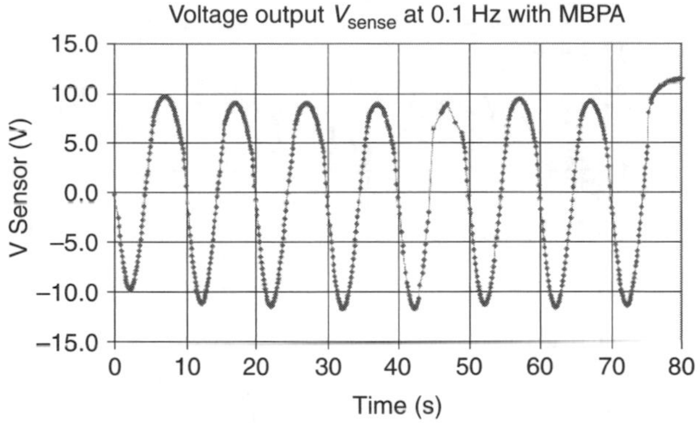

FIGURE 4.12 Top: Voltage sensitivity of a 25 μm thick capacitive PVDF foil sensor with Ag top and bottom contacts and a graphite absorbing layer on top. The pyroelectric response of the foil sensor was measured in the low-frequency range by a high-impedance parameter analyzer (MBPA) and at medium and higher frequencies by a lock-in amplifier. The inserts show an AFM micrograph of the phase signal of an Ag electrode on top of the PVDF foil and the equivalent circuit diagram of the measuring setup. Bottom: Voltage output of the PVDF sensor measured with the MBPA at a modulation frequency of 0.1 Hz.

For comparison, the voltage sensitivity and output voltage of a 2 μm thick spin-on PVDF-copolymer film are depicted in Fig. 4.13. The ferroelectric copolymer films are fabricated and characterized as described in Sec. 4.2.4 and afterward subdued to a stepwise poling procedure (see Sec. Setup for the Measurement of the Pyroelectric Response). Due to the much higher film capacitance (~ 600 pF) compared to the foil, the cutoff frequencies are expected to be one order of

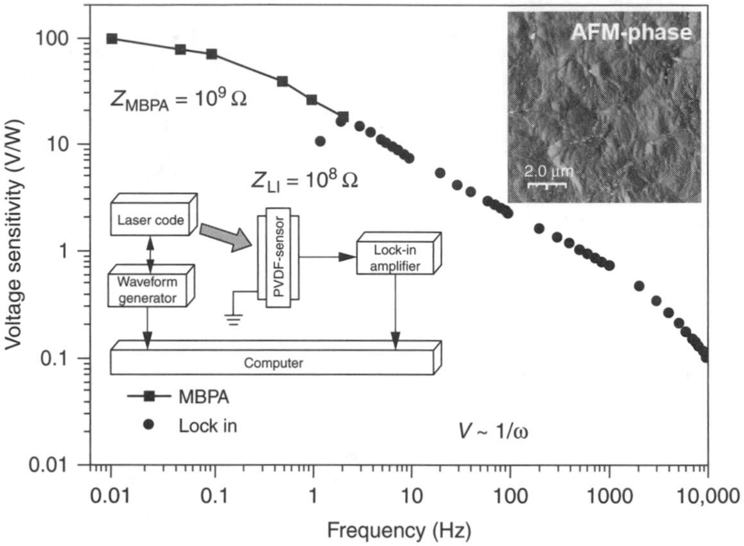

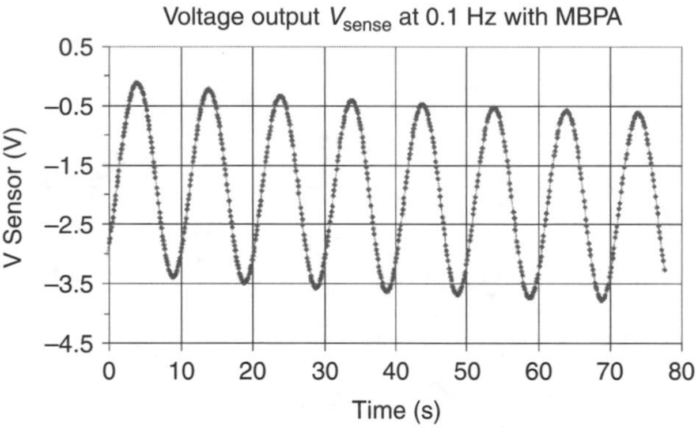

FIGURE 4.13 Top: Voltage sensitivity of a 2 μm thick capacitive PVDF-TrFE thin-film sensor with Ag top and bottom contacts and a graphite absorbing layer on top (PVDF:TrFE = 55:45). The pyroelectric response of the thin-film sensor was measured in the low-frequency range by a high-impedance parameter analyzer (MBPA) and at medium and higher frequencies by a lock-in amplifier. The inserts show an AFM micrograph of the phase signal of the surface of the PVDF layer and the equivalent circuit diagram of the measuring setup. Bottom: The voltage output of the PVDF sensor measured with the MBPA at a modulation frequency of 0.1 Hz is shown.

magnitude lower. The calculated values are $f_{co} = 2.6$ Hz for the lock-in measurement and $f_{co} = 0.3$ Hz for the parametric analyzer MBPA, both of which are consistently reproduced in the voltage sensitivity measurements. The overall voltage sensitivity at low frequencies (0.01 Hz) is about $R_V = 100$ V/W which, normalized to the ferroelectric polymer layer thickness, results in an $R_{V''} = R_V/d = 50$ MV/(W·m) that is significantly larger than the associated value $R_{V'} = 20$ MV/m obtained for the commercial PVDF film. That illustrates the sufficiently large crystallinity of the ferroelectric copolymer thin films and the effectiveness of the poling procedure for a parallel orientation of the dipoles.

Modelling of the Pyroelectric Response—Heat Distribution Models Next to the pyroelectric coefficient of the layer, the design of the sensor device is important to improve the pyroelectric current and voltage responses. The performance of the sensor is especially influenced by the substrate (material and thickness), the thickness of the pyroelectric layer, and the absorber structures. To determine the different influences qualitatively and quantitatively, two models for the thermal conduction in a sensor element were developed. The knowledge of the actual temperature change in the pyroelectric layer is necessary not only for the improvement of the design of the sensor element, but also for the determination of the pyroelectric coefficient of a given material.

The first model is a one-dimensional model that enables calculation of the temperature variations at any position in vertical direction from the surface of the sensor through the different material layers of the sample. This model is then used to calculate the average temperature change in the pyroelectric layer to compare various sensor designs, and the influences from different parameters of the sensor design on the pyroelectric response are investigated. Parameters under investigation are the substrate material and thickness, thickness of the pyroelectric layer, and surface area of the sensor. By solving the basic heat distribution equation, Eq. (4.26), for a set of adjacent layers, the average temperature variation in the pyroelectric layer can be calculated. The as-calculated temperature variations are then used to model the expected pyroelectric responses in the frequency range from 10^{-4} to 10^6 Hz by using an appropriate equivalent circuit for the measurement setup. The as-obtained current and voltage responses for different sensor designs (different area, substrate material, substrate thickness, thickness of pyroelectric layer) are compared with the results obtained from pyroelectric measurements performed in the way described in Sec. 3.3. The model helps thus to design a sensor with a maximum pyroelectric response in the targeted frequency range as well as to tailor the frequency dependence of current and voltage response.

The second model is based on a finite element method (FEM) using the MATLAB Partial Differential Equation (PDE) toolbox. Solving the heat transfer equation in two dimensions helps to calculate lateral resolution limits, which are important for designing an array of close-packed sensor elements. Consideration is also given to the time resolution limits of the sensor device which occur because the heat from a previous event has to be conducted away before a subsequent excitation occurs.

One-Dimensional Model

In accordance with the publication of Setiadi and Regtien,[31] a one-dimensional heat distribution equation has been used, enabling one to calculate the amplitude and time dependence of the temperature variations in the sensor. The first model takes into account the different specific heat capacities, thermal conductivities, and densities of the different materials in the thin-film system. A sketch of a modeled layer system is given in Fig. 4.14. The excitation of heat waves is done with a laser, intensity-modulated with a sine function. To extract the pyroelectric current and voltage responses, a suitable equivalent circuit for the sensor element had to be taken into account. To compare the results with the experimental data, the frequency-dependent measured values for C_p and R_p (obtained with an LCR meter) from the respective sensor elements were used as an input for the model.

The principal equation for heat conduction within the nth layer is

$$\frac{\partial T(x,t)}{\partial t} = \frac{\delta_n}{c_n \cdot d_n} \frac{\partial^2 T_n(x,t)}{\partial x^2} \tag{4.26}$$

where δ_n = heat conductivity
c_n = specific heat
d_n = mass density of the nth layer

As the intensity of the incident light is assumed to vary according to a sine wave ($P_i = P_o \cdot e^{i\omega t}$), producing heat variations in the sensor element, T_n can be written as

$$T_n(x,t) = T_n(x) \cdot e^{i\omega t} \tag{4.27}$$

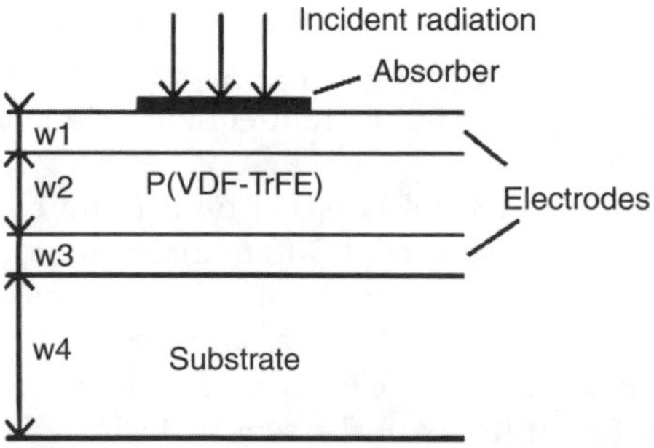

FIGURE 4.14 Layer setup for the one-dimensional model.

The second important equation is the spatial heat current density J which for the nth layer is defined as

$$J_n(x) = -\delta_n \frac{\partial T_n(x)}{\partial x} \qquad (4.28)$$

The problem can be solved analytically with respective boundary conditions at each layer, basically yielding the average temperature $\overline{T}_n(t)$ for the nth layer.

According to Fig. 4.14, the pyroelectric layer is the second layer and is given by

$$I_{\text{pyro}} = \frac{dQ}{dt} = \frac{dQ}{dT} \cdot \frac{dT}{dt} = p_{\text{pyro}} A \omega \overline{T} \qquad (4.29)$$

with $\overline{T}$ being the average temperature of the pyroelectric layer that is represented by a complicated matrix equation[32, 33] accounting for the material constants (δ, c, d) and for the heat input which is described by the absorption coefficient η, referring to the amount of light absorbed by the sensor and the heat radiation transfer coefficient g_H, taking into account the heat loss per unit area at the front and back sides of the sensor.

By means of the equivalent circuit, the voltage response can be calculated from the differential equation

$$p_{\text{pyro}} A \frac{\partial \overline{T}}{\partial t} = C \frac{\partial V}{\partial T} + \frac{V}{R} \qquad (4.30)$$

with the solution

$$V_{\text{pyro}}(t) = \left| \frac{pA\omega R}{\sqrt{1+\omega^2 R^2 C^2}} \cdot \overline{T} \right| = I_{\text{pyro}} \cdot \frac{R}{\sqrt{1+\omega^2 R^2 C^2}} \qquad (4.31)$$

This solution has already been given in the beginning of Sec. Phenomenology of the Pyroelectric Response and describes the absolute value of the pyroelectric voltage response that can be measured by a lock-in amplifier.

Comparison with Experiment Before using the model to test the influence of certain material parameters, a comparison between the experimental and the modeled results for one example is given (with the material properties specified in Fig. 4.15). The shape of the curve can be explained very well by the one-dimensional multilayer model over four orders of magnitude in frequency. Some of the

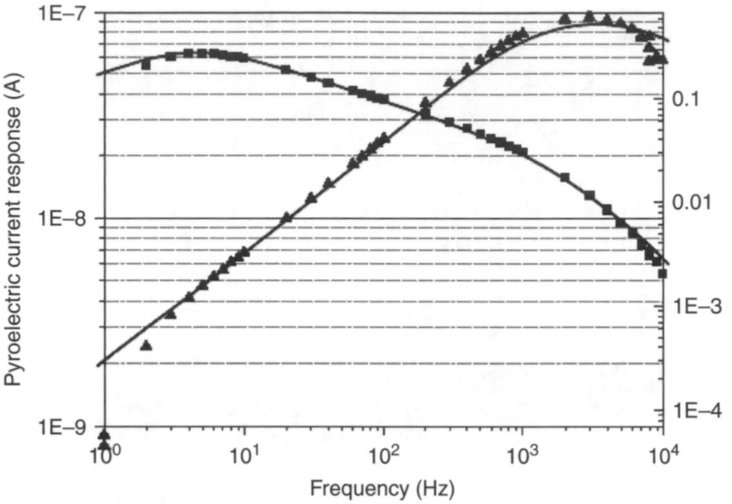

FIGURE 4.15 Comparison of the calculated and modeled current and voltage response for a sample similar in composition and fabrication route to that of Fig. 4.10. The lines correspond to the calculated voltage and current response. The parameters used for the calculation are P(VDF-TrFE) layer thickness $d = 3.3$ μm, input laser intensity of $P = 13$ kW/m², pyroelectric coefficient $p = 30$ μC/(K·m²), absorber thickness 1.8 μm, absorption coefficient $\eta = 0.9$, $R_i = 100$ MΩ, and $C_i = 75$ pF.

input parameters for the calculation are given by the sample geometry and are determined independently by ellipsometry (layer thickness values) or are inherent to the measurement setup (specified values for input resistance and capacitance of lock-in amplifier plus cable capacitance) or are determined by the excitation setup as the laser intensity.

The parameters that are not known exactly for this specific sample are the pyroelectric coefficient, which was assumed to be 30 μC/(K·m²) corresponding to measured values of similarly constructed samples and the absorbed intensity that is transferred to heat (90% absorption was estimated corresponding to independent NIR transmission measurements of the absorption layer). However, these values shift the obtained responses only linearly. The steeper drop of the measured values in the high-frequency range of the current response is due to the influence of the black absorber layer whose layer thickness seems to be underestimated in the model.

Dependence of the Voltage Response on the Sensor Area Even if the current response scales proportional to the surface area of the sensor, which follows directly from Eq. (4.29), the obtained voltage response shows a different behavior (see Fig. 4.16 for a sensor with 400 nm thick pyroelectric layer on glass substrate). According to Eq. (4.31) the

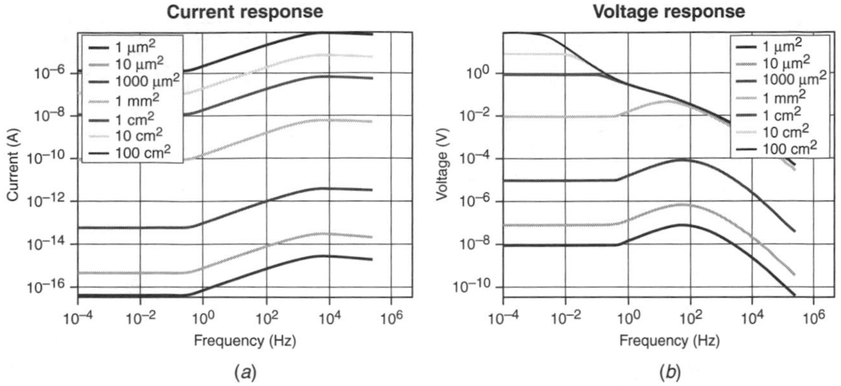

FIGURE 4.16 The calculated current and voltage response for different sensor areas. Substrate: glass 175 μm; pyroelectric layer thickness: 400 nm. Calculated for the values of $R_i = 100$ MΩ and $C_i = 75$ pF. (See also color insert.)

voltage response also depends on the sensor capacitance connected in parallel with the capacitance of the readout electronics. When the sensor area is very small, the low capacitance leads to an overall drop of the response over the whole frequency regime as in this case the cutoff frequency is much larger ($\omega << \omega_c$). Bigger sensor elements (≥ 1 mm^2 for the sensor configuration in Fig. 4.16), on the other hand, can only increase the signal in the low-frequency range ($\omega << \omega_c$), while the response remains constant at high frequencies where the linear dependence of current and capacitance on the area cancel each other due to

$$V_{\text{pyro}} = \frac{I_{\text{pyro}}}{\omega C} \qquad \text{at } \omega >> \omega_c \tag{4.32}$$

For an explanation of the detailed shape of the calculated results, the graphs have to be compared with the temperature lift calculated in the pyroelectric layer, given in Fig. 4.17*b* for a 175 μm thick glass substrate. In the low-frequency range, the temperature decreases linearly with the frequency. In the range from 1 Hz to 10 kHz the slope of the temperature lift declines slower than $1/\omega$ and thus produces an enhancement in the current response. In the high-frequency regime, the temperature is not reaching the pyroelectric layer anymore and thus the obtained response starts to decrease again. To explain the voltage response, the lowpass filter effect of the device connected to the lock-in amplifier has to be considered in addition to the temperature effects.

Comparison of Different Substrates and Their Thicknesses It is widely known in pyroelectric technology that the substrate has a strong influence on the pyroelectric response, since it is a heat sink for the

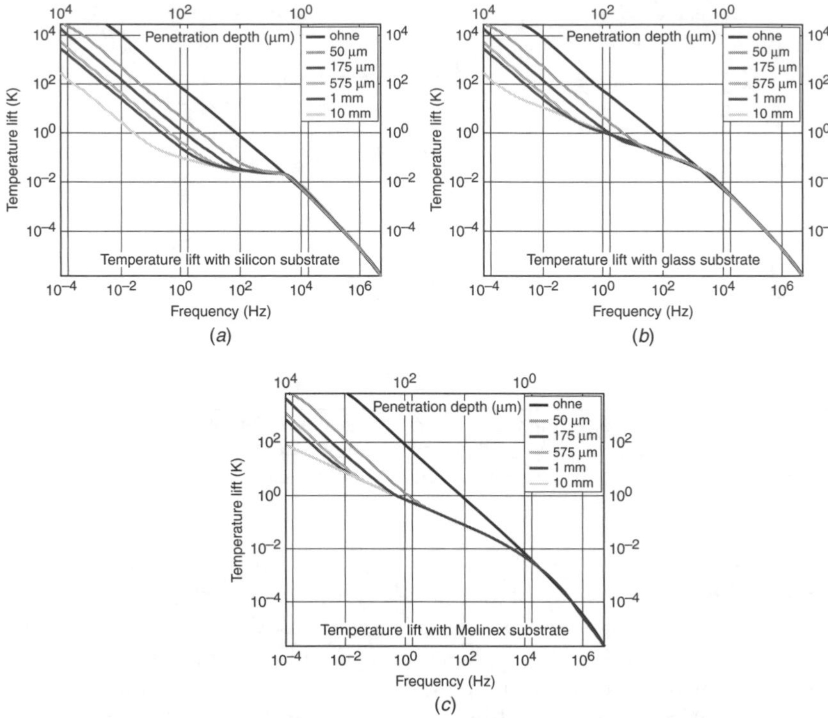

Figure 4.17 Calculated temperature lift for different substrates and the multilayer model depicted in Fig. 4.14. The substrates are (*a*) silicon, (*b*) glass, and (*c*) PET-foil Melinex. The penetration depth of the thermal wave and its assignment to the frequency are also indicated (red axis). (See also color insert.)

sensor element. This, however, is a big advantage for sensors using P(VDF-TrFE) because they are not automatically connected to highly thermal conductive substrates. Ceramics such as lead zirconate titanate (PZT), crystals such as triglycine sulfate (TGS), and other pyroelectrics are normally fabricated on a silicon substrate, which strongly decreases the average temperature lift in the sensor element (compare to Fig. 4.17*a*). A comparison of different substrates is thus very important. Since one big advantage of the PVDF copolymer is that it can be processed successfully on flexible substrates, a comparison of the thermal influences of different substrates and their thicknesses was made. To compare the results with the work of Setiadi and Regtien,[31] a calculation with a silicon substrate was done as well. In Fig. 4.17 the temperature lift (induced by the absorption of the laser) in the sensor element with respect to the ambient temperature is given. It is obvious that the silicon substrate has the worst responses in the low-frequency range because the substrate acts as a heat sink. The different colors in the plot are calculations for different thicknesses of the substrate. Of course, all substrate-related considerations are only important in the low-frequency range, when the penetration depth of the thermal wave is sufficiently high to reach down to the substrate. The *penetration depth* is defined as the distance in which the temperature has fallen to $1/e$ of its maximum value. In the high-frequency range (above 1 kHz), the penetration depth is shorter than the thickness of the pyroelectric layer. Since it is the average of the temperature lift $\bar{T}$ in the pyroelectric layer that is important, $\bar{T}$ decreases with a faster excitation, due to a shorter penetration depth λ at higher frequencies according to $\lambda = \sqrt{2K/\omega}$ with K being the thermal diffusivity.

The blue line in all diagrams of Fig. 4.17 corresponds to a calculation without any substrate, hence it does not change from one substrate to another. The frequency where the influence of the substrates is starting to become visible is around 10 kHz, and this frequency is the same for all substrates. But the extent to which the temperature lift is lowered by the substrate depends on its heat conductivity. The highest response is hence obtained for the PET foil (Melinex), having a heat conductivity in the range of the P(VDFTrFE) layer. Silicon is the worst substrate in the interesting frequency range between 0.1 and 100 Hz with respect to the magnitude of the voltage and current response because of its high thermal conductivity. With increasing thickness of the substrate the lowering of the temperature lift is extended to smaller frequencies. In Fig. 4.18*a* to *c* the different current and voltage responses are plotted. It is shown that they strongly depend on the substrate thickness in the low-frequency range. Again the detailed characteristics of these curves are determined and explained by the temperature lifts calculated and displayed in Fig. 4.17.

Comparison of Different Pyroelectric Layer Thicknesses The thickness of the pyroelectric layer has an influence on the response functions of the sensor as well (see Fig. 4.19*a* to *c*). At low frequencies, when due to the large penetration depth the whole sample is excited, the pyroelectric current is equal for all thicknesses of the pyroelectric layer. At higher frequencies the pyroelectric layer is activated only partly, resulting in a lower average temperature in the pyroelectric layer. It is interesting that for the high thermally conductive silicon substrate, the thin pyroelectric layer does not give the highest response, because the heat is conducted away by the substrate. However, for each layer thickness on the silicon substrate, there is a frequency range where the current response is maximized.

Influence of Impedance and Capacitance of the Measurement Circuit As expected from Eq. (4.31) the voltage response is strongly influenced by the input resistance of the measurement instrument or, strictly speaking, by the cutoff frequency ω_c (see Sec. 4.2.4), whereas the current response is independent of ω_c and only determined by the

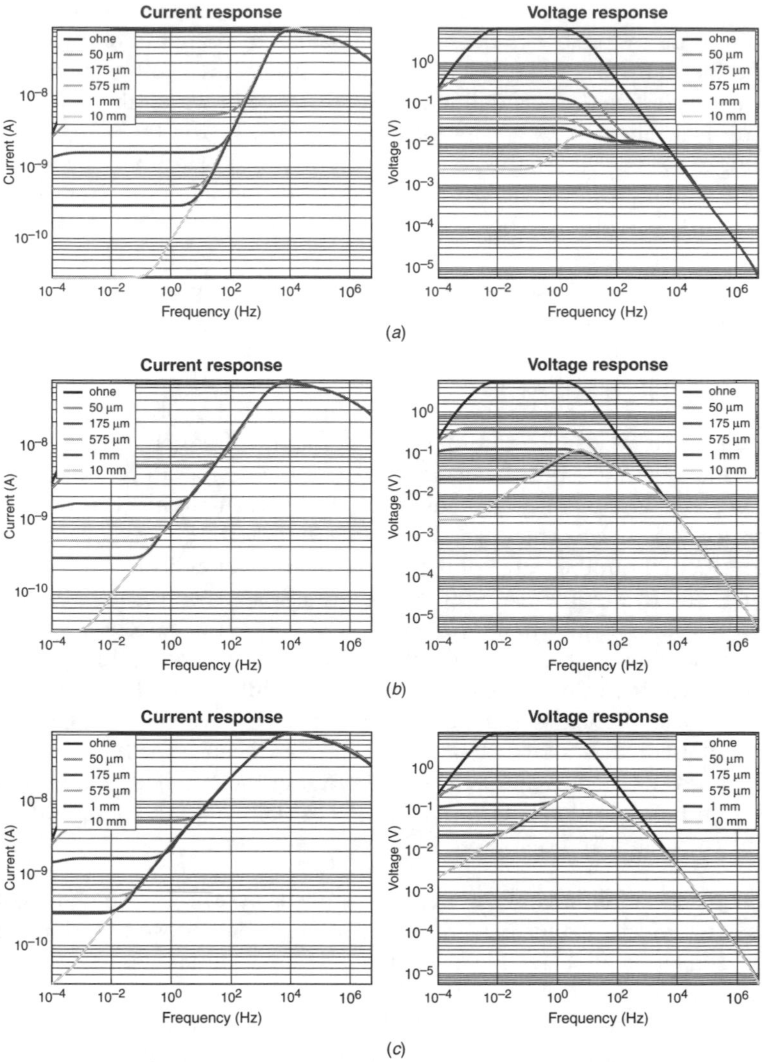

FIGURE 4.18 Current and voltage response in dependence of the substrate thicknesses and the material for (*a*) silicon (*b*) glass, and (*c*) PET foil (Melinex from DuPont Teijin), calculated for the values of $R_i = 100\ \text{M}\Omega$ and $C_i = 75$ pF. (See also color insert.)

average temperature in the pyroelectric layer and the product of frequency, area, and pyroelectric coefficient [see Eq. (4.29)]. For frequencies below the cutoff, the voltage response is proportional to the current response according to $V_{pyro} = I_{pyro}R$ with R being the overall resistance of the equivalent circuit. According to that, the voltage response is

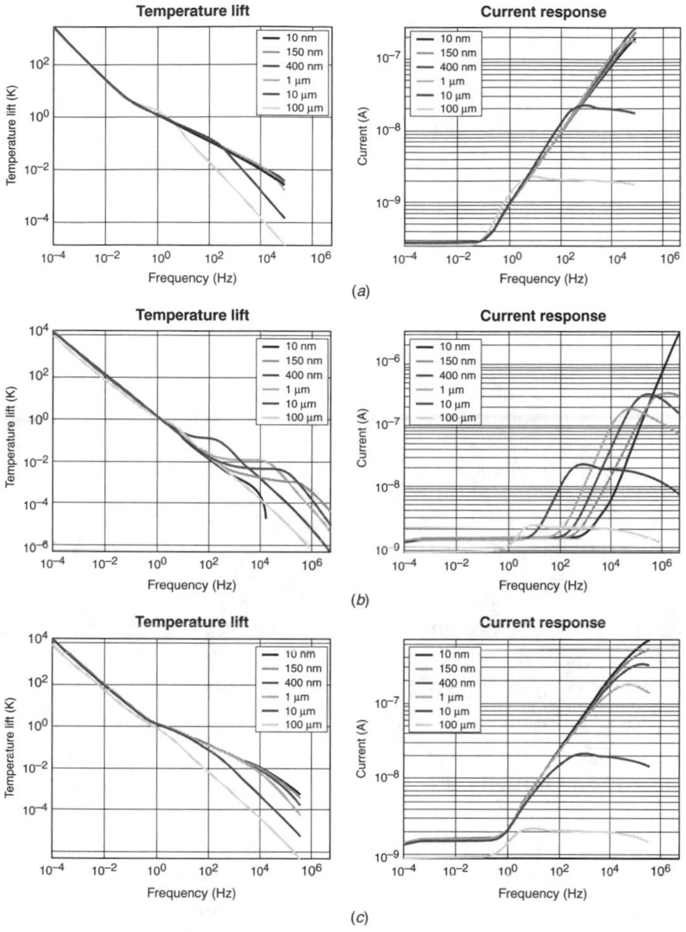

FIGURE 4.19 Temperature lift and current response with different thicknesses of the pyroelectric layer on 200 µm thick substrate of (*a*) glass, (*b*) silicon, and (*c*) PET, calculated for the values of R_i = 100 MΩ and C_i = 75 pF. (See also color insert.)

basically frequency-independent below 1 Hz, and this level increases with the resistance (Fig. 4.20).

Two-Dimensional Model

The one-dimensional model gives a relatively good picture of the thermal penetration of the device and is, by the use of an appropriate equivalent circuit, able to explain the measured pyroelectric responses very well. However, since it is a one-dimensional model, no information on the lateral resolution can be obtained from this approach. A two-dimensional model of the heat transfer in the sensor element was therefore developed to give further insight in the lateral thermal distribution.

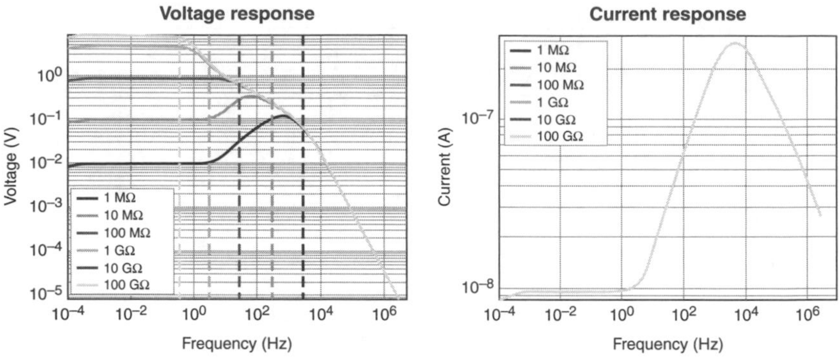

FIGURE 4.20 Frequency dependence of the voltage and current response for a sensor with 3 μm thick pyroelectric layer on 175 μm glass substrate and 75 pF input capacitance for different input resistance values of the measurement instrument (lock-in amplifier, parametric analyzer, etc.). The dotted lines indicate the respective cutoff frequencies $f_c = \omega_c/(2\pi)$. All lines collapse into one for the current response. (See also color insert.)

This becomes important for the fabrication of an array of close-packed sensor elements. The obtainable resolution of such an array, when used for thermal imaging, is obviously dependent on the amount of sensor elements, able to be addressed separately. But even if the electrode structure is fabricated with a high spatial density, the heat conduction between adjacent sensor elements is a resolution-limiting factor. This model uses the finite element method (FEM) to numerically calculate the thermal conduction in two dimensions, since no analytic solution can be provided in this case. The basis of the model is to define a geometry and to solve the heat-transfer equation on that geometry. Boundary conditions specify the behavior of the system at the edges of the observed region, as well as the heat entries and losses at these boundaries.

The Finite Element Method The basic idea of the finite element method is to solve a continuous differential equation numerically on a discrete grid and in discrete time steps. For each grid cell i a value for the solution $u(x_i, t_i)$ is calculated by the use of the solution in the previous time step $u(x_i, t_{i-1})$ and the solutions from adjacent cells $u(x_{i\pm1}, t_i)$. Boundary conditions specify the time behavior of the system at the edge of the specified geometry $u(x_0, t)$. Due to limited computer power possibilities, a dimensional reduction has to be taken into account. In this case a two-dimensional analysis of the problem seems to be sufficient, since the heat flux has the same properties in the x and y directions. Another important question concerns whether the surrounding environment should be part of the model or whether the interactions are put in the boundary conditions.

Since only a part of the sensor element is modeled, the edges where the real sensor is extended in reality are described by appropriate boundary conditions. The boundaries, where a heat transfer to the environment takes place, were taken into account in the boundary conditions as well, using the heat transfer coefficients from the one dimensional model.

Calculations of Different Heat Distributions and Comparison with One-Dimensional Model The FEM model gives the simulated temperature for every triangle in the mesh at each time *t*. The mesh is automatically generated by the MATLAB PDE toolbox and is used to discretize the problem (see Fig. 4.21*a*).

On the basis of this geometry, the calculated temperature distribution for a thermal excitation occurring at electrode 1 is displayed in Fig. 4.21*b*. The thermal conduction to the adjacent sensor elements is more clearly seen in the inset. An excitation temperature of 0.1 K is present at the surface of the top electrode.

In Fig. 4.22 the average temperature lifts in the sensor elements under various conditions are plotted. The upper line always corresponds to

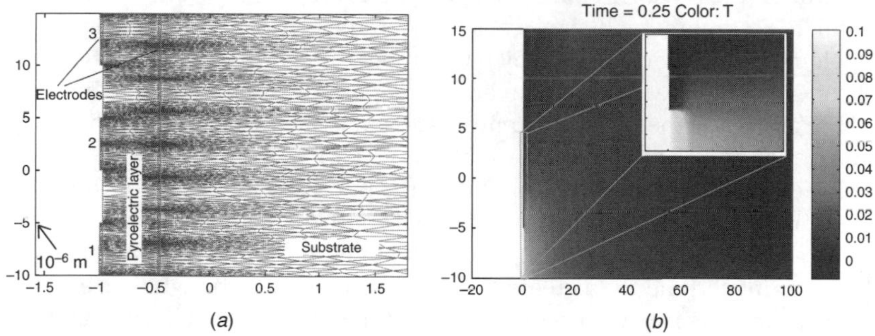

Figure 4.21 (*a*) Section of the geometry of three adjacent sensor elements on one 100 μm substrate. Only a part of the substrate is shown. The lines indicate the mesh. (*b*) Plot of the calculated temperature distribution. (See also color insert.)

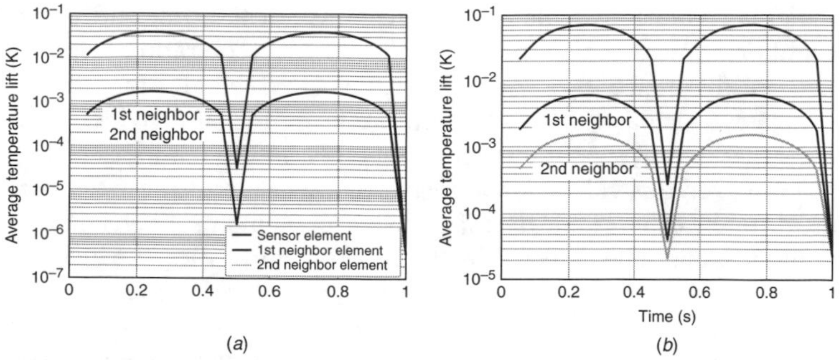

Figure 4.22 Difference in the temperature lift for adjacent sensor elements. (*a*) P(VDF-TrFE) pyroelectric layer on silicon substrate. (*b*) P(VDF-TrFE) pyroelectric layer on PET-foil substrate, excited element, 1st neighbor and 2nd neighbor.

the activated sensor element. The adjacent sensor elements are not activated, but due to the heat flux from the excited sensor element a thermal variation is also present in these elements. The middle line corresponds to the first neighbor and the lower line to the second neighbor. The different subfigures represent the temperature lift for various material compositions.

A direct quantitative comparison between the one- and the two-dimensional models is complicated. In the one-dimensional model, the starting point is the intensity of the laser, and the absorption at the top of the sensor element is already a part of the modeling process. The starting point in the two-dimensional model is the temperature lift at the front side of the sensor element. Since the temperature lifts for the different layers can be computed with the one-dimensional model, this value can be taken as a starting point for the two-dimensional model. In Fig. 4.22*a*, a calculation of a P(VDF-TrFE) layer on a silicon substrate is made, whereas in Fig. 4.22*b* PET is used as the substrate. The uppermost lines correspond to the time development of the average temperature in the sensor element excited. The average temperature was calculated by summing over all cells in the pyroelectric layer of the sensor element and dividing them by the number of cells. In the case of the silicon substrate the maximum average temperature is simulated to be 0.03 K (assuming the excitation temperature at the top side of 0.1 K and the frequency of 1 Hz). In the case of the PET foil substrate the maximum temperature lift is 0.06 K. This corresponds to the result of the one-dimensional model. The high thermal conductive substrate is acting as a heat sink, lowering the average temperature in the sample. It takes too much time to calculate this temperature over the whole frequency range as it was done in the one-dimensional model, but the result at the frequency of 1 Hz corresponds to the values obtained in the one-dimensional model.

Considerations on the Lateral Resolution Since the heat flux limits the spatial resolution of a sensor array, the heat flux to adjacent sensor elements has been calculated. The response of the first and second neighbor element is also displayed in Fig. 4.22. The ratio of the temperature difference $T_{\text{neighbor}}/T_{\text{excited}}$ is given in Table 4.3 for different substrates calculated. The determination of the allowed crosstalk between adjacent sensor elements depends on the readout electronics. However, this value can be seen as a lower limit for the temperature difference detectable by adjacent sensor elements. Even if smaller differences could be detected by the readout electronics, the heat conductivity becomes the resolution limiting factor for the device. It becomes obvious that a low thermally conductive substrate is advantageous for PVDF as the pyroelectric layer. However, the best results are obtained when using a high thermally conductive pyroelectric layer (see Fig. 4.22). The temperature lift in the pyroelectric layer is the highest, and the thermal crosstalk is low. This result is in contradiction

Pyroelectric layer—substrate	T_{exceed} (K)	$T_{neighbor}$ (K)	Ratio 1st	Ratio 2nd
PVDF–silicon	0.03	0.002	6.6×10^{-2}	6.6×10^{-2}
PVDF–PET	0.06	0.005	8.3×10^{-2}	2.5×10^{-2}
High thermal conductive layer–silicon	0.1	0.001	1×10^{-3}	4×10^{-3}

TABLE 4.3 Thermal Crosstalk Between Adjacent Sensor Elements

to the statement in several publications, that PVDF is advantageous in array applications because of its poor heat conductivity.[6] According to the fact that the thermal crosstalk between the excited element and the second neighbor is only 2.5% for a distance separation of 10 μm, virtually no thermal crosstalk is expected for elements that are separated by about 50 to 100 μm.

4.2.5 Description of Transistor Part

Why Do We Need an Organic Thin-Film Transistor and What Are the Requirements?

One very important aspect in capacitive pyroelectric sensors is the fact that they are high-impedance devices with a resistance R_p that is typically in the gigaohm range. If such a sensor is connected in parallel to a low resistive load R_L ($R_L << R_p$), the voltage output signal will be reduced adequately to the reduction of the overall resistance R according to

$$R = (R_L^{-1} + R_p^{-1})^{-1} \approx R_L \qquad \text{and} \qquad V_{\text{pyro},L} = V_{\text{pyro}} \cdot \frac{R_L}{R_p} \tag{4.33}$$

This would result in the collapse of the output signal, and therefore, it is very beneficial to have an impedance transforming element inserted between sensor and load.

An ideal device to achieve this is a field-effect transistor because of its high-impedance input and its low-impedance output normally connected to the load. For large-area flexible physical sensor technologies, organic thin-film transistors (OTFTs) are ideal candidates for impedance-transforming elements because they can be processed on large areas at reasonable price and the substrates can be flexible.

According to this, OTFTs have to fulfill several requirements to be applicable as impedance-transforming or, perhaps, even signal-amplifying elements for polymer-based large-area sensor technologies. First, they should operate at reasonably low voltages (< 5 V) because the output signals typically are in the region $0.1 < V_{pyro} < 1$ V. So if a clear impact of the sensor signal on the OTFT current is desired, the threshold voltage of the transistor should not be an order of magnitude larger. Second, the OTFT input really needs to have a high-impedance input (~ gigaohms) in order not to unwillingly downscale the sensor signal. That means basically that the gate dielectric has to be very dense with low leakage currents and sufficiently high breakdown strengths. Third, the overall performance and stability should be sufficient for the targeted application (that could be an interesting point in the case of automotive industry driven applications). And last but not least, the fabrication process should be compatible with large-area processing on flexible substrates, thus arguing for printing and large-area evaporation techniques.

Low-Voltage OTFTs

Reducing the threshold voltage and also the subthreshold swing is essential for operating OTFTs at low-voltage levels. When combined with very low gate leakage currents, OTFTs may also become a key element in high-end sensor applications, such as flexible touchpads and screens or thermal imaging tools for night vision, surveillance, or for the detection of undesired heat loss paths in buildings.

The aforementioned transistor parameters critically depend on not only the thickness and the dielectric properties of the gate insulator, but also the trapped charge densities at the interface between these materials.[33] The selection of semiconductors and gate insulators with excellent interface properties is currently the challenge in the quest for improving the performance of OTFTs.

Figure 4.23*a* shows the structure of low-voltage organic transistors with high dielectric constant (high-*k*) oxide–polymer nanocomposites. Al_2O_3 or ZrO_2 was chosen as high-*k* dielectric materials, combined with poly-alpha-methylstyrene (PαMS) or poly-vinyl-cinnamate (PVCi) to

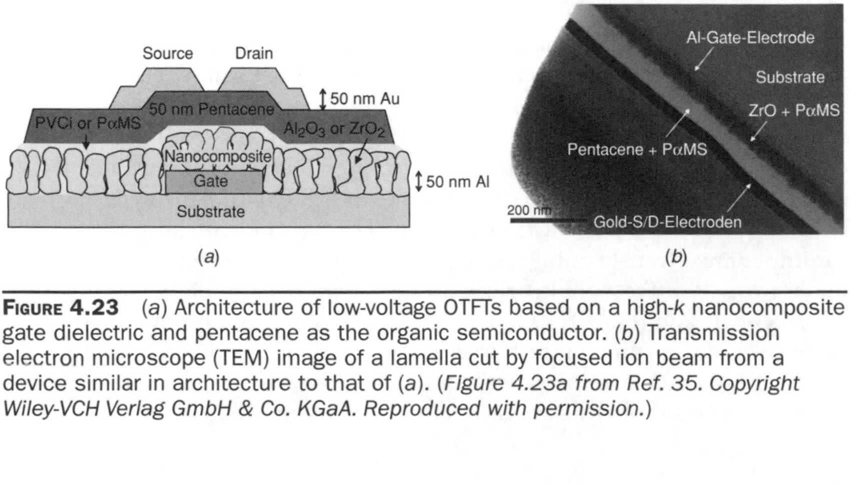

FIGURE 4.23 (*a*) Architecture of low-voltage OTFTs based on a high-*k* nanocomposite gate dielectric and pentacene as the organic semiconductor. (*b*) Transmission electron microscope (TEM) image of a lamella cut by focused ion beam from a device similar in architecture to that of (*a*). (*Figure 4.23a from Ref. 35. Copyright Wiley-VCH Verlag GmbH & Co. KGaA. Reproduced with permission.*)

form a smooth and dense nanocomposite gate dielectric.[35] Pentacene is used as the organic semiconductor material, the gate electrode is based on Al, while Au source and drain electrodes are employed. For the substrate glass or PET film, Melinex is used. The devices are fabricated according to the following procedure: the gate is formed by thermal evaporation of aluminum through shadow masks on a glass or PET substrate and the metal-oxide layer is fabricated by reactive sputtering of Al or Zr under high vacuum condition. Prior to the active organic semiconductor, a thin layer (~ 5 to 20 nm) of an appropriate hydroxyl-free polymer (PαMS or PVCi) was applied to the metal-oxide dielectric layer by spin-coating, thus forming a dense metal-polymer nanocomposite double layer as gate dielectric. Finally, for completion of the transistor device, 50 nm of pentacene is applied by thermal evaporation and structured via shadow masks at a rate of 0.1 nm/min and a substrate temperature of $T_S = 25°C$.

According to variable spectroscopic ellipsometry measurements done on the as-produced nanocomposite gate dielectrics but fabricated on silicon wafers, the measured layer thickness and optical constants can be modeled only if a mixed structure with club-shaped metal-oxide crystallites and interspaces filled by the polymer is assumed.[34] The club-shaped metal-oxide film growth with interspaces is clearly seen in the TEM micrograph (Fig. 4.23*b*).[34]

In Fig. 4.24 atomic force microscopy (AFM) images of a ZrO_2/PαMS nanocomposite gate dielectric-based transistor are displayed. The bare ZrO_2 metal-oxide surface is displayed in Fig. 4.24*a*, the nanocomposite in Fig. 4.24*b*, and the pentacene layer grown on top of the nanocomposite dielectric in Fig. 4.24*c*. The AFM images clearly reveal that the rough (surface rms-roughness = 1.5 nm) and less dense ZrO_2 layer, which is composed of regularly clubbed grains (see Fig. 4.23*b*), smoothens by forming the nanocomposite (rms-roughness = 0.4 nm). The substrate roughness critically influences the growth dynamics of pentacene molecules on top of dielectric surfaces,[36] grain sizes typically

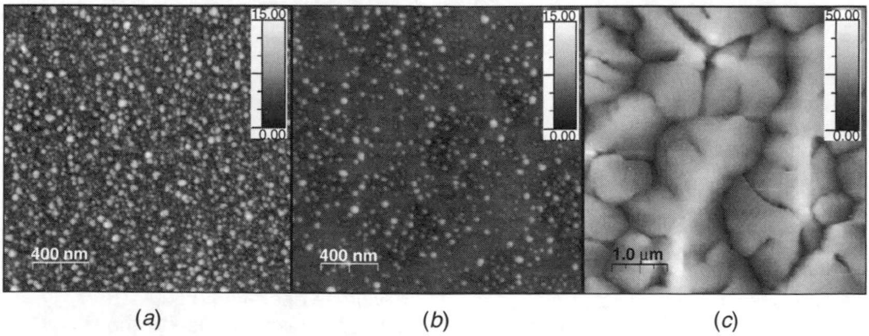

Figure 4.24 Atomic force height images of (*a*) bare ZrO_2 metal oxide surface, (*b*) the nanocomposite, and (*c*) pentacene grown on top of the nanocomposite. (*From Ref. 35. Copyright Wiley-VCH Verlag GmbH & Co. KGaA. Reproduced with permission.*)

increase with decreasing surface roughness. For rms-roughness values below 0.5 nm the pentacene morphology is characterized by dentritic crystallites of several microns height composed of well-separated monolayer-high terraces (see Fig. 4.24*c*).

It turned out that the nanocomposite-pentacene interface is a much higher-quality interface than the SiO_2-pentacene one showing up as (1) very low trap densities in the subthreshold region close to the theoretical limit (small swing or sharp turn-on), (2) low threshold voltages, (3) high charge carrier mobilities resulting in reasonably high drain currents at low voltages, and (4) low leakage currents resulting in high input impedance. In addition it can be shown (see Fig. 4.25) that the interface retains similar high quality if the whole device is fabricated on flexible substrates.[35]

The superior performance of pentacene transistors with high-*k* nanocomposite gate dielectrics is clearly indicated in the transfer characteristics shown Fig. 4.25 of a ZrO_2/PαMS pentacene OTFT with a gate capacitance $C_i = 120$ nF/cm^2 fabricated on PET. Note that in the on regime no hysteresis is evident between forward and backward gate voltage sweeps. The off current, however, reveals the typical gate field behavior with a trianglelike hysteresis.[37] The on/off ratio is deduced from the transfer characteristic $I_{ds}(V_{gs})$ as $I_{on/off} = 10^6$ (Fig. 4.25) with the following definitions: $I_{off} = I_{ds}(0\text{ V})$ and

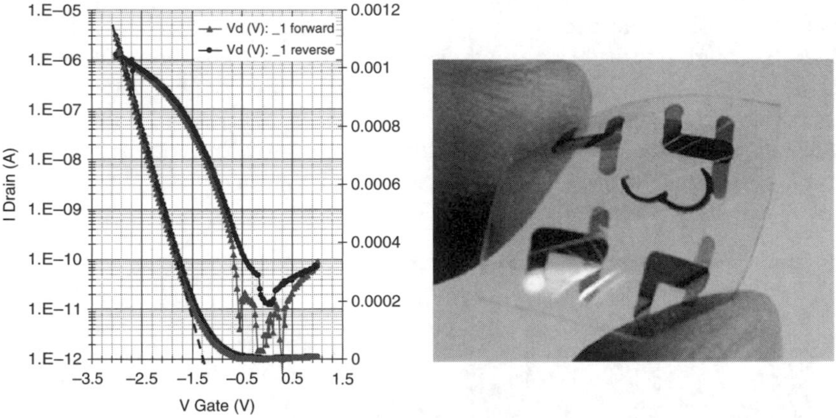

(*a*) (*b*)

FIGURE 4.25 (*a*) Transfer characteristics of a low-voltage OTFT with a nanocomposite gate dielectric based on ZrO_2 and PVCi fabricated on PET film (Melinex). The device has a gate capacitance of $C_i = 0.12$ µF/cm^2, proving low-voltage operation ($V_T = -1.3$ V) with a subthreshold swing $S \sim 100$ mV/dec (forward sweep), which is close to the theoretical limit of 82 mV/dec. The curve with triangles corresponds to the forward gate voltage sweep, whereas the curve with dots corresponds to the reverse sweep direction. The line is an extrapolation of the linear part of $\sqrt{I_{ds}}$ to $I_{ds} = 0$. (*b*) Photo image of this device.

$I_{on} = I_{ds}(-3\text{ V})$. The subthreshold swing defined as the inverse of the maximum slope of the current in the subthreshold regime is around $S = 100$ mV/dec in this device for forward sweep direction. From the square root dependence of the drain current as a function of the gate voltage, the threshold voltage V_T is determined to be about $V_T = -1.3$ V (Fig. 4.25), demonstrating that the device can be controlled and operated in the low-voltage regime, similar to pentacene OTFT devices based on TiO_2 with a PαMS capping layer.[38] It is interesting to mention that in all nanocomposite OTFTs the charge carrier mobility μ does not depend on the gate field for voltages above $V_{gs} = -2$ V, reaching values around $\mu = 0.4 - 1.2\ \text{cm}^2/(\text{V}\cdot\text{s})$. The high mobility values nicely correlate with the typical morphology of the polycrystalline pentacene layers grown on the nanocomposite dielectric surface, where the very large size of the crystallites is related to a smaller density of transport hindered by grain boundaries (Fig. 4.24*c*). Finally the gate leakage current is below 1×10^{-9} A, showing that the prepared transistors reveal a sufficiently high input impedance (in the GΩ regime) for sensor applications.

Very nice performance is also obtained for OTFTs with Al_2O_3-based nanocomposites. In Fig. 4.26 gate dielectrics (6 nm Al_2O_3 + 10 nm PαMS) and top-contact source and drain electrodes, made by e-beam evaporation of gold via shadow masks, are shown. From the subthreshold characteristics at $V_D = -3$ V of the as-produced OTFT, a subthreshold swing of about $S = 100$ mV/dec is extracted for the forward sweep of the gate voltage and an onset voltage $V_{on} = -1.2$ V. The channel length of this device is $L = 100$ μm, and the charge carrier mobility at $V_G = -5$ V is $\mu = 0.6\ \text{cm}^2/(\text{V}\cdot\text{s})$. These values emphasize the excellent performance of the low-voltage pentacene-based OTFT, which easily can be operated in the range $V_{op} < 3$ V. Due to the very thin gate dielectric layer thickness the gate leakage is higher than that of

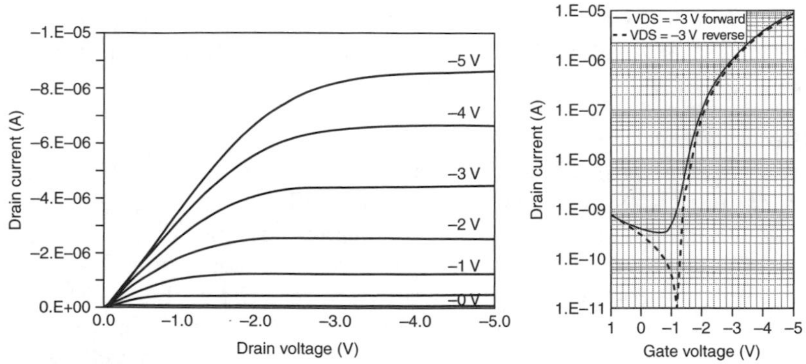

FIGURE 4.26 (Left) Output and (right) subthreshold characteristics of a pentacene-based OTFT with 6 nm Al_2O_3 + 10 nm PαMS as gate dielectric and Au source and drain electrodes.

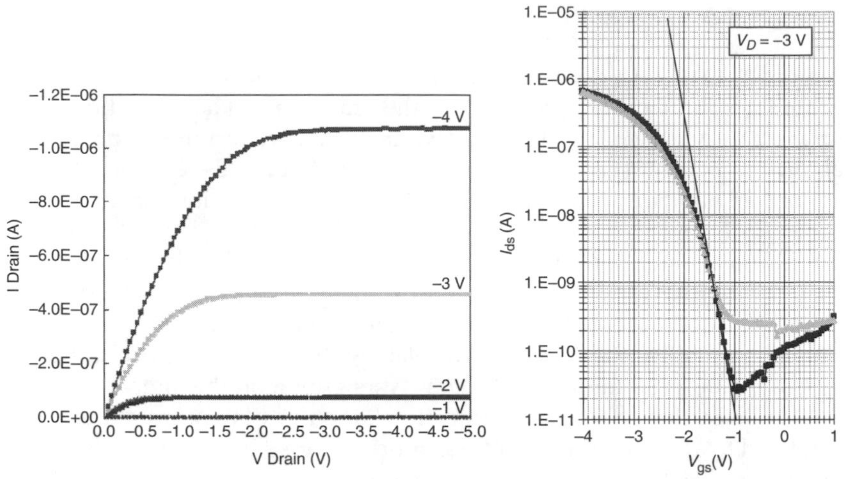

FIGURE 4.27 (Left) Output and (right) subthreshold characteristics of a pentacene-based OTFT with 30 nm Al_2O_3 + 10 nm PαMS as gate dielectric and Au-source and drain electrodes.

ZrO_2-based layers, and consequently the input impedance is somewhat smaller (~150 MΩ @ $V_D = -3$ V).

Therefore we also produced a thicker Al_2O_3 layer and, similar to the other device, a PαMS layer of about 10 nm, that resulted in an overall area-related gate capacitance of 80 nF/cm^2. The output and subthreshold characteristics of this device are depicted in Fig. 4.27. The drain current level of this device is smaller than that of the one with thinner Al_2O_3 which is partly due to the smaller gate capacitance and partly due to the slightly worse pentacene morphology that results in a smaller charge carrier mobility of about $\mu = 0.2$ cm^2/(V · s). The other relevant parameters are $V_{on} = -1$ V and $S = 200$ mV/dec and are extracted from the subthreshold characteristic displayed in Fig. 4.27. Clearly, the smaller gate leakage of ~1 to 2×10^{-10} A results in an input impedance of the order of about 5 to 6 GΩ.

Integrated Sensor

The fully flexible pyroelectric sensor element shown schematically in Fig. 4.28*a* illustrates the potential of such OTFTs in new high-end applications of organic electronics.[35] The sensing principle is based on the pyroelectric effect in a ferroelectric P(VDF-TrFE) copolymer. The sensor element is composed of a 2 μm thick pyroelectric P(VDF-TrFE) copolymer film that is fabricated by spin coating. The pyroelectric element is sandwiched between 70 nm thick Al electrodes directly integrated on glass or on a flexible PET substrate. The sensing capacitor is fabricated prior to the deposition of the OTFT, and both devices are integrated via the bottom electrode of the sensor serving as the

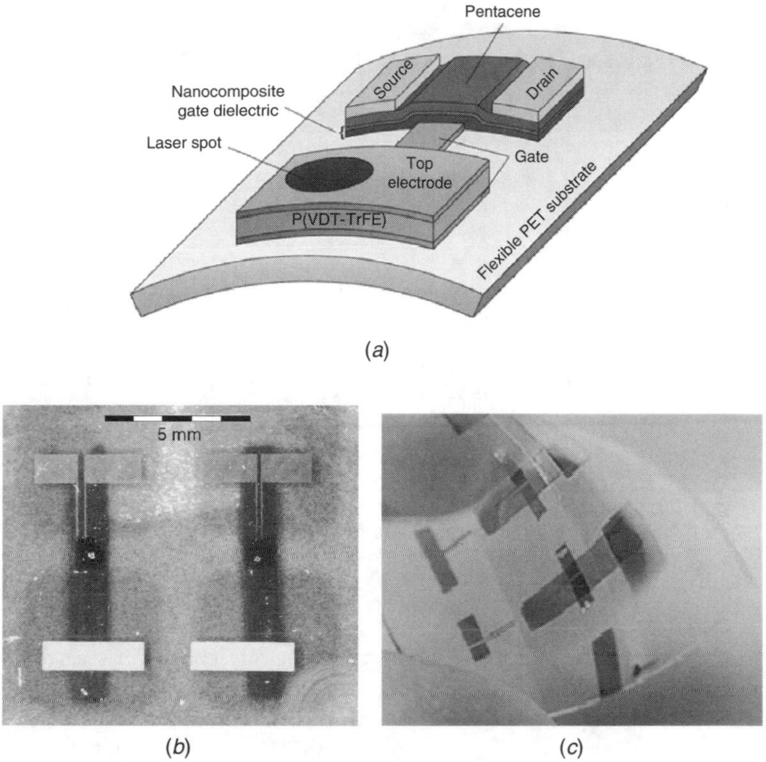

FIGURE 4.28 (*a*) Schematic view of the fully flexible sensor circuit, (*b*) microscopic image of the integrated optothermal sensor element, and (c) photo of its operation as a light-activated switch. (*From Ref. 35. Copyright Wiley-VCH Verlag GmbH & Co. KGaA. Reproduced with permission.*)

gate electrode of the OTFT. Figure 4.28*a* shows the scheme of the circuit, Fig. 4.28*b* a micrograph of an integrated optothermal sensor, and Fig. 4.28*c* a photo of the operation of such devices as an optothermal switch.

When intensity-modulated light of an infrared laser diode and a modulation frequency of 0.01 Hz are impinging on the top electrode, the transistor is switched on with an on/off ratio over up to four decades, as revealed in Fig. 4.29.[35] The on/off ratio depends on the threshold voltage and the input impedance of the OTFT; thus the large on/off ratio achieved here is a direct consequence of the superior transistor performance and the very small subthreshold swing, guaranteeing a sharp switch-on of the transistor. The on-off switching is stable over hours.[35] Rather than a laser diode, a simple laser pointer can be used for the excitation with somewhat smaller on/off ratios. The simple preparation of the circuit element lends itself to easily scale up to array sensors useful for thermal imaging of infrared scenes.

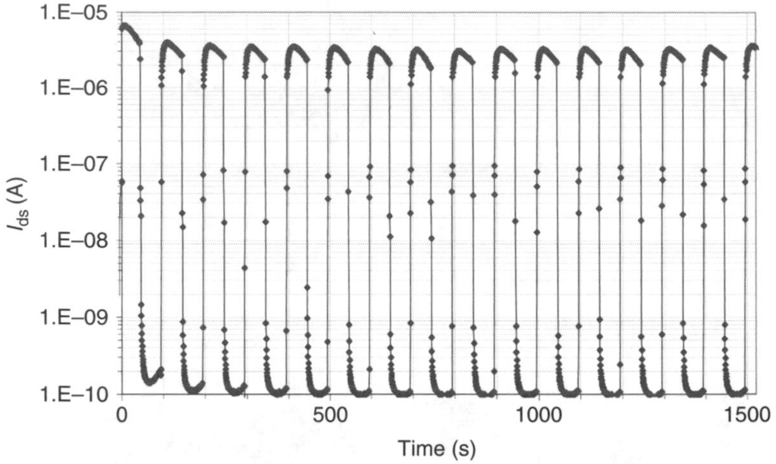

FIGURE 4.29 Demonstration of on/off cycles of the sensor circuit with light from an intensity-modulated laser diode impinging on the element. (*From Ref. 35. Copyright Wiley-VCH Verlag GmbH & Co. KGaA. Reproduced with permission.*)

Apart from direct integration the readout characteristics of a pyroelectric sensor by an OTFT were also investigated, which is especially interesting in the context of comparing commercial PVDF foil sensors (having no direct integrated OTFT) with sol-gel-derived PVDF copolymer sensors.

The drain current variations of the OTFT induced by the IR intensity modulations of the laser diode on the PVDF-foil capacitor are shown in Fig. 4.30 (top) for $V_G = -0.5$ V and in Fig. 4.30 (bottom) for $V_G = 0$ V both recorded at a driving voltage $V_D = -3$ V. The waveform of the modulations was square at $V_G = -0.5$ V and sine at $V_G = -0$ V, and in both cases the modulation frequency was 0.1 Hz. The switching behavior of the integrated IR sensor is quite stable as can be seen from the long-term cycle measurements plotted in Fig. 4.30. After about 1000 s (100 cycles) not only the ratio between off-and-on current, but also the absolute current values are completely stable.

As can be extracted from Fig. 4.30, the modulated drain current reproduces the desired laser diode controlled switch-on/switch-off characteristic of the OTFT, which is driven from the off state at $I_{off} = 6 \times 10^{-10}$ A to the on state at $I_{on} = 1 \times 10^{-7}$ A with a frequency of 0.1 Hz. Accounting for the measurement circuitry and the input impedance of the OTFT, this corresponds to an output modulation voltage of the sensor element of about $V_{sense;eff} = \pm 1.5$ V and a switch of the total gate voltage $V_{G;tot}$ from 1 to −2 V that is also confirmed by a comparison with the subthreshold characteristics depicted in Fig. 4.26. Note that the effective sensor output voltage $V_{sense;eff}$ is much smaller than the value extracted by the parametric analyzer which is basically due to

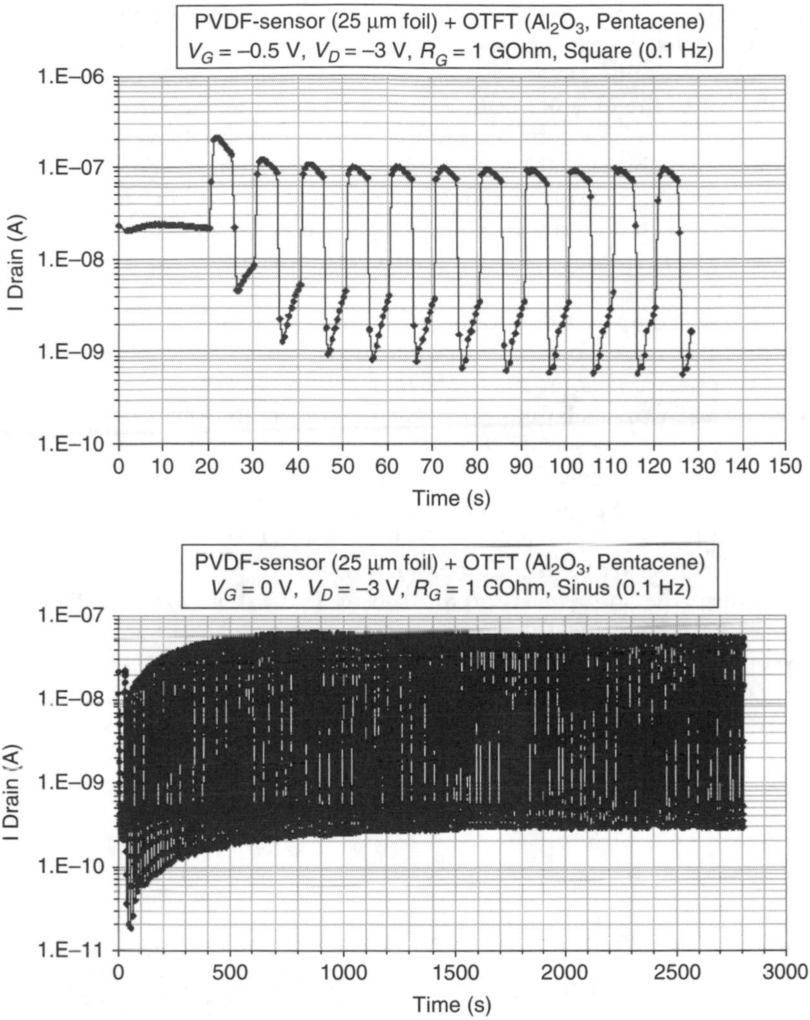

Figure 4.30 Top: Drain current modulations of an OTFT at $V_D = -3$ V and $V_G = -0.5$ V connected to a PVDF-foil capacitance induced by an intensity-modulated laser diode (square waveform). Bottom: Long-term behavior of the I_D modulation for an integrated PVDF-sensor solely controlled by intensity-modulated laser diode (square waveform, $V_D = -3$ V and $V_G = -0$ V).

the fact that the input impedance of the OTFT is of the order of 150 MΩ, which is significantly smaller than that of the parametric analyzer (1 GΩ).

If no voltage is applied from the analyzer sources, the OTFT is driven solely by the output voltage of the foil sensor element. In consideration of the measurement circuitry and the input impedance of the OTFT itself, in that case the total gate voltage $V_{G;tot}$ varies from 1.5 to −1.5 V, resulting in a variation of the drain current between $I_{off} = 2 \times 10^{-10}$ A to the on state at $I_{on} = 3 \times 10^{-8}$ A. Due to the maximum negative $V_{G;tot} \sim -1.5$ V compared to $V_{G;tot} \sim -2$ V that is achieved, if an additional gate bias of −0.5 V is applied, the on current is somewhat smaller in this case. This is mainly due to the very steep transition from off-to-on state intrinsic to the Al_2O_3-based OTFTs, meaning that a small increase in the $V_{mod;eff}$ can induce a significant increase in the on current level.

We have compared that to the output of a pyroelectric thin-film capacitor that is read out by an organic Al_2O_3-based OTFT similar to that of Fig. 4.27. This organic transistor device was connected to a pyroelectric PVDF-TrFE thin-film capacitor of 2 μm thickness, characterized by a voltage sensitivity as depicted in Fig. 4.13 to form an integrated organic pyroelectric thin-film sensor. Stimulation of the sensor resulted in a periodic modulation of the drain current between off state at $I_{off} = 1 \times 10^{-10}$ A and on state at $I_{on} = 4 \times 10^{-8}$ A that is shown to be stable at least over a period of 1 h (Fig. 4.30).

Because the impedance of the sensor (2 GΩ) is significantly smaller than that of the OTFT (5 to 6 GΩ), only a small reduction of the voltage output V_{sens} is expected ($V_{sens;eff} = 0.75V_{sens}$), as soon as sensor and OTFT are integrated. From an inspection of Figs. 4.27 and 4.31, it is seen that $V_{sens;eff}$ oscillates between −0.4 and −2.5 V, corresponding to $V_{sens;eff} \sim 0.7V_{sens}$ in the case of an integrated pyroelectric PVDF-TrFE thin-film sensor which is very close to the expected value.

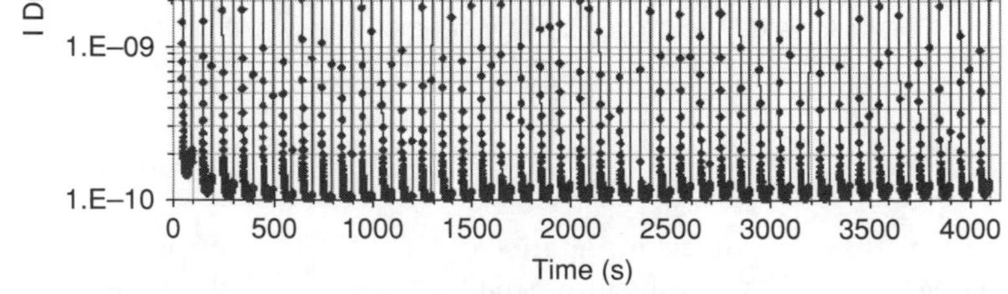

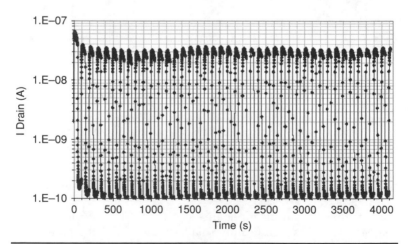

FIGURE 4.31 Long-term behavior of the ID modulation for an integrated PVDF-TrFE thin-film sensor solely controlled by an intensity-modulated laser diode (square waveform) at an operation voltage of $V_D = -3$ V.

To obtain an on-to-off output current ratio above 10^3 for the integrated organic pyroelectric thin-film sensors, an OTFT with a subthreshold swing around 0.1 V/decade and an onset voltage around 1 V is needed. The results reported in the beginning of this section on the integrated optothermal device clearly meet the aforementioned specifications.

Abbreviations and Acronyms

CPE	constant phase element
DUT	device under test
EC	equivalent circuit
(O)FET	(organic) field-effect transistor
IS	impedance spectrum
MIS	metal-insulator semiconductor
MPP	*N,N'*-Dimethyl-3,4,:9,10-perylenbis(carboximid)
OLED	organic light-emitting diode
OPD	organic photo diode
PVP	polyvinylphenol

Acknowledgment

This work was supported by the European Commission under Contract No. 215036 as a specific targeted research project with the acronym 3PLAST.

References

1. L. Young, "Anodic Oxide Films," *Trans. Faraday Soc.*, vol. 51, 1955, pp. 253–267.
2. U. Rammelt and C. A. Schiller, "Impedance Studies of Layers with a Vertical Decay of Conductivity or Permittivity," *ACH—Models in Chemistry*, vol. 137, 2000, pp. 199–212.
3. M. Shur, *Physics of Semiconductor Devices*, Prentice-Hall, Upper Saddle River, N.J., 1990.
4. S. M. Sze, *Physics of Semiconductor Devices*, Wiley, New York, 1982.
5. H. Schoen, "Charakterisierung organischer MIS-Strukturen durch gleichspannungsabhängige Impedanzspektroskopie," Thesis, TU-Graz, October 2006.
6. S. B. Lang, *Physics Today*, August 2005.
7. A. J. P. Martin, *Proc. Phys. Soc.*, 53:186 (1941).
8. J. B. Bergmann, J. H. McFee, and G. R. Crane, *Appl. Phys. Lett.*, 8:203 (1971).
9. A. M. Glass, J. H. McFee, and J. B. Bergmann, Jr., *J. Appl. Phys.*, 42:5219 (1971).
10. E. Yamaka, *Natl. Tech. Re.*, 18:141 (1972).
11. J. B. Bergmann, Jr., and G. R. Crane, *Appl. Phys. Lett.*, 21:497 (1972).
12. S. Bauer and S. B. Lang, *IEEE Trans. Diel. El. Ins.*, 3:647 (1996).
13. H. Meixner, *Ferroelectrics*, 115:279 (1991).
14. S. Webster and T. D. Binnie, *Sensors and Actuators*, A49:61 (1995).
15. P. C. D. Hobbs, *Proc. SPIE*, 4563:42 (2001).

16. D. Setiadi, P. M. Sarro, and P. P. L. Regtien, *Sensors and Actuators*, 52(A):103 (1996).
17. D. Setiadi, P. P. L. Regtien, and P. M. Sarro, *Microelectr. Eng.*, 29:85 (1995).
18. N. Neumann and R. Köhler, *SPIE*, 2021:35 (1993).
19. T. Someya, Y. Kato, T. Sekitani, S. Iba, Y. Noguchi, Y. Murase, H. Kawaguchi, et al., *PNAS*, 102:12321 (2005).
20. T. Furukawa, *Adv. Colloid. Interface Sci.*, 183:71–77 (1997).
21. R. W. Newsome, Jr., and E. Y. Andrei, *Phys. Rev. B*, 55(11):7264 (1997).
22. S. B. Lang, *Sourcebook of Pyroelectricity*, Gordon and Breach, New York, 1974.
23. M. E. Lines and A. M. Glass, *Principles and Applications of Ferroelectrics and Related Materials*, Clarendon Press, Oxford, 1977.
24. S. Ducharme, S. P. Palto, and V. M. Fridkin, Ferroelectrics and Dielectric Thin Films, *Handbook of Thin Film Materials*, vol. 3, Chapter 11, Academic Press, San Diego, 2002.
25. M. Okabe, R. Wada, M. Tazaki, and T. Homma, *Polymer J.*, 35(10):798–803 (2003).
26. M. Tazaki, R. Wada, M. Okabe, and T. Homma, *Polymer Bull.*, 44:93–100 (2000).
27. M. Zirkl, Ph.D. thesis, Karl-Franzens-Universität Graz, May 2007.
28. M. Zirkl, B. Stadlober, and G. Leising, *Ferroelectrics*, 353:173 (2007).
29. N. Neumann, R. Köhler, and G. Hoffmann, *Ferroelectrics*, 118:319–324 (1991).
30. S. Bauer-Gogonea and S. Bauer, in *Handbook of Advanced Electronics and Photonic Materials and Devices*, H. S. Nalwa (ed.), vol. 10, Academic Press, 2001.
31. D. Setiadi and P. P. L. Regtien, *Ferroelectrics*, 173:309 (2002).
32. J. Groten, Diploma thesis, Karl Franzens University Graz, 2008.
33. For a recent review see A. Maliakal in *Organic Field-Effect Transistors*, Z. Bao and J. Locklin (eds.), CRC Press, Boca Raton, FL, 2007, p. 229.
34. A. Fian, A. Haase, B. Stadlober, G. Jakopic, N. B. Matsko, W. Grogger, and G. Leising, *Anal. Bioanal. Chem.*, 390:1455 (2008).
35. M. Zirkl, A. Haase, A. Fian, H. Schön, C. Sommer, G. Jakopic, G. Leising, et al., *Adv. Mat.*, 17:2241 (2007).
36. B. Stadlober, M. Zirkl, M. Beutl, G. Leising, S. Bauer-Gogonea, and S. Bauer, *Appl. Phys. Lett.*, 86:242902 (2005).
37. U. Haas, A. Haase, V. Satzinger, H. Pichler, G. Jakopic, G. Leising, and B. Stadlober, et al., *Phys. Rev. B*, 73:235–339 (2006).
38. L. A. Majewski, R. Schroeder, and M. Grell, *Adv. Mater.*, 17:192 (2005).

CHAPTER 5

Progress and Challenges in Organic Light-Emitting Diode-Based Chemical and Biological Sensors

Ruth Shinar

Microelectronics Research Center and
Department of Electrical and Computer Engineering
Iowa State University, Ames

Yuankun Cai and Joseph Shinar

Ames Laboratory-USDOE and
Department of Physics and Astronomy
Iowa State University, Ames

5.1 Introduction

Remarkable advances in the development of chemical and biological sensors have occurred over the past 30 years.[1–8] One of the major current efforts focuses on development of low-cost field-deployable multianalyte sensors and sensor networks. Several organic-electronics-based thrusts within this effort, which are under different stages of development, are described in several chapters of this volume. The paradigm described in this chapter is that of photoluminescence (PL)-based sensors with small-molecular organic light-emitting diodes (SMOLEDs) as the PL excitation source. As detailed below, such an OLED-based sensing platform is very promising.

PL-based chemical and biological sensors, which are often utilized for monitoring a single analyte, are sensitive, reliable, and suitable for the wide range of applications in areas such as environmental, medical, food and water safety, homeland security, and the chemical industry.[1–8] The sensors are typically composed of an analyte-sensitive luminescent sensing component, a light source that excites the PL, a photodetector (PD), a power supply, and the electronics for signal processing. Recent compact light sources include diode lasers and LEDs,[9] but their integration with the sensing element, microfluidic architectures, and/or the PD is intricate, and not nearly as simple and potentially low-cost as that of the OLED-based platform. Based on the need and challenges, our goal is to develop a flexible, lightweight, compact, portable, and low-cost platform of sensor arrays, eventually miniaturized and usable for multianalytes, in which the light source, the sensing component, and eventually a thin-film PD are structurally integrated in the uniquely simple design described below. Indeed, the results reported to date on the OLED-based platform are very promising for developing miniaturized sensor arrays for the above-mentioned applications, including for high-throughput multianalyte analysis.[10–20]

This chapter reviews the recent advances in the development of such PL-based chemical and biological sensors, where an array of OLED pixels serves as the excitation source. This pixel array is structurally integrated with the sensing element to generate a compact, eventually miniaturized, sensor module. Advanced integration includes additionally an array of p-i-n, thin-film PDs that are based on hydrogenated amorphous Si (a-Si:H), a-(Si,Ge):H, or nanocrystalline Si. Organic PDs are also suitable for this advanced structural integration.

OLED attributes that are attractive for sensing applications include their simple and easy fabrication in any two-dimensional shape, including on flexible plastic substrates, their uniquely simple integration with a sensing component, and compatibility with microfluidic structures. Additionally, OLEDs consist of individually addressable pixels that can be used for multiple analytes, they can be operated at an extremely high brightness, they consume little power and dissipate little heat, and their cost will likely drop to disposable levels.

SMOLEDs are typically fabricated by thermal vacuum evaporation of the small molecules; polymer OLEDs (PLEDs) are typically fabricated by spin-coating or inkjet printing of the polymer-containing solutions.[21] They have dramatically improved over the past decade,[22–26] and commercial products incorporating them are proliferating rapidly. Their inherent advantages include the ability to fabricate them on glass and flexible plastic;[27] electrophosphorescent red-to-green OLEDs with external quantum efficiencies $\eta_{ext} > 17\,\%$[28] and blue OLEDs with $\eta_{ext} \approx 11\%$ have been reported.[25] They can be easily fabricated in sizes ranging from a few square microns to several square millimeters, and pixels as small as 60 nm in diameter have been reported.[29] These developments and their rapid commercialization present an opportunity to develop a new platform of integrated (micro)sensor arrays.

Figure 5.1 shows two examples of the envisioned integrated OLED-based sensing platform. Figure 5.1*a* demonstrates a simplified array operated in the "back detection" mode. That is, alternating OLED and thin-film PD pixels are fabricated on one side of a common substrate. The sensor component is fabricated on the opposite side of that substrate, or on a separate substrate that is attached back-to-back to the OLED/PD substrate. The OLED's electroluminescence (EL) excites the PL of the sensing component, which is then monitored by the PD pixels located in the gaps between the OLED pixels. In measuring analyte-induced changes in the PL intensity I, suitable measures to reduce interfering light, such as optical filters (not shown) above the OLED (e.g., a bandpass filter) and PD pixels (e.g., a long-pass filter), will be needed. Such measures will minimize the contribution of the long-wavelength tail of the EL and that of background light monitored by the PD. They are particularly essential when the Stokes shift between the absorption and emission of the analyte-sensitive component is small. However, as shown later, by pulsing the OLED,

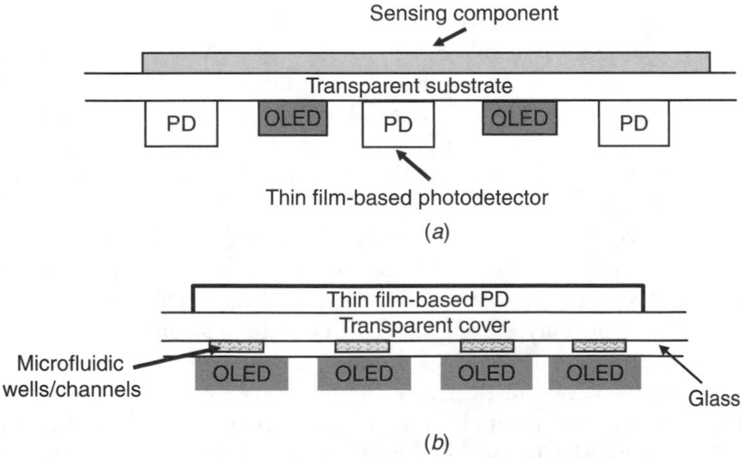

FIGURE 5.1 Simplified (not to scale) structural integration of three components of a PL-based sensor: OLED excitation source, sensing component, and a thin-film PD. (*a*) "Back detection" mode with the OLED and PD pixels on one side of a common substrate and (*b*) "front detection" mode with the sensing component between the excitation source and the PD arrays. Optical filters or other means of reducing interfering light are not shown. (*Figure 5.1a reprinted from Ref. 18. Copyright 2006, with permission from Elsevier.*)

analyte-induced changes in the PL decay time τ can be measured during the off period of the OLED to monitor specific analytes; this approach eliminates the need for such filters. Figure 5.1*b* shows an array operated in the "front detection" mode, where the PD array is fabricated on a separate substrate and placed in front of the analyte. As such, the PD array can serve as a cover for microfluidic wells or channels containing surface-immobilized or solutions of recognition elements. In both cases, compact arrays with flexible designs can be generated to address the various sensing needs whether in the gas or liquid phase.

In developing the OLED sensing platform, we focused first on OLED/sensing component integration, to generate the compact sensor module. As detailed below, this module was evaluated for monitoring various gas- and liquid-phase analytes, including multiple analytes in a single sample. The second step was directed toward advanced integration with additionally a PD.

5.2 Structurally Integrated OLED/Sensing Component Modules

The OLEDs used for the platform development consist of a transparent conducting anode, typically indium tin oxide (ITO) coated on a glass or plastic substrate, organic hole- and electron-transporting layers (HTLs and ETLs, respectively), an emitting layer, and a metal cathode.[30] They are easily fabricated using thermal evaporation in a low-vacuum ($\sim 1 \times 10^{-6}$ torr) glove box. Details on their fabrication and encapsulation, for enhanced long-term stability, are provided elsewhere.[13–17] The organic layers consisted of a 5 nm thick copper phthalocyanine (CuPc) hole injecting layer, which is also believed to reduce the surface roughness of the ~140 nm thick treated ITO,[31] and a 50 nm thick *N,N'*-diphenyl-*N,N'*-bis(1-naphthyl phenyl)-1,1'-biphenyl-4,4'-diamine (NPD) HTL. For blue OLEDs, with peak emission at ~460 to 470 nm, the 40 nm thick emitting layer was 4,4'-bis(2,2'-diphenylvinyl)-1,1'-biphenyl (DPVBi), or perylene (Pe)-doped 4,4'-bis(9-carbazolyl) biphenyl (Pe:CBP),[32] typically followed by a 4 to 10 nm thick tris(quinolinolate) aluminum (Alq_3) electron transport layer. For green OLEDs with peak emissions at ~535 and ~545 nm, the ~40 nm thick emitting layer and ETL were Alq_3 and rubrene-doped Alq_3, respectively. In all cases, an 8 to 10 Å CsF or LiF buffer layer was deposited on the organic layers,[33, 34] followed by the ~150 nm thick Al cathode. The total thickness of the OLEDs, excluding the glass substrate, was thus < 0.5 μm. Under forward bias, electrons are injected from the low-workfunction cathode into the ETL(s). Similarly, holes are injected from the high-workfunction ITO into the HTL(s). Due to the applied bias, the electrons and holes drift toward each other and recombine in the emitting layer. A certain fraction of the recombination events results in radiative excited states. These states provide the EL of the device.

Recent developments demonstrated the viability of the structurally integrated OLED array/sensing film, where these components are fabricated on two separate glass slides that are attached back-to-back. This geometry is unique in its ease of fabrication, and it eliminates the need for components such as optical fibers, lens, and mirrors, resulting in a compact and potentially very low cost OLED/sensor film module, whose ~2 mm thickness is determined by that of the substrates.[10–18] Additionally, the nearly ideal coupling between the excitation source and the sensor film enables operation at relatively low power, which minimizes heating that is a critical issue for sensor materials and analytes involving heat-sensitive bio(chemical) compounds.

The PD, usually a photomultiplier tube (PMT) or a Si photodiode, is typically positioned either in front of the sensing film ("front detection") or behind the OLED array ("back detection"). In the latter configuration, the PD collects the PL that passes through the gaps between the OLED pixels. As the PD is not structurally integrated with the other components, in this configuration the OLED array is sandwiched between the PD and the sensing component, unlike the envisioned structure of Fig. 5.1*a*, where the OLED and PD pixels are fabricated on the same side of a common substrate. The resulting sensor probe, which includes the OLED array, the sensor film, and the PD, is very compact when using, e.g., a surface-mount Si photodiode, which is < 2 mm thick.

The OLED pixel array enables fabrication of a new platform of compact multianalyte sensor arrays. The pixels are individually addressable and can be associated with different sensor films, and their emission can vary from single color (at any peak wavelength from 380 to 650 nm) to multicolor combinatorial arrays.[31, 35, 36] As the volume of manufactured OLEDs increases, they will eventually become disposable. Thus, the development of OLED-based field-deployable, compact, low-cost, user-friendly, and autonomous chemical and biological sensor arrays is promising.

5.3 Sensors Based on Oxygen Monitoring

In the examples provided below, typically 2 to 4 OLED pixels, 2×2 mm^2 each, were utilized as the excitation source; 0.3×0.3 mm^2 pixels were also successfully employed. Importantly, the OLEDs were often operated in a pulsed mode, enabling monitoring of the effect of the analytes not only on the PL intensity I, but also on the PL decay time τ. The τ-based detection mode is advantageous since moderate changes in the intensity of the excitation source, dye leaching, or stray light have essentially no effect on the measured value of τ. Thus, the need for a reference sensor or frequent sensor calibration is avoided. Additionally, as τ is measured during the off period of the OLED, no filters are needed to block any OLED light from the PD, and the issue of background light stemming from the tail of the broad EL band is practically eliminated.

5.3.1 Advances in Monitoring Gas-Phase and Dissolved Oxygen

The first effort to develop the OLED-based sensor platform focused on O_2 sensors, which command a large commercial market.[37] The well-known method for monitoring gas-phase and dissolved O_2 (DO) is based on the collisional quenching of the phosphorescence of oxygen-sensitive dyes such as Ru, Pt, or Pd chelates by gas-phase O_2 and DO. In a homogeneous matrix, the O_2 concentration can be determined ideally from changes in I under steady-state conditions or from τ using the Stern-Volmer (SV) equation

$$I_0/I = \tau_0/\tau = 1 + K_{SV}[O_2] \tag{5.1}$$

where I_0 and τ_0 are the unquenched values and K_{SV} is a temperature- and film-dependent constant.

Although this PL-based sensing is well established, extensive studies of optical O_2 sensors are still continuing in an effort to enhance the sensors' accuracy, reliability, limit of detection, and operational lifetime; reduce their cost and size, and develop an O_2 sensor for in vivo applications.[38] The large market for O_2 sensors is also an incentive for continued research in the field. The low-cost field-deployable, compact oxygen sensors will therefore be very attractive for the various biological, medical, environmental, and industrial applications.[37–40]

Among the structurally integrated OLED/dye-doped O_2 sensor film modules evaluated to date, those based on Pt and Pd octaethylporphyrine (PtOEP and PdOEP, respectively), embedded in polystyrene (PS) and excited by Alq_3-based OLEDs, exhibited the highest sensitivities.[17] Their dynamic range and sensitivity, defined as $S_g \equiv \tau_0/\tau(100\%\ O_2)$ in the gas phase and as the ratio S_{DO} of τ measured in a deoxygenated solution to that of an oxygen–saturated solution, from 0 to 60°C, are comparable to the best O_2 sensors reported to date. This demonstrates the viability of the platform for high-sensitivity monitoring.

The linear SV relation was confirmed for DO monitored by dyes in solution.[41] It was observed also for the solid [DPVBi OLED]/[Ru dye sensor film] matrix,[17] and as seen in Figs. 5.2 and 5.3 for DO measurements with the oxygen-sensitive dyes embedded in a PS film. The pulse width used for obtaining the data shown in Figs. 5.2 and 5.3 was 100 μs. We note that S_{DO} ~14 of DO in water measured with the OLED-based sensors is among the highest reported for PtOEP:PS.

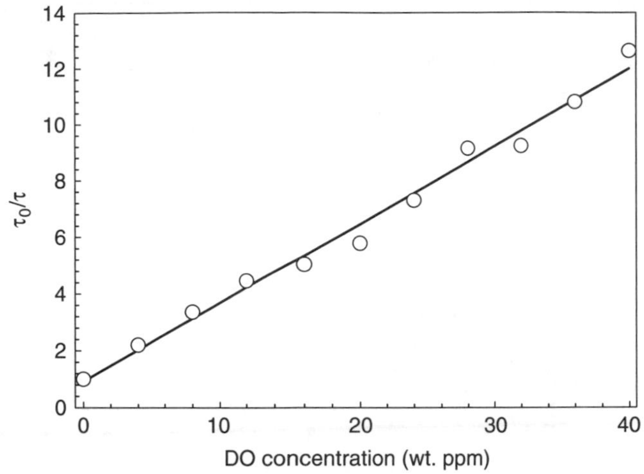

FIGURE 5.2 Calibration line for a PtOEP-based DO sensor at 23°C.

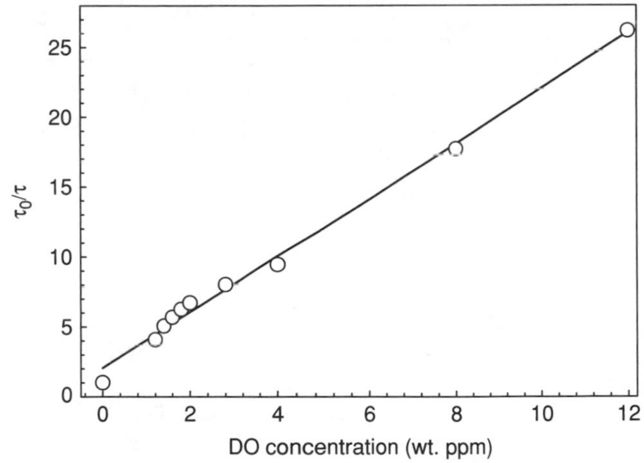

FIGURE 5.3 Calibration line for a PdOEP-based DO sensor at 23°C.

Deviations of SV plots from linearity are common, as shown in Figs. 5.4 and 5.5, which show such plots of PtOEP- and PdOEP-based gas-phase sensors. Changing the preparation conditions of the sensing film can affect the linearity of the SV plot.

For the results shown in Fig. 5.4, the [Alq_3 OLED]/[PtOEP:PS] sensor module was turned on once every 24 h, and τ was determined vs. the O_2 level, which varied from 0 to 100%. In this pulsed mode, the

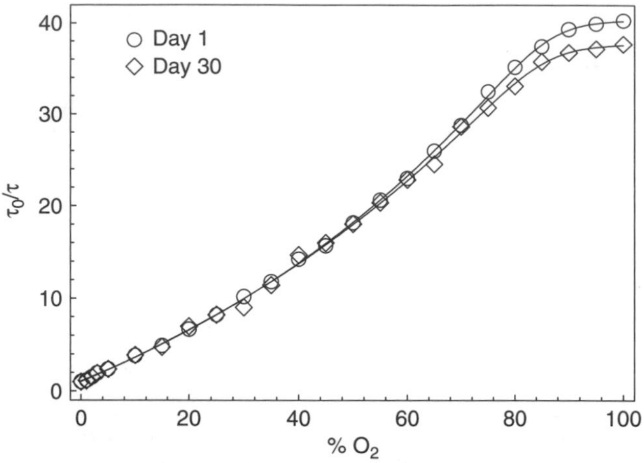

FIGURE 5.4 Stern-Volmer plots of a gas-phase O_2 sensor obtained at 23°C using an [Alq_3 OLED]/[PtOEP-doped PS film] sensor module measured daily for 30 days. The film was prepared by drop casting a solution containing PtOEP:PS at a ratio of 1:15.

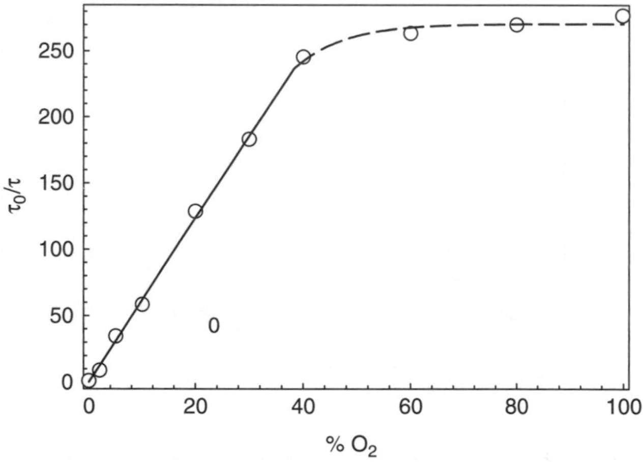

FIGURE 5.5 The SV plot at 23°C of a PdOEP-based gas-phase O_2 sensor excited by a rubrene-doped Alq_3 OLED.[17] The sensing film was prepared by drop-casting; it is doped with TiO_2 particles. The SV plot is linear up to ~40% O_2. (*Reprinted from Ref. 17. Copyright 2007, with permission from Elsevier.*)

OLED bias pulse amplitude was 20 V, its width was 100 μs, and its repetition rate was 25 Hz; τ was determined by averaging the PL decay curves over 1000 pulses. As clearly seen, the SV plots deviate from linearity, and the sensitivity decreased slightly from ~40 on day 1 to ~37 on day 30, with the main changes occurring at high O_2 levels.

The results indicated that the overall accuracy was generally ±0.5%. Over a testing period of 30 days, τ measured in air using the Alq_3/PtOEP:PS module gradually decreased from 20.2 ± 0.1 to 19.7 ± 0.05 μs. Hence, the relative error decreased from ~0.5 to ~0.25% over this period. These measurements, and others, demonstrated that these sensor module lifetimes are well over 30 days.

The results shown in Fig. 5.5 are for a film doped with PdOEP and 360 nm diameter titania (TiO_2) particles.[42] Such doping enhances the PL of the O_2 sensing films. When excited by an OLED, the dye PL intensity increases up to ~10-fold, depending on the TiO_2 concentration and the excitation source. The enhanced PL is attributed to light scattering by the embedded TiO_2 particles, due to their high $n \sim 2.8$ index of refraction relative to that of the polymer matrix ($n \sim 1.5$), and possibly by voids in the film, whose $n = 1$ is much lower than that of the matrix. The particles and voids scatter the EL, increasing its optical path and consequently its absorption and the PL. The particles can also result in an increase in the PL outcoupling, reducing waveguiding to the film edges.

As an example of the effect of additionally doping the sensor film with titania, Figs. 5.6 and 5.7 show the effect of the particles on the gas- and liquid-phase PL decay curves of PtOEP:PS and PdOEP:PS excited by the Alq_3 OLED in 100% Ar and O_2 environments at room temperature.

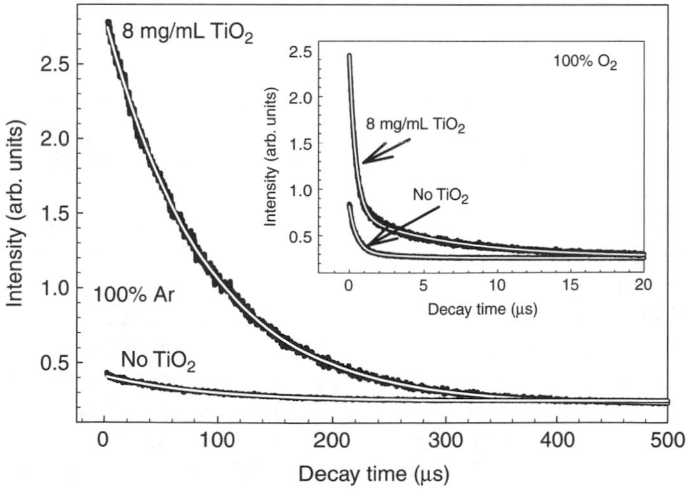

FIGURE 5.6 The gas-phase PL decay curves in Ar and O_2 for PtOEP:PS with titania doping (higher intensities) and without it (lower intensities); the exponential fitting (for Ar) and biexponential fitting (for O_2) are also shown. *(From Ref. 42. Copyright Wiley-VCH Verlag GmbH & Co. KGaA. Reproduced with permission.)*

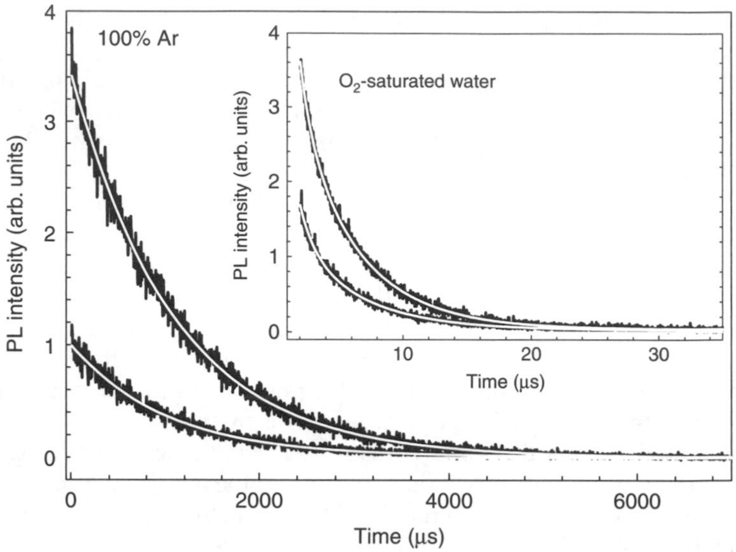

FIGURE 5.7 The PL decay curves in Ar- and O_2-saturated water for PdOEP:PS with TiO_2 doping (higher intensities) and without it (lower intensities); the exponential fitting (for Ar) and biexponential fitting (for O_2) are also shown. (*From Ref. 42. Copyright Wiley-VCH Verlag GmbH & Co. KGaA. Reproduced with permission.*)

The strong enhancement in the PL intensity is clearly seen. It improves the signal-to-noise S/N ratio, without any change in the response time or the long-term stability of the sensor films. The TiO_2 particles, however, affect τ, independently of their concentration above 1mg/mL.

The improved S/N improves the analyte limit of detection (LOD), shortens the data acquisition time, and reduces the needed excitation intensity. This reduces potential dye photobleaching and increases the exciting OLED lifetime.

The PL intensity could be further increased by using additionally Al mirrors to back-reflect the EL into the sensor film. Figure 5.8 shows such an example.

Preliminary results indicated also that films made of blended polymers may be of value. As an example, as shown in Fig. 5.9, blending PS with poly(dimethylsiloxane) (PDMS) resulted in an increased PL intensity. This may be a result of enhanced light scattering within the blended film associated with its microstructure. The PS:PDMS films are expected to result in improved LOD due to enhanced permeability to O_2; PDMS is more permeable to O_2 in comparison with PS, which is only moderately permeable to it.

The foregoing results indicate that different OLED/sensor films modules can be used for monitoring O_2 over different ranges of concentration, some (e.g., depending on the PS:dye ratio, or film thickness)

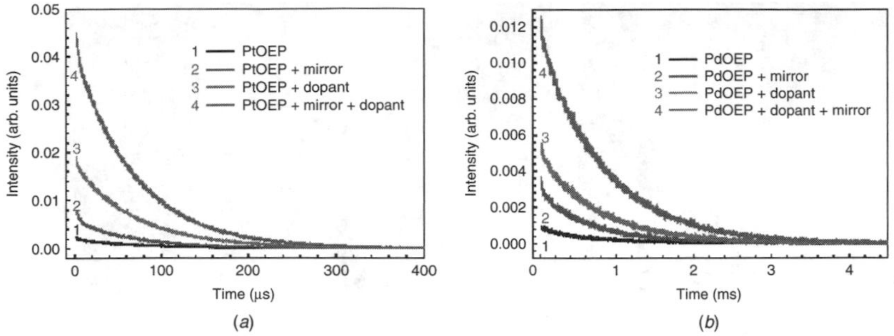

FIGURE 5.8 Effect of the titania dopant embedded in the PS film and Al mirrors on the PL decay of (*a*) PtOEP:PS and (*b*) PdOEP:PS films in water in equilibrium with Ar (~0 ppm DO). The films were prepared by drop-casting 60 μL of toluene solution containing 1 mg/mL TiO_2, 1 mg/mL dye, and 50 mg/mL PS. (See also color insert.)

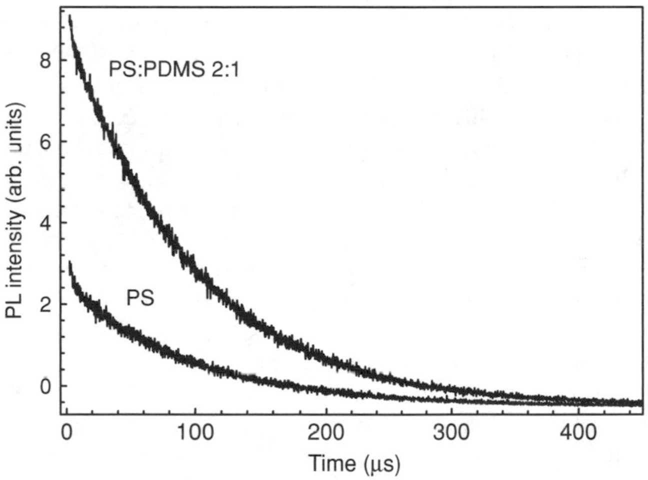

FIGURE 5.9 Enhancement of the PL intensity of a PtOEP-based gas-phase O_2 sensor in an Ar atmosphere at 23°C by using a 2:1 (by weight) PS:PDMS polymer blend.

with linear $1/\tau$ vs. $[O_2]$ calibrations. Thus, OLED/sensing film arrays, which are simple to fabricate and are of small size, are particularly promising for real-world applications. The use of such arrays will additionally improve the accuracy in analyte monitoring via redundant measurements. For example, one such simple array could comprise two sensing films: a 1:10 PtOEP:PS film that exhibits near linear SV plot over the whole 0 to 100% range (data not shown), which would be excited by Alq_3 OLED pixels, and a PdOEP:PS film, that is very sensitive to low levels of O_2 and exhibits a linear behavior up to ~40% O_2

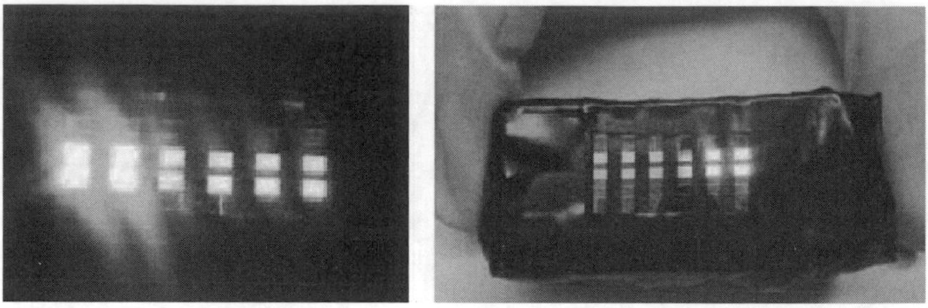

FIGURE 5.10 Mixed OLED platforms of green (outer four pixels in the arrays) and blue OLEDs. The array on the right is encapsulated. The pixel size is 2 × 2 mm^2. (See also color insert.)

(Fig. 5.5), which would be excited by rubrene-doped Alq_3 OLED pixels. The arrays of mixed OLED pixels can also include adjacent blue and green pixels based on, e.g., DPVBi or Alq_3, respectively. Such pixels will enable the use of Ru dye- or PtOEP- and PdOEP-based sensors, respectively. Such a mixed array of blue OLEDs (peak EL at ~460 nm) and green OLEDs (peak EL at ~530 nm) is shown in Fig. 5.10.

In such arrays, typically 2 to 4 pixels excite a given sensing film. Thus, through consecutive or simultaneous excitation of such small groups of OLED pixels, O_2 can be detected by different sensing films that exhibit linear calibration plots and sensitivities suitable for different ranges of O_2 levels. This approach provides the basis for sensor (micro)arrays for multianalyte detection, including using an array of OLEDs emitting at various wavelengths. Such arrays were recently fabricated using combinatorial methods.[35, 36]

In summary, the example of the O_2 sensor demonstrates that the use of OLEDs' as excitation sources in PL-based chemical sensors is promising. The ease of OLED fabrication and OLED/sensing component integration result in compact modules, which are expected to be inexpensive and suitable for real-world applications. The example of oxygen sensing demonstrates the advantageous decay time detection mode. The results also demonstrate the promise of the OLED pixel platform for developing sensor arrays for multiple analyses of a single analyte or for detection of multiple analytes, as shown next. In evaluating the OLED-based platform for oxygen sensing, a new assessment of PtOEP-based sensors in terms of PtOEP aggregation and the effects of film composition and measurement temperature on the PL lifetime was obtained.[17]

We note that OLED-based O_2 sensors are currently evaluated for potential commercialization. The main issues that face such devices are the OLED's long-term stability and large-scale availability. Figure 5.11 shows a structurally integrated OLED/PtOEP:PS/Si photodiode sensor operating in the back detection geometry.

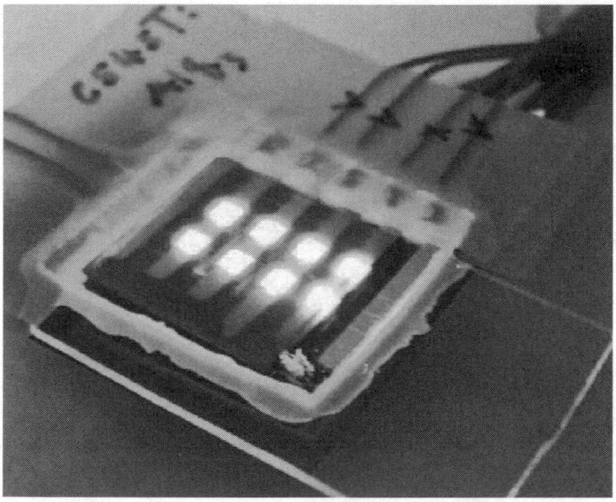

FIGURE 5.11 Top view of a structurally integrated OLED/sensor film/PD probe in a back detection geometry. (See also color insert.)

5.3.2 Multianalyte Sensing

Sensor arrays for detection of multiple analytes in a single sample have been reported extensively. The sensing transduction mechanisms included electrochemical,[43, 44] piezoelectric,[45, 46] electrical resistance,[47, 48] and optical.[49–55] Such wide-range studies are driven by the need for high-throughput, inexpensive, and efficient analyses of complex samples. Sensor arrays are often fabricated by using photolithography and soft lithography;[44, 48, 56–58] inkjet, screen, and pin printing;[59] and photodeposition.[49, 60, 61] These techniques frequently involve labor-intensive multistep fabrication and require sophisticated image analysis and pattern recognition codes, which often require relearning. The use of OLEDs as single- or multicolor excitation sources in sensor (micro)arrays would drastically simplify fabrication, miniaturization, and use of the PL-based sensors for sequential or simultaneous monitoring of multiple analytes in a single sample.

As mentioned, the OLED-based arrays are unique in their ease of fabrication and integration of the excitation source with the sensing component. The excitation source of individually addressable OLED pixels can be based on a single-color OLED or possibly on multicolor pixels fabricated in a combinatorial approach that results in adjacent OLED pixels that emit at wavelengths ranging from blue to red.[35] OLED pixels of nanometer size, reported recently,[29] should be suitable for future sensor micro/nanoarrays for a wide range of applications.

The OLED-based multianalyte sensor for DO, glucose, lactate, and ethanol, all present in a single sample, was based on the successful integration of the OLED/O_2 sensor. This multiple bioanalyte sensor is obviously important for various clinical, health, industrial, and environmental applications.

The glucose, lactate, and ethanol sensing methods were similar, so we describe the method for glucose only. The glucose level was determined from the DO level following the enzymatic oxidation of glucose by glucose oxidase (GOx) (the GOx was embedded in a thin sol-gel film or dissolved in solution):

$$\text{glucose} + O_2 + \text{GOx} \rightarrow H_2O_2 + \text{gluconic acid} \tag{5.2}$$

Hence, in the presence of glucose, the PL quenching of the O_2-sensitive dye is reduced (i.e., I and τ increase) due to O_2 consumption during the enzymatic oxidation.[13, 16, 62, 63] A similar reaction in the presence of lactate oxidase (LOx) or alcohol oxidase (AOx) results in lactate or ethanol oxidation, respectively.

The multianalyte measurements were performed in sealed cells, since the DO in such cells could not be replenished from the air, and, consequently, its final level yielded the initial analyte level directly. Indeed, the responses from cells open to air were more complex, clearly due to such replenishment.[16] The O_2-sensitive dye, embedded in a PS film, was deposited on the bottom of the reaction cells. Each cell contained a buffered solution of GOx, LOx, or AOx. The total volume of each cell was 100 to 200 μL.

Measurements in sealed cells also resulted in a low LOD of ~0.01 to 0.02 m*M* for glucose, lactate, and ethanol. In our previous studies on the OLED-based glucose sensor, the GOx was embedded in a sol-gel film, rather than dissolved in solution. The sol-gel film was deposited on the PtOEP:PS film, and the monitored sample was simply dropped on the sol-gel film. While the dynamic range was a much higher 0 to 5 mg/mL, the LOD was also higher. In reports on other PL-based glucose sensors, the excitation sources have included Ar^+ lasers (operated at 20 to 40 mW), Hg lamps, or a fluorimeter with a 150 W Xe lamp light source.[4, 62, 63] The glucose concentration c_{Gl} was usually monitored via changes in I.

The OLED sensor array was assembled in a back-detection geometry and operated in the τ mode. The analytes were monitored both sequentially, using a single PD, and simultaneously, using a compatible array of Si photodiodes. The sequential and simultaneous modes yielded similar results. Calibration curves were obtained by using a modified SV equation suitable for these analytes in sealed cells, where there is no supply of DO beyond the initial concentration. The modified SV equation was based on the following considerations: Let the initial analyte, initial DO, and final DO levels be $[\text{analyte}]_{\text{initial}}$, $[\text{DO}]_{\text{initial}}$, and $[\text{DO}]_{\text{final}}$, respectively. Then if

$$[\text{analyte}]_{\text{initial}} \leq [\text{DO}]_{\text{initial}} \tag{5.3}$$

and all the analyte is oxidized, then

$$[\text{DO}]_{\text{final}} = [\text{DO}]_{\text{initial}} - [\text{analyte}]_{\text{initial}} \tag{5.4}$$

This leads to the modified SV relation

$$I_0/I = \tau_0/\tau = 1 + K_{\text{SV}} \times ([\text{DO}]_{\text{initial}} - [\text{analyte}]_{\text{initial}}) \tag{5.5}$$

Therefore, $1/\tau$ vs. $[\text{analyte}]_{\text{initial}}$ will ideally be linear with a slope equal to $-K_{SV}$, which, as expected, was found to be film-dependent. Equation (5.5) is also valid for containers open to air, if the oxidation of the analyte [Eq. (5.2)] is much faster than the rate at which gas-phase oxygen diffuses into the solution.

The results shown below and published elsewhere[64] were in excellent agreement with Eq. (5.5). And although that equation appears to limit the dynamic range to $[\text{DO}]_{\text{initial}} \sim 8.6$ wt ppm ~ 0.25 m*M* in equilibrium with air at 23°C, it is only the dynamic range in the final test solution, which may be diluted. Thus, through dilution, the actual dynamic range is wider and covers the concentration range of the various applications.

Figure 5.12 shows the schematic of the OLED array designed for simultaneous monitoring of four analytes. The OLED pixels are defined by the overlap between the mutually perpendicular ITO and Al stripes. There is no crosstalk between the OLED pixels; 5×5 mm^2 Si photodiodes were assembled in an array compatible with the OLED pixel array and placed underneath it. The reaction cells, whose base is the PtOEP:PS film, were on top. Three of these reaction cells contained each an enzyme that specifically catalyzes the oxidation of one of the analytes.

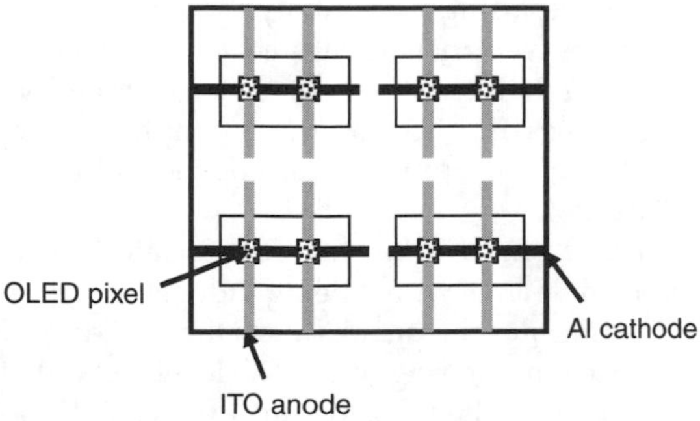

Figure 5.12 Schematic of the OLED array designed for simultaneous monitoring of four analytes. The vertical lines are the ITO anode stripes, and the horizontal lines are the Al cathode stripes. The (square) OLED pixels are defined by the overlap between the ITO and the Al stripes. (*Reprinted from Ref. 64. Copyright 2008, with permission from Elsevier.*)

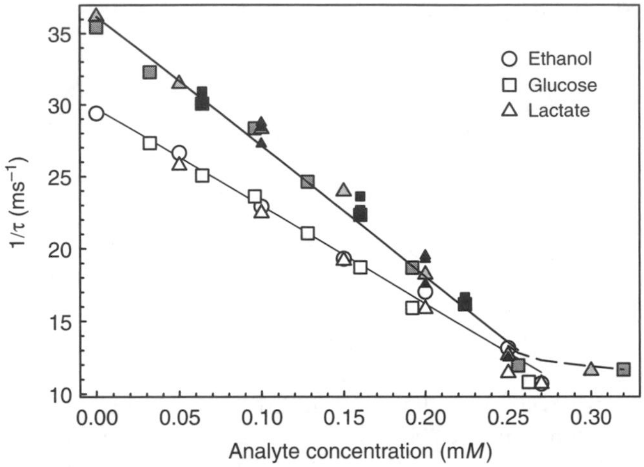

FIGURE 5.13 $1/\tau$ vs. analyte concentration for ethanol, glucose, and lactate for different sensor films. The open symbols represent a consecutive measurement, where comparable concentrations of the three analytes in a mixture were monitored with a single PD. The gray symbols show the monitoring of single analytes (glucose and lactate) to generate a calibration curve. The black symbols represent simultaneous measurement of glucose and lactate with all three analytes in a single sample, using an array of PDs. (*Reprinted from Ref. 64. Copyright 2008, with permission from Elsevier.*)

Figure 5.13 shows the calibration lines of $1/\tau$ vs. analyte concentration for ethanol, glucose, and lactate. Data of both consecutive and simultaneous measurements are presented. In the consecutive measurement, comparable concentrations of the three analytes in a mixture were monitored with a single PD. In the simultaneous measurement, calibration lines using single analytes were first generated, and the analyte mixtures were then monitored simultaneously using a PD array. As seen, though different sensor films generate somewhat different calibration lines, Eq. (5.5) and the assumptions associated with it are confirmed, and both approaches using the OLED array result in an LOD of ~0.02 m*M*.

For comparable sensor films, the data for all three analytes are presented by a single line, with τ being independent of the analyte, whether glucose, lactate, or ethanol, due to the similar oxidation reactions and the actual monitoring of the DO level. Yet only the analyte corresponding to the oxidase enzyme in any given solution affects τ; i.e., there is no interference between the analytes. For example, in the solution containing LOx and any combination of 0.15 m*M* or 0.35 m*M* ethanol or glucose, 0.1 m*M* lactate yielded $45.3 \leq \tau \leq 46.1$ μs, and 0.2 m*M* lactate yielded $68.0 \leq \tau \leq 69.1$ μs. Note that that the data points deviate by < 2% from the best-fit line, i.e., the predicted values.

To summarize, the results show that the use of the compact OLED-based sensor array is a viable approach for simultaneous monitoring of these multiple analytes. To reduce the array dimensions further, efforts are underway to integrate thin-film amorphous or nanocrystalline Si photodiode arrays with the OLED/sensing element arrays.[18] Some of these efforts are reviewed in Sec. 5.5.

5.3.3 Sensors for Foodborne Pathogens

Control of food pathogens at the various stages of food processing, transport, and distribution remains a challenge. As an example, remediation of *Listeria monocytogenes*, a ubiquitous foodborne bacterium typically associated with postprocessing contamination of ready-to-eat meats, has been studied extensively,[65] with more recent work focused on developing novel naturally produced antimicrobials,[66] addressing consumer demand to minimize synthetic chemicals in food processing. Concurrently, there is a need for compact and reliable field-deployable sensors that will monitor such pathogens.

Compounds derived from bacteria, such as the iron-chelating siderophores and bacteriocins, are among the potential natural candidates to control food pathogens.[67] Bacteriocins have received high consideration by the food industry;[68, 69] however, only purified nisin, a polycyclic antibacterial peptide, is being allowed for use as a food preservative. The search for other natural compounds, with the potential to safely and easily obliterate food toxins, continues.

Other environmental bacteria that may produce antimicrobial compounds are methanotrophs, i.e., aerobic methane-oxidizing bacteria. Methanobactin (mb) is a novel copper-binding chromopeptide recently isolated from different methanotrophs. DiSpirito, Shinar, and coworkers have recently obtained evidence regarding the Cu-chelated mb's (Cu-mb)[70] bactericidal activity toward *L. monocytogenes* and *Bacillus subtilis*,[71, 72] a gram-positive bacterium commonly found in soil.[71]

In principle, the effect of Cu-mb on the bacteria can be studied by following the bacteria's respiration. To that end, the antimicrobial activity of Cu-mb on the respiration of *B. subtilis* was studied by monitoring DO in sealed containers using the OLED-based sensor. The *B. subtilis*, a nontoxic bacterium, is often used as a genetic model system. The advantages of the OLED-based DO sensor for this application are its compact size and flexible design and thus its potential use in food processing and packaging.

Figure 5.14 shows the effect of Cu-mb on the respiration and survival of ~5 mg/mL *B. subtilis* in the presence of glucose. As seen, the bacteria in a sealed container consume the DO in about 20 min, after which the DO level is reduced to 0 ppm; however, the bacteria survive. Upon addition of sufficient Cu-mb, the level of DO reduces faster to a constant level. In the presence of 250 μM Cu-mb, the level

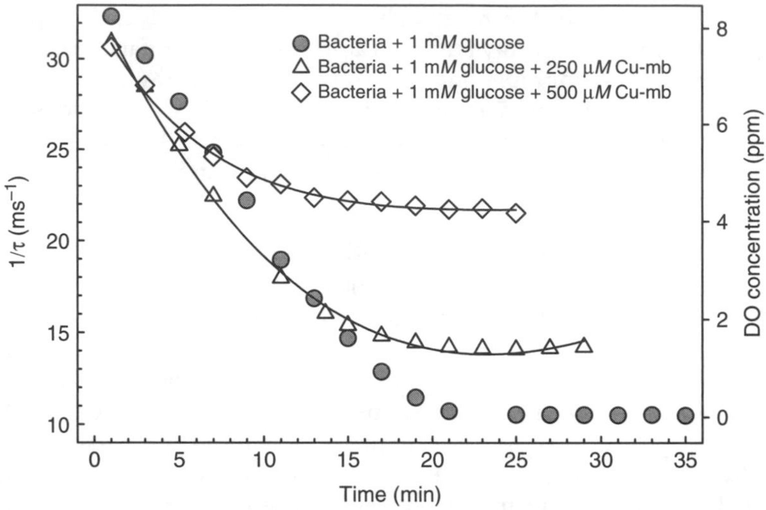

FIGURE 5.14 The effect of Cu-mb on the respiration and survival of ~5 mg/mL *B. subtilis* bacteria.

of DO reached a constant level smaller than 2 ppm in ~15 min. In the presence of 500 μ*M* Cu-mb, the level of DO reached a constant level of ~4 ppm after ~10 min. These constant, non-zero DO levels, which remain in the solution, indicate obliteration of the bacteria by Cu-mb; this process is faster with increasing Cu-mb concentration, as expected, with a higher remaining level of DO.

The foregoing initial results on monitoring the state of *B. subtilis* cultures using an OLED-based DO sensor demonstrate the potential power of a wide network of autonomous remote-controlled OLED-based DO sensors, which would monitor foodborne pathogens and food spoilage at key processing, transport, and distribution points of the food industry.

5.4 OLED Sensing Platform Benefits and Issues

The foregoing review demonstrated the promise of the OLED-based platform for monitoring oxygen and other analytes that can be monitored via oxygen. The advantages of the OLED arrays include their high brightness, design flexibility, and compatibility with glass or plastic substrates and consequently with microfluidic architectures as well. Moreover, their fabrication is facile, and their integration with the sensing elements is uniquely simple. Hence, they should eventually yield low-cost, miniaturized, field-deployable multianalyte sensor arrays for monitoring a wide variety of analytes. The continuing advances in OLED performance and their expanding commercialization will facilitate the realization of the OLED-based sensor platform.

The foregoing review also highlighted the advantages of operating PL-based sensors in the τ mode, as this mode eliminates the need for frequent sensor calibration, a reference sensor, and optical filters that block the EL from the PD. As the EL decay time of fluorescent OLEDs is typically < 100 ns, they are particularly suitable for pulsed excitation of oxygen-sensitive phosphorescent dyes, whose radiative decay time τ_{rad} is typically > 1 μs. Consequently, the PtOEP and PdOEP dyes, with a large PL quantum yield and τ_{rad} of ~100 and ~1000 μs, respectively, were particularly rewarding.

With absorption bands around ~380 and ~540 nm (the latter closely overlapping the EL of Alq_3-based OLEDs) and PL peaking at ~640 nm, the PtOEP and PdOEP dyes also have a large Stokes shift. This enables their use not only in the τ mode, but also in the *I* mode, as the EL tail at the PL band is minimal.

Another example of a successful OLED-based sensor is that of hydrazine, which is highly toxic and volatile, but a powerful monopropellant used in space shuttles.[15] It is also a common precursor for some polymer synthesis, plasticizers, and pesticides. Due to its extreme toxicity, the American Conference of Governmental Industrial Hygienists recommended that the threshold limit value (TLV) for hydrazine exposure, i.e., the time-weighted average concentration of permissible exposure within a normal 8 h workday, not exceed 10 ppb in air.[73] The OSHA recommended skin exposure limit is 0.1 ppm (0.1 mg/m^3), and the immediate threat to life is less than 60 ppm.[74]

The hydrazine sensor is based on the reaction between N_2H_4 and anthracene 2,3-dicarboxaldehyde.[15, 75] The reactants are not emissive, but the reaction product fluoresces around 550 nm when excited around 475 nm by, e.g., a blue DPVBi-based OLED; the signal is proportional to the N_2H_4 level.

The OLEDs used for the N_2H_4 sensor were operated in a dc mode at 9 to 20 V, or in a pulsed mode at up to 35 V. The PD was a PMT. At ~60 ppb, the PL was detected after ~1 min. Therefore, at ~1 ppb, the PL would be detected in ~1 h. Thus, this capability exceeds the OSHA requirements by a factor of ~80.

A third example of an OLED-based sensor on which preliminary evaluations have been reported is that of *Bacillus anthracis*, via the lethal factor (LF) enzyme it secretes.[15] The need for a compact, low-cost, field-deployable sensor for rapid, on-site detection of this bacterium, which would eliminate the need to send samples for diagnosis to a remote site, is obvious. The detection of LF, which is one of the three proteins secreted by the live anthrax bacterium, is based on its property of cleaving certain peptides at specific sites.[76–78] As the LF cleaves the peptide, which is labeled with a fluorescence resonance energy transfer (FRET) donor-acceptor pair, with the donor and acceptor on opposite sides of the cleaving site and the two cleaved segments separate, the PL of the donor, previously absorbed by the acceptor, becomes detectable by the PD.

A suitable peptide labeled with a rhodamine-based dye as the donor and a Molecular Probes QSY7 dark quencher as the acceptor was synthesized. Preliminary results using green ITO/CuPc/NPD/Alq_3/CsF/Al OLEDs yielded a maximal PL intensity increase of ~100% at 37°C and ~ 40 μ*M* peptide upon 15 min exposure to 25 n*M* LF. This PL increase is low relative to that observed by UV laser excitation,[76] pointing to an issue associated with OLED excitation in conjunction with the specific dye employed, which has a Stokes shift of only ~20 nm. Indeed, using a green inorganic LED with a 530 nm bandpass filter, and a 550 nm longpass filter in front of the PMT, the LF increased the PL six fold following peptide cleavage.[79]

To increase the change in the output of the PD resulting from cleavage of the peptide by LF, other donor-acceptor pairs are being examined, together with microcavity OLEDs that emit a much narrower EL band. Other approaches to reduce the contribution of the EL tail, e.g., the use of crossed polaroids between the OLED and the PD,[80] are also under investigation.

5.5 OLED/Sensing Component/Photodetector Integration

We are currently developing a compact PL-based O_2 sensor to evaluate a fully integrated platform, where the PD is a p-i-n or n-i-p structure based on thin films of hydrogenated amorphous Si (a-Si:H) and related materials, or nanocrystalline Si (nc-Si).[18] Similar to OLEDs, a-(Si,Ge):H-and nc-Si-based PDs are easily fabricated on glass or plastic substrates.

The composition of the layers in the PDs was first chosen so as to minimize their sensitivity at the Alq_3-based OLED EL (i.e., at ~535 nm) and maximize it at the PtOEP and PdOEP emission bands, i.e., at ~ 640 nm. To this end, a-Si:H and a-(Si,Ge):H PDs were evaluated by measuring their quantum efficiency (QE) vs. wavelength.[18] The a-(Si,Ge) PDs are attractive due to their better match with the dyes' PL, and lower response at the EL band. However, they are inferior overall due to their higher dark current and lower speed.

In monitoring the oxygen concentration, using the thin-film PDs, via the *I* mode, a lock-in amplifier was used with a pulsed OLED. The detection sensitivity, however, was relatively low due to issues related to the PDs' fabrication, and to an electromagnetic (EM) noise stemming from the pulsed OLED and the device wiring when using the lock-in detection.

One PD-related issue was its high dark current. It was suspected that this issue is a result of boron diffusion from the p to the i layer during growth of p-i-n devices. To minimize this diffusion, a SiC

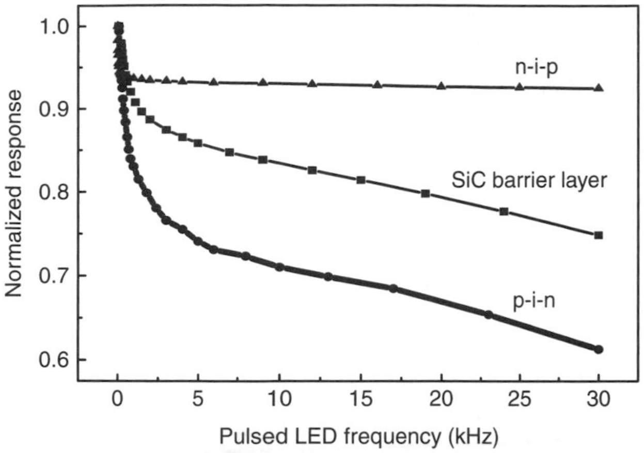

FIGURE 5.15 The normalized response of VHF-grown PDs vs the LED pulse frequency. The p+ and i layers were a-Si/nc-Si/a-Si and a-Si, respectively. (*Reprinted from Ref. 64. Copyright 2008, with permission from Elsevier.*)

barrier layer was grown at the p-i interface. This indeed resulted in a decrease in the dark current, by about two orders of magnitude. An additional reduction by an order of magnitude was achieved by etching to generate a mesa structure. A reduced dark current in n-i-p structures, where the p layer is grown last, further confirmed the suspected effect of B diffusion on the dark current. Reducing the dark current improved the response measured by the PD. This behavior is shown in Fig. 5.15. The response, however, strongly decreased with increasing pulse frequency at frequencies smaller than 5 kHz, which is currently not clear.

The PDs were further improved by using nc-Si based devices. The p layer in these devices was bandgap-graded SiC/a-Si/nc-Si for improved bandgap matching. Unlike the a-Si:H layer, the nc-Si layer was only lightly doped; the series resistance was reduced by post-growth anneal to diffuse boron into it from the a-Si:H layer. The i layer was nc-Si. The speed of this device was the highest of all, ~250 μs, but still too slow for monitoring O_2 in the τ mode.

Figure 5.16 shows SV plots obtained following the above mentioned recent improvements in the OLEDs and PDs. A detection sensitivity S_g of ~7 was obtained with an Alq_3 excitation source, a PtOEP:PS film, and an n-i-p a-(Si,Ge)-based PD, in which boron diffusion from the p layer to the i layer was eliminated. A sensitivity of ~26 was obtained with a PdOEP:PS film. As seen in Fig. 5.16, a higher S_g ~ 47 was achieved for an integrated sensor of [coumarin-doped Alq_3 OLED]/[PdOEP-doped polystyrene sensor film]/[VHF-grown SiC/a-Si/nc-Si (p+ layer):nc-Si (i layer) PD].

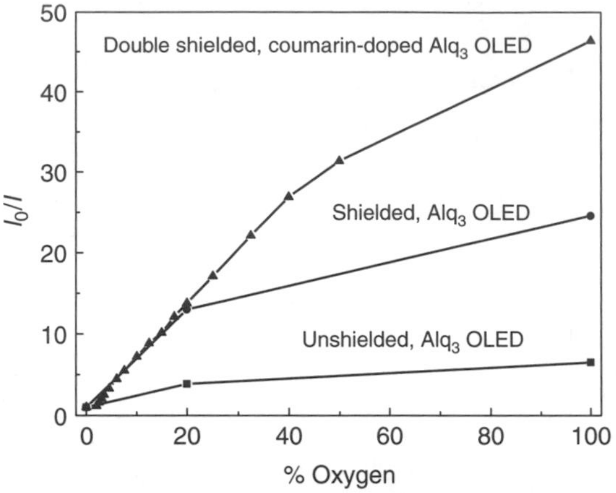

FIGURE 5.16 SV plots using unshielded and shielded structurally integrated O_2 sensors: OLED/PdOEP:PS/thin film Si-based PD. (*Reprinted with permission from Ref. 81, "Integrated Photoluminescence-Based Sensor Arrays: OLED Excitation Source/Sensor Film/Thin-Film Photodetector," SPIE 2007.*)

The latter higher sensitivity was achieved in part by shielding the electromagnetic noise generated by the pulsed current through the OLED and wiring when using lock-in detection.[81] In one approach, the shielding was achieved by placing a grounded ITO/glass, as shown in Fig. 5.17, over the OLED. This sensitivity is still lower than the sensitivity of the [Alq_3-based OLED]/[PdOEP:PS] module obtained with a PMT, and the thin-film PD could only monitor O_2 in the *I* mode (due to the PD's slow speed), but it still demonstrates the potential ultimate success of the three-component integrated platform. As the measurement of τ is beneficial over *I*, the development of the thin-film PDs focuses currently on obtaining a detailed understanding of the factors affecting their speed, in an attempt to shorten their response time.

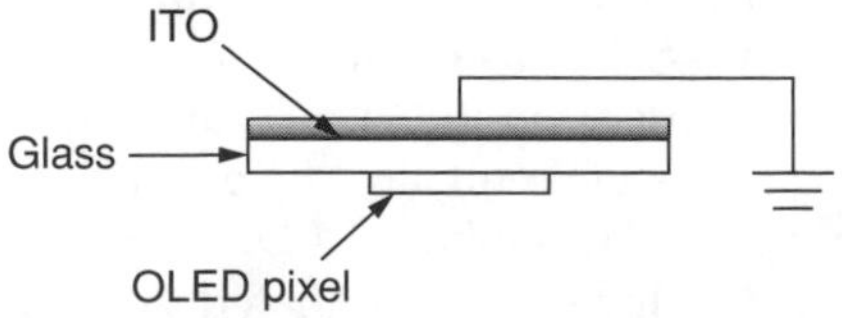

FIGURE 5.17 OLED shielding configuration (not to scale), which resulted in significantly improved S_g. (*Reprinted with permission from Ref. 81, "Integrated Photoluminescence-Based Sensor Arrays: OLED Excitation Source/Sensor Film/Thin-Film Photodetector," SPIE 2007.*)

5.6 Concluding Remarks

Recent advances in the development of the OLED-based luminescent chemical and biological sensor platform were reviewed. The advantages of this platform include its ease of fabrication, uniquely simple integration with the sensing elements, and consequent promise as a compact field-deployable low-cost monitor for various analytes. In addition, the OLEDs could be efficient light sources in sensor (micro)arrays for detection of multiple analytes in microfluidic architectures.

The viability of the OLED-based platform was demonstrated for gas-phase and dissolved oxygen, hydrazine, and simultaneous monitoring of multiple analytes, such as mixtures of dissolved O_2, glucose, lactate, and ethanol. Initial results on the structural integration of the OLED excitation source, the thin-film sensing element, and a Si-based thin-film PD were also reviewed. This effort is motivated by its potential to lead to badge-size monitors.

For the sensors based on monitoring oxygen, the fluorescent OLEDs are operated in a pulsed mode. This enables their use not only in the PL intensity I mode, but also in the PL decay time τ mode, as τ is determined by the analyte concentration. Since $\tau > 1$ μs, it is much longer than the ~100 ns EL decay time τ_{EL} of typical fluorescent OLEDs. Indeed, even some phosphorescent OLEDs, where $\tau_{EL} \sim 1$ μs, can be used to excite the sensor films and monitor the O_2 in the τ mode. This mode is strongly preferable over the I mode: (1) It removes the need for a reference sensor and frequent sensor calibration, since τ is independent of moderate changes in the sensing film, light source intensity, and background light. (2) It removes the need for optical filters that block the EL from the PD, as the PL decay curve of the O_2 sensing dye is monitored after the EL is turned off.

Some current and future efforts to enhance the OLED-based sensor platform include these: (1) Applications of microcavity, stacked (tandem), and/or mixed layer phosphorescent OLEDs.[82, 83] These would be brighter, more efficient, and longer-lived than the conventional fluorescent OLEDs; the microcavity OLEDs' narrow EL spectra could be tailored to the absorption spectrum of the sensing element, resulting in drastic improvement of the sensor platform. However, preliminary studies indicated, as expected, that the relatively thicker microcavity OLEDs require longer pulses to obtain the maximal EL and therefore PL (see Fig. 5.18). Increased PL intensity can improve the limit of detection; however, the use of longer pulses may adversely affect the OLED lifetime. (2) Development of the platform for various biological analytes, notably foodborne pathogens, which are a major ongoing issue in the global food supply chain. (3) Improvement of nc-Si-based PDs to shorten their response time and evaluation of organic PDs (OPDs) for integration with the OLED/sensor film modules, as Peumans et al. have demonstrated fast OPDs.[84]

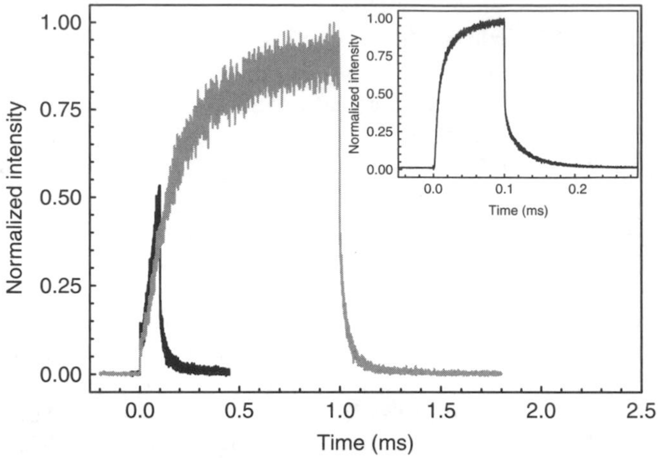

FIGURE 5.18 Normalized intensities of the OLED EL tail + sensor film PL measured during 0.1 and 1 ms pulses of a microcavity OLED; the OLED structure was [26.7 nm Ag] / [5 nm CuPc] / [85 nm NPD] / [85 nm Alq_3] / [1 nm LiF] / [136 nm Al]. The inset shows the normalized intensity measured during a 0.1 ms pulse of a conventional ITO/ [5 nm CuPc] / [50 nm NPD] / [50 nm Alq_3] / [1 nm LiF]/ [136 nm Al] OLED. The measurement was performed in the gas phase under ambient conditions with a PtOEP:PS sensing film with a 600 nm long-pass filter. As such, the intensities during the pulses show a combination of the EL tail and the PL. The FWHMs of the EL conventional and microcavity OLEDs were 94 and 19 nm, respectively.

References

1. O. S. Wolfbeis, L. Weis, M. J. P. Leiner, and W. E. Ziegler, *Anal. Chem.*, 60:2028 (1988).
2. B. H. Weigl, A. Holobar, W. Trettnak, I. Klimant, H. Kraus, P. O'Leary, and O. Wolfbeis, *J. Biotech.*, 32:127 (1994).
3. P. Hartmann and W. Ziegler, *Anal. Chem.*, 68:4512 (1996).
4. Z. Rosenzweig and R. Kopelman, *Sens. Actuat. B*, 35–36:475 (1996).
5. B. D. MacCraith, C. McDonagh, A. K. McEvoy, T. Butler, G. O'Keeffe, and V. Murphy, *J. Sol-Gel Sci. Tech.*, 8:1053 (1997).
6. A. K. McEvoy, C. M. McDonagh, and B. D. MacCraith, *Analyst*, 121:785 (1996).
7. V. K. Yadavalli, W.-G. Koh, G. J. Lazur, and M. V. Pishko, *Sens. Actuat.* B, 97:290 (2004).
8. M. Vollprecht, F. Dieterle, S. Busche, G. Gauglitz, K-J. Eichhorn, and B. Voit, *Anal. Chem.*, 77:5542 (2005).
9. E. J. Cho and F. V. Bright, *Anal. Chem.*, 73:3289 (2001).
10. W. Aylott, Z. Chen-Esterlit, J. H. Friedl, R. Kopelman, V. Savvateev, and J. Shinar, U.S. Patent No. 6,331, 438, December 2001.
11. V. Savvate'ev, Z. Chen-Esterlit, C.-H. Kim, L. Zou, J. H. Friedl, R. Shinar, J. Shinar, et al., *Appl. Phys. Lett.*, 81:4652 (2002).
12. B. Choudhury, R. Shinar, and J. Shinar, in *Organic Light Emitting Materials and Devices VII*, edited by Z. H. Kafafi and P. A. Lane, *SPIE Conf. Proc.*, 5214:64 (2004).
13. B. Choudhury, R. Shinar, and J. Shinar, *J. Appl. Phys.*, 96:2949 (2004).

14. R. Shinar, B. Choudhury, Z. Zhou, H.-S. Wu, L. Tabatabai, and J. Shinar, in *Smart Medical and Biomedical Sensor Technology II*, edited by Brian M. Cullum, *SPIE Conf. Proc.*, 5588:59 (2004).
15. Z. Zhou, R. Shinar, B. Choudhury, L. B. Tabatabai, C. Liao, and J. Shinar, in *Chemical and Biological Sensors for Industrial and Environmental Security*, edited by A. J. Sedlacek, III, S. D. Christensen, R. J. Combs, and T. Vo-Dinh, *SPIE Conf. Proc.*, 5994:59940E-1 (SPIE, Bellingham, Wash., 2005).
16. R. Shinar, C. Qian, Y. Cai, Z. Zhou, B. Choudhury, and J. Shinar, in *Smart Medical and Biomedical Sensor Technology III*, edited by B. M. Cullum and J. C. Carter, *SPIE Conf. Proc.*, 6007:600710-1 (SPIE, Bellingham, Wash., 2005).
17. R. Shinar, Z. Zhou, B. Choudhury, and J. Shinar, *Anal. Chim. Acta*, 568:190 (2006).
18. R. Shinar, D. Ghosh, B. Choudhury, M. Noack, V. L. Dalal, and J. Shinar, *J. Non Crystalline Solids*, 352:1995 (2006).
19. S. Camou, M. Kitamura, J.-P. Gouy, H. Fujita, Y. Arakawa and T. Fujii, *Proc. SPIE*, 4833:1 (2002).
20. O. Hofmann, X. Wang, J. C. deMello, D. D. C. Bradley, and A. J. deMello, *Lab on a Chip*, 5:863 (2005).
21. J. Shinar (ed.), *Organic Light-Emitting Devices: A Survey*, Springer-Verlag, New York, 2003.
22. C. Adachi, M. A. Baldo, S. R. Forrest, S. Lamansky, M. E. Thompson, and R. C. Kwong, *Appl. Phys. Lett.*, 78:1622 (2001).
23. C. Adachi, R. C. Kwong, P. Djurovich, V. Adamovich, M. A. Baldo, M. E. Thompson, and S. R. Forrest, *Appl. Phys. Lett.*, 79:2082 (2001).
24. R. H. Friend, R. W. Gymer, A. B. Holmes, J. H. Burroughes, R. N. Marks, C. Taliani, D. D. C. Bradley, et al., *Nature*, 397:121 (1999).
25. www.universaldisplay.com
26. www.cdt.co.uk
27. G. Gustafsson, Y. Cao, G. M. Treacy, F. Klavetter, N. Colaneri, and A. J. Heeger, *Nature*, 357:447 (1992); C. C. Wu, S. D. Theiss, G. Gu, M. H. Lu, J. C. Sturm, S. Wagner, and S. R. Forrest, *IEEE Electron. Dev. Lett.*, 18:609 (1997); H. Kim, J. S. Horwitz, G. P. Kushto, Z. H. Kafafi, and D. B. Chrisey, *Appl. Phys. Lett.*, 79:284 (2001); E. Guenther, R. S. Kumar, F. Zhu, H. Y. Low, K. S. Ong, M. D. J. Auch, K. Zhang, et al., in *Organic Light-Emitting Materials and Devices V*, edited by Z. H. Kafafi, *SPIE Conf. Proc.*, 4464:23 (SPIE, Bellingham, Wash., 2002); T.-F. Guo, S.-C. Chang, S. Pyo, and Y. Yang, ibid., p. 34.
28. C. Adachi, M. A. Baldo, M. E. Thompson, and S. R. Forrest, *J. Appl. Phys.*, 90, 5048 (2001).
29. J. G. C. Veinot, H. Yan, S. M. Smith, J. Cui, Q. Huang, and T. J. Marks, *Nano Lett.*, 2:333–335 (2002); F. A. Boroumand, P. W. Fry, and D. G. Lidzey, *Nano Lett.*, 5:67 (2005); H. Yamamoto, J. Wilkinson, J. P. Long, K. Bussman, J. A. Christodoulides, and Z. H. Kafafi, *Nano Lett.*, 5:2485 (2005).
30. J. Shinar and V. Savvate'ev, in *Organic Light-Emitting Devices: A Survey*, J. Shinar (ed.), Springer-Verlag, New York, 2003, Chap. 1.
31. L. Zou, V. Savvate'ev, J. Booher, C.-H. Kim, and J. Shinar, *Appl. Phys. Lett.*, 79:2282 (2001).
32. V. G. Kozlov, G. Parthasarathy, P. E. Burrows, S. R. Forrest, Y. You, and M. E. Thompson, *Appl. Phys. Lett.*, 72:144 (1998).
33. L. S. Hung, C. W. Tang, and M. G. Mason, *Appl. Phys. Lett.*, 70:152 (1997).
34. S. E. Shaheen, G.E. Jabbour, M. M. Morell, Y. Kawabe, B. Kippelen, N. Peyghambarian, M.-F. Nabor, et al., *J. Appl. Phys.*, 84:2324 (1998).
35. K. O. Cheon and J. Shinar, *Appl. Phys. Lett.*, 83:2073 (2003).
36. J. Shinar, R. Shinar, and Z. Zhou, *Appl. Surf. Sci.*, 254:749 (2007).
37. Frost & Sullivan, *Market Engineering Research for the U.S. Water and Wastewater Dissolved Oxygen Instrumentation Market, 1997*; Frost & Sullivan, *World Process Analytical Instrumentation Markets*, Apr. 19, 2005.
38. P. A. S. Jorge, P. Caldas, C. C. Rosa, A. G. Oliva, and J. L. Santos, *Sens. Actuat. B*, 103:290 (2004).
39. C. Preininger, I. Klimant, and O. S. Wolfbeis, *Anal. Chem.*, 66:1841 (1994).

40. Y. Amao, T. Miyashita, and I. Okura, *Analyst*, 125:871 (2000).
41. Z. J. Fuller, D. W. Bare, K. A. Kneas, W.-Y. Xu, J. N. Demas, and B. A. DeGraff, *Anal. Chem.*, 75:2670 (2003).
42. Z. Zhou, R. Shinar, A. J. Allison, and J. Shinar, *Adv. Func. Mater.*, 17:3530 (2007).
43. H. Frebel, G.-C. Chemnitius, K. Cammann, R. Kakerow, M. Rospert, and W. Mokwa, *Sens. Actuat. B*, 43:87–93 (1997).
44. M. S. Wilson and W. Nie, *Anal. Chem.*, 78:2507–2513 (2006).
45. I. Sugimoto, M. Nakamura, and H. Kuwano, *Sens. Actuat. B*, 10:117–122 (1993).
46. E. T. Zellers and M. Han, *Anal. Chem.*, 68:2409 (1996).
47. M. S. Freund and N. S. Lewis, *Proc. Natl. Acad. Sci. USA*, 92:2652–2656 (1995).
48. A. Carbonaro and L. Sohn, *Lab on a Chip*, 5:1155 (2005).
49. L. Li and D. R. Walt, *Anal. Chem.*, 67:3746 (1995).
50. T. A. Dickinson, J. White, J. S. Kauer, and D. R. Walt, *Nature*, 382:697 (1996).
51. D. R. Walt, T. Dickinson, J. White, J. Kauer, S. Johnson, H. Engelhardt, J. Sutter, et al., *Biosens. Bioelec.*, 13:697–699 (1998).
52. K. L. Michael, L. C. Taylor, S. L. Schultz, and D. R. Walt, *Anal. Chem.*, 70:1242–1248 (1998).
53. Maria Dolores Marazuela and Maria Cruz Moreno-Bondi, *Anal. Bioanal. Chem.*, 372:664–682 (2002).
54. E. J. Cho and F. V. Bright, *Anal. Chem.*, 74:1462 (2002).
55. E. J. Cho, Z. Tao, E. C. Tehan, and F. V. Bright, *Anal. Chem.*, 74:6177 (2002).
56. S. P. A. Fodor, R. P. Rava, X. C. Huang, A. C. Pease, C. P. Holmes, and C. L. Adams, *Nature*, 364:555–556 (1993).
57. M. Chee, R. Yang, E. Hubbell, A. Berno, X. C. Huang, D. Stern, J. Winkler, et al., *Science*, 274:610–614 (1996).
58. G. McGall, J. Labadie, P. Brock, G. Wallraff, T. Nguyen, and W. Hinsberg, *Proc. Natl. Acad. Sci. USA*, 93:13555–13560 (1996).
59. A. V. Lemmo, J. T. Fisher, H. M. Geysen, and D. J. Rose, *Anal. Chem.*, 69:543–551 (1997).
60. B. G. Healey and D. R. Watt, *Anal. Chem.*, 69:2213–2216 (1997).
61. Y.-H. Liu and T. H. Pantano, *Anal. Chim. Acta*, 419:215–225 (2000).
62. O. S. Wolfbeis, I. Oehme, N. Papkovskaya, and I. Klimant, *Biosens. & Bioelec.*, 15:69 (2000).
63. H. Xu, J. W. Aylott, and R. Kopelman, *Analyst*, 127:1471 (2002).
64. Y. Cai, R. Shinar, Z. Zhou, C. Qian, and J. Shinar, *Sens. & Actuat. B*, 134:727–735 (2008).
65. http://en.wikipedia.org/wiki/Listeria_monocytogenes
66. C. L. Johnson dissertation, "Methanobactin: A Potential Novel Biopreservative for Use against the Foodborne Pathogen *Listeria Monocytogenes*," Chapters 2 and 3 and references therein, Iowa State University, 2006.
67. B. Ray, in *Food Biopreservatives of Microbial Origin*, B. Ray and M. Daeschel (eds.), CRC Press, Boca Raton, Fla., 1991, pp. 25–26.
68. T. Abee, L. Krockel, and C. Hill, *Int. J. Food Microbiol.*, 28:169–185 (1995).
69. J. Cleveland, T. J. Montville, I. F. Nes, and M. L. Chikindas, *J. Food Microbiol.*, 71:1–20 (2001).
70. D. W. Choi, J. D. Semrau, W. E. Antholine, S. C. Hartsel, R. C. Anderson, J. N. Carey, A. M. Dreis, et al., *J. Inorg. Biochem.*, 102:1571–1580 (2008).
71. http://en.wikipedia.org/wiki/Bacillus_subtilis#cite_note-Brock-0.
72. Y. Cai, R. Shinar, D. W. Choi, A. DiSpirito, and J. Shinar, in Iowa State University Institute for Food Safety and Security (IFSS) 2d Annual Symposium, "Food Safety and Public Health: Production, Distribution, and Policy," April 12, 2007.
73. American Conference of Governmental Industrial Hygienists (ACGIH), *1999 TLVs and BEIs. Threshold Limit Values for Chemical Substances and Physical Agents, Biological Exposure Indices*, Cincinnati, Ohio, 1999. See also http://www.epa.gov/ttn/atw/hlthef/hydrazin.html#ref12
74. National Institute for Occupational Safety and Health (NIOSH), *Pocket Guide to Chemical Hazards*, U.S. Department of Health and Human Services, Public

Health Service, Centers for Disease Control and Prevention, Cincinnati, Ohio, 1997; http://www.cdc.gov/niosh/npg/npg.html
75. S. Rose-Pehrsson and G. E. Collins, U.S. Patent 5,719,061, Feb. 17, 1998.
76. R. T. Cummings, S. P. Salowe, B. R. Cunningham, J. Wiltsie, Y. W. Park, L. M. Sonatore, D. Wisniewski, et al., *Proc. Natl. Acad. Sci*, 99:6603 (2002).
77. F. Tonello, P. Ascenzi, and C. Montecucco, *J. Biol. Chem.*, 278:40075 (2003).
78. B. E. Turk, T. Y. Wong, R. Schwarzenbacher, E. T. Jarrell, S. H. Leppla, R. J. Collier, R. C. Liddington, et al., *Natl. Struc. & Mol. Biol.*, 11:60 (2004).
79. Z. Zhou, R. Shinar, L. Tabatabai, C. Liao, and J. Shinar, unpublished results.
80. E. Kraker, A. Haase, B. Lamprecht, G. Jakopic, C. Konrad, and S. Köstler, *Appl. Phys. Lett.*, 92:033302 (2008).
81. D. Ghosh, R. Shinar, Y. Cai, Z. Zhou, V. L. Dalal, and J. Shinar, in *Organic-Based Chemical and Biological Sensors*, edited by R. Shinar and G. G. Malliaras, *SPIE Conf. Proc.*, 6659:66590E-1 (SPIE, Bellingham, Wash., 2007).
82. L. S. Liao, K. P. Klubek, and C. W. Tang, *Appl. Phys. Lett.*, 84:167 (2004).
83. T.-Y. Cho, C.-L. Lin, and C.-C. Wu, *Appl. Phys. Lett.*, 88:111106 (2006).
84. P. Peumans, V. Bulovic´, and S. R. Forrest, *Appl. Phys. Lett.*, 76:3855 (2000).

CHAPTER 6

An Introduction to Organic Photodetectors

Jingsong Huang
Molecular Vision Ltd.
BioIncubator Unit, London, UK

John deMello
Chemistry Department
Imperial College London, London, UK

6.1 Introduction

In this chapter, we review the use of organic photodiodes (OPDs) as sensitive photodetectors, and we compare their performance characteristics with conventional vacuum tube and solid-state detectors. Huge research efforts have been invested in organic light-emitting diodes, solar cells, and transistors, but organic photodetectors have received far less attention to date. Although they are structurally similar to organic solar cells, their performance criteria are quite different. Materials systems and device geometries that work well for solar energy applications may perform poorly when applied to photodetectors, yielding poor spectral characteristics, excessive noise, or sluggish response. Organic photodetectors should not therefore be viewed as a simple offshoot of existing solar cell research, but as technological devices in their own right with their own unique set of technical challenges. This chapter is written with this viewpoint in mind, and offers a tutorial-style introduction to organic photodetectors that is intended to provide an accessible starting point for the would-be OPD researcher. We review the fabrication and operating principles of OPDs and the preferred techniques for reading them electronically. We introduce the principal figures of merit for photodetectors, and we investigate how OPDs match up against other technologies. Finally, we consider some promising application areas where OPDs look set to find important uses. Other unexpected applications are bound to emerge in the coming years, and in writing this chapter, we hope to stimulate exactly this outcome by encouraging further research in this hugely important area.

The optical detector is one of the cornerstones of modern technology, and it plays a critical role in numerous applications including imaging, communications, data retrieval, proximity and motion detection, environmental monitoring, and chemical analysis, to name but a few. In crude terms, optical detection can be divided into two broad categories: data communications ("data comms"), where light is used to carry an encoded signal, and sensing, where it is the properties of the light itself that are of interest. The frequency, intensity, and spectral characteristics of the optical signal vary tremendously from one application to the next. In fiber-optic communications systems, photodetectors are used to receive infrared (IR) signals at rates of up to 100 GHz. In optical disc drives, they retrieve data at up to 200 MHz by detecting visible laser light reflected from the pits of a spinning disc. They are also used in lower bandwidth applications, such as remote controls for electronic equipment, optical "trip switches" for home security systems, proximity detectors for motor vehicles, and orientation/position sensors in optical-mouse devices. In industry, photodetectors play a critical role in process control where they are used to ensure the correct positioning of components, to monitor product throughput, and to provide real-time information for the feedback control of robotic systems. In chemical and biological analysis, they are used to monitor changes in absorption, fluorescence, chemiluminescence, or refractive index due to the presence of specific analytes. In environmental monitoring, optical detectors are used to determine the concentration of air- or waterborne particulates by monitoring the frequency of optical scattering events—a technique that can also be used to determine particle size via the autocorrelation of the scattering signal.

One-dimensional sensors based on arrays of CMOS photodiodes or charge coupled devices (CCDs) are used for optical-scanning applications and for optical detection in spectrographs. Two-dimensional image sensors based on the same technologies form the basis of digital cameras and play a critical role in quality-assured manufacturing where (as the "eyes" of machine-vision systems) they are used to seek out defective products. IR-sensitive image sensors are used in night vision and thermographic imaging systems and even in on-board missile guidance systems. Two-dimensional amorphous selenium and silicon sensors, meanwhile, are rapidly replacing photographic film as the preferred detectors in X-ray imaging systems due to the immediacy of image replication, superior dynamic range, and easier archiving and data retrieval.

The above examples cover just a small fraction of the uses for photodetectors, and in choosing which detector to use for a given application, it is important to take into account both the optical demands of that application and wider issues relating to the circumstances in which the detector will be used. In data comms, the intensity of the optical signal is usually quite high, and the speed of response tends to be the overriding consideration. In sensing applications, other issues such as sensitivity, linearity, dynamic range, and spectral range are often more important. In some cases, tolerance to harsh operating conditions may be required, including resilience to corrosive chemicals or extremes of temperature. In others, issues such as the size, weight, power consumption, or cost of the photodetectors may be important. It is rarely the case that a single technology will meet all of an engineer's design criteria perfectly, and selecting a suitable detector frequently comes down to finding an acceptable compromise between many competing criteria.

It is for the above reasons that the emergence of a new detection technology is important. OPDs will offer their own mix of advantages and disadvantages, making them superior for some applications and inferior for others. It is likely that new applications will arise for which conventional detectors are either technically or commercially unsuited, opening up completely new technological opportunities. We shall speculate in later sections about what these opportunities might be, but we start by reviewing the main photodetectors in current use.

6.2 Conventional Photodetectors

Photomultiplier tubes (PMTs) are vacuum tube devices that use the photoelectric effect to convert photons to electrons (Fig. 6.1).[1, 2]

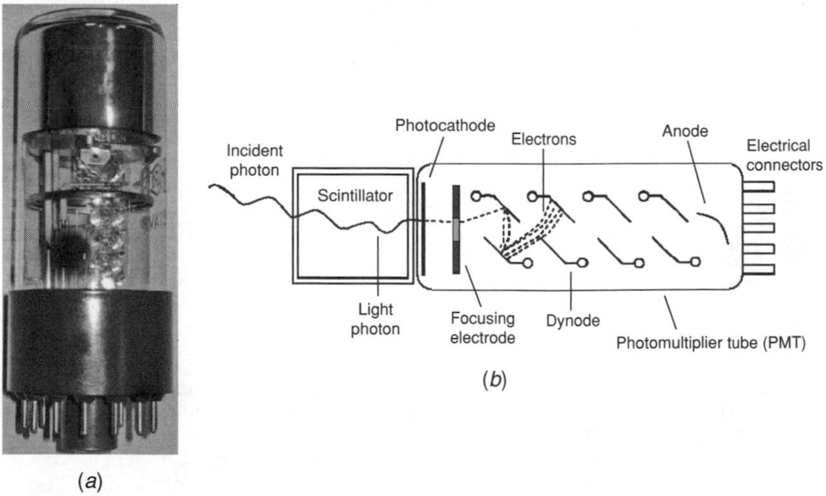

FIGURE 6.1 (*a*) Typical photomultiplier tube in vacuum housing. (*b*) Schematic of PMT, indicating current gain via secondary electron emission at dynodes. (*Picture courtesy of wikipedia*.)

Incident photons strike a photocathode, which then emits electrons into the surrounding vacuum. The emitted electrons are directed by a focusing electrode onto a series of electrodes known as dynodes that are held at successively higher potentials. When the electrons strike the dynodes, secondary emission occurs, causing the number of electrons (and hence the current) to multiply rapidly. The cascade of dynodes generates a virtually noise-free gain that can exceed 1 million with a bandwidth of more than 1 GHz. As a result, PMTs are an excellent choice for measurements that must be made at high speed or in low light levels.

Photodiodes are solid-state devices that are normally based on either p-n or p-i-n type architectures (Fig. 6.2).[2] In a p-n junction, the absorption of photons with energy greater than the semiconductor band gap generates an electron in the conduction band and a hole in the valence band. A high electric field exists in the depletion layer of the p-n junction which drives the photogenerated electrons and holes in opposite directions toward their respective electrodes, from where they are then extracted into the external circuit in the form of a current. The p-i-n devices behave in a similar way, except a lightly doped layer of near-intrinsic semiconductor is "inserted" between the p- and n-doped regions. This increases the width of the (high field) depletion layer and so increases the photoactive thickness of the device, resulting in improved efficiency. In addition, since the electric potential is dropped over a greater distance, the junction capacitance is reduced, which, as we will see later, allows for faster performance.

In normal operation, p-n and p-i-n photodiodes generate at most one electron-hole pair for every absorbed photon, but they can be made to exhibit internal gain if they are operated at sufficiently high reverse biases. In this mode of operation, the photogenerated electrons and holes collide with atoms in the semiconductor crystal, generating

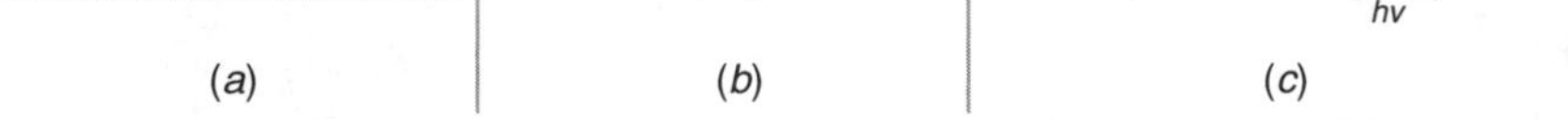

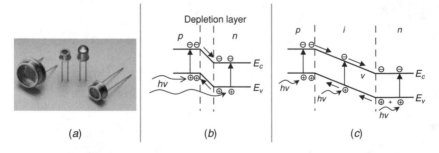

Figure 6.2 (*a*) Typical Si photodiodes. (*b*) Band diagram of p-n type photodiode; the main photoactive region is defined by the high field depletion zone which drives electrons and holes to the cathode and anode, respectively. (*c*) Band diagram of p-i-n type photodiode; the central intrinsic layer increases the thickness of the high field region, resulting in an extended photoactive region and improved photosensitivity. The increased thickness also reduces the capacitance of the device, resulting in faster response. (*Image (a) courtesy of Hamamatsu Photonics KK., All Rights Reserved.*)

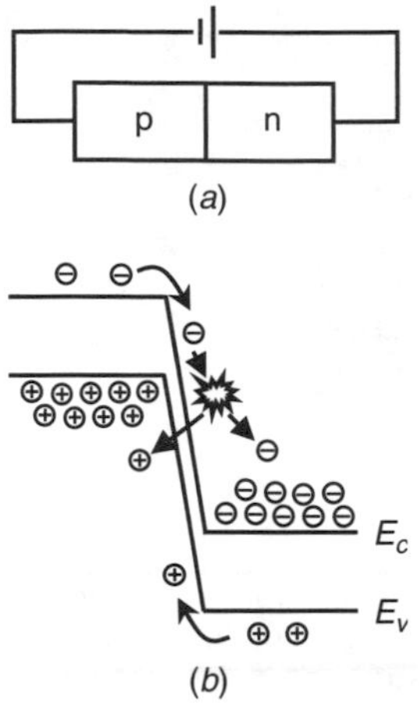

FIGURE 6.3 (a) Typical silicon avalanche photodiode. (b) Schematic indicating gain mechanism in an APD; under a strong applied bias, an electron or a hole collides with an atom of the lattice, creating an additional electron-hole pair.

secondary carriers which amplify the current; such devices are known as avalanche photodiodes or APDs (Fig. 6.3).[3] The overall gain G is determined by the field-dependent *impact ionization coefficients* $\alpha_n(E)$ and $\alpha_p(E)$, which represent the average distance traveled by an electron and hole before generating a new electron-hole pair via impact ionization. If the width of the depletion zone is W, it can be shown[2, 4] that

$$G = \frac{(1-\alpha_p/\alpha_n)\exp[\alpha_n W(1-\alpha_p/\alpha_n)]}{1-\alpha_p/\alpha_n \exp[\alpha_n W(1-\alpha_p/\alpha_n)]} \tag{6.1}$$

Unlike cascade processes in PMTs, impact ionization generates significant noise (although not nearly as much as would be introduced by an equivalent external amplifier). This is so because the electron-hole pairs collide with the crystal atoms at random locations throughout the depletion zone and so undergo different amounts of multiplication. In a PMT, by contrast, gain occurs only at the discrete locations defined by the dynodes. The excess noise factor F in an APD is equal to[2, 4]

$$F = G\left(\frac{\alpha_p}{\alpha_n}\right) + 2\left(1-\frac{1}{G}\right)\left(1-\frac{\alpha_p}{\alpha_n}\right) \tag{6.2}$$

This expression is largest when $\alpha_n = \alpha_p$ in which case the noise factor is equal to the gain G (and there is thus no signal-to-noise benefit to

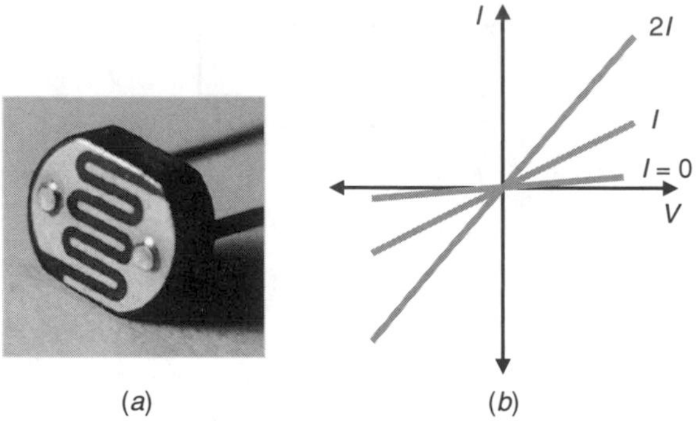

FIGURE 6.4 (*a*) Typical photoresistor. (*b*) The conductivity of a photoresistor increases in approximate proportion to the intensity of light, providing a simple low-cost light sensor.

be derived from the internal amplification); F is minimized if either α_n or α_p is equal to zero, in which case the noise factor reduces to 2. Intricate device architectures have been developed to maximize the difference in the ionization coefficients and so approach the ideal situation $F = 2$ (which, for a typical gain of 100, would yield a 50-fold improvement in the signal-to-noise ratio). The relative complexity of APDs and their stringent manufacturing requirements lead to high fabrication costs, with high-end avalanche photodiodes often selling for several thousand U.S. dollars.

Photoresistors (Fig. 6.4)[2]—also known as photoconductors or photoconductive cells—are closely related to photodiodes in the sense that they too use a semiconductor as the photoactive medium. However, unlike photodiodes, they contain no junction and so provide no driving force for charge separation in the absence of an applied electric field. The density of photogenerated charge and hence the photoconductivity increases in direct proportion to the intensity of incident light. Photoresistors are typically fabricated by a simple two-step process, in which a thick layer of photosensitive material such as CdS is first deposited onto a substrate, and interdigitated metal electrodes are then deposited by thermal evaporation on top of the active layer. Photoresistors have the benefit of being tolerant to high operating voltages of several hundred volts, enabling direct AC usage and, due to their simplicity of fabrication, are extremely low-cost devices. However, they are inherently less sensitive than photodiodes since they require an applied voltage to operate (which generates a dark current and associated noise, see Sec. Shot Noise), and they also tend to respond slowly on a >10 ms time scale. Photoresistors made from mercury telluride and cadmium telluride offer a means of accessing the 2 to 15 μm range of the optical spectrum (which is inaccessible to photodiodes), although the costs of these devices tend to be much higher than those based on CdS.†

Just as all diodes are photoactive to some extent, so too are all transistors. Phototransistors[2] operate in a similar way to photodiodes except, due to their transistor operation, they exhibit current gains of up to a few thousand. (Note that this is a conventional amplification effect as opposed to the low-noise internal gain exhibited by PMTs and APDs.) In operation, a voltage is applied between the collector and the emitter, and when the transistor is illuminated, a photocurrent flows between the base and the collector, which in turn causes an amplified current to flow between the collector and the emitter. Photodarlingtons are closely related devices, comprising a photoactive input transistor followed by a secondary transistor, which can provide increased levels of current gain upward of several hundred thousand. When phototransistors and photodarlingtons are used in conjunction with suitable load resistors, their built-in gain allows them to drive TTL and CMOS logic gates directly, which simplifies circuitry considerably. However, the need for an applied bias between the base and the collector gives rise to a constant (dark) current even in the absence of light. This is usually several nanoamperes and sets an approximate lower limit for the detectable photocurrent. As well as having higher dark currents, phototransistors tend to exhibit slower response speeds and worse linearity than photodiodes, and suffer from significant device-to-device variability. In virtually all cases, superior performance is achieved by separating the detection and gain stages, using a photodiode for the former and a discrete amplifier for the latter (although this clearly entails some additional cost and circuit complexity).

The last kind of photodetector in common use is the charge-coupled device (CCD) (Fig. 6.5)[2] which forms the basis of virtually all digital cameras and image sensors. The term CCD formally refers to the specific architecture that is used to shuttle charges around the sensor array rather than the individual photosensors themselves. CCDs were originally conceived in the 1960s as a new kind of memory circuit, but were soon applied to other tasks such as signal processing and imaging. Today they are no longer of interest as memory elements (having been surpassed by a variety of superior technologies), but they have evolved into the current solution of choice for imaging applications. In essence an imaging CCD is an array of metal-insulator-semiconductor photocapacitors, which collect and maneuver photogenerated charges. The CCD is usually "built" on a grounded p-type silicon substrate (the semiconductor) with a surface

---

†*Note*, many countries are phasing out the heavy metals commonly used in photoresistors for environmental reasons.

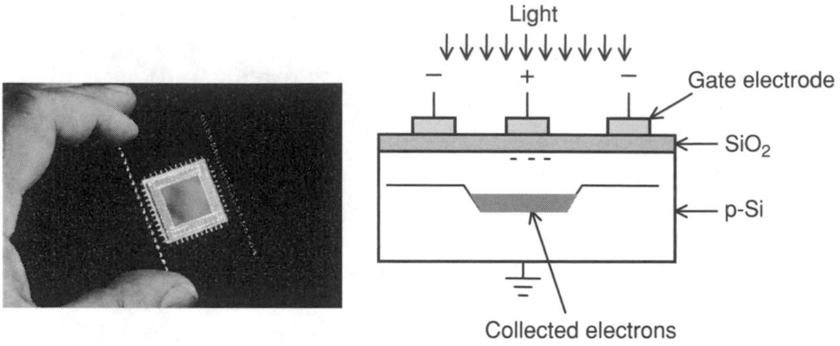

(*a*) (*b*)

FIGURE 6.5 (*a*) Typical charge-coupled device. (*b*) Schematic of a light-sensitive pixel; electrons and holes are photogenerated in the p-Si layer and driven to the p-Si/SiO_2 interface and ground electrode, respectively. The gate electrode of the pixel is held at a higher potential than the gate electrodes of the adjacent pixels, and so electrons are trapped at the p-Si/SiO_2 interface and accumulate in direct proportion to the cumulative photon exposure.

coating of silicon oxide (the insulator), on top of which sits an array of conductive polysilicon "gate" electrodes (the "metal") (Fig. 6.5*b*). By setting the potential of a given gate electrode positive with respect to its immediate neighbors, a potential well is constructed in which photogenerated electrons can collect. The positive charge on the gate electrode drives (intrinsic) holes away from the semiconductor/oxide interface, creating a depletion zone. When the pixel is illuminated, electrons and holes are generated in the depletion zone and driven in opposite directions by the electric field—electrons to the semiconductor/oxide interface and holes to the ground rail. The electrons accumulate at the interface in direct proportion to the number of incident photons. Hence, if the charge on the photocapacitor is measured after a given time, the integrated photon count can be determined. This is done by shuttling the charges pixel by pixel toward a single charge-to-voltage amplifier which processes the data from each pixel in sequence. The major cause of noise in CCDs is thermal generation of electron-hole pairs which causes additional electrons to accumulate at the semiconductor/oxide interface. The rate of photogeneration must substantially exceed the rate of thermal generation if a reliable measurement is to be obtained, and for this reason a thermoelectric cooler is frequently employed to reduce the temperature of the CCD.

The optimum detector for a given application is governed by many considerations, including spectral response, sensitivity, dynamic range, speed, active area, and cost. PMTs offer high levels of virtually noise-free gain and are therefore the preferred choice for ultralow light level detection. They are also currently the only choice for very large area applications, where they offer photoactive surfaces of up to 5000 cm^2. On the negative side, PMTs require high operating voltages of several thousand volts, have relatively narrow spectral ranges, and are bulky, fragile, and expensive due to the need for vacuum housing. APDs are compact solid-state alternatives to PMTs, although they exhibit markedly lower gains and bandwidths. They are often used when light levels are too high for photomultiplier tubes but too low for photodiodes. They too are very expensive. In contrast to PMTs and APDs, photodiodes have no internal gain and, for low light level measurements, usually require an external amplifier to increase the signal to a manageable level. As we shall see later, external amplification is a major cause of measurement noise, and photodiodes are therefore best suited to detection in fairly high light level conditions where amplifier noise is negligible. On the positive side, photodiodes are more robust than APDs and PMTs, have very low operating voltages and wide dynamic range, and are fairly cheap due to their simple fabrication. Phototransistors offer an integrated alternative to photodiode/amplifier combinations and are a good choice for fairly high light level applications where simplicity of design and low cost are important considerations, and high linearity is not required. Photoresistors are ideal for ultralow cost applications that do not require high levels of sensitivity, and for this reason tend to be the solution of choice for the toy industry where economy of fabrication is paramount. CCDs are the dominant technology for one- and two-dimensional sensor arrays, although photodiode arrays are gaining rapidly in importance. At present, there is little to distinguish CCD and diode arrays in terms of performance and cost. CMOS detectors have a slight edge in terms of responsivity and speed, whereas CCDs have a marginal advantage in terms of dynamic range and pixel uniformity.

The above discussion has highlighted the main types of photodetectors in common use today, and it is natural to ask at this point whether there is really a need for a new detection technology at all. OPDs are photodiodes with no internal gain and so—in their current form at least—are unable to compete with PMTs and APDs in terms of sensitivity. They are also high-capacitance devices based on low-mobility materials and consequently exhibit slow response times compared to PMTs, APDs, and conventional inorganic photodiodes. On the other hand, their favorable processing characteristics may open up new applications that are currently unviable, e.g., in large-area light detection. The largest single-element photodetectors are 20 in PMTs, and the largest imaging arrays are ~1600 cm^2 amorphous silicon panels. Using printing techniques, it should be possible to create organic photovoltaic (OPV) devices of much larger area at greatly reduced cost. In addition, by fabricating OPV devices on flexible substrates, conformal photodetectors can be created that adapt to the shape of the surface on which they are mounted. A critical area where

OPV devices may in the future have an advantage over other technologies is in spectral range. OPDs with a spectral range 350 to 1000 nm have already been reported,[5] which is better than typical multialkali-based PMTs (300 to 850 nm)[†] and approaches that of UV-enhanced silicon photodiodes (190 to 1100 nm). We are confident that OPDs with greatly enhanced spectral ranges will be developed in due course. Finally—and this is a point that we will consider in some depth in the following pages—organic photodiodes may one day become the solution of choice for low bandwidth low light-level optical detection, offering comparable performance to the silicon photodiode at a price point close to that of the photoresistor.

6.3 OPV Devices

The vast majority of organic photodetectors can be classified as photodiodes, in that they (1) generate a photocurrent even in the absence of an applied bias and (2) exhibit no internal gain (i.e., at most one electron is extracted into the external circuit for each absorbed photon). In addition, there have been a few reports of organic phototransistors (OPTs)[6, 7] which, like their inorganic counterparts, combine photosensitivity with current amplification. OPTs have been less well studied than even organic photodiodes and will not concern us here, although they may turn out to have important applications in image sensors due to their ease of fabrication.

6.3.1 Device Architectures

There are two main device architectures for organic photodiodes: the discrete heterojunction (Fig. 6.6*b*) and the bulk heterojunction (Fig. 6.6*c*). The simplest discrete heterojunction devices consist of a bilayer of a hole transporting material (the donor) and an electron transporting material (the acceptor) sandwiched between two electrodes. The simplest bulk heterojunction devices consist of a single blended layer of the donor and acceptor, again sandwiched between two electrodes. The key requirement in both cases is that the frontier orbitals of the two organic materials be substantially offset to encourage electron transfer from the donor to the acceptor or hole transfer in the opposite direction (Fig. 6.6*a*). The result is a partitioning of the holes and electrons into the donor and acceptor phases, respectively, from where they can be transported to the relevant electrodes.

The charge separation step does not in itself guarantee exciton dissociation since the electron-hole pairs may remain coulombically

[†]Note that Ag-O-Cs photocathodes are sensitive from 400 to 1200 nm, but they have low peak quantum efficiencies of just 0.36% compared to ~20% for alkali-based photocathodes. However, PMTs based on Ag-O-Cs can still be a viable choice for low level light detection due to the low-noise nature of the gain process.

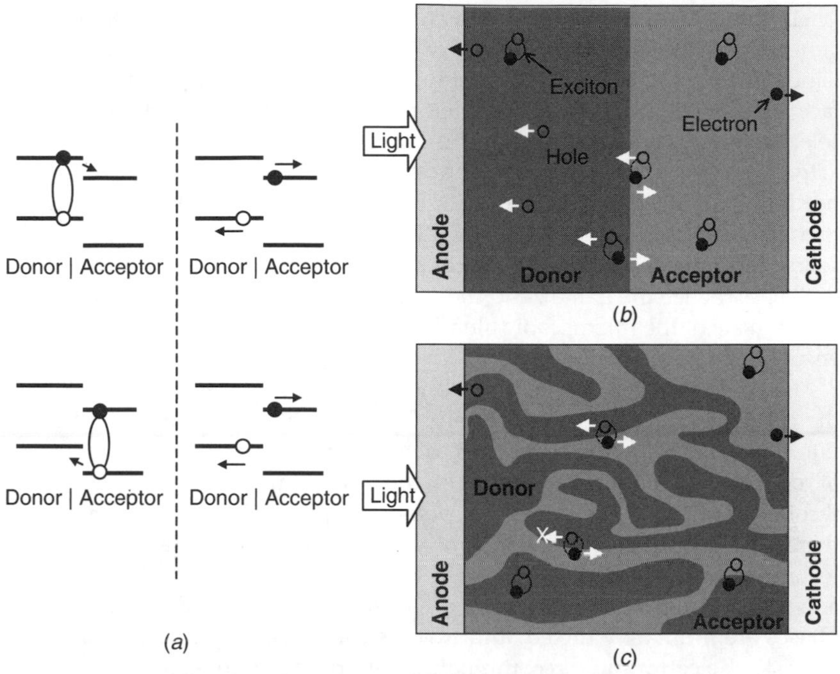

Figure 6.6 (*a*) Energy level diagrams for organic donor and acceptor materials. Photoexcitation of the donor creates a tightly bound electron-hole pair known as an exciton. The electron from the exciton can lower its energy by passing to the lower-lying LUMO level of the acceptor, splitting the exciton; excitons in the acceptor are split when a hole transfers to the higher-lying HOMO level of the donor. (*b*) Schematic of a discrete heterojunction device; excitons created close to the donor/acceptor interface are split into free electrons and holes that are then transported to the cathode and anode by the acceptor and donor layers, respectively. (*c*) Schematic of a bulk heterojunction device, in which the donor and acceptor materials are blended together on a nanometer length scale; all excitons are created close to an interface, resulting in a high yield of free carriers; for efficient operation, continuous pathways must exist from the point of generation to the electrodes, otherwise charges become trapped at dead ends and eventually recombine.

bound in the form of intermolecular charge-transfer excitons even after partitioning has occurred, leaving them susceptible to eventual geminate recombination. To achieve complete dissociation, the electrons and holes must (at a pictorial level) gain sufficient kinetic energy in the charge-transfer process to overcome their residual attraction, which is thought to require an offset† of at least 0.5 eV in the energies of the relevant frontier orbitals.[8]

†Unlike solar cell applications where one wishes to avoid excessive energy loss during the charge-transfer process to maintain high power conversion effciencies, in photodetector applications it is beneficial to make this offset as large as possible so as to maximize the free carrier yield.

In the case of discrete heterojunction devices, only those excitons generated within a diffusion range of the interface yield free carriers (since the others are lost by recombination before reaching the interface). Hence, to ensure high device efficiencies, it is important to select materials in which the diffusion range is comparable to the active layer thickness (~100 nm).[9] (Additional blocking layers may be needed to prevent excitons from diffusing to the electrodes where they are liable to be quenched.) In the case of bulk heterojunction solar cells, donor/acceptor interfaces are formed throughout the film thickness (hence the name), and the main requirement is to match the length scale of the microscopic blend to the exciton diffusion range to ensure efficient dissociation.[10, 11]

Once dissociated, the electron and hole must be transported to the cathode and anode, respectively, from where they are extracted into the external circuit in the form of an electric current. In the case of discrete heterojunction devices, the holes and electrons pass through pure donor and acceptor phases, respectively, and they are therefore channeled efficiently to the electrodes with minimal risk of recombination (except when they are in close proximity to the donor/acceptor interface). In the case of bulk heterojunction devices, the two phases are intimately mixed, and there is consequently an appreciable risk of electron-hole recombination before the charges reach the electrodes. To minimize this risk, continuous percolation pathways are required from the point of generation to the electrodes in order to shuttle the charge carriers rapidly to the electrodes before they have an opportunity to recombine (see, however, the discussion in Sec. 6.2). In spite of the apparent difficulty of avoiding recombination in such circumstances, a suitably optimized bulk heterojunction device can exhibit short-circuit quantum efficiencies approaching 100%.

6.3.2 Device Fabrication

The first successful demonstration of an OPV device based on the discrete heterojunction architecture was reported by Tang[12] in 1986. The active layers consisted of successive vacuum-deposited films of copper phthalocyanine (the donor, Fig. 6.7*a*) and a perylene tetracarboxylic di-imide (the acceptor, Fig. 6.7*b*). The device (which, like many of the OPV devices reported to date, was developed as a solar cell rather than a photodetector) exhibited a power conversion efficiency of 1 percent under simulated AM2 illumination, a record value for an organic solar cell at the time and a value that was to remain unchallenged for several years to come.

The most extensively investigated discrete heterojunction devices are those based on organic/fullerene bilayers.[11, 13] The first fullerene to be tested in an OPV application was C_{60} (Fig. 6.7*c*)—a material in which the 60 electrons from the p_z orbitals give rise to a delocalized π system similar to that in conjugated molecules and polymers.[14] C_{60} has an

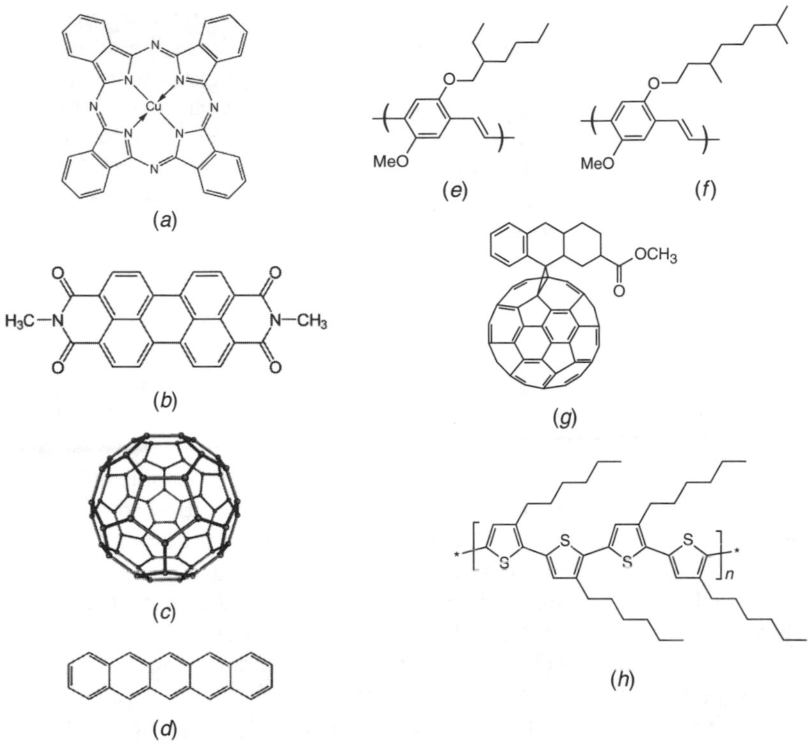

Figure 6.7 Chemical structure of some common PV materials. (*a*) Copper phthalocynanine, (*b*) perylene tetracarboxylic diimide, (*c*) C_{60}, (*d*) pentacene, (*e*) MEH-PPV, (*f*) MDMO-PPV, (*g*) PCBM, and (*h*) P3HT. Perylenes, C_{60}, and its soluble derivative PCBM are widely used acceptor materials in OPV devices due to their low-lying LUMO levels and good electron transport properties. MEH-PPV, MDMO-PPV, and P3HT are widely used donor materials.

optical gap in the range of 1.5 to 1.8 eV, although this transition is only weakly allowed in the solid state so C_{60} is a relatively poor light absorber. However, what makes C_{60} particularly interesting for OPV applications is its excellent performance as an electron acceptor (which derives from its deep LUMO level at ~4.4 eV)[15] and its high electron mobility of 2.0 to 4.9 $cm^2/(V \cdot s)$.[16] In contrast to many organic materials, electron transport dominates the conduction process in C_{60}.

C_{60} may be readily deposited onto a polymer or molecule film by vacuum sublimation, enabling heterojunction devices to be fabricated with relative ease. The properties of the C_{60}/organic interface have been well studied for many organic materials, and it has been shown that proximate excitons in the donor are efficiently dissociated on a subpicosecond time scale by transfer of an electron to the C_{60} (with an associated quenching of the luminescence efficiency).[17, 18] Heterojunction OPV devices with C_{60} acceptor layers have been reported with a wide range of polymer and molecular donors. Sariciftci et al. [19] demonstrated the first such device in 1993, using soluble MEH-PPV (Fig. 6.7*e*) as the donor material, and they reported a 20-fold improvement in photocurrent relative to devices without C_{60}. Although C_{60} is the most studied of the molecular acceptor materials, alternative acceptors have also been investigated in the context of heterojunction OPV devices. Perylenes (Fig. 6.7*b*) have relatively high electron affinities and are known for their photoconductive properties, having been widely exploited in xerographic applications. Indeed, in many xerographic devices a bilayer photoconductor is used, with the perylene in contact with a hole transport layer. In contrast to most organic dyes, perylenes are very stable and are commonly used as colorants for paints and plastics. Moderately efficient double-layer cells with peak external quantum efficiencies greater than 10% have been fabricated using bilayers of pentacene (Fig. 6.7*d*) and perylenes.[20]

Applying the heterojunction approach to solution-processable polymers is more of a challenge, since the process of depositing the second layer is liable to dissolve and wash away the first layer. One solution is to find materials that are soluble in different "orthogonal" solvents, which allows the acceptor layer to be deposited without disturbing the predeposited donor layer.[21] An alternative strategy is to use a donor material that is prepared via a thermal conversion route, such as poly(p-phenylenevinylene) (PPV), which is subsequently rendered insoluble by curing.[22] There is increasing interest in the ability of certain materials to self-organize into relatively complex structures, a phenomenon with the potential to significantly reduce manufacturing costs by in effect allowing devices to "build" themselves. For instance, if two polymers with differing polarities are dissolved in a common solvent, they may sometimes stratify into discrete layers when deposited onto a suitably treated substrate, allowing a bilayer structure to be formed in a single deposition step.[23] Alternatively, heterojunction OPV devices may be fabricated by "lamination" of two organic layers.[24] In this approach, the thin films of the donor and acceptor layers are deposited onto separate anode- and cathode-coated substrates, respectively, and the layers are subsequently fused together under heat and pressure. Adhesion between the organic layers may be promoted by mixing a small quantity of the electron acceptor into the hole acceptor and vice versa before deposition.

The above techniques notwithstanding, the vast majority of solution-processed solar cells and photodetectors are based around the bulk heterojunction architecture, in which a composite layer of the two materials is cast directly from solution.[25–27] For polymer systems, the donor and acceptor polymers can be simply mixed together in the same solvent and deposited by, e.g., spin-coating or printing as with a simple single-layer device. For small-molecule devices, the distributed architecture can be achieved through codeposition of donor and acceptor materials in vacuum,[28] although some solution-processable small-molecule materials are also available.[29, 30] The performance of bulk heterojunction devices is intimately related to the morphology of the donor/acceptor blends. A fine-structured mixing of the donor and acceptor materials is preferred for efficient exciton dissociation, as this enhances the total area of the donor/acceptor interface, and the excitons consequently have less far to diffuse to reach a dissociation site. However, the blend morphology also influences the transport of charges back to the collecting electrodes. Ideally, the donor and acceptor materials should form an interpenetrating network, in which complete transport pathways from the dissociation site to the relevant electrode exist for both electrons and holes.[31] Breaks or islands in the network can form trapping sites for charges (see Fig. 6.6), which can be detrimental to device performance since they increase the probability of electron-hole recombination.

The morphology can be influenced through a variety of processing techniques. In the case of small-molecule devices, the morphology depends primarily on the strength of interactions between individual deposited molecules and their thermal energy, which together determine the ease of molecular rearrangement. The morphology may be controlled to some extent by varying the vapor pressure, the substrate temperature, and the deposition rates. In the case of solution-processed polymer devices, the phase separation arises during the deposition of the solution.[23, 32, 33] As the solvent evaporates and the interchain distances decrease, the interactions are increased, and the mixture that exists in solution begins to de-mix to form domains rich in one material or the other. This phase separation is eventually arrested when the interchain interactions become sufficiently strong to prevent further rearrangement. The solvent boiling point and polarity, as well as the ambient conditions during deposition, can all influence the nature of the morphology, as can substrate surface treatments. For example, the surface can be treated to attract the donor material to the (anode-coated) substrate during deposition of the polymer blend solution, ensuring a proper connection of the hole accepting donor regions to the anode.

The majority of bulk heterojunction devices comprise a small-molecule acceptor dispersed in a polymer donor; all-polymer systems are still relatively uncommon due to the relative scarcity of good electron transporting polymers. The most widely used small-molecule acceptors are based on perylenes and solubilized fullerene derivatives such as [6,6]-phenyl-C61-butyric acid methyl ester (PCBM, Fig. 6.7*g*), whereas the donors are often substituted poly(phenylene-vinylene)s (PPVs) and polythiophenes. Initial interest in PPVs was driven mainly by their use in organic light-emitting diodes and the considerable flexibility they provide in terms of chemical structure engineering. The two most widely used PPV derivatives to date have been poly[2-methoxy-5-(3′,7′-dimethyloctyloxy)-1,4-phenylene vinylene] (MDMO-PPV)

(Fig. 6.7*e*) and poly[2-methoxy, 5-(2-ethylhexoxy)-1,4-phenylene vinylene] (MEH-PPV) (Fig. 6.7*f*).[18, 26, 34] In combination with fullerenes, PPV-based devices have been reported to reach external quantum efficiencies of 66% and power conversion efficiencies of 3% under AM 1.5 solar simulation.[35] However, MDMO-PPV and MEH-PPV have relatively low glass transition temperatures of 45 and 65°C, respectively,[36, 37] which can lead to poor thermal stability under intense illumination or in hot ambient conditions. They also exhibit relatively low charge mobilities of ~ 10^{-5} and 10^{-4} $cm^2/(V \cdot s)$, which may exacerbate recombination losses under low field conditions.[19, 38] For these reasons, polythiophenes tend to be the donor of choice in bulk heterojunction devices. The archetypal polythiophene, poly(3,hexylthiophene) [P3HT] (Fig. 6.7*h*), has a high glass transition temperature and a high charge mobility of up to 10^{-2} $cm^2/(V \cdot s)$ due to its crystalline morphology.[39] The P3HT:PCBM donor/acceptor combination is the most widely studied material system to date, and so far it has yielded the best across-the-board device performance (for solar applications) in terms of long-term stability and power conversion efficiencies of 4 to 5%.[39–43] The P3HT:PCBM system will be used as an exemplar of many of the concepts discussed in this chapter. Due to its advanced state of development, P3HT:PCBM can be considered to be a benchmark material against which others should be judged.

The most widely used anode material is the degenerate n-type semiconductor indium tin oxide (ITO), which combines high conductivity with good transparency in the visible part of the spectrum. Indium tin oxide, however, has a spiked morphology that can result in non-uniform current flow though the device and premature aging. It is therefore usual to coat the indium tin oxide layer with a thick layer of a conducting polymer such as poly(3,4 ethylene-dioxythiphene): polystryrene sulfonate (PEDOT:PSS) that acts as an ameliorating layer for the ITO, resulting in substantially improved operating lifetimes. The cathode is usually a thermally deposited metal of low to moderately low work function such as Ca or Al.

6.3.3 Current-Voltage Characteristics

Figure 6.8*a* shows a simple schematic for an OPV device that is connected to a load resistance *R*. We consider the idealized case of a simple single-layer bulk heterojunction solar cell, in which the blended material can be treated as a simple composite semiconductor, in which the HOMO level is derived from the donor and the LUMO level from the acceptor. This approximation allows us to ignore the microscopic structure of the active layer and to analyze device operation using simple ideas from conventional semiconductor physics. There are two extreme situations we can consider: short-circuit and open-circuit. In short-circuit, the load resistance is zero and so presents no obstacle to the flow of charge. This results in the maximum possible photocurrent, known as

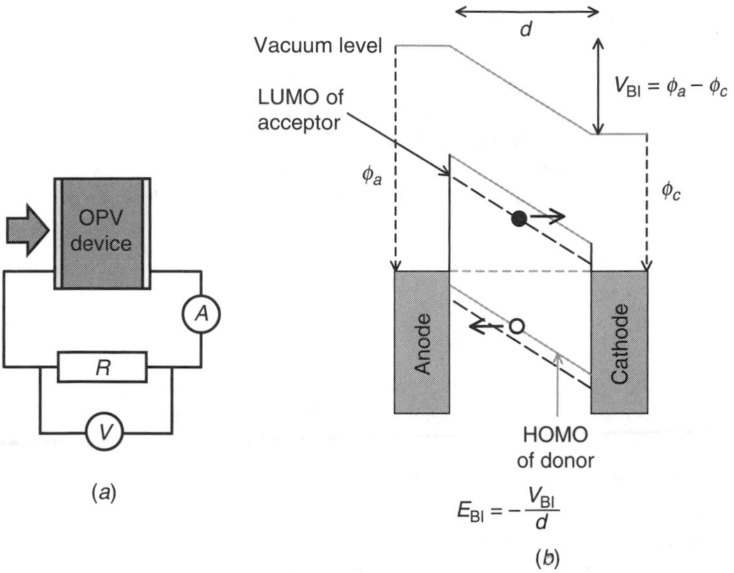

FIGURE 6.8 (*a*) OPV device driving a load resistance *R*; under illumination, the photodiode generates a photovoltage V_{photo} and a photocurrent I_{photo}. (*b*) Energy level diagram for a single-layer solar cell; the electric field exerts a force on the charge carriers that tends to drive the electrons and holes to the cathode and anode, respectively. In a simple device, the electric field strength is determined by the work function difference between the electrodes and the thickness of the device.

the short-circuit photocurrent J_{SC}, and a photovoltage V_{photo} of zero (since the two electrodes are shorted together). The aligned Fermi levels of the two electrodes create a negative internal field $E_{BI} = -V_{BI}/d$, known as the built-in field,[44, 45] that helps drive electrons toward the cathode and holes toward the anode (Fig. 6.8*b*), resulting in a negative photocurrent.

Under weak illumination, the size of the photocurrent is directly proportional to the intensity of the incident light—a property that is essential for most light-sensing applications. The linearity comes about because, at steady state, the rate of free carrier generation by photoexcitation must be exactly balanced by the rate of free carrier loss by extraction into the external circuit. Hence, if the incident light level doubles, so too must the photocurrent to compensate. This equality holds true only in the low-intensity regime where internal losses due to electron-hole recombination in the bulk are negligible (and the charge density is too low to affect the electric field distribution inside the device). In well-optimized devices, recombination effects are appreciable only at high illumination levels exceeding[46]

100 mW/cm^2, so the linear range can extend over many orders of magnitude.

Negligible recombination does not imply a quantum efficiency of 100 percent. The most obvious reason is that only a fraction γ of the incident photons lead to the successful generation of free carriers (due to incomplete absorption and non-dissociative exciton decay channels). A second, more subtle, reason is that some of the photogenerated carriers reach the "wrong" electrodes. To generate the maximum (negative) photocurrent, the photogenerated charges should be extracted only by their "parent" electrodes—electrons by the cathode and holes by the anode. In reality, the electrons and holes move by a combination of diffusion and drift. The diffusive trajectories of the electrons and holes resemble random walks onto which the electric field superimposes ordered drift: diffusion drives charges indiscriminately to both electrodes whereas drift drives the electrons and holes systematically to their parent electrodes. The randomizing effects of diffusion mean that only a fraction α_e of electrons reach the cathode and only a fraction α_h of holes reach the anode. The remaining charges migrate to the "wrong" electrode where they are extracted into the external circuit, giving rise to a positive photocurrent that partially cancels the negative one. Hence, if photons strike the photodiode at a rate $\Re$, the short-circuit photocurrent will equal

$$\frac{I_{SC}}{e} = -\gamma\left(\frac{\alpha_e + \alpha_h}{2}\right)\Re + \gamma\left[\frac{(1-\alpha_e)+(1-\alpha_h)}{2}\right]\Re \tag{6.3}$$

where the first term on the right-hand side corresponds to the negative photocurrent generated by charges that reach the "correct" electrode and the second term corresponds to the positive photocurrent generated by charges that reach the wrong electrode. (The factor 2 in the denominator is required to avoid double counting since for every photogenerated electron-hole pair, at most one electron can be extracted into the external circuit.) Equation (6.3) can be rewritten in the simpler form

$$\frac{I_{SC}}{e} = -\gamma(\alpha_e + \alpha_h - 1)\Re \tag{6.4}$$

which, dividing through by $e\Re$, yields the following expression for the short-circuit quantum efficiency η:

$$\eta(E_{BI}) = -\gamma(E_{BI})[\alpha_e(E_{BI}) + \alpha_h(E_{BI}) - 1] \tag{6.5}$$

In writing this last equation, we have made explicit the dependence of γ, α_e, and α_h on the built-in field strength E_{BI}. The larger the built-in field, the more effective it is in driving the photogenerated charges toward their parent electrodes, and hence the closer in value are α_e and α_h to unity. In addition, in some material systems, an electric field is needed to promote exciton dissociation,[8] causing γ to increase monotonically with increasing field strength. The short-circuit quantum efficiency is therefore optimized by maximizing the built-in field, which requires the use of electrodes with widely differing work functions.[44, 45]

If a load resistance R is now connected between the terminals of the photodiode, a positive voltage

$$V_{\text{photo}} = R \times I_{\text{photo}} \tag{6.6}$$

will appear between the anode and the cathode, where I_{photo} is the observed photocurrent. The appearance of the photovoltage reduces the electric field strength inside the solar cell to a value

$$E = \frac{V_{\text{photo}} - V_{\text{BI}}}{d} \tag{6.7}$$

I_{photo} is always smaller (less negative) than the short-circuit photocurrent I_{SC}, which can be understood by dividing I_{photo} into two parts:

$$I_{\text{photo}} = I_{\text{ph}} + I_{V\,\text{photo}} \tag{6.8}$$

The first part I_{ph} arises from the continuous generation and extraction of free carriers into the external circuit. By direct extension of the argument presented above, I_{ph} is given by

$$\frac{I_{\text{ph}}(E)}{e} = -\gamma(E)[\alpha_e(E) + \alpha_h(E) - 1]\Re \tag{6.9}$$

The second part, $I_{V\text{photo}}$, is due to the photovoltage V_{photo} between the two electrodes. The photovoltage is indistinguishable in its effects from an externally applied positive bias, and so causes electrons and holes to be injected from their parent electrodes into the bulk of the device. This gives rise to a positive current that opposes the negative current I_{ph}. In most devices, under normal operating conditions, $I_{V\text{photo}}$ is identical[46] to the dark current I_{dark}.

$$I_{V\,\text{photo}}(E) = I_{\text{dark}}(E) \tag{6.10}$$

which implies

$$I_{\text{photo}} = I_{\text{ph}} + I_{\text{dark}} \tag{6.11}$$

As the load resistor is progressively increased in value from zero to infinity, the photovoltage increases from its short-circuit value of zero to a maximum value known as the open-circuit voltage V_{OC}. The internal field strength therefore diminishes progressively from a maximum (negative) value E_{BI} to a minimum (negative) value E_{OC} where

$$E_{OC} = \frac{V_{OC} - V_{BI}}{d} \tag{6.12}$$

The increasing photovoltage causes I_{Vphoto} (which is positive) to increase steadily in value, and the decreasing field-strength causes I_{ph} (which is negative) to decrease steadily in magnitude. The observed photocurrent I_{photo} therefore decreases steadily in magnitude from a peak value of I_{SC} at short circuit (where the photovoltage is zero) to a value of zero at open circuit (where the infinite load resistance presents an unsurpassable obstacle to current flow).

The effects of a load resistance can be replicated using a simple voltage source (Fig. 6.9*a*). An applied bias of zero ($R = 0$) is equivalent

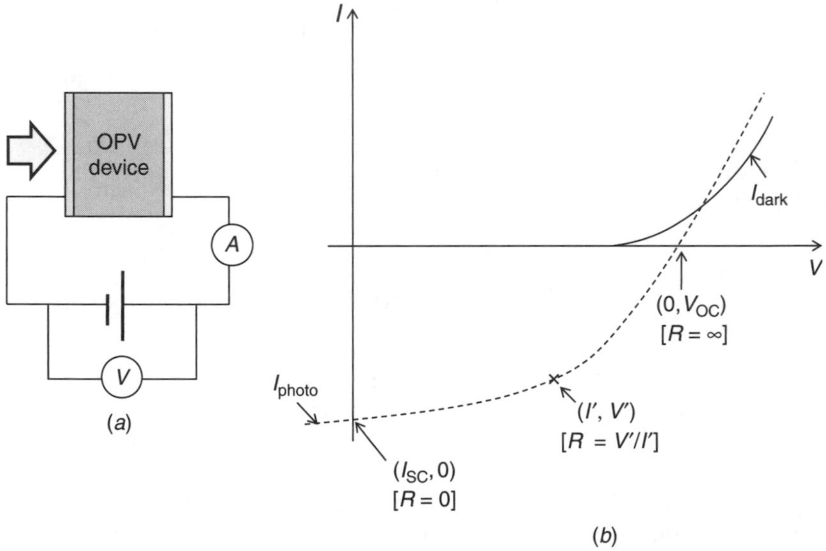

FIGURE 6.9 (*a*) Schematic of an organic photodiode connected to a variable voltage source. (*b*) Model current-voltage curve for a device in the dark and under illumination. The short-circuit current I_{SC} occurs at $V = 0$ (i.e., when the load resistance is zero), and the open-circuit voltage V_{OC} occurs at $I = 0$ (i.e., when the load resistance is infinite). A general point (I', V') on the current-voltage curve corresponds to the current that would be obtained if the device were connected to a load resistance $R' = V'/I'$.

to short circuit, whereas an applied bias of V_{OC} is equivalent to open circuit ($R = \infty$); intermediate applied biases ($0 < V < V_{OC}$) are equivalent to intermediate load resistances ($0 < R < \infty$). In Fig. 6.9*b*, we show the photocurrent as a function of applied bias for a typical device under a fixed illumination level. A general point (I', V') on the photocurrent-voltage curve corresponds to the photocurrent I' and the photovoltage V' that would be obtained if the photodiode were connected to a load resistance $R' = V'/I'$ (keeping the light intensity the same). Also shown on the plot is the dark current I_{dark} which increases quickly with voltage.

The photocurrent curve in Fig. 6.9*b* extends beyond the limited range $0 < V < V_{OC}$ that we considered in the discussion above. Inside this range, the power

$$P = I_{photo} \times V_{photo} \tag{6.13}$$

is negative, which indicates that power is dissipated by the photodiode in the external circuit. This is the relevant regime for solar cells, where we are required to dissipate power harnessed from the sun in an external load resistance. In the case of photodiodes, however, we can apply a bias of any desired size using an external voltage source. If the photodiode is subjected to a reverse bias, we obtain an internal field strength that is larger in magnitude than the built-in field. This has three important benefits. First, from Eq. (6.9), it enhances the value of I_{ph}, giving rise to improved photosensitivity. Second, it sweeps charges more rapidly from the device, shortening response times. Third, it reduces the steady-state charge density inside the device, reducing the amount of electron-hole recombination and so extending the linear range to higher intensities.

These benefits, however, come at an important cost since the application of a reverse bias also generates a (negative) voltage-dependent dark current. The dark current sets a baseline beneath which it is difficult to measure a photocurrent since small fractional drifts in the dark current (e.g., due to temperature changes) can mask the smaller photocurrent; hence, if the dark current is in the nanoampere range, it is difficult to measure photocurrents very much smaller than this. In addition, as we discuss later, biasing the device generates noise in the photodiode which degrades the signal-to-noise ratio (see Sec. Shot Noise). The short-circuit quantum efficiency of a well-optimized organic device can average 25% over its full spectral range. Hence, at best a fourfold increase in quantum efficiency can be obtained by applying a reverse bias. The dark current, on the other hand, can increase by several orders of magnitude when a sizable bias is applied, meaning the slight increase

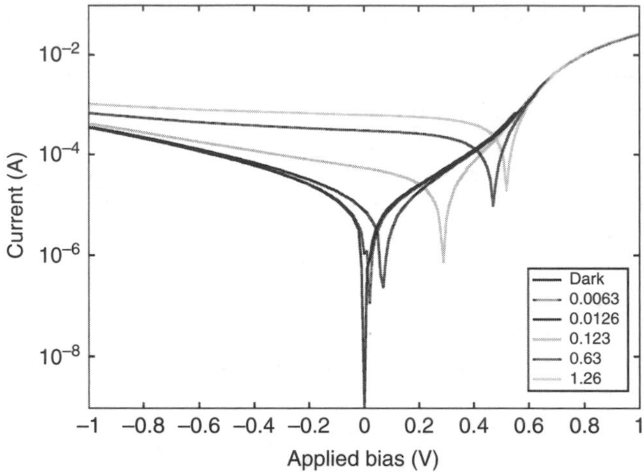

FIGURE 6.10 Current-voltage curves for an ITO/PEDOT:PSS/P3HT:PCBM/Al bulk heterojunction device under varying illumination levels (in milliwatt). (See also color insert.)

in quantum efficiency is strongly outweighed by the large increase in dark current.†

The current-voltage characteristics of a typical ITO/PEDOT:PSS/P3HT:PCBM/Al bulk heterojunction device under white light illumination are shown in Fig. 6.10 for a wide range of intensity levels. To enable all curves to be shown on a single plot, the absolute magnitude of the current is plotted against voltage using a logarithmic y axis. The open-circuit voltage for each light level occurs at the point where the corresponding current-voltage curve dips toward zero. The current is negative to the left of the open-circuit voltage and positive to the right. In Fig. 6.11, the short-circuit photocurrent I_{SC} is plotted as a function of the illumination level; I_{SC} increases linearly with the incident power up to more than 1 mW. Also shown on the plot is the open-circuit voltage V_{OC} which, by contrast, has a strongly sublinear dependence on the incident power. This sublinear behavior can be understood by remembering that

$$I_{ph}(V_{OC}) = -I_{V\text{photo}}(V_{OC}) \tag{6.14}$$

Now I_{ph} varies fairly slowly with voltage, whereas $I_{V\text{photo}}$ shows a strong exponential-like dependency. Hence, if the light level is

†We note in passing that the device can also be subjected to a forward bias, but this results in much larger dark currents than reverse bias and so far worse sensitivity. Hence, photodiodes should only ever be operated in short circuit or reverse bias if an increased speed of response or dynamic range is needed.

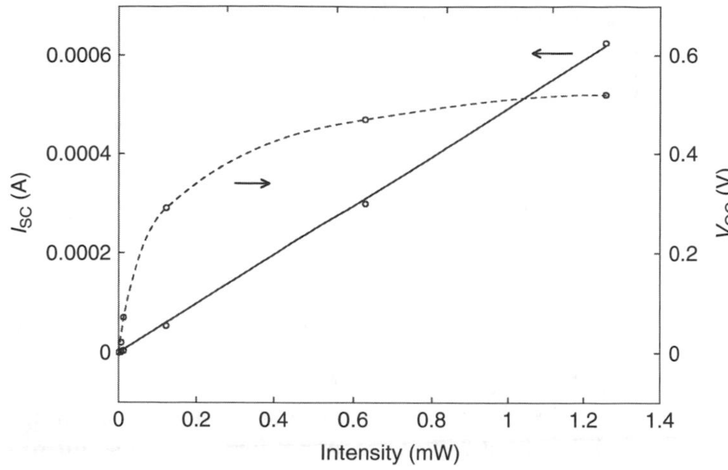

FIGURE 6.11 The short-circuit photocurrent and open-circuit voltage as a function of light intensity for the ITO/PEDOT:PSS/P3HT:PCBM/Al bulk heterojunction device shown in Fig. 6.10.

increased by an order of magnitude, giving rise to an approximate 10-fold increase in I_{ph}, then V_{OC} need change by only a very small amount to generate a compensating increase in $I_{V_{photo}}$. The sublinear intensity dependence of V_{OC} is far less useful than the linear response of I_{SC}. Moreover, V_{OC} inherits from $I_{V_{photo}}$ a strong temperature dependence that makes it far more susceptible to drifts in temperature than I_{SC}. Consequently in light-sensing applications, I_{SC} is virtually always the preferred sense parameter.

In the interests of simplicity, the discussion above has focused on simple single-layer bulk heterojunction devices. The situation is slightly different in the case of discrete heterojunction devices since asymmetry in the generation profiles of the electrons and holes generates sizable concentration gradients at the heterojunction that tend to drive the charges to their respective parent electrodes even in the absence of an electric field.[47] Nonetheless similar arguments to those provided above apply, and the resultant photocurrent-voltage curves are qualitatively similar in appearance. Importantly, our conclusion that I_{SC} is the preferred sense parameter due to its linear response and superior temperature stability also applies to discrete heterojunction devices. This same message also extends to more complex multilayer architectures.

6.3.4 The Equivalent Circuit

A photodiode—organic or otherwise—may be represented conceptually by an equivalent circuit comprising an (infinite impedance) current source I_S in parallel with a diode D, a shunt resistor R_{sh} and a

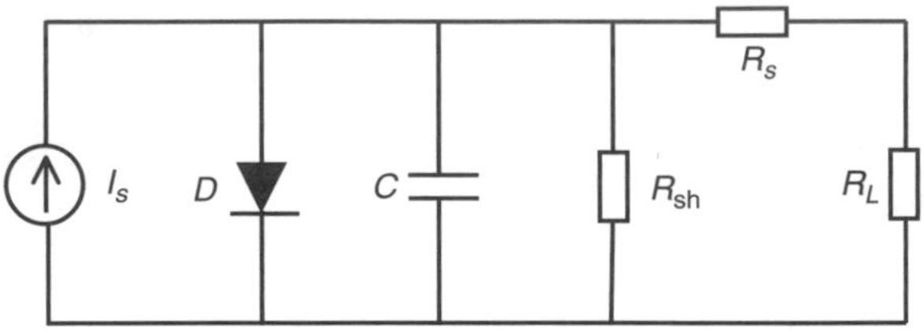

FIGURE 6.12 Equivalent circuit representation of a photodiode.

capacitor C (Fig. 6.12). Also shown is a series resistance R_s due to the resistance of the electrodes, although in many circumstances R_s can be neglected since it is normally just a few tens of ohms. The current source accounts for I_{ph} in Eq. (6.11), and the parallel combination of the diode and the shunt resistor accounts for I_{dark}. The shunt resistance is normally determined by measuring the dark current under a small reverse bias of 10 mV.

$$R_{sh} = \frac{-0.01\text{V}}{I_d(-0.01\text{V})} \tag{6.15}$$

Typical values for the shunt resistances of organic devices range from a few kilohms to a gigaohm or more, compared to around 50 GΩ for a very good quality Si device (Hamamatsu S4797-01). The actual value of the shunt resistance depends on the device architecture, and the care taken in fabrication; poorly made OPV devices tend to exhibit lower shunt resistances due, for example, to spikes of indium tin oxide or filaments of the thermally evaporated cathode that bridge the two electrodes and so allow charge to bypass the (high-impedance) active materials. In a carefully fabricated device, the shunt resistance is determined by the intrinsic transport properties of the active layer material. As we shall see, the shunt resistance has a very significant influence on the photodetector sensitivity.

The capacitance of an organic photodiode can be estimated from the standard formula for the geometric capacitance

$$C = \frac{A\varepsilon_r\varepsilon_0}{d} \tag{6.16}$$

where A = area of the electrodes
ε_r = relative permittivity
ε_0 = permittivity of free space
d = width of the active layer[48]

The *capacitance density* is defined as the capacitance per unit area, and it is a convenient area-independent measure of device capacitance. Using typical values of $\varepsilon_r = 3$ and $d = 100$ nm, we obtain for the geometrical capacitance density a value of ~300 pF/mm^2. This compares with about 30 pF/mm^2 for a typical silicon photodetector (Hamamatsu S4797-01) and about 3 pF/mm^2 for a PIN-type silicon photodetector (Hamamatsu S5821). There is some scope for lowering the capacitance of OPV devices by using thicker films, but this probably allows for a 5- to 10-fold reduction at best (since the use of excessively thick films frustrates the extraction of charge carriers and tends to reduce device efficiencies). Organic photodiodes are therefore relatively high-capacitance devices, which have important implications for their noise characteristics and speed of response.

6.4 Device Characteristics

6.4.1 Spectral Response

The spectral response of a photodetector is normally characterized in terms of either its "photosensitivity" or its quantum efficiency, both of which are ordinarily measured under short-circuit conditions. The *photosensitivity* $S(\lambda)$ at an illumination wavelength λ is defined as the ratio of the photocurrent $I(\lambda)$ to the incident power $P(\lambda)$:

$$S(\lambda) = \frac{I(\lambda)}{P(\lambda)} \tag{6.17}$$

In situations where the photodiode is illuminated by a broadband excitation source, the resultant photocurrent I is given by

$$I = \int_{\lambda_{min}}^{\lambda_{max}} P(\lambda)S(\lambda)\,d\lambda \tag{6.18}$$

The quantum efficiency $\eta(\lambda)$ is the fraction of incident photons that are successfully converted to a photocurrent in the external circuit

$$\eta(\lambda) = \frac{I(\lambda)/e}{\Re(\lambda)} \tag{6.19}$$

where $\Re(\lambda)$ is the rate at which photons of wavelength λ impinge on the device. Clearly, $\Re(\lambda) = P(\lambda)/(hc/\lambda)$. The quantum efficiency and the photosensitivity are therefore related by the following identity:

$$\eta(\lambda) = \frac{I(\lambda)/e}{\Re(\lambda)} = \frac{I(\lambda)/e}{P(\lambda)/(hc/\lambda)} = \frac{hc}{e\lambda}\frac{I(\lambda)}{P(\lambda)} = \frac{hc}{e\lambda}S(\lambda) \tag{6.20}$$

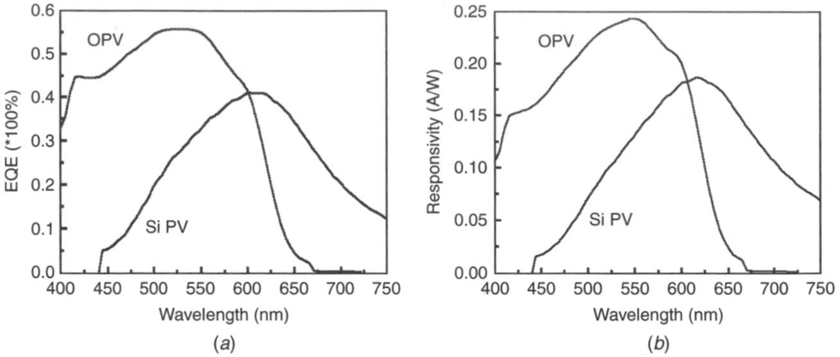

Figure 6.13 (*a*) External quantum efficiency spectral response curves and (*b*) photosensitivity spectral response curves for a Si device and an ITO/PEDOT: PSS/P3HT:PCBM/Al device.

Usually $S(\lambda)$ and $\eta(\lambda)$ are determined by reference to a calibrated photodetector: if, at the illumination wavelength λ, the calibrated photodetector has a photosensitivity $S_{cal}(\lambda)$ and a quantum efficiency $\eta_{cal}(\lambda)$ and the currents from the test device and calibrated detector are $I(\lambda)$ and $I_{cal}(\lambda)$, respectively, then:

$$\frac{S(\lambda)}{S_{\text{cal}}(\lambda)} = \frac{I(\lambda)}{I_{\text{cal}}(\lambda)} \qquad \frac{\eta(\lambda)}{\eta_{\text{cal}}(\lambda)} = \frac{I(\lambda)}{I_{\text{cal}}(\lambda)} \tag{6.21}$$

The spectral response characteristics of a typical ITO/PEDOT: PSS/P3HT:PCBM/Al device are shown in Fig. 6.13 in terms of both photosensitivity and quantum efficiency. Also shown are the spectral response characteristics of a typical Si photodiode (OSRAM, SFH2430) for comparison. The peak efficiencies of the organic device (0.24 A/W, 55.7%) are comparable to those of the silicon device (0.18 A/W, 41.0%), although the active range is much narrower due to the narrow absorption range of the P3HT:PCBM system. (Note that although the silicon photodiode we have chosen is fairly typical, superior devices with substantially higher quantum efficiencies are available; likewise, superior organic devices can also be fabricated.)

6.4.2 Rise Time and Cutoff Frequency

The speed of response is a critical consideration in data communications and time-resolved sensing applications, and it is usually characterized in terms of either the *rise time* or *cutoff frequency*. The rise time of a photodiode is conventionally defined as the time required for the output to change from 10 to 90% of its final level in response to a step increase in the intensity (Fig. 6.14). The fall time is similarly

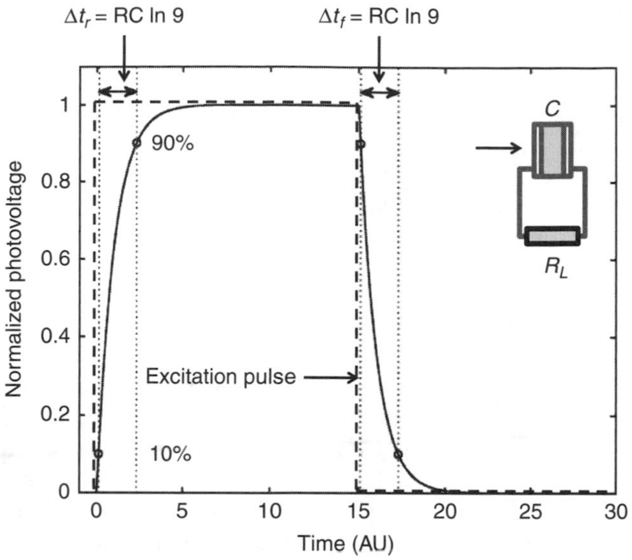

FIGURE 6.14 The theoretical response of an RC-limited photodiode to a square wave light source. The rise and fall times are related to the RC time constant by $\Delta t = \mathrm{RC} \ln 9$.

defined as the time required for the output to change from 90 to 10% of its initial level in response to a step decrease in the intensity. The rise and fall times depend on the wavelength of the incident light (which affects the generation profile) and the value of the load resistance R_L (which is conventionally chosen to be 50 Ω); higher load resistances result in longer response times due to an increased RC time constant (where the relevant capacitance C is the photodiode capacitance).

For an RC-limited process, the time-dependent photovoltage in response to a step increase in the light level from zero is given by

$$V(t) = V_0\left[1 - \exp\left(-\frac{t}{\mathrm{RC}}\right)\right] \tag{6.22}$$

where R = load resistance
C = photodiode capacitance
V_0 = final steady-state voltage

The 10 and 90% points are reached at times t_1 and t_2, respectively, where

$$0.1V_0 = V_0\left[1 - \exp\left(-\frac{t_1}{\mathrm{RC}}\right)\right] \rightarrow t_1 = \mathrm{RC}(\ln 10 - \ln 9) \tag{6.23}$$

and

$$0.9V_0 = V_0\left[1-\exp\left(-\frac{t_2}{\mathrm{RC}}\right)\right] \rightarrow t_2 = \mathrm{RC}\ln 10 \tag{6.24}$$

The rise time Δt_r is therefore given by

$$\Delta t_r = t_2 - t_1 = \mathrm{RC}\ln 9 \tag{6.25}$$

A similar calculation on the falling edge yields the same value for the fall time.

The 3 dB cutoff frequency is defined as the frequency at which the amplitude of the photodiode signal falls to $1/\sqrt{2}$ of its DC value (Fig. 6.15), and may be straightforwardly determined by illuminating the photodiode with sinusoidal light of fixed amplitude and sweeping its frequency f. The cutoff frequency may alternatively be determined by recording the photodiode's impulse response, i.e., its transient behavior in response to an extremely short light pulse. A delta function contains equal portions of all possible excitation frequencies. The Fourier transform of the impulse response therefore gives the gain-frequency response of the photodiode,[49] from which the cut-off frequency can be directly determined.

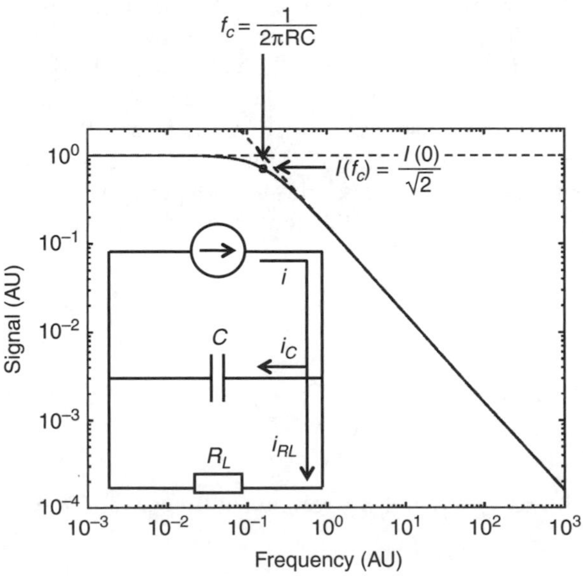

FIGURE 6.15 The theoretical signal vs. frequency response of an RC-limited photodiode. The 3 dB cutoff frequency f_c occurs when the frequency of the incident light is equal to $1/(2\pi\mathrm{RC})$.

The rise time and cutoff frequency are closely related as can be seen from the following analysis. The current i generated by the photodiode is divided into two parts i_c and i_{RL} that flow through the internal capacitance and the external load resistor, respectively (inset to Fig. 6.15):

$$i = i_C + i_{RL} \tag{6.26}$$

(Here, we are assuming $R_L << R_{SH}$ and so we can ignore the negligible current that flows through the shunt resistance.) The photodiode and load resistor are connected in parallel and are consequently subjected to the same potential difference, allowing us to write

$$i_C Z_C = i_{RL} R_L \tag{6.27}$$

where $Z_C = 1/(2\pi jfc)$ is the complex impedance of the capacitance. Hence, we can write

$$\frac{i_c}{2\pi jfC} = i_{RL} R_L \tag{6.28}$$

from which it follows that

$$i_C = i_R 2\pi jfCR_L \tag{6.29}$$

Substituting into Eq. (6.26), we obtain

$$i_{RL} = \frac{1}{1+2\pi jfCR} \rightarrow |\, i_{RL} \,| = \frac{1}{\sqrt{1+(2\pi fCR)^2}} \tag{6.30}$$

The 3 dB cutoff frequency f_C occurs when the signal falls by a factor $\sqrt{2}$, i.e., when $2\pi fCR_L = 1 \rightarrow f = 1/(2\pi CR)$. Hence, in the RC-limited regime, the rise time and the cutoff frequency are related by

$$\Delta t_r = \mathrm{RC} \ln 9 = \frac{\ln 9}{2\pi f_C} = \frac{0.35}{f_C} \tag{6.31}$$

6.4.3 Intrinsic Photodiode Noise Characteristics

Thermal Noise

Above we discussed how photodiodes can be modeled as a current generator in parallel with a diode, a shunt resistor, and a capacitor. In the immediate vicinity of $V = 0$, the slope of the current-voltage curve is linear, meaning it is possible to drop the diode from the equivalent circuit representation and use the shunt resistance alone to account

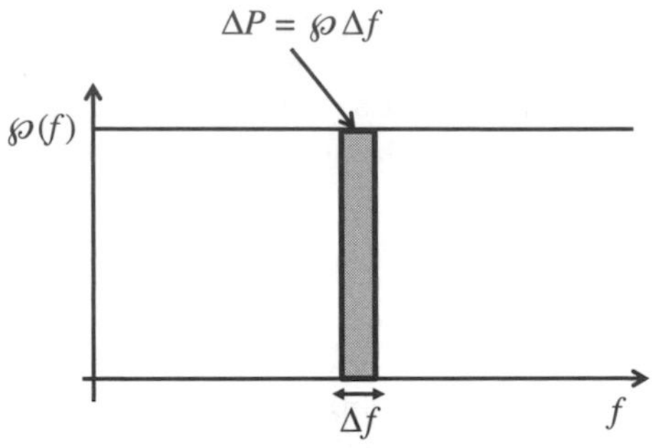

FIGURE 6.16 The noise spectral density for a white noise source. The power is distributed evenly over all frequencies up to some (extremely high) maximum cutoff frequency; if the amplitude of the power spectrum (known as the power spectral density) is $\wp$, then the power ΔP contained within a frequency range Δf is given by $\Delta P = \wp \Delta f$.

for $I_{V\text{photo}}$. The electrons inside any resistor undergo random motion as they are constantly buffeted backward and forward by atomic collisions, leading (in short circuit) to an overall current of zero. This motion results in a constantly changing charge distribution and so induces a randomly fluctuating voltage across the terminals of the resistor, known as thermal or Johnson noise. A process that is random in time† is evenly distributed in the frequency domain. It follows that thermal noise is "white" in its characteristics; i.e., the time-averaged power per unit frequency, known as the power spectral density $\wp(f)$, is the same at all frequencies (Fig. 6.16).‡ The power P dissipated by a resistor is related to its voltage V by $P = V^2/R$. Hence, we can write for the mean squared voltage

$$\langle V^2 \rangle = R\langle P \rangle = R\,\wp(f)\,\Delta f = \widetilde{\sigma_V^2}\,\Delta f \tag{6.32}$$

where $\widetilde{\sigma_V^2} = R\,\wp(f)$ is the voltage variance per unit frequency and Δf is the frequency range of the measurement. Since $\wp(f)$ is constant for a white noise source, so too is $\widetilde{\sigma_V^2}$, and the mean squared voltage is therefore directly proportional to the bandwidth Δf. Note, in this chapter, we will use tilde (~) notation to indicate bandwidth-corrected parameters. The voltage variance $\widetilde{\sigma_V^2} = \widetilde{\langle V^2 \rangle}$ has units V^2/Hz (volts squared per hertz) whereas the corresponding standard deviation $\widetilde{\sigma_V}$ has units $V/\sqrt{Hz}$ (volts per square-root hertz).

†Or, more precisely, one whose autocorrelation is a delta function.

‡In reality, this can be true only up to some maximum cutoff frequency, or else the integrated power would be infinite. In practice, though, the cutoff frequency is sufficiently high as to be of no practical relevance to our discussion.

Unlike a resistor, a capacitor generates no noise.[50] The voltage across the capacitor is fixed the instant the two electrodes are disconnected and thereafter undergoes no further thermal fluctuations, since the isolated plates cannot spontaneously acquire additional charge through thermal motion. It is only when the plates of the capacitor are connected by a resistor that fluctuations can arise. It follows that the only circuit element that contributes directly to the intrinsic noise characteristics of the photodiode is the shunt resistance R_{sh}. It is possible to derive the following expression for the voltage variance per unit frequency due to a resistor R (see Appendix and Usher[50]):

$$\widetilde{\sigma_V^2} = 4Rk_BT \tag{6.33}$$

The thermal noise can be alternatively expressed in terms of $\widetilde{\sigma_I^2} = \widetilde{\sigma_V^2}/R^2$, the mean-squared current per unit frequency flowing through the resistor

$$\widetilde{\sigma_I^2} = \frac{4k_BT}{R} \tag{6.34}$$

The current noise is therefore smallest for large values of the shunt resistance. Importantly, biasing the resistor has virtually no effect on the voltage and current fluctuations since it has the minor effect of superimposing a very small constant drift velocity on the (much larger) instantaneous velocities of the charge carriers. The thermal noise of a discrete resistor is therefore largely independent of the applied bias. In the specific case of a photodiode, the shunt resistor is in parallel with a diode, and both "components" contribute thermal noise (whose variances combine in a sum of squares manner). In short circuit, the diode resistance is much higher than the shunt resistance, and the shunt resistance is therefore the dominant source of thermal current noise; at sufficiently high applied biases, the diode resistance drops substantially, causing it to become the dominant source of thermal noise.

Shot Noise

The second major source of noise is shot noise, which is due to statistical fluctuations in the flowing current caused by the discrete nature of the electrons. Consider an experiment in which we determine the number N of particles that flow through the external circuit in a measurement time Δt. N will fluctuate randomly from one measurement to the next, but we can assign a mean μ_N and a variance $\widetilde{\sigma_N^2}$ for the number of electrons detected per measurement. The mean current I is related to μ_N by

$$I = \frac{e\mu_N}{\Delta t} \tag{6.35}$$

and the current variance σ_I^2 is related to σ_N^2 by

$$\sigma_I^2 = \left(\frac{e\sigma_N}{\Delta t}\right)^2 = \frac{e^2\sigma_N^2}{\Delta t^2} \tag{6.36}$$

In situations where the motion of the individual electrons is completely uncorrelated, the number of electrons detected in the measurement time Δt will be described by a Poisson distribution,[50] for which the mean μ equals the variance σ^2. Therefore the current variance can be written

$$\sigma_I^2 = \frac{e^2\sigma_N^2}{\Delta t^2} = \frac{e^2\mu_N}{\Delta t^2} = \frac{e^2}{\Delta t^2}\frac{I\Delta t}{e} = \frac{eI}{\Delta t} \tag{6.37}$$

Since each measurement takes a time Δt to complete, the bandwidth B, i.e., the highest frequency that can be measured (the Nyquist Frequency) is given by

$$B = \frac{1}{2\,\Delta t} \tag{6.38}$$

The mean-squared variation in the current per unit frequency is therefore given by

$$\widetilde{\sigma_I^2} = \frac{\sigma_I^2}{B} = 2eI \tag{6.39}$$

which, from Eq. (6.11), can be written in the expanded form

$$\widetilde{\sigma_I^2} = 2eI_{\text{dark}}(V) + 2eI_{\text{ph}}(V) \tag{6.40}$$

The first term in Eq. (6.40) is zero when $V = 0$, meaning the shot noise is minimized (and hence the signal-to-noise ratio is maximized) for measurements obtained under short-circuit conditions. This mode (known as the *photovoltaic* mode) is the preferred mode of operation for high-precision measurements. Under an applied bias (the *photoconductive* mode), charges are extracted more quickly, allowing for improved speed and linearity, but this comes at the expense of increased shot noise. The signal-to-noise ratio (SNR) for a signal dominated by shot noise is given by

$$\text{SNR} = \frac{\widetilde{\sigma_1}}{I} = \frac{\sqrt{2eI}}{I} \propto \sqrt{I} \tag{6.41}$$

Shot noise is therefore most important at low photocurrents where the statistical fluctuations in the flowing electrons are most evident. In passing, we point out that Eq. (6.39) was obtained using Poisson statistics and consequently applies only when the electrons behave as independent particles. Interactions between particles can lead to more or less noise than predicted by the shot expression. An important example in inorganic semiconductor devices is the so-called flicker noise, which has a spectral energy density that varies as f^{α} where $\alpha \leq 1$; its precise origin is poorly understood and remains a subject of considerable debate.[50] The effect of flicker noise is to increase the noise level above the shot limit at frequencies less than a few hundred hertz, and whenever possible, it is advisable to work at appreciably higher frequencies. We are aware of no studies investigating the existence of flicker noise in organic photodiodes.

Noise Equivalent Power

The noise equivalent power (NEP) is a useful means of characterizing the sensitivity of a detector. The NEP is the minimum detectable power and is formally defined as the incident power required to achieve a signal-to-noise ratio of 1. In a system dominated by shot and thermal noise, the total noise current per square root of bandwidth is

$$\tilde{\sigma}_I = \sqrt{2eI_{\text{dark}}(V) + 2eI_{\text{ph}}(V) + \frac{4k_B T}{R}} \tag{6.42}$$

To generate a photocurrent of equal size would require an incident power $P(\lambda) = \tilde{\sigma}_I(\lambda) / S(\lambda)$, and hence we obtain for the noise equivalent power per square root of hertz

$$\widetilde{NEP} = \frac{1}{S(\lambda)}\sqrt{2eI_{\text{dark}}(V) + 2eI_{\text{ph}}(V) + \frac{4k_B T}{R}} \tag{6.43}$$

6.5 Measuring a Current

The simplest way to measure a small current is to pass it through a large "sense" resistor and measure the associated voltage drop across the resistor (Fig. 6.17*a*). The current can then be determined by simple application of Ohm's law. There are several problems with this approach, however. To obtain a reasonably sized voltage from a small current, a very large sense resistor is required; e.g., a gigaohm resistance is required to generate a millivolt from a picoampere. The high resistance combines with the high capacitance of the photodiode to create a large RC time constant, which results in a sluggish response. For instance, a photodiode capacitance of 100 pF coupled to a gigaohm sense resistance yields a time constant of 100 ms, making it difficult to measure signals above 10 Hz. A faster response can be obtained by reducing the sense resistor, but the increased speed comes at the expense of reduced gain. The use of a large sense resistor also limits

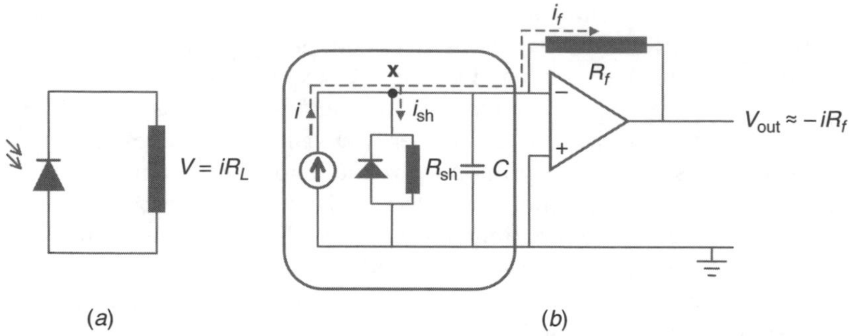

FIGURE 6.17 (*a*) Using a sense resistor to measure a current. The voltage *V* across a known resistor *R* is measured, and the current *i* is determined from Ohm's law. Using a sense resistor has several disadvantages, including poor linearity and slow response. (*b*) Using a transimpedance amplifier to measure a current. The amplifier generates an output voltage $V_{out} = iR_f$, where *i* is the current generated by the current source of the photodiode and R_f is the gain resistor. Using a transimpedance amplifier overcomes many of the drawbacks associated with using a sense resistor (see main text).

the linear range since even a fairly small photocurrent can cause a significant voltage drop across the resistor. This has the effect of biasing the photodiode and so reducing the internal field strength, which in turn lowers the quantum efficiency [see Eq. (6.9)]. So 300 pA through a gigaohm sense resistance would result in a 0.3 V bias across the photodiode—a significant fraction of the built-in potential (which is typically no more than 1 V). The (positive) biasing of the photodiode causes a diminishing quantum efficiency with increasing light intensity, resulting in a problematically sublinear response.

6.5.1 The Transimpedance Amplifier

To avoid the above problems, it is common to use a *transimpedance* amplifier (Fig. 6.17*b*) to convert the photocurrent to an easily measured voltage, an approach that beneficially achieves large signal gain, while also maintaining a fast speed of response and good linearity (until the amplifier reaches saturation). The effectiveness of OPV devices as photodetectors is closely related to their performance in transimpedance circuitry, so it is useful to review here some basic aspects of operational amplifier (op-amp) theory. Our discussion is deliberately brief, and the interested reader is referred to standard texts for further information.[51] It will be sufficient for our purposes to use ideal op-amp theory, in which it is assumed that the input impedances of the two op-amp terminals are infinite and that no potential difference exists between them.

Ideal Op-Amp Behavior

In Fig. 6.17*b* we show a circuit for a transimpedance amplifier, where we have represented the photodiode in equivalent circuit form and, for now, we ignore the photodiode capacitance. The current source of the photodiode generates a current i which is split at node **x** into two parts i_{sh} and i_f that pass through the shunt and feedback resistances, respectively. (Note that no current is able to flow into the inverting terminal of the op-amp due to its infinite input impedance.) Clearly,

$$i = i_{sh} + i_f \tag{6.44}$$

from which it follows directly that

$$i = \left(\frac{V_- - 0}{R_{sh}}\right) + \left(\frac{V_- - V_{out}}{R_f}\right) \tag{6.45}$$

where V_- is the potential at the inverting terminal. After some rearrangement, this yields

$$V_{out} = -iR_f + \left(1 + \frac{R_f}{R_{sh}}\right)V_- \tag{6.46}$$

In ideal op-amp theory, feedback ensures there is zero potential difference between the inverting and non-inverting terminals. Hence, applying $V_- = V_+ = 0$, we obtain

$$V_{out} = -iR_f \tag{6.47}$$

The input current is therefore converted to an output voltage by a transimpedance gain equal in value to the feedback resistor R_f.

In reality, although we have assumed $V_- = V_+$, a very small potential difference V_{out}/A will exist between the input terminals since the output of an op-amp is related to the potentials at its input terminals by the gain equation

$$V_{out} = A(V_+ - V_-) \tag{6.48}$$

where A is the open-loop gain. From the perspective of the photodiode, the transimpedance amplifier therefore resembles a load resistance R_{eff} where

$$R_{eff} = \frac{V_- - V_+}{i} = \frac{-V_{out}/A}{i} = \frac{-(-iR_f)/A}{i} = \frac{R_f}{A} \tag{6.49}$$

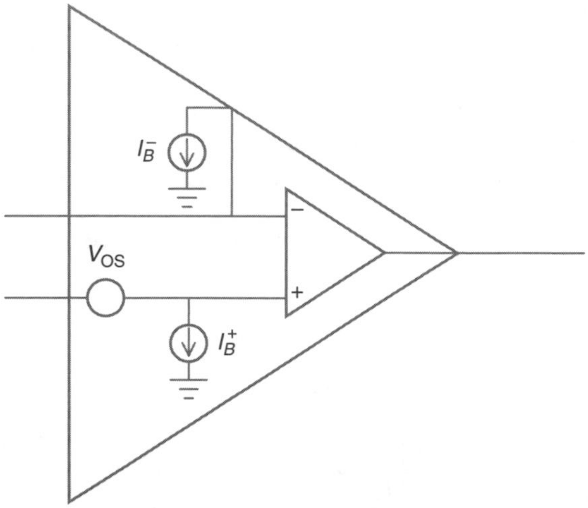

Figure 6.18 Equivalent circuit for a real op-amp. Small input bias currents I_B^- and I_B^+ flow into the inverting and non-inverting inputs of the amplifier, and an offset voltage V_{OS} gives rise to an output voltage even when the input terminals are shorted. These current and voltage offsets lead to systematic errors in the output voltage.

(Note that in this calculation we have ignored the tiny fraction of the current that passes through the large shunt resistance of the photodiode.) In a good-quality op-amp, the open-loop gain can be 100 million or more, so the effective resistance is much smaller than the feedback resistance. This enables a large current-to-voltage gain to be obtained without significantly biasing the photodetector and, hence, without compromising linearity or unduly sacrificing speed of response.†

Amplifier Offsets

In the above discussion we assumed ideal op-amp behavior, but real op-amps exhibit internal noise and offsets that must be taken into account when performing low light level measurements. The behavior of a real op-amp can be modeled using the equivalent circuit in Fig. 6.18. Two current sources are shown at the input terminals of the op-amp which reflect the fact that real op-amps require small steady "input bias" currents to flow at their input terminals to

†The cutoff frequency is now determined by the amplifier response and in particular by the size of the feedback capacitance (if any) that is used in parallel with the feedback resistor to reduce noise and improve stability; see Sec. Amplifier Stability.

operate.[51] The input bias current i_B^- at the inverting terminal is split between the feedback resistor and the shunt resistance of the photodiode in exactly the same way as the current i generated by the photodiode. Therefore i_B^- adds directly to i and experiences the same transimpedance gain R_f, giving rise to a steady systematic offset $\Delta V_{ia}^{\text{out}}$ in the output voltage:

$$\Delta V_{ia}^{\text{out}} = i_B^- R_f \tag{6.50}$$

Also shown in the equivalent circuit is a small voltage source (or "input offset voltage") v_{OS} at the noninverting terminal, which accounts for the fact that real op-amps give out a constant output voltage when their input terminals are shorted. Referring to Eq. (6.46) and writing $V_- = V_+ = v_{\text{OS}}$, the effect of the input offset voltage is to generate a systematic offset $\Delta V_{va}^{\text{out}}$ in the output voltage:

$$\Delta V_{va}^{\text{out}} = \left(1 + \frac{R_f}{R_{\text{sh}}}\right) v_{\text{OS}} \tag{6.51}$$

The amplifier offsets can be minimized by selecting op-amps with the smallest possible values of i_B^- and v_{OS} and by ensuring that the shunt resistance R_{sh} of the photodiode is as high as possible. In addition, most op-amps allow any remaining offset to be trimmed to zero using an externally applied voltage, although this is an imperfect solution due to natural drift in the offset voltage and bias current over time.

Amplifier Noise

The input bias current and the amplifier offset voltage are subject to noise fluctuations. If we define the *input noise current* $\widetilde{\sigma_{Ia}}$ to be the noise per square root of frequency on the input bias current, then by analogy with Eq. (6.50), the corresponding noise on the output voltage is given by

$$\sigma_{Ia}^{\text{out}} = \sigma_{Ia} R_f = \widetilde{\sigma_{Ia}} \sqrt{B} R_f \tag{6.52}$$

where B is the measurement bandwidth. And if we define the *input noise voltage* $\widetilde{\sigma_{Va}}$ to be the noise per square root of frequency on the offset voltage, then by analogy with Eq. (6.51), the corresponding noise on the output voltage is given by

$$\sigma_{Va}^{\text{out}} = \left(1 + \frac{R_f}{R_{\text{sh}}}\right) \sigma_{Va} = \left(1 + \frac{R_f}{R_{\text{sh}}}\right) \widetilde{\sigma_{Va}} \sqrt{B} \tag{6.53}$$

In fact this last equation isn't strictly correct because we have overlooked the capacitance C_d of the photodiode.[†] This can be taken into account by replacing the shunt resistance R_{sh} by the complex impedance $Z_{sh} = R_{sh} \,||\, Z_C$, where $Z_C = 1/(j2\pi f C_d)$. The capacitance introduces a frequency dependence into the gain, and after integrating over all frequencies (see the Appendix), the root mean squared noise voltage at the output becomes

$$\sigma_{Va}^{\text{out}} = \widetilde{\sigma_{Va}}\sqrt{B}\sqrt{\left(1+\frac{R_f}{R_{\text{sh}}}\right)^2 + \frac{4\pi^2}{3}C_d^2 R_f^2 B^2} \tag{6.54}$$

The transimpedance circuitry introduces one other noise source into the final amplified signal: thermal noise from the feedback resistor, which appears at the output with a root mean squared value:

$$\sigma_{Rf}^{\text{out}} = 4k_B T R_f \sqrt{B} \tag{6.55}$$

Combining all three noise contributions, we obtain a total root mean squared noise voltage at the output of

$$V_N = \sqrt{\left(\sigma_{Ia}^{\text{out}}\right)^2 + \left(\sigma_{Va}^{\text{out}}\right)^2 + \left(\sigma_{Rf}^{\text{out}}\right)^2}$$

$$V_N = B^{1/2}\sqrt{\left(1+\frac{R_f}{R_{\text{sh}}}\right)^2 \widetilde{\sigma_{Va}^2} + \frac{4\pi^2}{3}C_d^2 R_f^2 B^2 \widetilde{\sigma_{Va}^2} + R_f^2 \widetilde{\sigma_{Ia}^2} + 4k_B T R_f} \tag{6.56}$$

Hence, dividing through by the amplified signal iR_f, we obtain for the signal-to-noise ratio

$$\text{SNR} = \frac{V_N}{iR_f} = \frac{B^{1/2}}{i}\sqrt{\left(1+\frac{R_f}{R_{\text{sh}}}\right)^2 \frac{\widetilde{\sigma_{Va}^2}}{R_f^2} + \frac{4\pi^2}{3}C_d^2 B^2 \widetilde{\sigma_{Va}^2} + \widetilde{\sigma_{Ia}^2} + \frac{4k_B T}{R_f}} \tag{6.57}$$

It follows from inspection of Eq. (6.57) that to maximize the signal-to-noise ratio, we must

- Maximize the photocurrent i by collecting as much of the incident light as possible and ensuring the photodiode has a high quantum efficiency.
- Select an op-amp with small values of $\widetilde{\sigma_{Ia}}$ and $\widetilde{\sigma_{Va}}$, i.e., one in which the input noise current and input noise voltage are

[†]Note that we didn't need to worry about the capacitance when dealing with the offset voltage because V_{OS} has a constant dc value.

both small; state-of-the-art op-amps have noise values in the $nV/\sqrt{Hz}$ and $fA/\sqrt{Hz}$ range.

- Use as large a feedback resistor as possible; although this increases the thermal noise from the feedback resistance, this drawback is outweighed by the improved signal gain.
- Minimize the measurement bandwidth B, ideally matching it to the signal bandwidth to eliminate extraneous noise at other frequencies; this can be done by adding a bandpass filter after the op-amp to reject unwanted frequencies.
- Maximize the shunt resistance of the photodiode, preferably ensuring it is at least as large as the feedback resistor, in order to minimize the effect of the amplifier voltage noise; fortuitously, maximizing the shunt resistance also minimizes thermal noise from the photodiode.
- Minimize the capacitance of the photodiode (again to minimize the effects of amplifier voltage noise).

To determine what can feasibly be detected using an organic photodiode, we estimate the output noise voltage based on typical performance characteristics for a high-precision low-noise op-amp and for an organic PV device of area 1 mm^2. We assume a noise voltage of 8 $nV/\sqrt{Hz}$ and a noise current 0.6 $fA/\sqrt{Hz}$—the data sheet specifications for the state-of-the-art AD795 op-amp from Analog Devices. We also assume a feedback resistance of 1 GΩ, which is about the highest that can be realistically used, and a shunt resistance of 1 GΩ and a capacitance of 400 pF for the photodiode. Inserting these values into Eq. (6.57) and assuming a modest measurement bandwidth of 1 kHz, we obtain a root mean squared noise voltage at the output of 0.4 mV. This is equivalent to a root mean squared current noise at the input of 0.4 pA, implying, on the basis of the amplifier performance alone, it should be possible to measure currents of ~1 pA and above. The photodiode has a thermal noise of 0.1 pA and shot noise of 0.02 pA (assuming a photocurrent of 1 pA). The amplifier characteristics are therefore the main determiner of sensitivity. The measurement noise is lowered to approximately 0.1 pA if the photodiode capacitance is reduced by a factor of 10. Increasing the shunt resistance by a factor of 10 leads to only a marginal reduction in the measurement noise.

Amplifier Stability

A final issue to note when using transimpedance amplifiers is stability. The photodiode capacitance and the feedback resistance together act as a low-pass filter that introduces a phase lag into the feedback loop, which approaches 90° at high frequencies. The op-amp itself introduces an extra phase lag that also approaches 90° at high frequencies, leading to a combined "lag" of almost 180°. Since the signal is fed back into the inverting input, this leads to positive feedback; in consequence, any high-frequency noise that contaminates the signal can force the amplifier into instability (sustained oscillations). The usual solution is to add a feedback capacitance in parallel with the feedback resistance, which reduces the gain at high frequencies (since the capacitor behaves as a low-impedance short) and so suppresses the high-frequency noise components, albeit at the expense of signal bandwidth. Stability issues can often be solved by careful amplifier selection, some amplifiers being more susceptible to oscillations than others.

6.5.2 The Charge Integrator

In situations where the photocurrent is significantly less than 1 pA, it may not be possible to achieve sufficient gain using a single-stage transimpedance circuit since 1 GΩ is normally considered the highest practical value for the feedback resistor (1 GΩ × 1 pA = 1 mV). One solution is to use a two-stage amplifier, comprising an initial current-to-voltage stage followed immediately by a voltage amplification stage. This has the advantage of providing improved bandwidth since substantially lower gains can be employed at each stage. However, the increased bandwidth comes at the expense of increased noise: if the feedback resistor at the current-to-voltage stage is reduced by a factor α, the signal is reduced by the same factor α, but from Eq. (6.33) the thermal noise is reduced only by a factor $\sqrt{\alpha}$. To achieve the same overall signal gain, the voltage gain at the second stage must be equal to α so the noise contribution from the first feedback resistor is increased by a factor $1/\sqrt{\alpha} \times \alpha = \sqrt{\alpha}$ compared to the single-stage process.

An alternative, lower-noise, solution is to use a (single-stage) current integrator in which the feedback resistor is replaced by a (noise-free) feedback capacitor (Fig. 6.19). In this case, at low frequencies where the impedance of the photodiode capacitance is high, we can write:

$$i(t) = i_{sh}(t) + i_f(t) - i_B^- \tag{6.58}$$

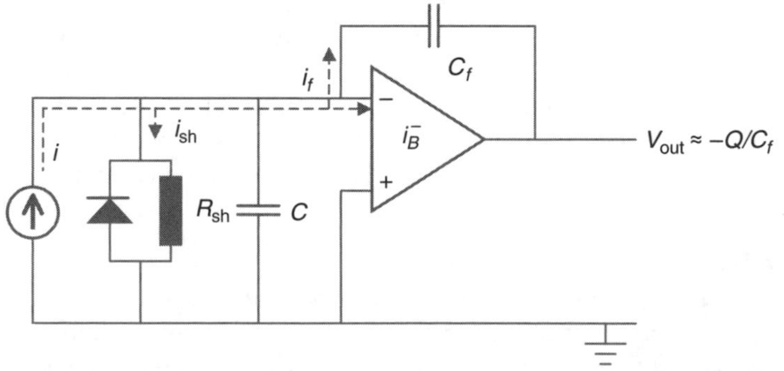

Figure 6.19 Simple circuit for an op-amp integrator, in which the feedback element is a capacitor. Using an integrator is usually the best way of measuring extremely small (sub-pA) currents.

where t = time

i = current generated by the current source of the photodiode

i_{sh} and i_f = currents that flow through the shunt resistance and the feedback resistance, respectively

i_B^- = input bias current, i.e., the current that flows into the inverting input of the op-amp

The current that flows through the feedback capacitor is proportional to the rate of change of the potential difference across it.

$$i_f(t) = C\frac{d}{dt}[V_- - V_{out}(t)] \tag{6.59}$$

Hence, from Eqs. (6.58) and (6.59), we can write

$$i(t) = \frac{V_- - 0}{R_{sh}} + C\frac{d}{dt}[V_- - V_{out}(t)] - i_B^- \tag{6.60}$$

which, on integrating over a measurement time τ, yields

$$\int_0^\tau i(t)dt = \int_0^\tau \left\{\frac{V_-}{R_{sh}} + C\frac{d}{dt}[V_- - V_{out}(t)] - i_B^-\right\}dt \tag{6.61}$$

The left-hand side can be equated with the total charge Q created by the photodiode current source. Hence, we can write

$$Q = \frac{\tau V_-}{R_{sh}} - C\Delta V_{out} - \tau i_B^- \tag{6.62}$$

where

$$\Delta V_{out} = V_{out}(\tau) - V_{out}(0) \tag{6.63}$$

which rearranges to give

$$\Delta V_{out} = -\frac{Q}{C} + \varepsilon\tau \tag{6.64}$$

where

$$\varepsilon = \frac{1}{C}\left(\frac{V_-}{R_{sh}} - i_B^-\right) \tag{6.65}$$

Ignoring ε for a moment, we can see that the voltage change ΔV_{out} is proportional to the total amount Q of charge generated by the current source of the photodiode over the measurement time τ. And ε is an error term arising from the non-ideal behavior of the amplifier (namely, the effects of the DC offset voltage and the input bias current) and increases linearly with the measurement time τ. Error ε is minimized by ensuring that (1) the shunt resistance is as high as possible; (2) the input bias current and dc offset voltage are as small as possible; and (3) the integration period is as short as possible, extending only over the time during which the device is illuminated.

To measure small currents, a small feedback capacitance should be used: a 1 fA input, a 1 pF feedback capacitance, and a 1 s measurement time will, for instance, result in a 1 mV output. In use the feedback capacitance is discharged immediately prior to performing the measurement. As we will see later, current sensing via charge integration is especially useful for photodiode arrays.

6.6 The State of the Art

The message from the above discussion is clear: for fast low-noise amplification, we require photodiodes with low capacitances, high shunt resistances, and high quantum efficiencies over the spectral range of interest. In this section, we consider what is currently achievable using organic photodiodes, and we speculate what might be possible in the future through improved device engineering and materials design.

6.6.1 Capacitance

As noted above, the capacitance densities of organic devices are relatively high, typically of the order of 300 pF/mm^2 compared with around 30 pF/mm^2 for standard Si devices and 3 pF/mm^2 for PIN devices. We are aware of no research that has specifically addressed the issue of minimizing the capacitance density of organic devices (while maintaining good overall device performance). However, since the capacitance density is inversely proportional to both the electrode separation and the relative permittivity of the intervening medium, the obvious solution is to increase either the thickness d or the relative permittivity ε_r of the active layers. (The latter could be achieved by grafting easily polarized side-chain groups onto existing OPV materials, assuming it is possible to do so without unduly compromising their charge transport properties).† The principal challenge in both cases is to maintain a high quantum efficiency in spite of the reduced internal field strength, e.g., through the use of high carrier mobility materials. At the time of this writing, an order of magnitude reduction in capacitance densities to ~30 pF/mm^2 seems quite realistic,

†Conjugated macromolecules with permittivities as high as 900 were studied and reported[52] in the 1960s, but we are not aware of any recent work that has sought to incorporate such materials into devices.

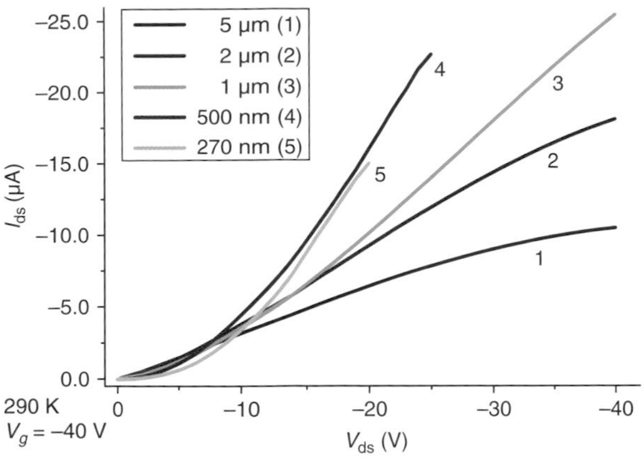

FIGURE 1.6 Drain current as a function of source-to-drain voltage for different channel lengths. The characterization was taken at room temperature and high density of charge carriers ($V_g = -40$ V, well beyond threshold voltages of each channel). For observation of the scaling behavior, the W/L ratios of all channels were kept at the same value of 10 in fabrication to exclude geometric factors. Clearly in the regime of $V_{ds} < V_g - V_{th}$, the current-voltage characteristic transitions from linear to superlinear upon scaling from micron to submicron channel length. For submicron channels there is an exponential dependence at very small V_{ds} due to the injection-limited transport through Schottky barrier at the metal-semiconductor contact. (*Reprinted with permission from Ref. 60. Copyright 2007, American Institute of Physics.*)

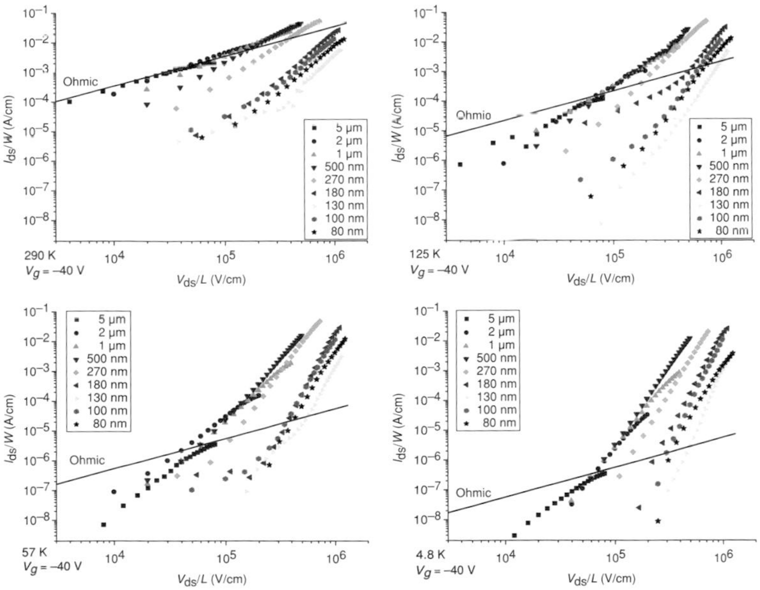

FIGURE 1.8 The current density vs. longitudinal field plots for various channel lengths at four different temperatures. The solid lines in these plots are the ohmic channel transport currents, calculated based on the mobility extracted from long-channel (5 μm) devices. These lines serve as the references to investigate the issues of contact injection-limited transport and field-dependent mobility. All the four figures are exactly on the same scale for the purpose of comparison.

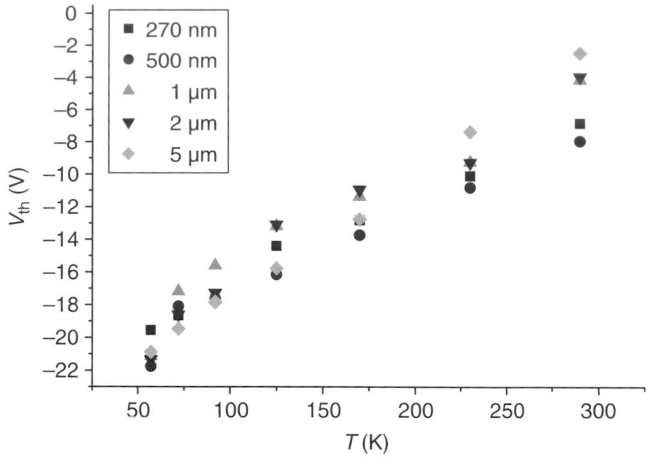

FIGURE 1.9 The temperature dependence of threshold voltage for scaled channel lengths. Threshold voltages V_{th} were extracted from the transconductance plots (I_d vs. V_g for $V_{ds} < V_g - V_{th}$; $(I_d)^{1/2}$ vs. V_g for $V_{ds} \geq V_g - V_{th}$) on high end of V_{ds} for each channel. V_{th} does not significantly change with longitudinal field. Channels of different lengths follow the same trend; namely, V_{th} shifts to high gate bias quasi-linearly with decreasing temperature. (*Reprinted with permission from Ref. 60. Copyright 2007, American Institute of Physics.*)

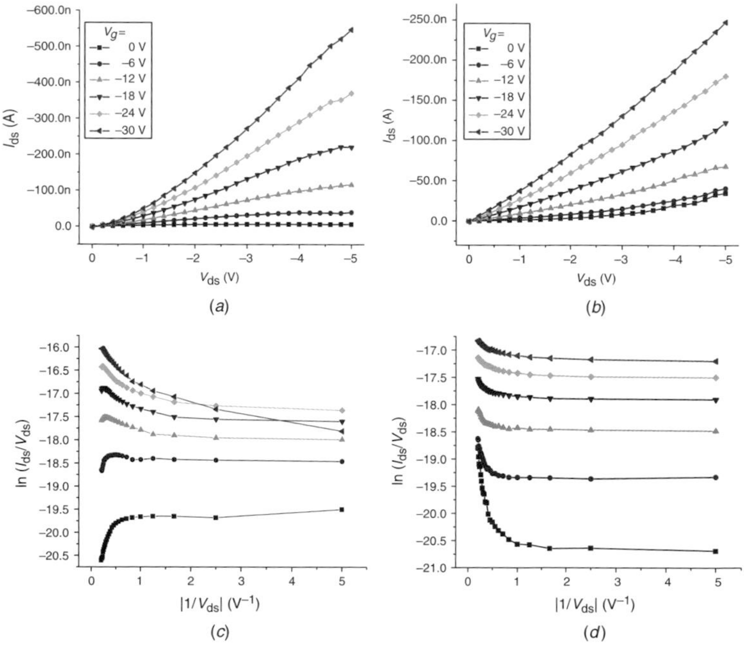

FIGURE 1.14 The DC characteristics of sub-10-nm pentacene FETs. (*a*) The DC *I–V* measurement of the device in Fig. 1.13*b*, with the side guards biased at the same potential as the drain. (*b*) The DC *I–V* measurement of a 19 nm channel device with the side guards biased; I_{ds}/I increases with increasing $|V_g|$. (c) $\ln(I_{ds}/V_{ds})$ vs. $1/|V_{ds}|$ plot of the data in (*a*). (*d*) $\ln(I_{ds}/V_{ds})$ vs. $1/|V_{ds}|$ plot of the data in (*b*). (*Reprinted with permission from Ref. 40. Copyright 2004, American Institute of Physics.*)

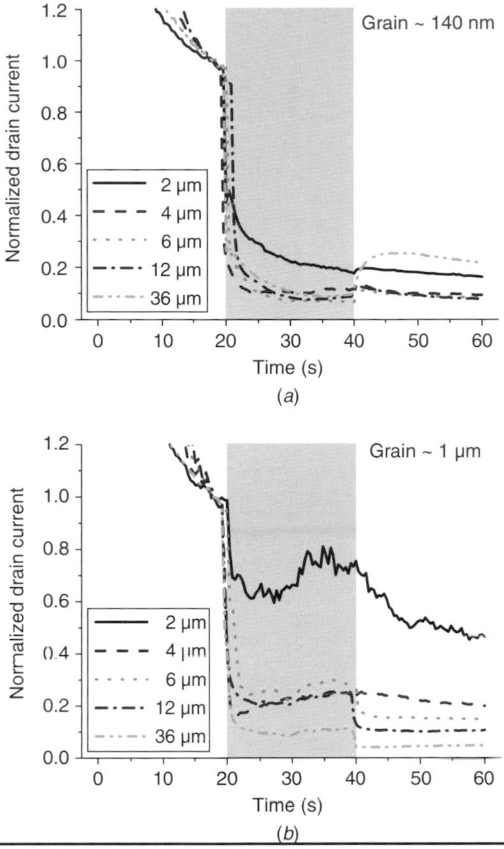

FIGURE 1.20 Sensing data of large-scale pentacene transistors upon exposure to 1-pentanol: normalized I_{ds} under the condition of $V_g = V_{ds} = -25$ V, $v = 45$ mL/min, and $d = 2$ mm for different microscale channel lengths, with average pentacene grain size of (*a*) 140 nm and (*b*) 1 μm, respectively. (*Reprinted with permission from Ref. 115. Copyright 2004, American Institute of Physics.*)

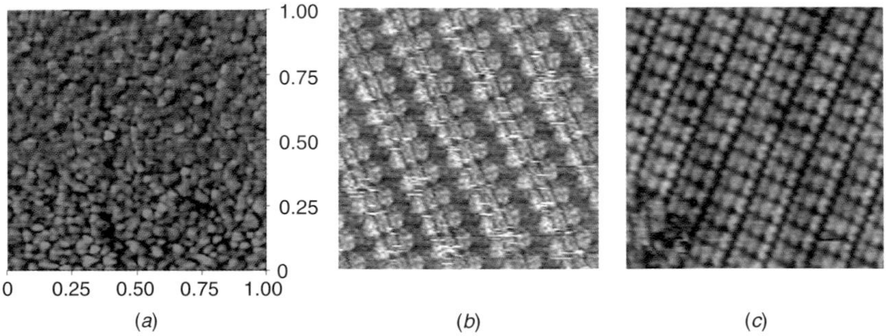

FIGURE 2.6 (*a*) AFM topographical image of D3ANT deposited on Si/SiO_2 (1 × 1 μm) and typical constant-current STM images of self-organized monolayers of D3ANT adsorbed (*b*) at the *n*-tetradecane–HOPG interface (16 × 16 nm²; $V_t = -333$ mV; $I_t = 27$ pA) and (*c*) at the *n*-tetradecane–Au(111) interface (16 × 16 nm²; $V_t = -62$ mV; $I_t = 83$ pA). (*Reproduced by permission of The Royal Society of Chemistry, Ref. 131.*)

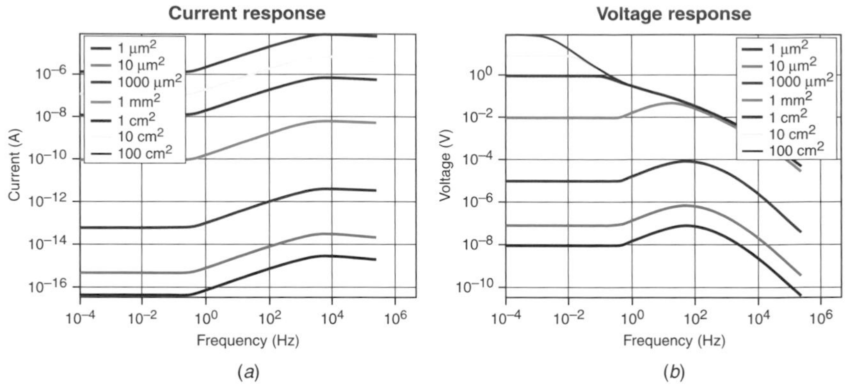

Figure 4.16 The calculated current and voltage response for different sensor areas. Substrate: glass 175 μm; pyroelectric layer thickness: 400 nm. Calculated for the values of $R_i = 100$ MΩ and $C_i = 75$ pF.

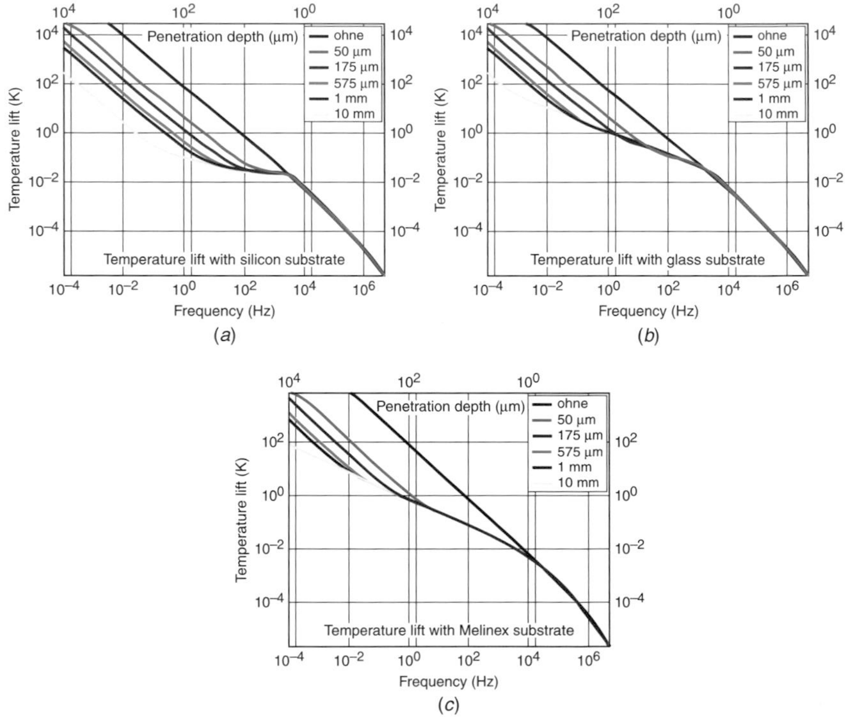

Figure 4.17 Calculated temperature lift for different substrates and the multilayer model depicted in Fig. 4.14. The substrates are (*a*) silicon, (*b*) glass, and (*c*) PET-foil Melinex. The penetration depth of the thermal wave and its assignment to the frequency are also indicated (red axis).

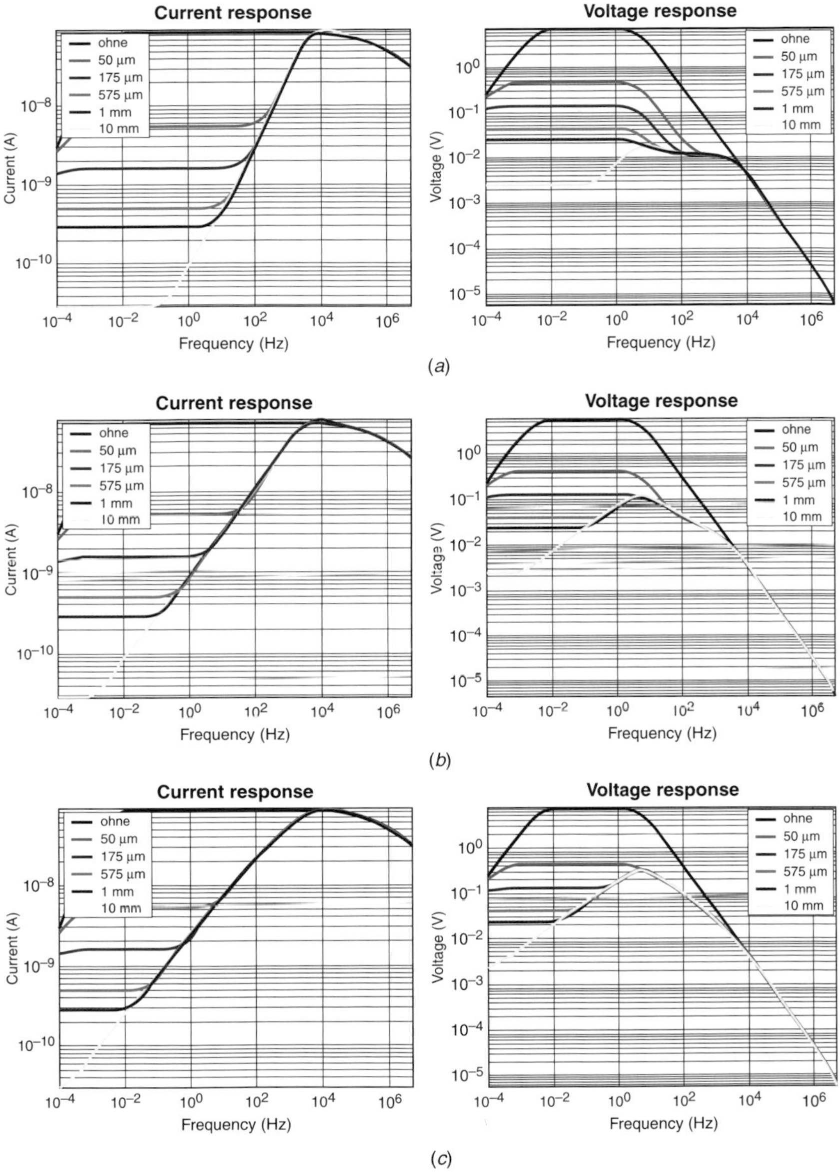

FIGURE 4.18 Current and voltage response in dependence of the substrate thicknesses and the material for (*a*) silicon (*b*) glass, and (*c*) PET foil (Melinex from DuPont Teijin), calculated for the values of $R_i = 100\ \mathrm{M\Omega}$ and $C_i = 75$ pF.

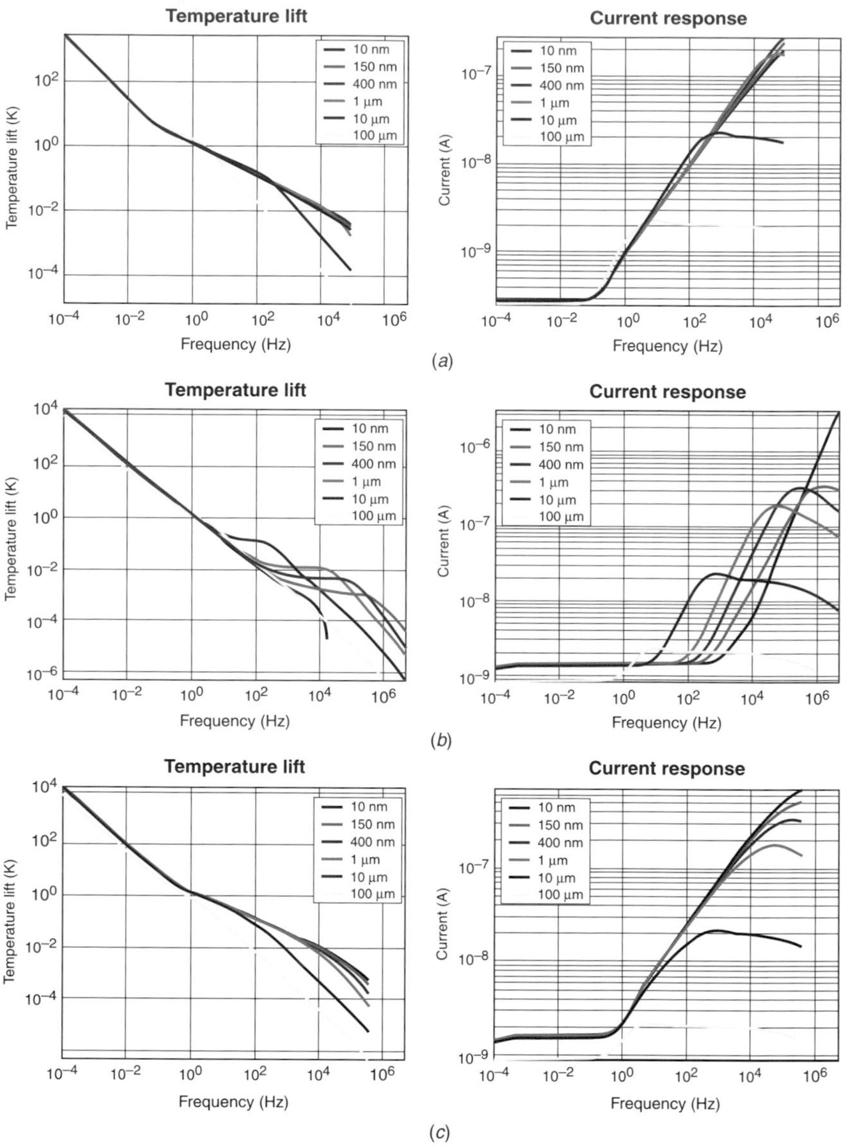

FIGURE 4.19 Temperature lift and current response with different thicknesses of the pyroelectric layer on 200 μm thick substrate of (*a*) glass, (*b*) silicon, and (*c*) PET, calculated for the values of R_i = 100 MΩ and C_i = 75 pF.

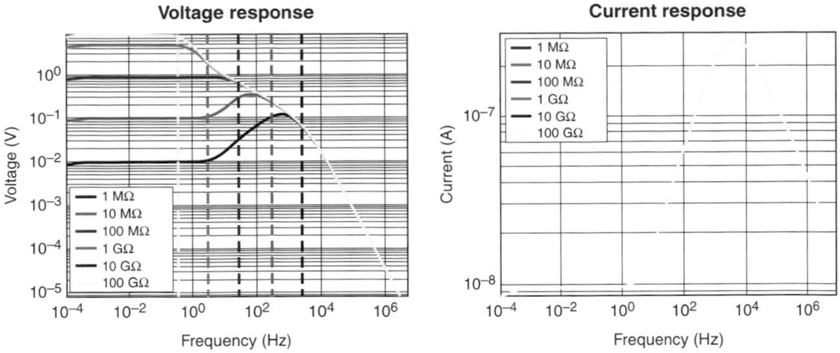

Figure 4.20 Frequency dependence of the voltage and current response for a sensor with 3 μm thick pyroelectric layer on 175 μm glass substrate and 75 pF input capacitance for different input resistance values of the measurement instrument (lockin amplifier, parametric analyzer, etc.). The dotted lines indicate the respective cutoff frequencies $f_c = \omega_c/(2\pi)$. All lines collapse into one for the current response.

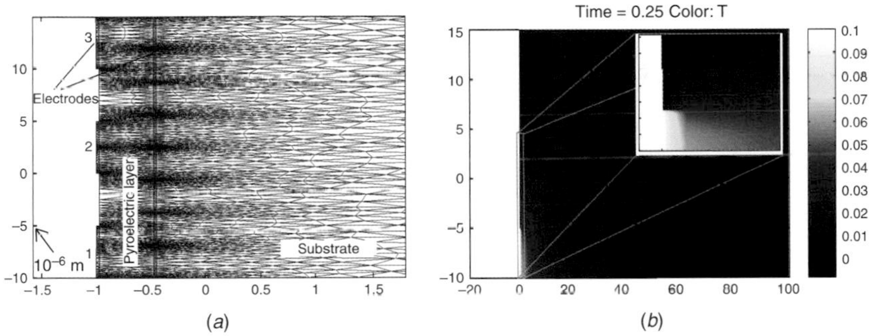

Figure 4.21 (*a*) Section of the geometry of three adjacent sensor elements on one 100 μm substrate. Only a part of the substrate is shown. The lines indicate the mesh. (*b*) Plot of the calculated temperature distribution.

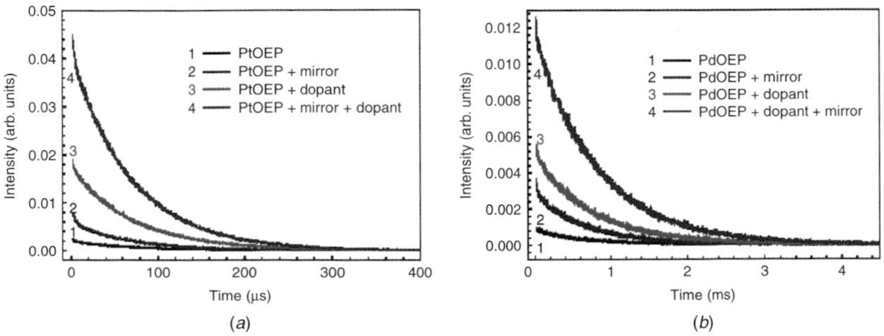

Figure 5.8 Effect of the titania dopant embedded in the PS film and Al mirrors on the PL decay of (*a*) PtOEP:PS and (*b*) PdOEP:PS films in water in equilibrium with Ar (~0 ppm DO). The films were prepared by drop-casting 60 μL of toluene solution containing 1 mg/mL TiO_2, 1 mg/mL dye, and 50 mg/mL PS.

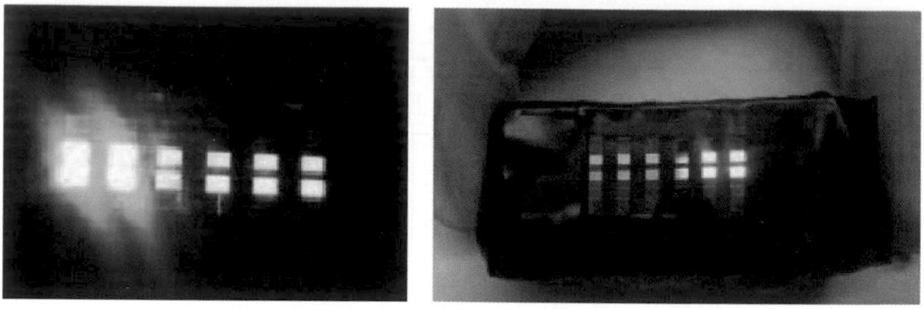

Figure 5.10 Mixed OLED platforms of green (outer four pixels in the arrays) and blue OLEDs. The array on the right is encapsulated. The pixel size is 2×2 mm^2.

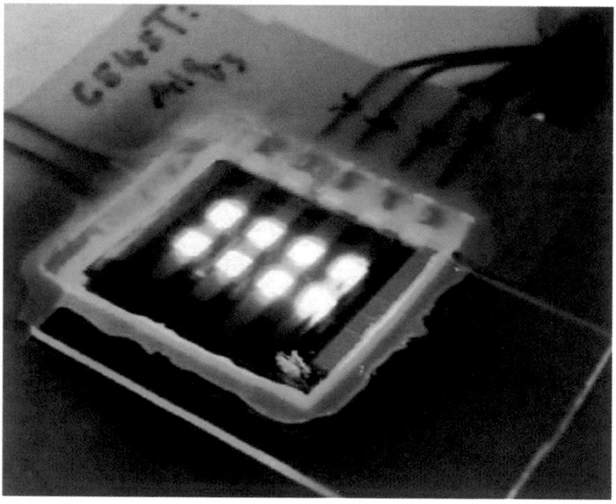

Figure 5.11 Top view of a structurally integrated OLED/sensor film/PD probe in a back detection geometry.

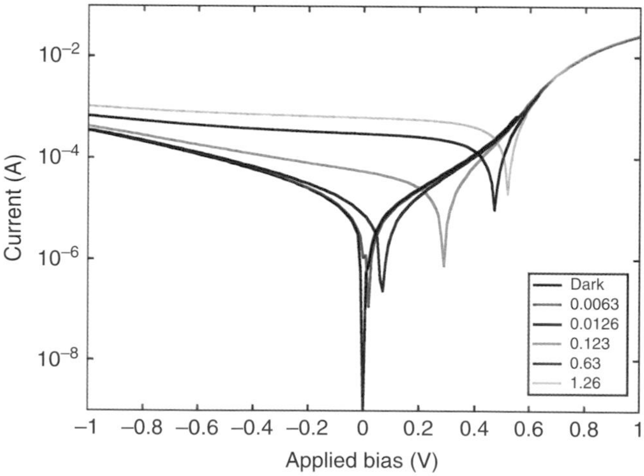

Figure 6.10 Current-voltage curves for an ITO/PEDOT:PSS/P3HT:PCBM/Al bulk heterojunction device under varying illumination levels (in milliwatt).

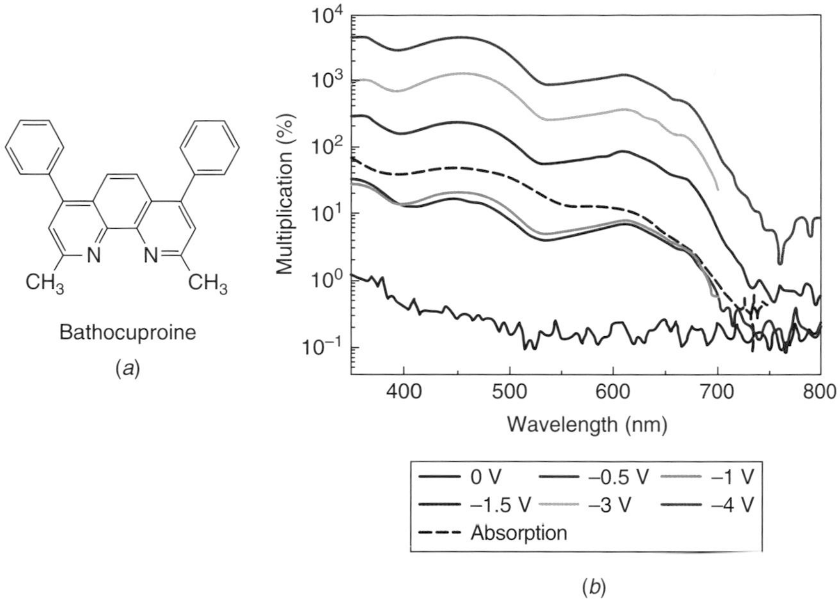

FIGURE 6.25 (*a*) The chemical structure of bathocuproine. (*b*) The apparent gain vs. emission wavelength in an ITO/PEDOT:PSS/pentacene/C60/bathocuproine/Al device under various reverse DC biases. (*Reprinted with permission from Ref. 61. Copyright 2007, American Institute of Physics.*)

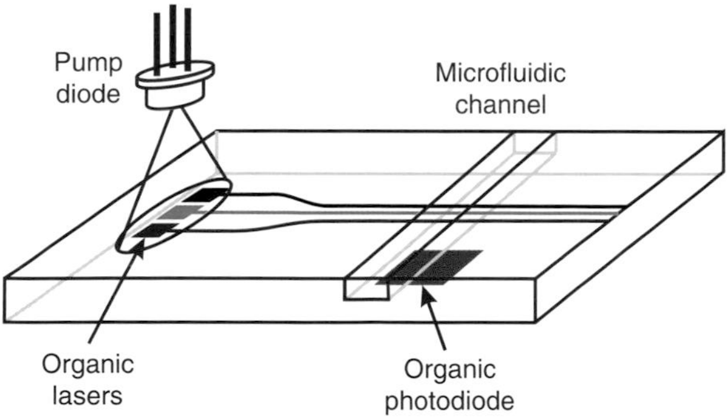

FIGURE 7.20 Scheme of a possible lab-on-a-chip design incorporating a microfluidic system for analyte preparation and handling, multiple laser sources, and photodetection.

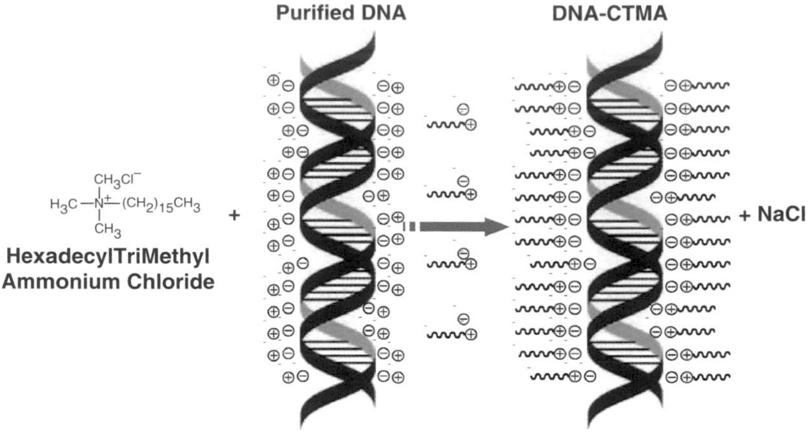

Figure 8.5 The purified DNA is initially soluble only in aqueous solutions and does not dissolve in any organic solvent. Purified DNA is modified through a cationic surfactant (hexadecyltrimethyl ammonium chloride—CTMA) cation exchange reaction to enhance solubility, processing, and stability.

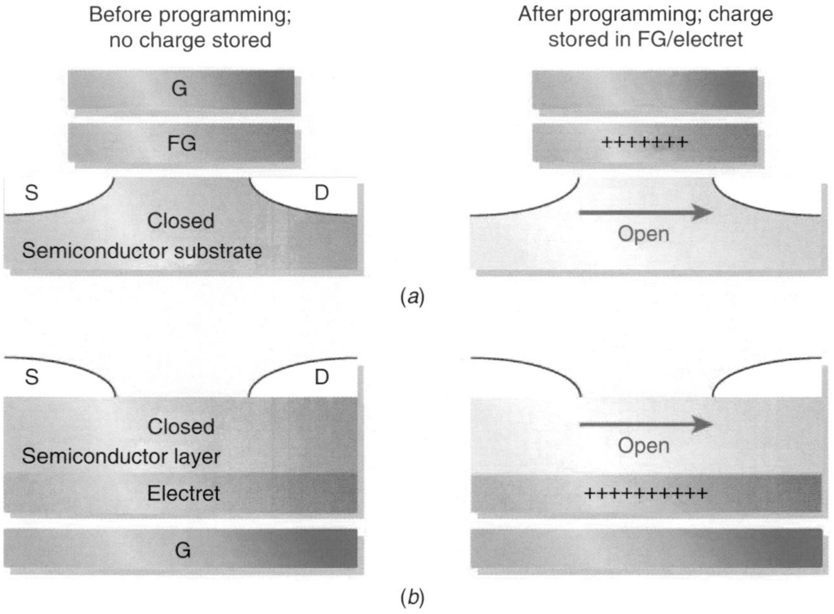

Figure 8.7 Methods against memory loss. The basic transistor is a device in which a small voltage applied at the control gate (G) modulates a much larger current flow from source (S) to drain (D) through a semiconductor substrate. (*a*) In flash memories, an amount of charge is trapped on a floating gate (FG) that modifies the control voltage required for current to flow from S to D. Whether current flows or not defines a boolean 1 or 0. The memory of this state persists as long as the charge remains trapped on the floating gate. (*b*) In Baeg and colleagues' organic device shown, the same principle is used, but the charge is trapped locally on a thin *electret* of chargeable polymer, rather than on an isolated floating gate.[73] (*Reproduced with permission from Ref. 77. Copyright 2004, Nature Publishing Group.*)

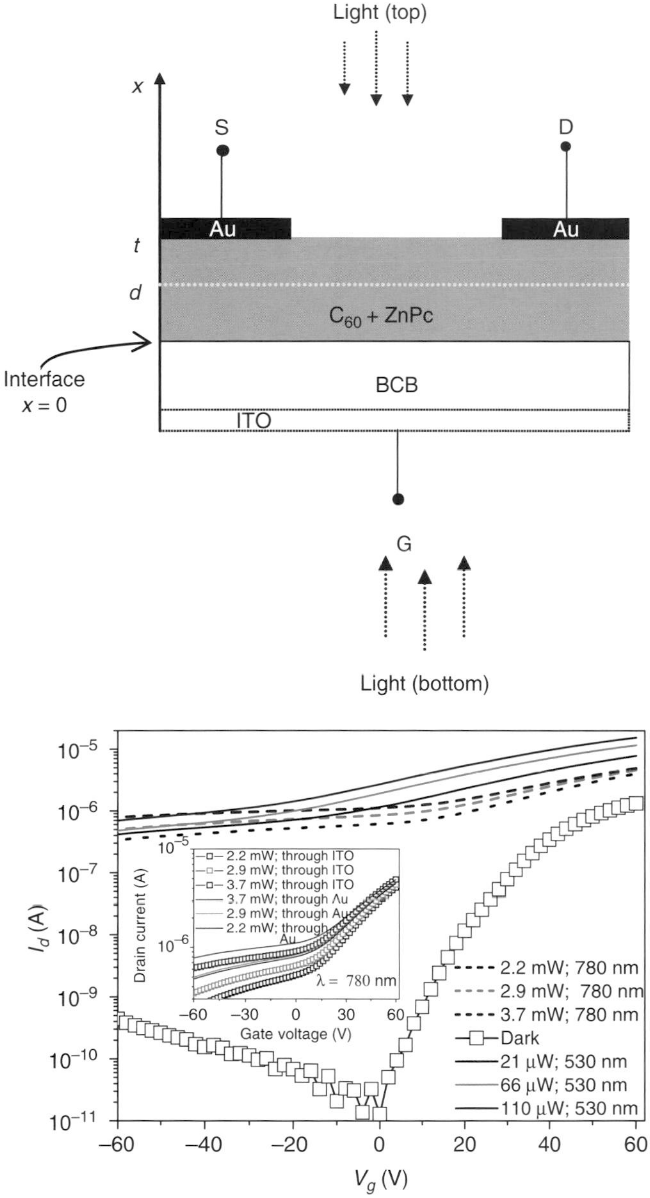

FIGURE 8.17 Top: Device scheme of the C_{60} and ZnPc-based phototransistors. Bottom: Photoresponse of the above device fabricated in the author's laboratory.

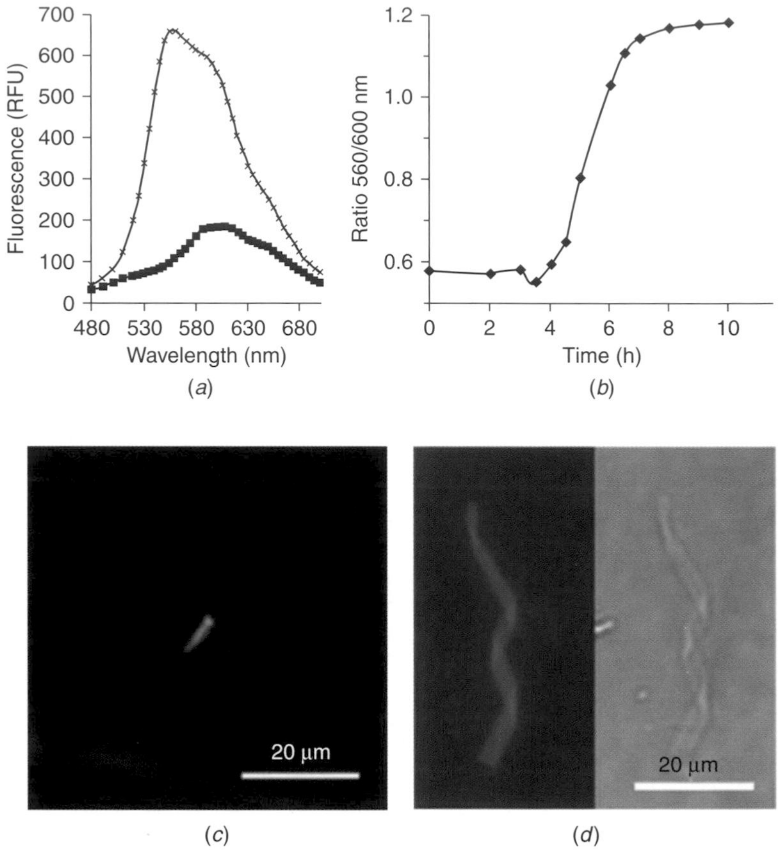

FIGURE 9.8 (*a*) Emission spectra (bottom) of PONT-native bovine insulin (■) and PONT-amyloid fibrillar bovine insulin (x). (*b*) Kinetics of insulin amyloid fibril formation monitored by PONT fluorescence. (c), (d) Images of self-assembled electroactive nanowires of PONT and insulin amyloid fibrils that are being formed when PONT is present during the fibrillation event. (*Parts (a) and (b) are reproduced with permission from Ref. 13. Copyright 2005 American Chemical Society Parts (c) and (d) are from Ref. 109. Copyright 2005 Wiley-VCH Verlag GmbH & Co. KGaA.*)

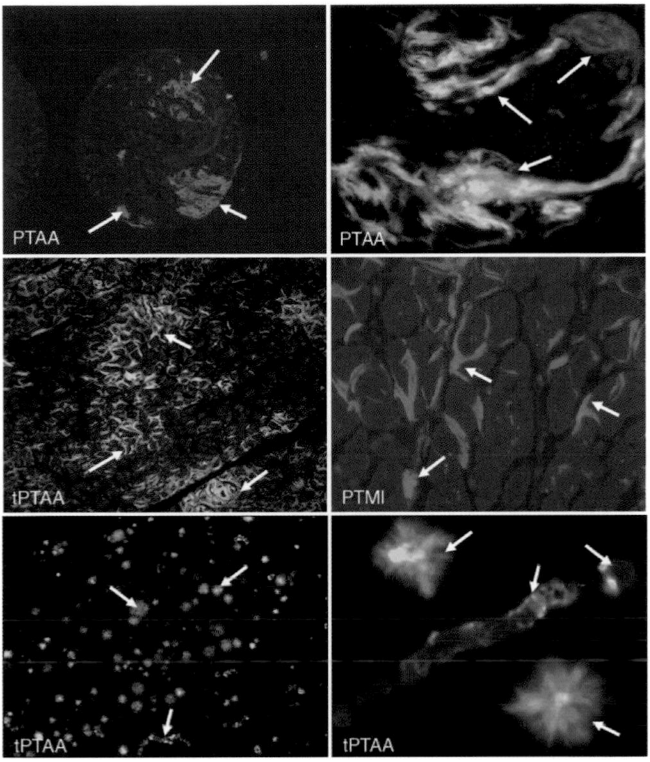

FIGURE 9.9 A plethora of amyloid deposits in tissue samples stained by different LCPs as indicated in the figure. Some typical amyloid deposits being stained by the LCPs are highlighted by arrows. Notably, PTAA and tPTAA bound to diverse amyloid deposits in the same tissue emit light with different colors, indicating that there is a heterogenic population of amyloid deposits in these samples (top right, bottom left and right).

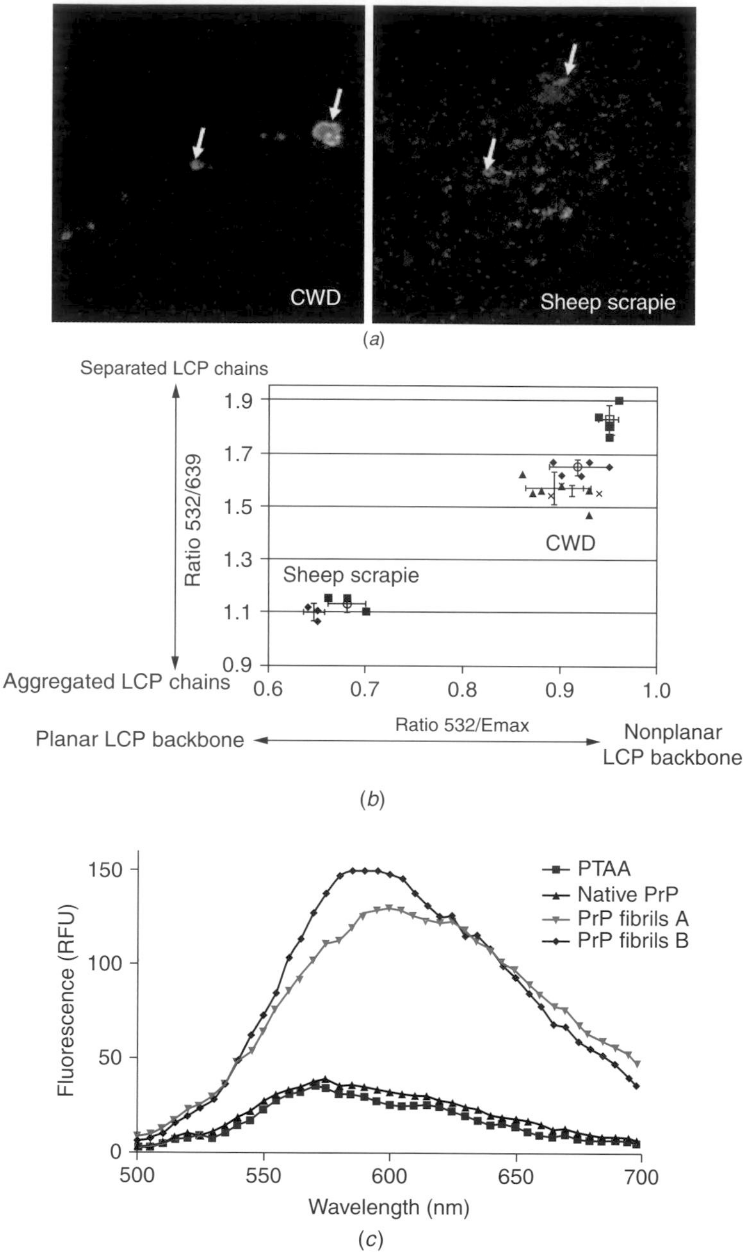

FIGURE 9.10 (*a*) Fluorescent images of prion deposits associated with distinct prion strains, chronic wasting disease (CWD, left) and sheep scrapie (right), which have been stained by PTAA. (*b*) Correlation diagram of the ratios, R532/639 and R532/E_{max}, of the intensity of the emitted light from PTAA bound to prion deposits originating from individual mice infected with CWD (four generations denoted with black symbols, ■, ♦, ▲, x) or sheep scrapie (two generations denoted with purple symbols, squares and diamonds). (*c*) Emission spectra of PTAA in buffer (black), PTAA bound to native (blue) or two different forms of fibrillar recombinant mouse PrP (green and purple). [*Part (c) reproduced with permission from Ref. 112. Copyright 2007 Nature Publishing Group.*]

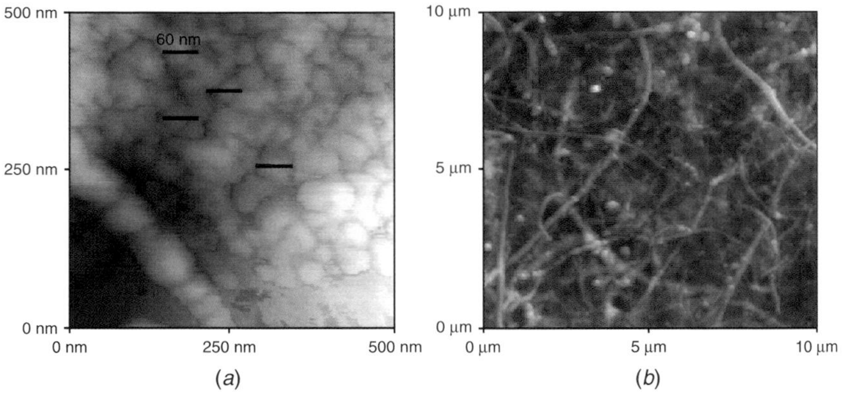

FIGURE 10.6 AFM images of electrophoretically deposited (*a*) polyaniline and (*b*) PANI/MWCNT-c films. (*Reprinted from Ref. 91. Copyright 2008 with permission from Elsevier.*)

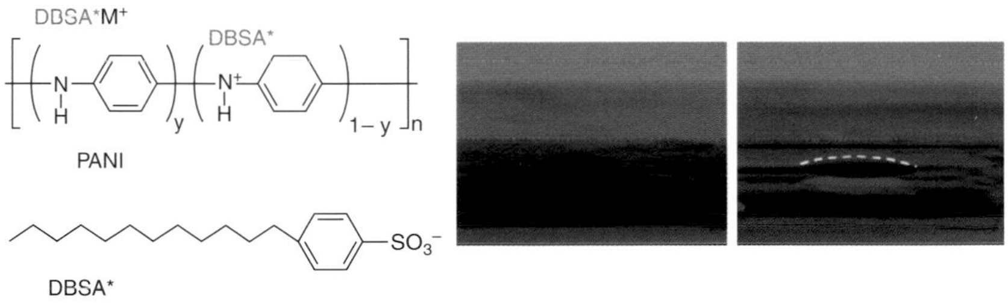

FIGURE 11.2 Left: The chemical structure of PANI and DBSA. Water droplets added to surface switches including PANI:DBSA, in which PANI is in its oxidized (middle) and reduced (right) state. (*From Ref. 10. Copyright 2004 Wiley-VCH Verlag GmbH & Co. KGaA. Reproduced with permission.*)

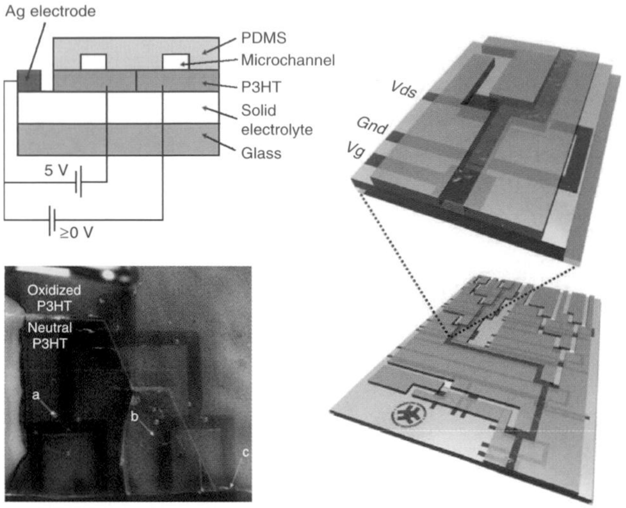

FIGURE 11.5 Top left: P3HT-based surface wettability switches used to control the flow of aqueous samples in microfluidic systems. Bottom left: Water is transported relatively faster along the microchannel of the Y-branches that include a floor of P3HT switched to the oxidized state (c). Once the water approaches the neutral P3HT (red color), it slows down considerably (a and b). Right: Electronic control of the gating of aqueous samples can be used in various lab-on-a-chip technologies to enable multiplexing of the sample analyte and different reagents. (*From Ref. 13. Reproduced by permission of the Royal Society of Chemistry, 2006.*)

FIGURE 11.7 Left: Chemical structure of PEDOT and PSS. Middle: The conjugated PEDOT:PSS-based ion pump made of patterned, adjacent PEDOT:PSS electrodes (A through D). Right: By addressing the electrodes ions (M^+) migrate from the source to the target electrolytes through the nonconducting PEDOT:PSS channel (pink).

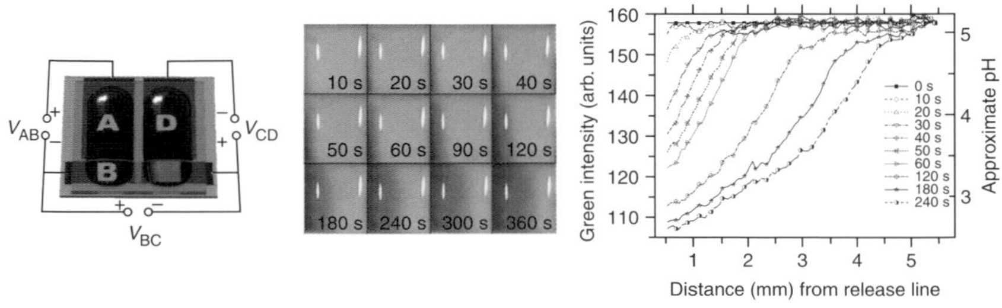

FIGURE 11.8 Left: The electronic ion pump with a pH paper on the C electrode as an indicator of the proton gradient. Middle: The proton gradient formed over time. Right: The associated measured pH gradients versus time. (*Reprinted from Ref. 20. Copyright 2008, with permission from Elsevier.*)

which would put organic devices on a par with standard pn-type Si photodiodes; achieving substantially greater reductions, although possible, will probably prove challenging with existing materials systems.

Anyway, regardless of what might be possible in the future, organic photodiodes in use today have typical capacitance densities of a few hundred pF/mm^2. In practical terms this means they are 10 to 100 times slower than Si photodiodes and very much slower than APDs and PMTs. The transient characteristics of a typical P3HT: PCBM photodiode of area 9 mm^2 are shown in Fig. 6.20, obtained using a 50 Ω load resistance. The rise and fall times are 0.45 and 0.5 μs, respectively from which it follows from Eq. (6.31) that the cutoff frequency is about 0.7 MHz. This compares with a rise time of 50 ns for a typical Si photodiode (Hamamatsu S6931) of similar area (6.6 mm^2) and with a comparable load resistance (100 Ω).

The OPV transients exhibit characteristic exponential charge and discharge profiles, indicating that the response speed is capacitance-limited in these devices (see fits to experimental data in Fig. 6.20). In large-area devices, the speed of response is determined by the RC time constant and therefore scales linearly with device area. When the device area is sufficiently small, the response time is no longer determined by the capacitance, but rather by the transit time τ, i.e., the

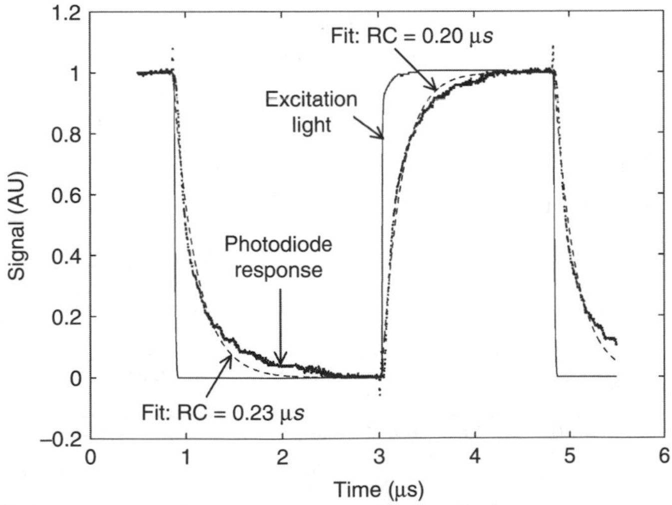

FIGURE 6.20 Transient response curve (black line) of a 9 mm^2 ITO/PEDOT: PSS/P3HT:PCBM/Al device in response to a stepped excitation source. The dotted line indicates a numerical fit to exponential charging and discharging curves. (*Data kindly provided by X. Wang, Imperial College London.*)

time required for the photogenerated charge carriers to pass through the thickness of the device. The transit time can be estimated from the drift velocity $v = \mu E$ of the slowest charge carriers where μ is the mobility and $E = (V - V_{BI})/d$ is the internal field strength. The transit time is therefore equal to

$$\tau = \frac{d}{v} = \frac{d}{[\mu E]} = \frac{d^2}{\mu \,|\, V - V_{BI} \,|} \tag{6.66}$$

In the non-RC-limited regime, a device is expected to reach steady state within a few transit times, and hence the speed of response is maximized by using thin devices under a significant reverse bias. Punke and coworkers[53] have reported the transient characteristics (Fig. 6.21) of ITO/PEDOT:PSS/P3HT:PCBM/Al photodiodes with an active layer thickness of 170 nm and an area of 0.2 mm^2. The devices were excited at 532 nm using a frequency doubled Nd:YAG pulsed laser with a pulse width of 1.6 ns. They exhibited a quantum efficiency of 12% at short circuit rising to 34% under –5 V operation. They had short (instrument-response limited) rise times of < 2 ns at all biases when connected to a 25 Ω load resistance. The fall times, however, were significantly longer and strongly bias-dependent, varying from 426 ns at short circuit to 36 ns under –5 V. They

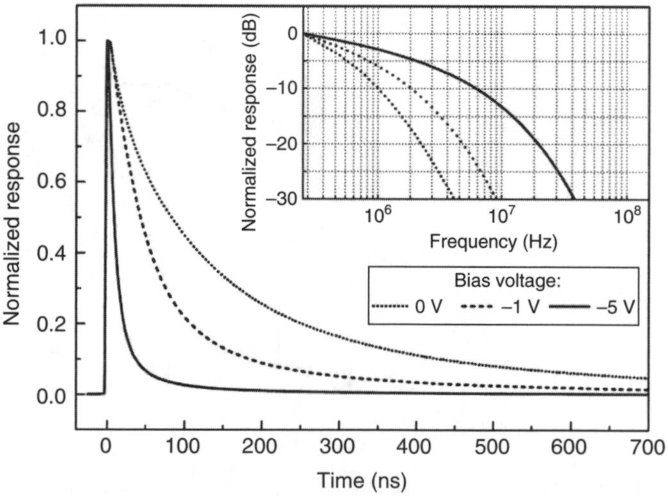

FIGURE 6.21 Normalized pulse curves for a 0.2 mm^2 ITO/PEDOT:PSS/P3HT:PCBM/Al device at various applied biases. The inset shows the corresponding signal vs.frequency curves determined from the Fourier transform of the pulse response curve. (*Reprinted with permission from Ref. 53. Copyright 2007, American Institute of Physics.*)

determined a 1 MHz cutoff frequency at –5 V from the Fourier transform of the impulse response (Fig. 6.21). This was limited by the slow turnoff characteristics of the device, and the fast turn-on characteristics suggest significantly faster response is in principle possible; based on the < 2 ns rise time, a cutoff frequency of order 100 MHz would be expected.

6.6.2 Shunt Resistance

As discussed above, the shunt resistance influences both the thermal noise of the photodiode and the output noise voltage from the amplifier, with it being important in both cases to maximize the shunt resistance in order to minimize the noise. In this section, we will consider how the shunt resistance of an OPV device can be maximized, although we will find it more convenient to couch our discussion in terms of the reciprocal parameter

$$G_{sh} = \frac{1}{R_{sh}} \tag{6.67}$$

where G_{sh} is the *shunt conductance*. Conventionally, a *shunt* is considered to be a conducting filament that bridges the two electrodes, thereby bypassing the (high-impedance) active materials and so allowing a sizable current to flow even at low applied biases.[54, 55] The term *shunt conductance*, however, is used here (and elsewhere) in a more general sense, and it simply refers to the conductance of the photodiode in the dark under (near) short-circuit conditions whatever its physical origin.

In a poorly fabricated device, physical shunts are indeed the major cause of the shunt conductance. The use of stringent fabrication procedures, however, can greatly reduce the shunt conductance; critical issues here include

- The amelioration of asperities in the (widely used) indium tin oxide anode, e.g., by using a thick surface coating of PEDOT:PSS
- The avoidance of pinhole defects in the active layer by using dust-free preparation and fabrication procedures (including rigorous filtering of all solutions)
- The avoidance of "harsh" deposition procedures for the top electrode; thermal evaporation at low temperatures is preferred
- Careful optimization of the processing conditions (such as solvent choice, active layer composition, spin speed, and annealing temperature), which strongly affect the film morphology

Failure to adequately address these issues can result in disastrously high shunt conductances of 10^{-4} S or more. With care, though, it is possible to reduce shunt conductances below the 10^{-8} S level, corresponding to shunt resistances in excess of 100 MΩ. At this level, the residual conductance is most probably due to current flow through the active layer rather than shunts. In many OPV devices, ohmic contacts are present at one or both electrodes (since this maximizes the built-in field and is therefore beneficial for the quantum efficiency [see Eq. (6.5)]. In such situations, appreciable injection can occur from the electrodes into the active layer materials even at very low biases. This is especially problematic for bulk heterojunction devices where the donor and acceptor materials can make continuous percolation pathways from one electrode to the other, providing effective shunts for injected holes and electrons (Fig. 6.22*a*). (In discrete heterojunction devices, injected charges are blocked at the heterojunction, resulting in low dark currents and high shunt resistances.) The simplest way to minimize the dark current in bulk heterojunction devices is to use a three-layer structure (Fig. 6.22*b*), in which pure regions of the donor and acceptor are located next to the anode and the cathode, respectively, and a uniform blended region exists in between. Any injected electrons (holes) that manage to pass through the bulk of the device are blocked on reaching the donor (acceptor) layer adjacent to the anode (cathode), resulting in extremely low dark currents

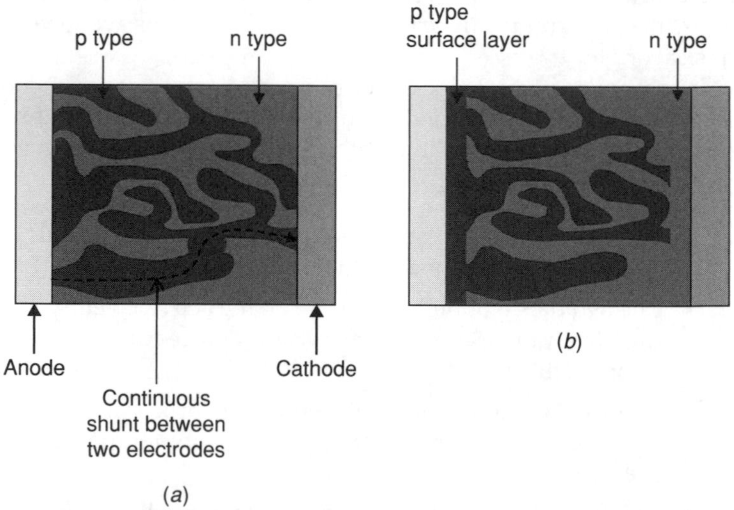

FIGURE 6.22 Schematic for (*a*) a standard *single-layer* bulk heterojunction photodiode and (*b*) a *three-layer* bulk heterojunction photodiode designed to minimize the shunt resistance by blocking leakage of injected charges.

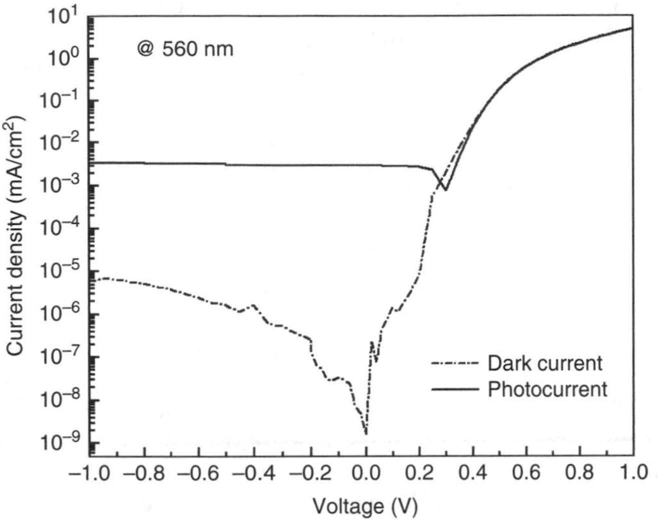

FIGURE 6.23 Current-voltage curves for an optimized multilayer ITO/PEDOT: PSS/P3HT:PCBM/Al device in the dark and in the light. The device exhibits a very high shunt resistance of 2 GΩ and remains fairly resistive even under substantial reverse bias (1.25 GΩ at –1 V).

and very high shunt resistances. The data in Fig. 6.23 show the current-voltage characteristics of a carefully optimized 15 mm^2 ITO/PEDOT:PSS/P3HT:PCBM/Al photodiode. The device is extremely resistive, exhibiting a short-circuit shunt resistance of ~ 2 GΩ. Importantly, the OPV device remains highly resistive even under reverse bias, yielding a dark current of just 0.8 nA under –1 V applied bias (~1.25 GΩ). The shunt resistance of this device compares very favorably with the majority of silicon devices, which tend to have shunt resistances of a few hundred megaohms or less. A resistance of 2 GΩ is still some way below the very highest shunt resistances (of 50 GΩ or more) quoted for top-end silicon photodiodes, but we believe such high values will also be attainable using organic devices with further optimization.

6.6.3 Spectral Response

One of the most appealing aspects of organic semiconductor technology is the ability to control the spectral properties of electronic devices through chemical design. In the case of organic photodiodes, the absorption spectrum of the active materials determines the wavelength range over which the cell is active, and this can be controlled by appropriate tuning of the active layer materials. This is an important advantage over traditional inorganic semiconductors, which afford only limited opportunities for tuning. Organic materials can be designed to absorb strongly over a wide spectral range to create broadband photodetectors or over a more restrictive range to create wavelength-selective devices. The most widely studied OPV system is the P3HT:PCBM donor/acceptor combination, which has a relatively narrow spectral range from about 300 to 650 nm (Fig. 6.13). In recent years—and driven primarily by the need to improve solar cell efficiencies—much research has focused on developing lower energy-gap materials that can harvest a wider part of the solar spectrum.[5, 56–59] In the context of photodiodes, these same materials systems are of interest for their wider spectral response. The donor-acceptor combination in Fig. 6.24*a*, reported by Wang and coworkers, provides one of the widest spectral ranges reported to date: both the donor polymer APFO-Green1 and the C_{70}-based acceptor molecule BTPF70 have absorption spectra that extend well beyond 1 μm.[5] The active range of optimized bulk heterojunction devices based on these materials spans 330 to 1000 nm (Fig. 6.24*b*), with the lower wavelength limit being due to absorption by the glass substrate. This is only marginally less than the spectral range of Si photodiodes whose long-wavelength sensitivity extends to about 1100 nm.

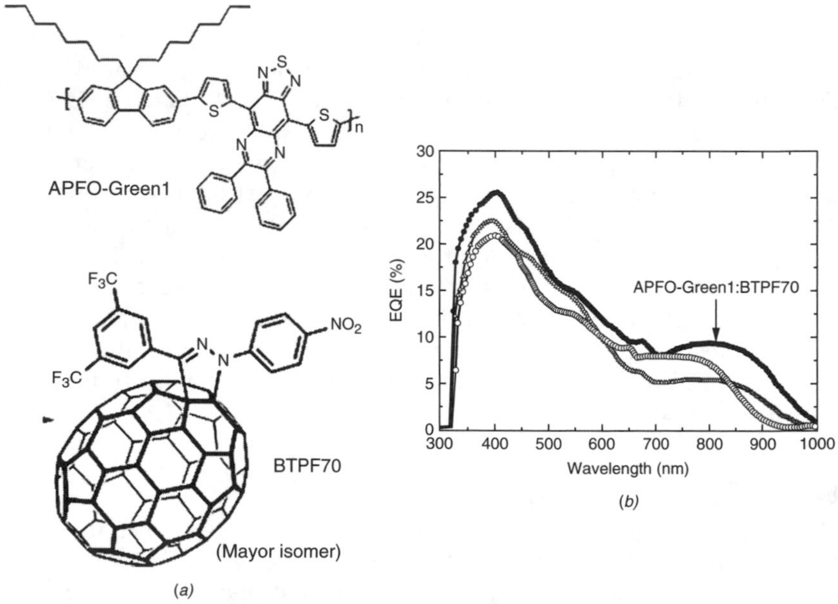

FIGURE 6.24 (*a*) Chemical structures and (*b*) spectral response curve for low band-gap materials developed by Wang and coworkers. Photodiodes based on this material combination have a similar spectral range to Si devices. (*Reprinted with permission from Ref. 5. Copyright 2004, American Institute of Physics.*)

In solar energy applications, there is an optimal band gap that, for a given illumination source, will yield the highest achievable power conversion efficiency: if the band gap is too high, few of the incident photons are absorbed; and if it is too low, much of the incident energy is wasted through internal conversion. The optimal band gap is variously estimated to lie between 1.0 and 1.5 eV, depending on the specific assumptions made (e.g., concerning the absorption spectrum, exciton binding energy, and light source). APFO-Green1 falls roughly in the middle of this range with a band gap of about 1.2 eV. In the case of photodiodes, no such optimum exists: the smaller the band gap ΔE, the lower the photocurrent onset energy and the wider the spectral range (assuming, of course, that a functioning device can be made). At the time of writing, we are aware of no published work that has focused on ultralow band gap materials ($\Delta E << 1$ eV). While such materials would be of little interest for solar energy applications (since they would yield extremely low power conversion efficiencies), they would be of great interest for photodiodes: for instance, a single device that could continuously harvest photons from < 300 nm up to 3 μm would offer something genuinely new that conventional photodiodes are unable to deliver. The inorganic photodiodes with the widest spectral range are based on InGaAs which, when suitably optimized, are sensitive from 250 to 1700 nm.† These are expensive high-end photodetectors used for ultrafast applications; a low-cost organic alternative that offered comparable or wider spectral range would find many applications. A new generation of "black" organic semiconductors is now under development, which can absorb continuously from 300 to > 2500 nm. Also of interest are carbon nanotubes, which can absorb continuously up to 2000 nm.[60] These materials have not yet been implemented in OPV devices—due to some processing issues—but could potentially offer much wider spectral ranges than could be achieved with other technologies.

6.6.4 Gain

As discussed in Sec. 6.1, PMTs and APDs are the most sensitive detectors available due to their internal gain processes. In the case of PMTs, the noise introduced by the gain process is virtually zero, and in well-optimized APDs it is little more than a factor of two. In both cases the resultant noise is far lower than would be incurred using an external amplifier of equivalent gain, and so it is possible to detect extremely low light levels right down to the single photon level. In OPV devices, under normal operating conditions, each incident photon can generate at most

†Tandem photodetectors (comprising a UV/Vis sensitive photodiode on top of an IR-sensitive photodiode) have also been reported, but do not appear to be in widespread use perhaps due to manufacturing complexity or cost.

one electron-hole pair and so can contribute at most one electron to the external circuit. The absence of gain limits sensitivity and means that OPV devices are best compared to conventional (gain-free) photodiodes. The question arises, whether it might be possible to achieve gain in organic photodiodes under any circumstances. There have in fact been several reports of OPV devices that, when operated under reverse bias, show anomalously high photocurrents (see Refs. 61 and 62 and references therein). In these devices, the change in current ΔI under illumination satisfies

$$\left|\frac{\Delta I(V)}{e}\right| = \left|\frac{I_{\text{light}}(V) - I_{\text{dark}}(V)}{e}\right| \gg \Re \tag{6.68}$$

where $\Re$ is the rate at which photons strike the device. In other words, under illumination, the current changes by an amount that massively exceeds $e\Re$, seemingly implying that more than one electron is generated for each absorbed photon. This is shown in Fig. 6.25 for an

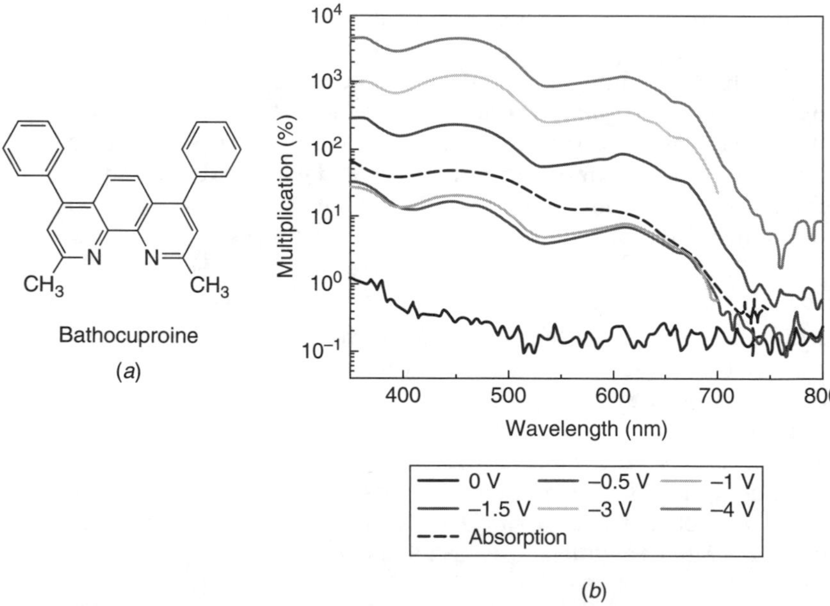

Figure 6.25 (*a*) The chemical structure of bathocuproine. (*b*) The apparent gain vs. emission wavelength in an ITO/PEDOT:PSS/pentacene/C60/bathocuproine/Al device under various reverse DC biases. (*Reprinted with permission from Ref. 61. Copyright 2007, American Institute of Physics.*) (See also color insert.)

ITO/PEDOT:PSS/C60/bathocuproine/Al device where the multiplication factor $M = \Delta I/e\Re$ is plotted against wavelength for a variety of applied biases.[61] At first glance, these data seem to suggest that at a reverse bias of –4 V, as many as 40,000 electrons are generated in the external circuit for each absorbed photon. Needless to say, this is not the case, and in reality the effect is photoconductive rather than photovoltaic. The behavior is encountered when there is strong trapping of one or both of the charge carriers. We consider here the situation in which both charge carriers are trapped. As previously described, the current I_{photo} under illumination can be divided into two parts: a photovoltaic part I_{ph} due to the continuous conversion of absorbed photons into electron-hole pairs and a photoconductive part $I_{V\text{photo}}$ due to the applied bias.

$$I_{\text{photo}} + I_{\text{ph}} + I_{V\,\text{photo}} \tag{6.69}$$

In the dark state under a reverse bias V_{rev}, the current is entirely photoconductive and is very low due to repeated trapping and (slow) detrapping of the electrons and holes. Under illumination, some of (or all) the trap sites are filled by the photogenerated charges, leading to an increase in the mobilities of the remaining untrapped charges. This in turn causes a reduction in the resistance R of the device and changes the photoconductive current by an amount

$$\Delta I_{V\,\text{photo}} = \frac{V_{\text{rev}}}{R(V_{\text{rev}},0)} - \frac{V_{\text{rev}}}{R(V_{\text{rev}},\Re)} \tag{6.70}$$

where the resistance R depends on both the bias V_{rev} and the rate $\Re$ at which photons strike the device: if $\Delta I_{V\,\text{photo}} > \Re/e$, the illusion of gain will be given, although in reality the effect is entirely photoconductive in origin.

Although photoconductive OPV devices can undergo large changes in current in response to very small light intensities, they have two main drawbacks: First, they tend to exhibit poor linearity since the effective charge mobilities need not increase linearly with the light intensity. Second, since by definition they have massively reduced impedance under illumination, they suffer from considerable thermal noise. They do not therefore exhibit the virtually noise-free cascade amplification processes found in PMTs and APDs and hence will never offer comparable levels of sensitivity. It may yet be possible to develop organic devices that exhibit genuine internal gain, but this is likely to require the development of new OPV materials that exhibit high carrier mobilities and are able to tolerate the high fields required for impact ionization.

6.7 Technology and Applications

In this final section, we consider some of the applications to which organic photodiodes have so far been applied. This list is in no sense intended to be exhaustive, and given the vast number of uses to which photodetectors have historically been put, it is inevitable that many new applications will be found in the coming years.

6.7.1 Printed and Flexible Devices

The ability to process conjugated polymers (and an increasing number of small molecules) from solution opens up considerable opportunities for cost savings, relative to conventional inorganic semiconductors. The usual method for "one-off" laboratory-scale device fabrication is spin-coating, a very reliable process that yields high-quality uniform films but involves significant materials wastage. The on-demand nature of printing methods results in efficient materials usage and offers an attractive means of making complex structures such as one- and two-dimensional sensor arrays. The most developed technique for organic semiconductor devices is inkjet printing, in which a small droplet of solvent containing the active materials is propelled from a chamber or *head* usually by piezoelectric expansion. Hoth and coworkers, for instance, reported a 2.9% organic solar cell (Fig. 6.26),[63] which—although still some way below the ~5% efficiencies achievable with spin-coated devices—indicates the potential of

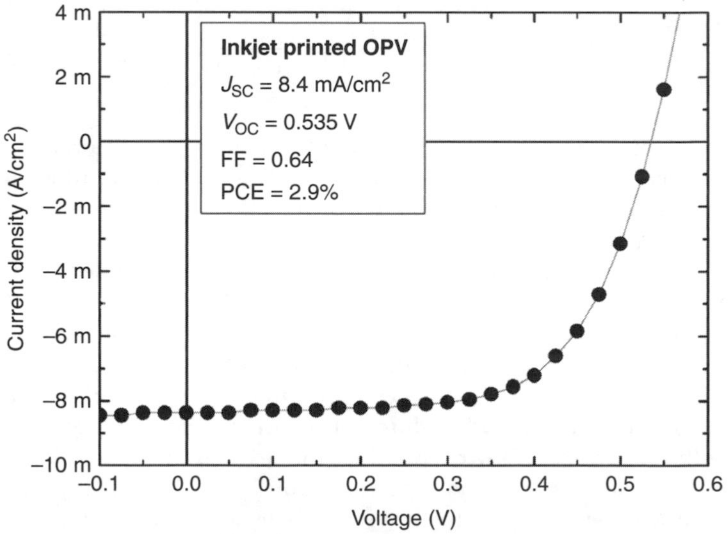

Figure 6.26 Current-voltage curve for an inkjet-printed ITO/PEDOT:PSS/P3HT:PCBM/Ca:Ag device. (*From Ref. 63. Copyright Wiley-VCH Verlag GmbH & Co. KGaA. Reproduced with permission.*)

the inkjet approach. In addition, a range of conventional contact-based printing and coating techniques are under development, e.g., gravure, flexographic, screen, and offset printing. These techniques have not been widely used for photodiode applications, although they are all being intensively investigated for solar energy applications. Contact methods, although still at an early stage of development, show considerable promise: Tobjork et al., for instance, fabricated a flexible solar cell with 0.74% power conversion efficiency using a gravure coating technique,[64] and Sakai et al. used screen printing to fabricate a bulk heterojunction solar cell reaching 2.35%.[65] The possibility of using printing methods to fabricate large continuous and pixelated photodetectors offers exciting opportunities. The largest single-element detector currently available is a 20 in photomultiplier tube, and the largest two-dimensional sensors are approximately 2000 cm^2. Printed OPDs offer the prospect of creating much larger devices at significantly lower cost.

The best OPV devices are currently fabricated on rigid glass substrates coated with ITO, and so they do not make full use of the processing advantages of organic materials. In particular, they do not exploit the potential for low-cost large-area reel-to-reel manufacturing, which requires the use of a conformable substrate. Accordingly, in recent years, many researchers have sought to fabricate organic devices on flexible ITO-coated plastic substrates.[66–69] Brabec and co-workers at Linz, for instance, have reported 6 cm × 6 cm PV devices on ITO-coated plastic substrates with solar power conversion efficiencies of 3%.[70] The use of such substrates, however, presents two significant technological problems. The first problem is that plastics, unlike glass, are highly permeable to oxygen and water, so additional barrier coatings must be incorporated into the substrates to prevent contamination (and associated degradation) of the active layer. These barrier coatings must balance the need for extremely low permeability (which, generally speaking, calls for thick and dense barrier coatings) with the conflicting needs for high transparency and flexibility (which require that the barrier layers be as thin as possible). One option is to use a single layer of material such as silicon oxide, which exhibits an extremely low permeability to oxygen and water. In practice, however, this approach is ineffective since unavoidable pinhole defects in the oxide layer provide pathways for moisture to reach the active layer. The most successful barrier coatings to date use a stack of different materials, e.g., alternating thin layers of polymer and ceramics, to create a highly impermeable film. Pinhole defects are still present in these composite films, but they generally extend across a single layer only and hence do not create continuous moisture pathways to the active layer. Today's barrier coatings still lack the degree of transparency and flexibility one ideally requires, and further development is needed in this regard.

The second problem is that plastics suffer from significant thermal shrinkage and are therefore only suited to low-temperature processing methods. ITO by contrast should ideally be deposited and annealed above 350°C to achieve optimal film quality.[71] The deposition of ITO on plastic substrates must be carried out at significantly lower temperatures, resulting in porous films with lower conductivity, reduced transparency and poorer substrate adhesion, which leads ultimately to reduced device quality.[71] These issues, together with the tendency of ITO to crack when the substrate is bent, have led researchers to seek alternative anode materials for flexible applications. A variety of materials systems have been proposed including other metal oxides,[72] thin metallic films,[73] polymer-metal composites,[71] polymer-fullerene composites,[74] and conducting polymers.[75–83] Conducting polymers are especially appealing because they can be deposited over large areas using printing methods, with obvious potential for reel-to-reel implementation. There are still considerable issues to be addressed before conducting polymers become a suitable replacement for ITO—most notably the need for improved transparency and conductivity—but materials such as polyaniline and PEDOT:PSS already show considerable promise as polymeric anodes. PEDOT:PSS is used extensively as an antistatic coating in the photographic film industry, where it is deposited from water-based solution by roll-to-roll coating.

In recent work, Huang et al.[81, 83] have reported efficient ITO-free P3HT:PCBM based organic photodiodes using flexible polyethylene-terephthalate (PET) substrates coated with PEDOT:PSS (Fig. 6.27). The devices were fabricated by depositing a patterned layer of PEDOT:PSS directly on a PET substrate, followed by a spin-coated film of the donor/acceptor blend, and finally a thermally evaporated aluminum cathode. The devices had high external quantum efficiencies of 58%, comparable to similar devices fabricated on rigid glass substrates, indicating that PEDOT:PSS is a credible replacement for ITO in organic photodetectors.[†] Importantly, the devices had low dark currents of 1.5 pA/mm^2 under near short-circuit conditions, slightly lower than comparable devices fabricated on ITO-coated glass. Moreover, they exhibited very low dark currents under reverse bias, rising to just 0.12 nA/mm^2 at a reverse bias of 1 V—nearly three orders of magnitude lower than a comparable device on ITO-coated glass.

[†]PEDOT:PSS is currently less suitable as an anode material for organic solar cells due to its relatively high resistivity (e.g., 1 Ω·cm for the Baytron P formulation[84]), which causes significant internal energy dissipation in the PEDOT:PSS contact. However, new materials systems such as vapor-phase polymerized PEDOT[85] (which eliminates the need for insulating PSS) are starting to challenge indium tin oxide in terms of conductivity and transparency. These efforts may ultimately lead to high-efficiency solar cells based on PEDOT or a similar conducting polymer.

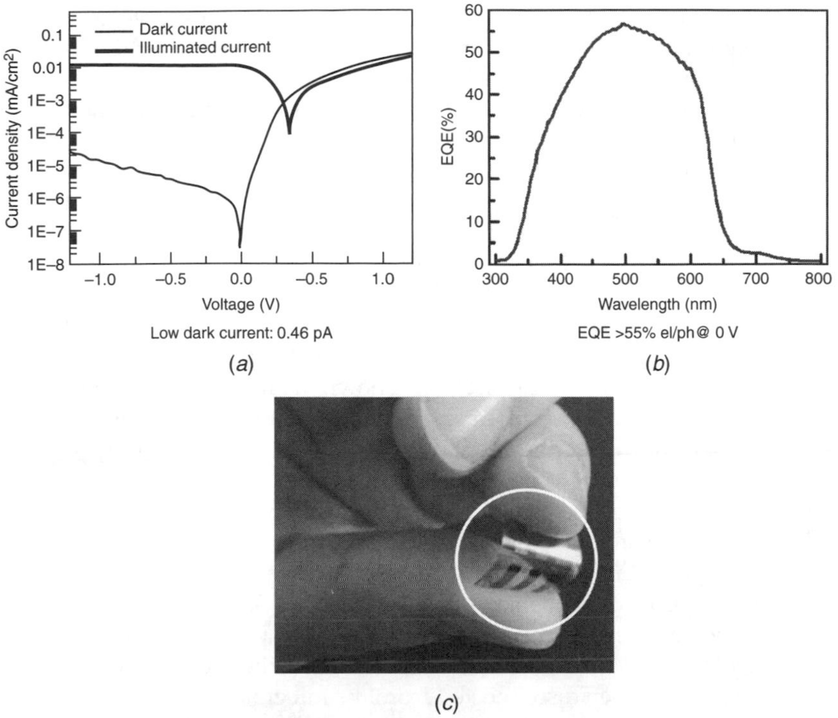

FIGURE 6.27 (*a*) Current-voltage curves, (*b*) spectral response curve, and (*c*) photograph of a flexible ITO-free photodiode fabricated on a PET plastic substrate. The structure of the device was PEDOT:PSS/P3HT:PCBM/Al. (*From Ref. 83—reproduced by permission of the Royal Society of Chemistry, 2007.*)

6.7.2 X-Ray Imaging

In X-ray imaging (XRI), a uniform source of X-rays is passed through an obstacle—often human tissue—and the transmitted light is detected using a two-dimensional panel of photodiodes. X-rays are difficult to focus, and the XRI panels must therefore be at least as large as the objects they are used to image. Crystalline silicon wafers are available only with diameters up to 300 mm, and for substantially larger applications, amorphous semiconductors must be used. The most widely used material for XRI is amorphous silicon (a-Si).[86] However, due to its low atomic mass, a-Si is unable to detect X-rays directly and must be used in conjunction with a scintillator screen that emits visible photons when struck by X-rays. Higher atomic mass materials such as amorphous selenium (a-Se) can detect X-rays directly and hence do not require a scintillator. Both a-Si and a-Se can be deposited onto large-area substrates by chemical vapor deposition, and panels of up to about 1600 cm^2 are currently available.[87] However, due to the complexity of the fabrication process, the panels are expensive, and there is significant interest in alternative technologies that would permit larger panels to be fabricated at substantially reduced cost.

OPV devices are especially appealing in this respect due to their excellent optoelectronic characteristics and their amenability to large-area, low-cost printing. The use of flexible plastic substrates raises the possibility of creating low-cost, large-area conformable panels that can adapt to the shape of the object or the radiation source, offering superior resolution and imaging capabilities. With these benefits in mind, Blakesley and Speller[87] undertook a conceptual feasibility study to determine whether OPV panels could provide a viable alternative to existing panel technologies.[87] They modeled the behavior of an OPV array on top of an a-Si backplane of thin-film transistors as shown in Fig. 6.28. The operating principle of the OPV panels is most easily understood by considering a single pixel, which comprises a photodiode and an associated transistor. The anode of the photodiode is held at a fixed negative potential V_b, and the cathode is connected to the source terminal of the transistor. The drain terminal is connected to the inverting input of an op-amp integrator. The non-inverting input of the op-amp is permanently grounded, and so the inverting input behaves as a virtual earth. To start, the pixel is reset by opening the gate, causing a reverse bias of approximately V_b to be applied to the photodiode and thereby charging the capacitor to this value. The pixel is then disconnected from the integrator by closing the gate. The pixel is illuminated, which generates a negative photocurrent that partially discharges the capacitor. The pixel is then reconnected to

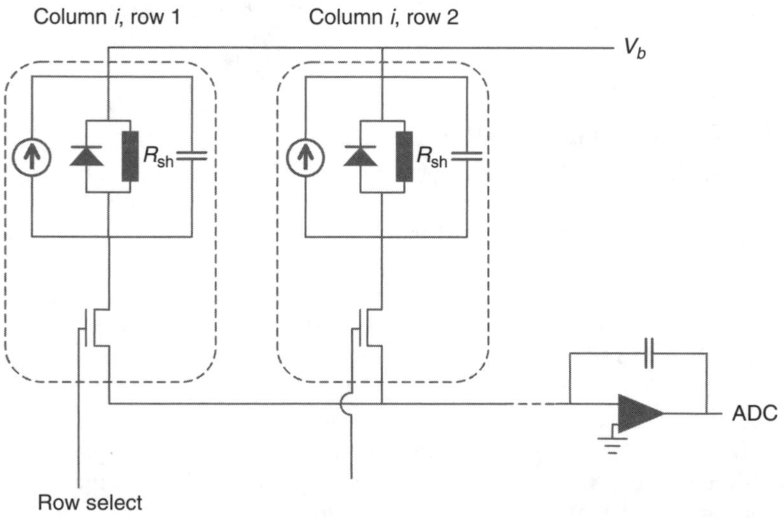

Figure 6.28 A simple passive pixel circuit for an organic photodiode array. (*Adapted with permission from Ref. 87, American Association of Physicist in Medicine, 2008.*)

the integrator by opening the gate, and the photodiode capacitor is recharged to a value V_b. This requires an amount Q of charge to flow that—assuming an ideal transistor with infinite off resistance and zero on resistance—is equal to the charge generated by the photodiode. The integrator generates an output voltage that is proportional to Q, which is then read using an analog-to-digital converter (ADC). The panel operates in a similar way, except there is one integrator/ADC for each column of the array, and the pixels are read on a row-by-row basis. The pixels are first reset by opening and closing their gates. They are then briefly illuminated with the X-ray image. The pixels in the first row are selected by opening their gates and read as described above. The first row is deselected, and the read process is repeated for each row in turn.

The dark current is an important source of error in photodiode sensor arrays. There is a delay between resetting each pixel and reading its charge, during which a dark current will flow; each photodiode is reverse-biased, so the dark current flows in the same direction as the photocurrent and leads to faster discharging of the pixels. To correct for this effect, the measurement can be repeated in the dark and the background count subtracted from the signal count. This procedure is valid, however, only if the dark currents are the same for the two measurements. In reality, the dark current changes continuously as the capacitor discharges and the device bias diminishes. A significant change in the device bias would also reduce the quantum efficiency, leading to non-linear behavior. The change ΔV in the device bias due to the photogenerated charge Q is given by

$$\Delta V = \frac{Q}{C} \tag{6.71}$$

Hence, a large pixel capacitance will minimize the change in the device bias and so enable an accurate dark current correction to be made. At the same time, too high a capacitance will generate excessive thermal noise and lead to extended readout times. Interestingly, Blakesley and Speller found a photodiode capacitance in the region of 1 to 20 pF to be optimal for a 100 μm × 100 μm pixel, which compares to about 4 pF for a typical organic photodiode of that size. This is one application where the high capacitance density of organic photodiodes is not an issue and, if anything, it is a little low (and so needs supplementing with an external capacitor). Blakesley and Speller further reported that OPV panels on a plastic substrate could offer significant improvements in terms of resolution and X-ray detection efficiency, and would be competitive with existing technologies providing the individual photodiodes had a quantum efficiency of 40 to 50% and a dark current density of less than 10 pA/mm^2. The authors concluded that "research in the future needs to focus on reducing dark currents to less than 10 pA/mm^2 while maintaining quantum

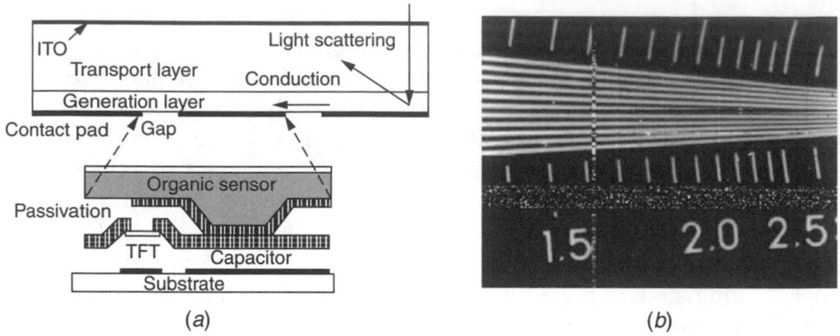

FIGURE 6.29 (*a*) Cross section of a pixel in an X-ray imaging circuit comprising an organic photodiode and an amorphous silicon driver. (*b*) Photograph of an X-ray image obtained with the sensor array. (*Reprinted from Ref. 88. Copyright 2002, with permission from Elsevier.*)

efficiencies of ideally 40–50% or greater." In practice, as we have shown above, these performance levels are quite achievable using devices available today, so organic photodiode X-ray imagers have very considerable promise.

There have been relatively few reports of OPD-based XRIs, but Street and coworkers have reported a prototype 512 × 512 array with a pixel size of 100 μm × 100 μm. Their system used a back plane of a-Si TFTs, which was coated with a simple continuous layer of the organic sensor material (see Fig. 6.29*a*). The address circuitry for each pixel was similar to Fig. 6.28 and comprised an a-Si thin film transistor, a 0.4 pF storage capacitor, the address lines, and a contact pad to the photodiode. The organic photodiode itself was based on a discrete heterojunction geometry, comprising a 300 nm vacuum-deposited layer of benzimidazole perylene (BZP) and a 10 μm blade-coated hole transport layer of tetraphenyldiamine (TPD) dispersed in a binder.† The pixels were completed with a semitransparent layer of evaporated Au/Pt. Finally, a GdO_2S_2:Tb scintillator screen (which emits at 550 nm) was placed on top of the array. Figure 6.29*b* shows an XRI image obtained using the panel; detailed performance characteristics were not provided.

The detectors in XRI systems experience varying amounts of X-ray exposure depending on factors such as the X-ray source intensity,

†The large thickness of the TPD layer was intended to minimize the dark current although, as shown above, with careful fabrication a conventional thin-film structure would suffice. The hole mobility in the TPD layer was reported to be about 10^{-5} $cm^2/(V\cdot s)$, which corresponds to a transit time of 3 ms under a 20 V reverse bias, which is broadly sufficient for video imaging purposes.

physical configuration, choice of substrate material, and frequency of use. X-ray exposure can cause a number of chemical changes in polymeric materials—cross-linking, chain scission, ionic desorption, etc.—and hence might be expected to degrade optoelectronic properties. Keivanidis and coworkers[89] investigated the aging of a variety of OPV devices under cumulative radiation doses of up to 500 Gy—typical of the total exposure experienced by a diode array inside a chest X-ray system over its full operational lifetime. The researchers considered three separate bulk heterojunction systems based on three commonly used donor-acceptor complexes: P3HT:PCBM, F8BT:PDI, and TFB: PDI (Fig. 6.30). The devices were fabricated by spin-coating a 70 nm layer of PEDOT:PSS onto ITO-coated glass substrates, followed by a 100 to 200 nm layer of the blended active material, and finally a 70 nm layer of evaporated aluminum. The spectral response curves were measured before and after exposure to 500 Gy X-ray radiation through the glass substrate. The P3HT:PCBM device showed a 17% reduction in quantum efficiency due to X-ray exposure, compared to 2% for F8BT:PDI and 3% for TFB:PDI. The dark currents of all three devices were unaffected by X-ray exposure. The reduction in the quantum efficiency of the P3HT:PCBM device was tentatively attributed to the effects of secondary electrons, which are released when X-rays

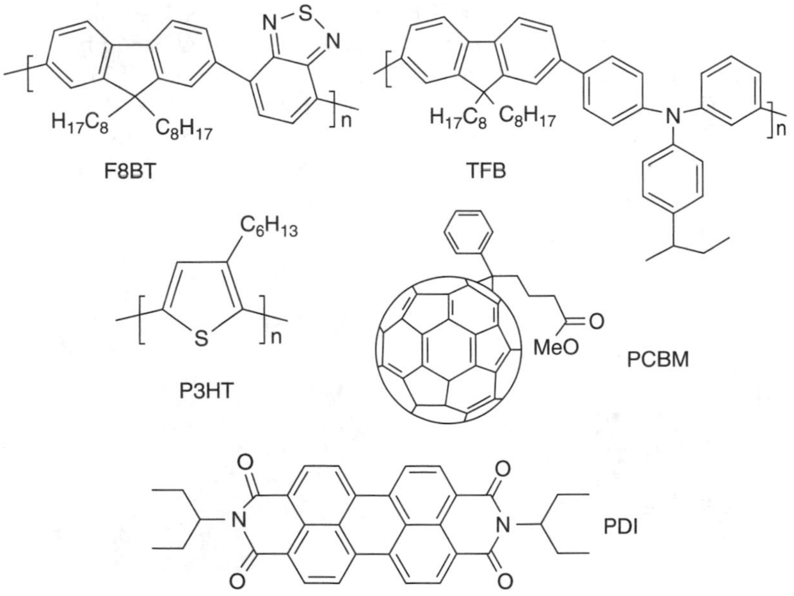

FIGURE 6.30 Chemical structures of three donor/acceptor complexes tested by Keivanidis and coworkers for stability against X-ray-induced degradation. (*Reprinted with permission from Ref. 89. Copyright 2008, American Institute of Physics.*)

interact with glass, and it was suggested that improved stability against might be achieved using plastic substrates. The other two material systems exhibited a remarkable tolerance to X-ray exposure, and it is apparent from these studies that organic materials are able to withstand the lifetime X-ray doses experienced in standard XRI applications.

Although we have focused here on XRI applications, there are a number of other applications where large-area imagers would be of value. The most obvious, perhaps, is document scanning where an extended 2D image sensor offers the possibility of faster image acquisition than conventional moving 1D scanners, while also removing the need for mechanical parts. Other applications where large-area scanners could be of great utility are X-ray crystallography and neutron detection, the latter being widely used in industry to detect corrosion. The above 2D sensors used amorphous silicon back panels to address the pixels, and hence did not make full use of the low-cost processing advantages offered by organic devices. Someya et al. have demonstrated a paper-thin image scanner on a plastic PEN substrate that combines organic photodiodes with organic transistors (Fig. 6.31).[90] The organic photodiodes were fabricated using bilayers of copper phthalocyanine and 3,4,9,10-perylene-tetracarboxylic-diimide (see Fig. 6.7). The transistors used pentacene as the active material and polyimide as the gate dielectric, and had a top-contact geometry with a channel length of 18 μm and a mobility of 0.7 $cm^2/(V \cdot s)$. They had reasonable on/off ratios of up to 10^5. The individual pixels exhibited good linearity up to 100 mW/cm^2. The array comprised 5184 pixels and had an effective sensing area of 5×5 cm^2 with a resolution of 36 dpi. Its thickness was 0.4 mm and it weighed just 1 g, making it especially attractive for portable electronic applications. The flexible array was able to distinguish black and white patterns (Fig. 6.31*b*) and could be used to accurately scan images on a curved surface.

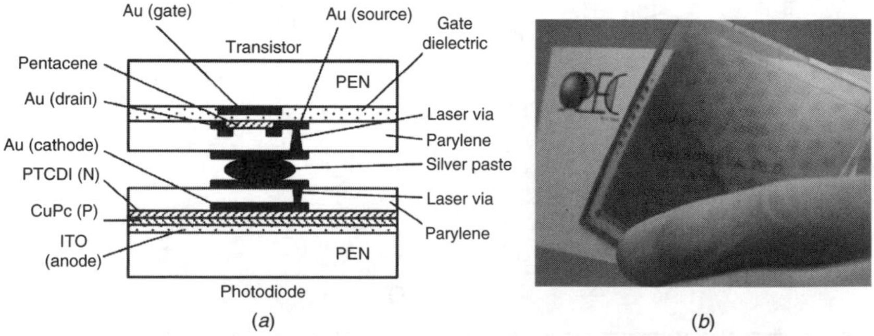

FIGURE 6.31 An all-organic scanner comprising organic transistors and organic photodiodes.(*Reprinted with permission from Ref. 90. Copyright 2005 IEEE.*)

Finally we note that the above discussion has focused on passive pixel sensors (PPSs), in which the transistor is used merely as a means of controlling the connection of the photodiode to an external amplifier. This kind of approach is appropriate for large-area applications where cost considerations mean it is impractical to provide each pixel with its own dedicated amplifier circuitry. However, PPS systems do not provide in situ amplification and can suffer from excessive noise arising from the capacitance of the data lines. In active pixel sensors (APSs), every photodiode has three or four dedicated transistors that play the dual role of pixel selection and in situ amplification. Tedde et al. have shown[91] that APS panels provide superior signal-to-noise performance and improved scalability to large-area panels, albeit at the expense of increased circuit complexity. Using a pixel circuit comprising an OPD on top of three a-Si TFTs, they were able to achieve a detection limit of 1 $\mu W/cm^2$ compared to 6 $\mu W/cm^2$ for a comparable PPS system. They reported good signal linearity and an in situ amplification of up to 10 using the APS pixel circuitry.

6.7.3 Diagnostics

The low-cost nature of organic semiconductor devices raises exciting opportunities in the areas of point-of-care and self-test diagnostics. These markets are notoriously price-sensitive and have historically been dominated by products with manufacturing costs in the cents to low dollars range. The current state of the art in this price bracket is the lateral-flow test,[92] which is the technology behind home fertility and pregnancy tests and involves a simple color change in response to an analyte of interest. Lateral flow tests are qualitative (or at best semiquantitative), typically offering results of the yes/no or high/medium/low variety. In many situations, however, quantitative data are required. The medical practitioner, for instance, may need to monitor the progression of a disease or the efficacy of a pharmaceutical treatment by determining the precise concentration of a specific biomarker. This currently requires that a blood or urine sample be sent to a remote laboratory where a quantitative assay is carried out using conventional analytical chemistry procedures. The ability to carry out such tests directly at the point of care would bring significant benefits: in particular, it would eliminate the need for patients to make repeat visits, thereby freeing up clinician's time, bringing forward the initiation of treatment, improving recovery prospects, and lowering treatment costs. Importantly such tests could also be deployed in developing countries that lack the centralized infrastructure needed for conventional testing, and hence would help address serious deficiencies in current health care provision.

There are relatively few technologies that offer a viable low-cost solution to quantitative point-of-care chemical analysis, but one promising approach is the use of microfluidic devices.[93] The microfluidics field aims to transform the analytical sciences, using techniques developed in the silicon processing industry, to engineer miniature devices on which chemical and biological processing can take place under precisely controlled conditions. Miniaturization of chemical processing provides significant advantages over conventional systems. These include (1) reduced analysis times, (2) cost reductions through downscaled fabrication and reduced consumption of reagents, (3) superior control of reaction conditions, (4) enhanced ability to carry out parallel processing, and (5) the ability to perform in-the-field or point-of-care measurements. Typical microfluidic devices are fabricated by forming continuous channels in silicon, glass, or plastic substrates and then sealing the channels using a second substrate as a lid (Fig. 6.32). All the standard chemical processing steps can be carried out in microfluidic devices, e.g., filtering, mixing, heating, cooling, separation.

The workhorse of clinical diagnostics is the immunoassay.[92] In simple terms, an immunoassay is a biochemical test that quantitatively measures the presence of a specific analyte in a biological fluid (usually serum, urine, or saliva) using the specific reaction of an antibody to its antigen. The final signal is usually optical, e.g., a qualitative color change or a quantitative absorption or emission signal.

FIGURE 6.32 Typical microfluidic devices. (*Picture courtesy of Wikipedia.*)

Basic blood chemistry panels comprising 10 to 20 quantitative tests are typically analyzed using automated benchtop instruments located in a centralized laboratory. The ability to replace such instruments with low-cost disposable devices would transform modern health care, allowing immediate on-the-spot testing. Immunoassays are readily integrated into microfluidic devices[94–96] and, importantly, can be implemented using capillarity-based fluid delivery schemes that passively draw the sample into the chip from an entry port without the need for external pumps. The microfluidic platform is therefore a promising basis for an entirely self-contained diagnostic device.

Microfluidic devices have shown themselves to be highly effective for laboratory-based research, where their superior analytical performance has established them as efficient tools for complex tasks in genetic sequencing, proteomics, and drug discovery applications. However, to date they have not been well suited to point-of-care or in-the-field applications, where cost and portability are of primary concern. Although the chips themselves are cheap and small, they are generally used in conjunction with bulky (off-chip) optical detectors, which are needed to quantify the optical signal. Although there have been some attempts to integrate optics within the chip structure itself, few have demonstrated the high levels of integration demanded of a point-of-care stand-alone system.

One promising option for creating integrated light sources and detectors is the use of organic semiconductor devices.[97–99] In use, the organic light-emitting diode (OLED) and photodetector are arranged in a face-on geometry on the top and bottom surfaces of a microfluidic chip with a channel in between (Fig. 6.33): biolabels present in the detection volume of the channel absorb photons from the LED which they may subsequently reemit as lower-energy photons.

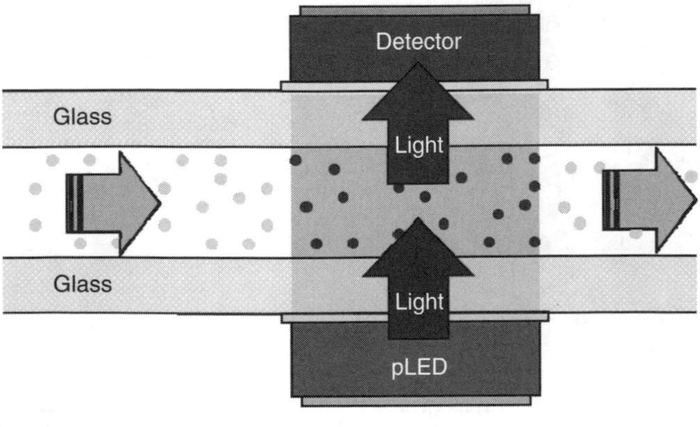

FIGURE 6.33 Detection geometry for integrated emission and absorption measurements in microfluidic devices.

Depending on the nature of the optical filtering employed, the photodiode detects the intensity of either the transmitted light from the OLED or the emitted light from the biolabel; either way, it is possible to deduce the concentration of the biolabel (and hence the analyte). Importantly, by splitting the sample stream, it is possible to perform multiple immunoassays in parallel, which can be monitored using separate LED/photodetector pairs. In an alternative configuration, the need for the OLED light source can be eliminated altogether by using a chemiluminescent assay that generates its own emission.[98] Chemiluminescence (CL) reactions typically involve the formation of a metastable reaction intermediate in an electronically excited state, which subsequently relaxes to the ground state with the emission of a photon. CL is particularly attractive for portable microfluidic assays, because the CL reaction acts as an internal light source, thereby lowering instrumental requirements and significantly reducing power consumption and background interference compared to fluorescence assays.

Figure 6.34*a* shows a prototype device developed by Molecular Vision Ltd. for absorption- or fluorescence-based detection of a single analyte. The prototype device comprises a filtered blue OLED light source in the lid. The base of the demonstrator contains an organic photodetector with integrated filtering plus transimpedance amplifier and readout electronics. In a fluorescence configuration—the most sensitive and versatile of the three detection methods—the system is able to detect fluorescent beads in the picomolar concentration range, which is sufficient for a wide range of diagnostic tests (Fig. 6.35). Figure 6.34*b*

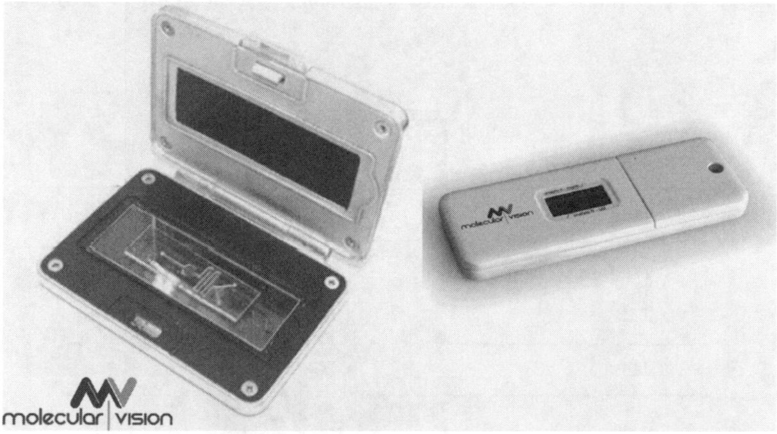

(*a*) (*b*)

Figure 6.34 (*a*) Early prototype device for emission- and absorption-based chemical assays. (*b*) Recent version offering capability for multiple-analyte testing. (*Reproduced with permission from Molecular Vision Ltd.*)

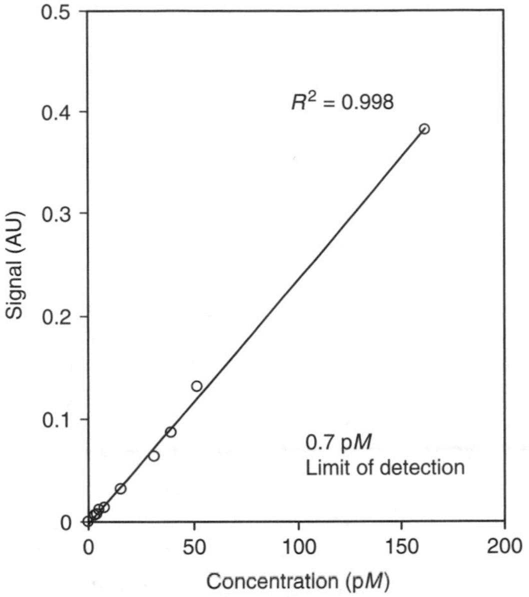

Figure 6.35 Dose-response curve for fluorescent beads obtained using the prototype device shown in Fig. 6.34*a*. (*Reproduced with permission from Molecular Vision Ltd.*)

shows a newer prototype device that offers equivalent performance capabilities in a smaller format and allows for the testing of multiple biolabels in parallel.

The above prototype devices confirm the feasibility of using organic devices for sensitive diagnostic testing, but conform to the familiar "cartridge plus reader" format, in which a disposable test device is plugged into a reusable reader that contains the optics and detection electronics. The real value of using organic devices will be realized in future work where the organic light sources and photodetectors will be directly printed onto the microfluidic chip itself. While adding only marginal size, weight, and cost to the microfluidic devices, this would improve detection efficiencies (by bringing the optics into closer proximity to the reaction channel), enable easy parallel interrogation of multiple-reaction channels, and eliminate the need for a separate reader. (In one possible format, the microfluidic device could be plugged into the data port of a mobile phone. The phone would provide power, data processing and display capabilities, removing the need for a dedicated reader.) The combined OLED/microfluidic/OPV architecture offers a promising route to low cost self-contained panel tests that would be difficult to implement at the necessary price point using standard inorganic components. There are numerous other potential applications of organic devices in chemical and biological analysis, many of which are described in other chapters of this book.

6.8 Conclusions

This chapter has focused on the emerging field of organic photodiodes. OPDs have received far less attention than the other major organic devices, but it is fast becoming clear that they hold very great promise as photodetectors, combining excellent optoelectronic performance with simple scalable processing and thin, lightweight, flexible form factors. There are already a few obvious application areas where these benefits have been recognized, and OPV devices are being used to significant advantage (albeit at a precommercial level at present), and many more are bound to emerge in the coming years. In writing this chapter it is our hope that more researchers will be encouraged to turn their attention to OPD development. The vast majority of organic photodetector research to date has been carried out as a curiosity-driven offshoot of standard solar cell research, rather than a discipline in its own right. This is unfortunate as the technical challenges involved in designing high-performance OPDs are quite distinct from those involved in designing efficient solar cells. New materials systems, device architectures, and fabrication techniques are urgently needed if OPD technology is to progress sufficiently to challenge the incumbent competition. These advances, however, are only likely to come about as the result of a dedicated research effort focused squarely on OPD optimization. The organic photodetector field offers some fascinating scientific challenges and tremendous technological opportunities, and we wholeheartedly encourage the interested researcher to get involved.

Appendix: Noise Analysis

Determining the Thermal Noise of a Resistor

In an RC network, the resistor is the only circuit element that contributes thermal noise.[51] Hence, to determine the noise characteristics of a resistor, we can consider a simple RC circuit, knowing that all noise in the circuit is directly attributable to the resistor. We model the resistor as a (mythical) noise-free resistor R in series with a noise source of unknown amplitude $\widetilde{\sigma_V}$ per square root of frequency. The situation is depicted in Fig. 6.36 where it is apparent that the resistor and capacitor act as a potential divider for the noise source.

In the case of a white noise source, a noise source of frequency f and bandwidth Δf will generate an amount $\widetilde{\sigma_V}\sqrt{\Delta f}$ of noise. Treating

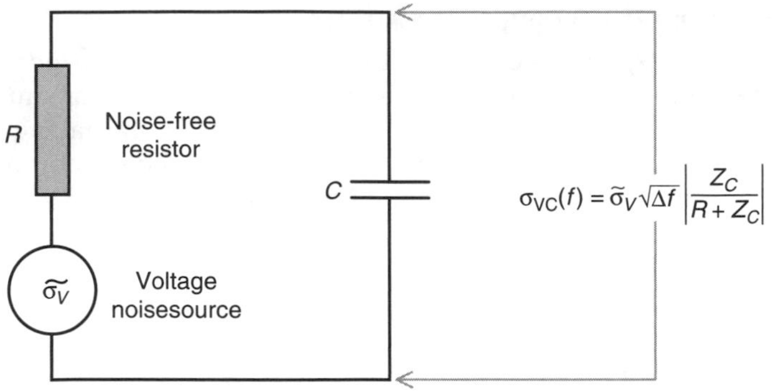

FIGURE 6.36 A noisy resistor can be represented by a noise-free resistor in series with a noisy voltage source. The noise due to the resistor can be determined by considering the behavior of an RC circuit.

the capacitor and resistor as a potential divider, the corresponding voltage σ_{VC} across the capacitor is given by

$$\sigma_{VC}(f) = \widetilde{\sigma_V}\sqrt{\Delta f}\left|\frac{Z_C}{R+Z_C}\right| = \widetilde{\sigma_V}\sqrt{\Delta f}\left|\frac{1/(2\pi jfC)}{R+1/(2\pi jfC)}\right| = \widetilde{\sigma_V}\sqrt{\Delta f}\left|\frac{1}{1+2\pi jfCR}\right|$$
$$= \frac{\widetilde{\sigma_V}\sqrt{\Delta f}}{\sqrt{1+(2\pi fCR)^2}}$$

Hence, squaring and integrating over all frequencies, we obtain for the total noise variance

$$\sigma_{VC}^2 = \int_0^\infty \frac{\widetilde{\sigma_V^2}\,df}{1+(2\pi fCR)^2} = \frac{\widetilde{\sigma_V^2}}{4CR}$$

Rearranging, we obtain

$$\widetilde{\sigma_V^2} = 8R\left(\frac{1}{2}C\sigma_{VC}^2\right) = 8R\langle U\rangle$$

where $\langle U\rangle$ can be identified as the time-averaged energy stored by the capacitor.[51] From the equipartition theory of classical thermodynamics,[51] $\langle U\rangle$ is equal to $1/2k_BT$. Hence, we obtain for $\widetilde{\sigma_V^2}$

$$\widetilde{\sigma_V^2} = 4k_BTR$$

Calculating the Output Noise Voltage

From Eq. (6.32), a noise component $\widetilde{\sigma_{Va}}$ of frequency f and bandwidth Δf will perturb the potential of the inverting input of the op-amp by an amount $\sigma_{Va}(f) = \widetilde{\sigma_{Va}}(f)\sqrt{\Delta f}$. Using Eq. (6.53) and taking into account the capacitance C_d of the photodiode, this noise component will appear at the output amplified by an amount $[1 + R_f/Z_{\text{sh}}(f)]$ where

$$|Z_{\text{sh}}(f)| = |R_{\text{sh}} \,||\, Z_{Cd}(f)| = \left|\frac{R_{\text{sh}}/j2\pi f C_d}{R_{\text{sh}} + 1/j2\pi f C_d}\right| = \frac{R_{\text{sh}}}{\sqrt{1+4\pi^2 f^2 C_d^2 R_{\text{sh}}^2}}$$

The output noise variance per unit frequency due to this noise component will therefore be

$$\sigma_{Va}^2(f) = \widetilde{\sigma_{Va}^2}\left|1 + \frac{R_f}{Z_{\text{sh}}(f)}\right|^2 \Delta f$$

which, after some rearrangement, yields

$$\sigma_{Va}^2(f) = \widetilde{\sigma_{Va}^2}\left[\left(1 + \frac{R_f}{R_{\text{sh}}}\right)^2 + 4\pi^2 f^2 C_d^2 R_{\text{sh}}^2\right]\Delta f$$

Hence, integrating over the full measurement bandwidth, we obtain for the total mean squared noise at the output

$$\sigma_{Va}^2 = \int_0^B \widetilde{\sigma_{Va}^2}\left[\left(1 + \frac{R_f}{R_{\text{sh}}}\right)^2 + 4\pi^2 f^2 C_d^2 R_{\text{sh}}^2\right] df$$

$$\sigma_{Va}^2 = \left[\left(1 + \frac{R_f}{R_{\text{sh}}}\right)^2 B + \frac{4\pi^2}{3} C_d^2 R_f^2 B^3\right]\widetilde{\sigma_{Va}^2}$$

References

1. S. Flyckt and C. Marmonier, *Photomultiplier Tubes: Principles and Applications*, Philips Photonics, Brive, Francem, 2002.
2. S. Donati, *Photodetectors*, Prentice-Hall PTR, 1999.
3. M. McClish, R. Farrell, R. Myers, F. Olschner, G. Entine, and K. S. Shah, *Nuclear Instruments & Methods in Physics Research Section a—Accelerators Spectrometers Detectors and Associated Equipment*, 567:36 (2006).
4. "Avalanche Photodiode: A User's Guide" http://optoelectronics.perkinelmer.com.content/applicationnotes/appapduserguide.pdf
5. X. J. Wang, E. Perzon, J. L. Delgado, P. de la Cruz, F. L. Zhang, F. Langa, M. Andersson, et al., *Appl. Phys. Lett.*, 85:5081 (2004).

6. T. P. I. Saragi, R. Pudzich, T. Fuhrmann, and J. Salbeck, *Appl. Phys. Lett.*, 84:2334 (2004).
7. T. P. I. Saragi, K. Onken, I. Suske, T. Fuhrmann-Lieker, and J. Salbeck, *Opt. Mat.*, 29:1332 (2007).
8. Z. E. Ooi, T. L. Tam, A. Sellinger, and J. C. deMello, *Energy Environ. Sci.*, 1:300 (2008).
9. Y. Terao, H. Sasabe, and C. Adachi, *Appl. Phys. Lett.*, 90:103515 (2007).
10. Z. R. Hong, B. Maennig, R. Lessmann, M. Pfeiffer, K. Leo, and P. Simon, *Appl. Phys. Lett.*, 90:203505 (2007).
11. S. Yoo, B. Domercq, and B. Kippelen, *Appl. Phys. Lett.*, 85:5427 (2004).
12. C. W. Tang, *Appl. Phys. Lett.*, 48:183 (1986).
13. J. Drechsel, B. Mannig, F. Kozlowski, M. Pfeiffer, K. Leo, and H. Hoppe, *Appl. Phys. Lett.*, 86:244102 (2005).
14. M. Paulsson and S. Stafstrom, *Syn. Met.*, 101:469 (1999).
15. C. Rogero, J. I. Pascual, J. Gomez-Herrero, and A. M. Baro, *J. Chem. Phys.*, 116:832 (2002).
16. K. Itaka, M. Yamashiro, J. Yamaguchi, M. Haemori, S. Yaginuma, Y. Matsumoto, M. Kondo, et al., *Adv. Mat.*, 18:1713 (2006).
17. B. Kraabel, C. H. Lee, D. McBranch, D. Moses, N. S. Sariciftci, and A. J. Heeger, *Chem. Phys. Lett.*, 213:389 (1993).
18. N. S. Sariciftci, L. Smilowitz, A. J. Heeger, and F. Wudl, *Science*, 258:1474 (1992).
19. N. S. Sariciftci, D. Braun, C. Zhang, V. I. Srdanov, A. J. Heeger, G. Stucky, and F. Wudl, *Appl. Phys. Lett.*, 62:585 (1993).
20. A. K. Pandey, K. N. N. Unni, and J. M. Nunzi, *Thin Solid Films*, 511:529 (2006).
21. K. Tada, M. Onoda, H. Nakayama, and K. Yoshino, *Syn. Met.*, 102:982 (1999).
22. J. H. Burroughes, D. D. C. Bradley, A. R. Brown, R. N. Marks, K. Mackay, R. H. Friend, P. L. Burns, et al., *Nature*, 347:539 (1990).
23. A. C. Arias, N. Corcoran, M. Banach, R. H. Friend, J. D. MacKenzie, and W. T. S. Huck, *Appl. Phys. Lett.*, 80:1695 (2002).
24. M. Granstrom, K. Petritsch, A. C. Arias, A. Lux, M. R. Andersson, and R. H. Friend, *Nature*, 395:257 (1998).
25. R. Pacios and D. D. C. Bradley, *Syn. Met.*, 127:261 (2002).
26. S. E. Shaheen, C. J. Brabec, N. S. Sariciftci, F. Padinger, T. Fromherz, and J. C. Hummelen, *Appl. Phys. Lett.*, 78:841 (2001).
27. D. Chirvase, J. Parisi, J. C. Hummelen, and V. Dyakonov, *Nanotechnology*, 15:1317 (2004).
28. S. Heutz, P. Sullivan, B. M. Sanderson, S. M. Schultes, and T. S. Jones, *Solar Energy Mat. Solar Cells*, 83:229 (2004).
29. Z. E. Ooi, T. L. Tam, R. Y. C. Shin, Z. K. Chen, T. Kietzke, A. Sellinger, M. Baumgarten, et al., *J. Mat. Chem. (in press)* 2008.
30. J. Morgado, E. Moons, R. H. Friend, and F. Cacialli, *Syn. Met.*, 124:63 (2001).
31. J. Parisi, V. Dyakonov, M. Pientka, I. Riedel, C. Deibel, C. J. Brabec, N. S. Sariciftci, et al., *Zeitschrift Fur Naturforschung Section a-a J. Phys. Sci.*, 57:995 (2002).
32. H. Hoppe, M. Niggemann, C. Winder, J. Kraut, R. Hiesgen, A. Hinsch, D. Meissner, et al., *Adv. Funct. Mat.*, 14:1005 (2004).
33. M. Campoy-Quiles, T. Ferenczi, T. Agostinelli, P. G. Etchegoin, Y. Kim, T. D. Anthopoulos, P. N. Stavrinou, et al., *Nature Mat.*, 7:158 (2008).
34. G. Yu, J. Gao, J. C. Hummelen, F. Wudl, and A. J. Heeger, *Science*, 270:1789 (1995).
35. M. M. Wienk, J. M. Kroon, W. J. H. Verhees, J. Knol, J. C. Hummelen, P. A. van Hal, and R. A. J. Janssen, *Angewandte Chemie—International Edition*, 42:3371 (2003).
36. S. Bertho, I. Haeldermans, A. Swinnen, W. Moons, T. Martens, L. Lutsen, D. Vanderzande, et al., *Solar Energy Mat. & Solar Cells*, 91:385 (2007).
37. T. W. Lee and O. O. Park, *Adv. Mat.*, 12:801 (2000).
38. W. Geens, S. E. Shaheen, B. Wessling, C. J. Brabec, J. Poortmans, and N. S. Sariciftci, *Organic Electr.*, 3:105 (2002).
39. Y. Kim, S. Cook, S. M. Tuladhar, S. A. Choulis, J. Nelson, J. R. Durrant, D. D. C. Bradley, et al., *Nature Mat.*, 5:197 (2006).

40. J. L. Li, F. Dierschke, J. S. Wu, A. C. Grimsdale, and K. Mullen, *J. Mat. Chem.*, 16:96 (2006).
41. W. L. Ma, C. Y. Yang, X. Gong, K. Lee, and A. J. Heeger, *Adv. Funct. Mat.*, 15:1617 (2005).
42. M. Reyes-Reyes, K. Kim, and D. L. Carroll, *Appl. Phys. Lett.*, 87:083506 (2005).
43. C. J. Brabec, J. A. Hauch, P. Schilinsky, and C. Waldauf, *Mrs Bull.*, 30:50 (2005).
44. I. H. Campbell, T. W. Hagler, D. L. Smith, and J. P. Ferraris, *Phys. Rev. Lett.*, 76:1900 (1996).
45. P. A. Lane, P. J. Brewer, J. S. Huang, D. D. C. Bradley, and J. C. deMello, *Phys. Rev. B*, 74:7 (2006).
46. Z. E. Ooi, R. Jin, J. Huang, Y. F. Loo, A. Sellinger, and J. C. deMello, *J. Mat. Chem.*, 18:1644 (2008).
47. C. M. Ramsdale, J. A. Barker, A. C. Arias, J. D. MacKenzie, R. H. Friend, and N. C. Greenham, *J. Appl. Phys.*, 92:4266 (2002).
48. S. A. DiBenedetto, I. Paci, A. Facchetti, T. J. Marks, and M. A. Ratner, *J. Phys. Chem. B*, 110:22394 (2006).
49. D. Donnelly, *Computing Sci. & Eng.*, 8:92 (2006).
50. M. J. Usher, *J. Phys. E: Sci. Instrum.*, 7:957 (1974).
51. W. Jung, *Op Amp Applications Handbook*, Analog Devices Series, Newnes, Oxford, UK, 2004.
52. R. Rosen and H. A. Pohl, *J. Poly. Sci. Part A-1*, 4:1135 (1965).
53. M. Punke, S. Valouch, S. W. Kettlitz, N. Christ, C. Gartner, M. Gerken, and U. Lemmer, *Appl. Phys. Lett.*, 91:071118 (2007).
54. P. J. Brewer, P. A. Lane, J. S. Huang, A. J. deMello, D. D. C. Bradley, and J. C. deMello, *Phys. Rev. B*, 71:6 (2005).
55. R. A. Collins, G. Bowman, and Sutherland R. R., *J. Phys. D, Appl. Phys.*, 4:L49 (1971).
56. J. Y. Kim, K. Lee, N. E. Coates, D. Moses, T. Q. Nguyen, M. Dante, and A. J. Heeger, *Science*, 317:222 (2007).
57. D. Muhlbacher, M. Scharber, M. Morana, Z. G. Zhu, D. Waller, R. Gaudiana, and C. Brabec, *Adv. Mat.*, 18:2884 (2006).
58. E. Perzon, F. L. Zhang, M. Andersson, W. Mammo, O. Inganas, and M. R. Andersson, *Adv. Mat.*, 19:3308 (2007).
59. Y. Yao, Y. Y. Liang, V. Shrotriya, S. Q. Xiao, L. P. Yu, and Y. Yang, *Adv. Mat.*, 19: 3979 (2007).
60. A. D. Grishina, L. Y. Pereshivko, L. Licea-Jimenez, T. V. Krivenko, V. V. Savel'ev, R. W. Rychwalski, and A. V. Vannikov, *High Energy Chem.*, 41:267 (2007).
61. J. S. Huang and Y. Yang, *Appl. Phys. Lett.*, 91 (2007).
62. J. Reynaert, V. I. Arkhipov, P. Heremans, and J. Poortmans, *Adv. Funct. Mat.*, 16:784 (2006).
63. C. N. Hoth, S. A. Choulis, P. Schilinsky, and C. J. Brabec, *Adv. Mat.*, 19:3973 (2007).
64. D. Tobjork, H. Aarino, T. Mäkelä, R. Österbacka, *Mat. Res. Soc. Symp. Proc.*, 1091:1091-AA05-45 (2008).
65. J. Sakai, E. Fujinaka, T. Nishimori, N. Ito, J. Adachi, S. Nagano, and K. Murakami, *Conference Record of the Thirty-First IEEE Photovoltaic Specialists Conference—2005*, p. 125.
66. M. A. De Paoli, A. F. Nogueira, D. A. Machado, and C. Longo, *Electrochimica Acta*, 46:4243 (2001).
67. F. Padinger, R. S. Rittberger, and N. S. Sariciftci, *Adv. Funct. Mat.*, 13:85 (2003).
68. M. Al-Ibrahim, H. K. Roth, U. Zhokhavets, G. Gobsch, and S. Sensfuss, *Solar Energy Mat. & Solar Cells*, 85:13 (2005).
69. M. Al-Ibrahim, H. K. Roth, and S. Sensfuss, *Appl. Phys. Lett.*, 85:1481 (2004).
70. C. J. Brabec, N. S. Sariciftci, and J. C. Hummelen, *Adv. Funct. Mat.*, 11:15 (2001).
71. K. Yamada, K. Tamano, T. Mori, T. Mizutani, and M. Sugiyama, *Proceedings of the 7th International Conference on Properties and Applications of Dielectric Materials, vols.* 1–3, 2003.

72. K. Morii, M. Ishida, T. Takashima, T. Shimoda, Q. Wang, M. K. Nazeeruddin, and M. Gratzel, *Appl. Phys. Lett.*, 89 (2006).
73. J. Lewis, S. Grego, B. Chalamala, E. Vick, and D. Temple, *Appl. Phys. Lett.*, 85:3450 (2004).
74. J. S. Moon (Ed. J. H. Park), *Sixth International Conference on the Science and Application of Nanotubes*, Sweden, abstract, p. 214, 2005.
75. G. Gustafsson, Y. Cao, G. M. Treacy, F. Klavetter, N. Colaneri, and A. J. Heeger, *Nature*, 357:477 (1992).
76. A. N. Krasnov, *Appl. Phys. Lett.*, 80:3853 (2002).
77. B. Y. Ouyang, C. W. Chi, F. C. Chen, Q. F. Xi, and Y. Yang, *Adv. Funct. Mat.*, 15:203 (2005).
78. F. L. Zhang, M. Johansson, M. R. Andersson, J. C. Hummelen, and O. Inganas, *Adv. Mat.*, 14:662 (2002).
79. V. K. Basavaraj, A. G. Manoj, and K. S. Narayan, *IEE Proc. Circuits, Devices and Systems*, 150:552 (2003).
80. G. P. Kushto, W. H. Kim, and Z. H. Kafafi, *Appl. Phys. Lett.*, 86:093502 (2005).
81. J. Huang, X. Wang, Y. Kim, A. J. deMello, D. D. C. Bradley, and J. C. deMello, *Phys. Chem. Chemical Physics*, 8:3904 (2006).
82. J. Huang, X. Wang, A. J. deMello, J. C. deMello, and D. D. C. Bradley, *J. Mat. Chem.*, 17:3551 (2007).
83. J. Huang, R. Xia, Y. Kim, X. Wang, J. Dane, O. Hofmann, A. Mosley, et al., *J. Mat. Chem.*, 17:1043 (2007).
84. A. Elschner, F. Jonas, S. Kirchmeyer, K. Wussow, *Proc. ASIA Display/IDW 2001*, 01:1427 (2001).
85. S. Admassie, F. L. Zhang, A. G. Manoj, M. Svensson, M. R. Andersson, and O. Inganas, *Solar Energy Mat. & Solar Cells*, 90:133 (2006).
86. J. T. Rahn, F. Lemmi, J. P. Lu, P. Mei, R. B. Apte, R. A. Street, R. Lujan, et al., *IEEE Trans. on Nucl. Sci.*, 46:457 (1999).
87. J. C. Blakesley and R. Speller, *Med. Phys.*, 35:225 (2008).
88. R. A. Street, J. Graham, Z. D. Popovic, A. Hor, S. Ready, and J. Ho, *J. Non-Crystalline Solids*, 299:1240 (2002).
89. P. E. Keivanidis, N. C. Greenham, H. Sirringhaus, R. H. Friend, J. C. Blakesley, R. Speller, M. Campoy-Quiles, et al., *Appl. Phys. Lett.*, 92:023304 (2008).
90. T. Someya, Y. Kato, S. Iba, Y. Noguchi, T. Sekitani, H. Kawaguchi, and T. Sakurai, *IEEE Trans. Elec. Dev.*, 52:2502 (2005).
91. S. Tedde, E. S. Zaus, J. Furst, D. Henseler, and P. Lugli, *IEEE Electron. Dev. Lett.*, 28:893 (2007).
92. J. Gordon and G. Michel, *Clin. Chem.*, 54:A204 (2008).
93. A. J. deMello, *Nature*, 442:394 (2006).
94. T. G. Henares, F. Mizutani, and H. Hisamoto, *Analytica Chimica Acta*, 611:17 (2008).
95. H. Huang and X. Zheng, *Sheng Wu Yi Xue Gong Cheng Xue Za Zhi*, 24:928 (2007).
96. T. Tachi, N. Kaji, M. Tokeshi, and Y. Baba, *Bunseki Kagaku*, 56:521 (2007).
97. O. Hofmann, X. H. Wang, A. Cornwell, S. Beecher, A. Raja, D. D. C. Bradley, A. J. deMello, et al., *Lab on a Chip*, 6:981 (2006).
98. X. H. Wang, O. Hofmann, R. Das, E. M. Barrett, A. J. deMello, J. C. deMello, and D. D. C. Bradley, *Lab on a Chip*, 7:58 (2007).
99. X. H. Wang, P. A. Levermore, O. Hofmann, J. C. deMello, A. J. deMello, and D. D. C. Bradley, *Proceedings of the First Shenyang International Colloquium on Microfluidics*, 165 (2007).

CHAPTER 7

Organic Semiconductor Lasers as Integrated Light Sources for Optical Sensors

T. Woggon,* M. Punke,* M. Stroisch,* M. Bruendel,‡ M. Schelb,§ C. Vannahme,† T. Mappes,§ J. Mohr,† U. Lemmer*

7.1 Introduction

The field of organic optoelectronic devices is steadily growing.[1] One of the most prominent applications is the use of organic light-emitting diodes (OLEDs) in small displays, e.g., in mobile phones or MP3 audio players.[2] Other research areas include organic solar cells and photodetectors, being of great interest for large-area, low-cost applications.[3–5] Organic devices can be fabricated by relatively low-cost manufacturing methods on a wide range of different substrates.[6]

*Light Technology Institute (LTI), Universität Karlsruhe (TH), Karlsruhe, Germany
†Institut für Mikrostrukturtechnik, Forschungszentrum Karlsruhe GmbH, Karlsruhe, Germany
‡Institut für Mikrostrukturtechnik, Forschungszentrum Karlsruhe GmbH, Postfach 3640, 76021 Karlsruhe, Germany and is now with the Robert Bosch GmbH, Stuttgart
§Institut für Mikrostrukturtechnik, Universität Karlsruhe (TH), Karlsruhe, Germany

Organic lasers are a special type of organic light-emitting devices.[7] In contrast to OLEDs, these devices emit a highly directional, monochromatic beam of light. By using different materials the whole visible wavelength range can be covered by the laser emission.[8–11] These lasers can also be fabricated by low-cost methods,[12,13] rendering them interesting for an application in optical sensor systems. In the last years the vision of a complete analysis system integrated "on a chip," a so-called lab-on-a-chip, has attracted an increasing number of users and research groups.[14–18] Possible applications for such systems are in the areas of medicine, drug discovery, and environmental monitoring. Instead of shipping a sample for an analysis to a central laboratory, one could use such small, portable systems for point-of-care diagnostics.[19,20] Optical detection schemes such as laser-induced fluorescence, absorbance detection, or evanescent field sensing are commonly used approaches in sensor systems. Today, these techniques are usually based on external laser sources and complicated coupling optics being bulky and expensive. The monolithic integration of laser light sources onto the chip will result in small, portable, and far more flexible systems.[21–25] In this chapter we discuss first the fabrication and properties of organic semiconductor lasers. Some focus is put on the aspects of low-cost replication. Then we address the integration of organic lasers into optical sensor chips and the choice of sensing principles.

7.2 Organic Semiconductor Lasers

The term *organic semiconductor laser* applies to a large variety of devices that are currently under development. Common to most of them is the use of the basic laser approach consisting of an optical active material placed inside a resonator which provides the feedback required for efficient laser operation. In contrast to inorganic semiconductors, their organic counterparts typically exhibit a large spectral range with optical gain under optical pumping. This provides the possibility to tune the emission wavelength of the laser by just modifying a wavelength selective resonator.

7.2.1 Distributed Feedback Resonators

Among different approaches for the resonator design[26] the distributed feedback (DFB) resonator scheme is one of the most popular realizations due to its rather simple and efficient design. Typically, the organic distributed feedback laser consists of a thin-film active material that is deposited on top of a periodical corrugated substrate. The active material is optically pumped, e.g., by an external pulsed UV laser. The layers sequence forms a slab waveguide (thickness < 400 nm) confining the optical wave in the substrate plane. The so-called distributed feedback mechanism of the laser relies on Bragg scattering induced by a periodic modulation of the refractive index induced by surface corrugations

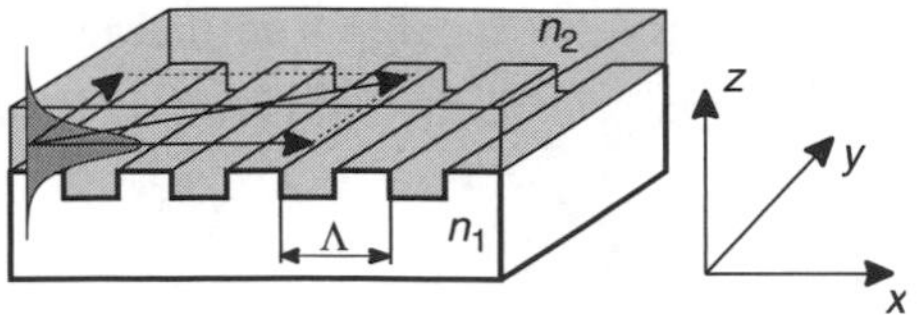

FIGURE 7.1 One-dimensional slab waveguide structure with a periodic variation of the refractive index.

(gratings) of the substrate. Figure 7.1 illustrates the distributed feedback mechanism in a slab waveguide structure with a one-dimensional, periodic variation of the refractive index.

The lattice constant Λ of the grating transforms to the reciprocal lattice constant **G** in *k*-space:

$$\mathbf{G} = \begin{pmatrix} 0 \\ \dfrac{2\pi}{\Lambda} \\ 0 \end{pmatrix}$$

Lasing operation occurs in plane and can be discussed by starting with a plane wave propagating in the slab with wave vector **k**:

$$\mathbf{k} = \begin{pmatrix} k_x \\ k_y \\ 0 \end{pmatrix} \qquad \text{with } |\mathbf{k}| = n_{\text{eff}} \cdot \frac{2\pi}{\lambda}$$

The effective refractive index n_{eff} represents the effective index of the propagating guided wave. For efficient feedback, which is essential for laser operation, constructive interference of a scattered wave is required. This is the case if the Bragg equation is fulfilled.

$$k_{x,m} = \frac{m}{2}|G| \qquad m \in \mathbb{N} \tag{7.1}$$

The positive integer factor m determines the scattering order. The scalar $k_{x,m=1}$ gives the smallest possible value of the wave vector component (corresponding to largest wavelength) in the direction of the refractive index modulation for which constructive interference is possible. In the special case of a wave traveling parallel to the reciprocal lattice vector **G**, Eq. (7.1) transforms to the well-known equation

$$\Lambda = \frac{m}{2}\frac{\lambda}{n_{\text{eff}}}$$

In general the constructive scattering of an incident wave with the vector $\mathbf{k}_i$ occurs if the Laue condition is fulfilled:

$$2\mathbf{k}_i \cdot \mathbf{G} = |\mathbf{G}|^2 \qquad \text{with } \mathbf{k}_i = \begin{pmatrix} k_{x,m} \\ k_y \\ 0 \end{pmatrix} \tag{7.2}$$

This formulation of the feedback condition states that Bragg scattering can occur if the incident photon **k** vector is located on a Bragg plane, being the perpendicular bisector of a line connecting the origin with the reciprocal lattice point **G**. A wave vector $\mathbf{k}_i$ which is located on a Bragg plane is scattered to $\mathbf{k}_d$ and vice versa. The wave vector $\mathbf{k}_i$ of the incident and $\mathbf{k}_d$ of the scattered wave correlate in the form

$$\mathbf{k}_d = \mathbf{k}_i + \mathbf{G} \tag{7.3}$$

Energy conservation results in the additional relation

$$|\mathbf{k}_i| = |\mathbf{k}_d| = \text{const}$$

Figure 7.2*a* shows first-order scattering ($m = 1$) of a wave vector $\mathbf{k}_i$. If the incident wave vector $\mathbf{k}_i$ complies with the condition in Eq. (7.2), it is scattered as stated in Eq. (7.3). Second order scattering ($m = 2$) is depicted in Fig. 7.2*b*. In this case there are two possibilities for the scattering process. The scattering condition (Bragg) determines only the component k_x of the wave vector. Therefore a nonzero z component of the wave vector results since energy conservation has to be met. This means the incident wave is scattered out of the waveguide.

Second-order laser resonators are advantageous for applications with free space optics due to their good light extraction efficiencies.

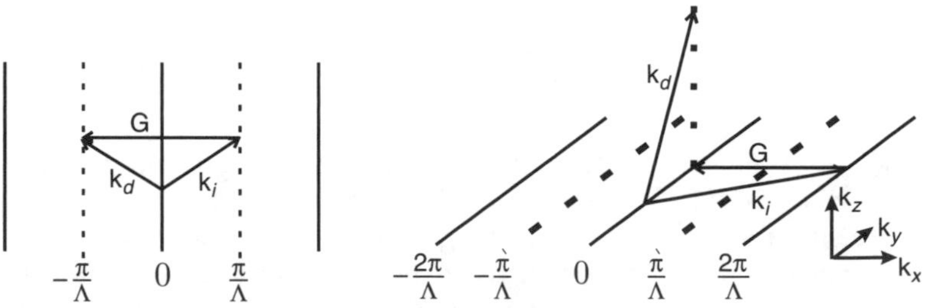

(*a*) First-order scattering (*b*) Second-order scattering

FIGURE 7.2 Feedback through scattering processes in a one-dimensional Bragg slab waveguide.

First-order resonators are of special interest for on-chip integration (see Sec. 7.4) and enable lower laser thresholds.

7.2.2 Organic Semiconductor Energy Transfer Systems

Material systems with a large spectral gap between the absorption of the excitation light and the laser emission are of particular interest for organic semiconductor lasers. This reduces the self-absorption of the emission, leading to a more efficient lasing operation. Such a system can be created by doping the active material, thus forming a so-called guest-host material.

The exciting radiation is absorbed in the host material. The absorbed energy is then transferred by a radiationless process to the guest material, lifting it into an electronically excited state (see Fig. 7.3). The predominantly utilized energy transfer mechanism for organic semiconductor lasers is the *Förster* resonance energy transfer.[8] This process is based on a dipole-dipole interaction and therefore requires a spectral overlap of the donor's emission with the acceptor's absorption. The effective radius of this process can be up to 10 nm.

The second important energy transfer system is the *Dexter* transfer. Here, an excited electron is transferred from the donor to the acceptor. In return, an electron in the ground state is moved from the acceptor to the donor. This particle exchange needs the atomic orbitals of acceptor and donor to overlap. Therefore, this process is typically only effective at distances of less than 1 nm.

7.2.3 Optical Pumping

There are several approaches for optically pumping an organic semiconductor laser. A high absorbance of the pump light in the laser material is a basic requirement. Additionally, short pulses and a small focus are advantageous. Depending on the material, the pulse duration has a critical influence on the threshold where lasing can be observed. This condition depends on the ratio of fluorescence lifetime to the pump pulse duration.[27]

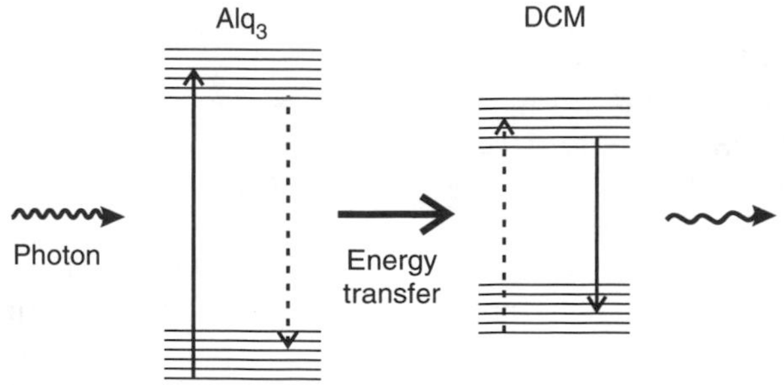

FIGURE 7.3 Energy transfer in dye-doped organic semiconductors.

Due to their short pulses and high energy densities, diode-pumped solid-state (DPSS) lasers were the first compact laser sources used for pumping organic semiconductor lasers.[10, 28] With steady advances in developing low-threshold organic lasers, a gallium nitride laser diode-pumped organic laser was presented in 2006.[29–31] In 2008 the first organic laser pumped by a blue light-emitting diode was demonstrated.[32]

7.2.4 Prospects for Organic Laser Diodes

With various optoelectronic components based on organic semiconductors being implemented successfully during the last decade, an organic laser diode device has become of particular interest. However, it could not be realized so far. Electrically pumped organic lasers are expected to provide numerous advantages over their inorganic counterparts such as tunability over the whole visible spectrum and the promise of low-cost fabrication.

The implementation of such a device seems to be straightforward. Electrodes are put on both sides of the slab waveguide formed by the organic laser material to inject charge carriers. The problem, however, are the strong optical losses which are introduced by coating the waveguide with a metal layer or even with a transparent metal oxide layer such as indium tin oxide (ITO).

To deal with this problem, the thickness of the organic layer can be increased in order to minimize the overlap of the guided mode with the absorbing electrodes. However, this results in very high operation voltages of the device.

Although high current densities for thin-film devices can be achieved without damaging the organic layer,[33–35] neither an electrically pumped organic laser nor an electrically pumped organic amplifier has been shown yet. The main reason for this behavior is the rising charge carrier density which causes additional losses such as bimolecular annihilation, induced absorption by polarons and triplet excitons, and field-induced exciton dissociation.[36]

7.3 Fabrication

There are basically two methods to realize DFB resonator gratings: either the lower or the upper active layer boundary has to be patterned with the requested topography of the distributed feedback grating. As shown in Fig. 7.4, this may be achieved either by direct patterning of the active laser material or by structuring the substrate beneath the active layer. The quality of the resonator structure affects the threshold and efficiency of the laser. For patterning of the active layer, photoisomerization,[37] direct embossing,[38, 39] or direct laser interference ablation (see Sec. 7.3.4) can be used.

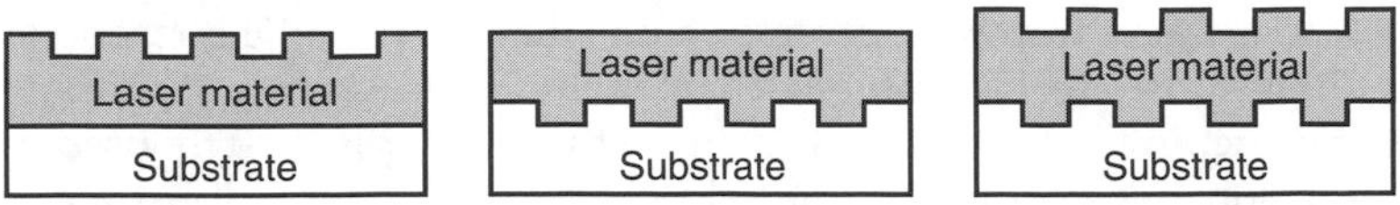

Figure 7.4 Different methods to pattern the active material layer of an organic semiconductor laser.

Patterning of the substrate might be advantageous since this process is independent of the properties of the active material. Hence, most fabrication techniques for organic semiconductor lasers are based on structured substrates.

To fabricate DFB resonators of first and second order for lasers in the visible range, grating periods of 100 to 400 nm are required.[40] In the following paragraph a (non comprehensive) overview of different fabrication methods of DFB resonators for organic semiconductor lasers is given. Figure 7.5 schematically illustrates the possible process strategies and methods.

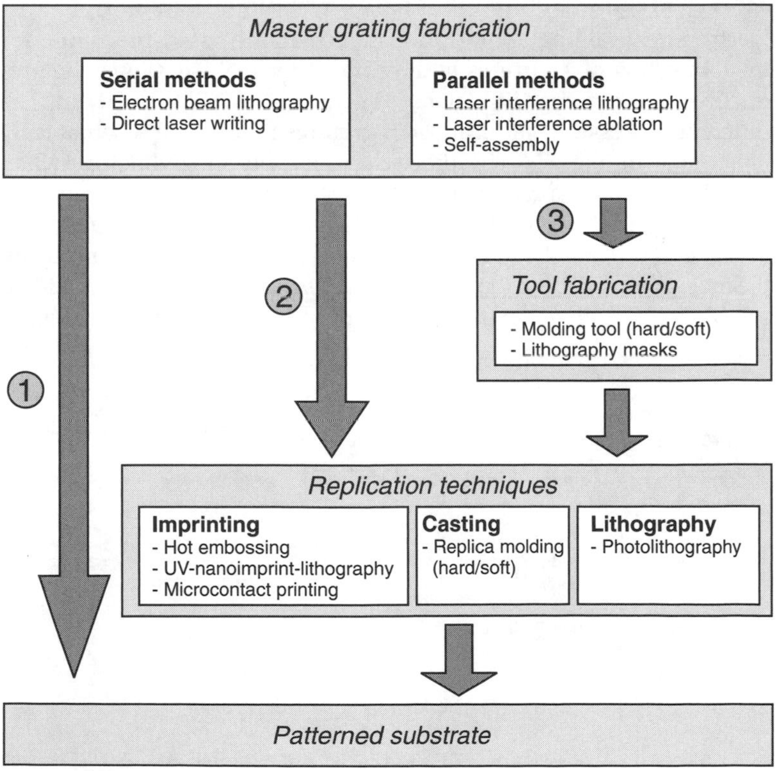

Figure 7.5 Overview of DFB resonator fabrication methods.

There are three main process chains that lead to a DFB resonator grating substrate. Each process chain includes structuring of a master substrate. In the first process chain (1) this master is directly used as laser substrate. However, the fabrication of a master substrate is expensive and time-consuming, thus this method is only used for prototyping. A second process chain (2) uses the replication of the master substrate. Different methods allow the multiple replication with the master substrate used as a molding tool. However, the master itself may be damaged during the process. This is avoided by introducing the fabrication of a replication tool (3) which can be optimized for long durability. Thus, the latter process chain is suited best for commercial applications.

The master substrate can be fabricated by serial or parallel processes. Serial processes such as electron beam lithography or direct laser writing allow for highest freedom in design of the structures. In comparison, parallel processes such as laser interference lithography or ablation are fast and applicable to large areas. In the following we will describe the most important fabrication methods for the master.

7.3.1 Master Fabrication: Electron Beam Lithography

Electron beam lithography (EBL) is a commonly used procedure for the fabrication of structures in the nanometer scale. It allows the production of structures with lateral dimensions of less than 20 nm. The different processes to fabricate DFB gratings by EBL are schematically illustrated in Fig. 7.6. In process A, a poly(methylmethacrylate) (PMMA) resist is spun onto the silica layer of a thermally oxidized silicon wafer, and the solvent content of the polymer layer is reduced by a prebake step. Then the resist is exposed to the electron beam, reducing the molecular weight of the PMMA in the exposed areas, thus making it soluble by a developer in the subsequent process step. After deposition

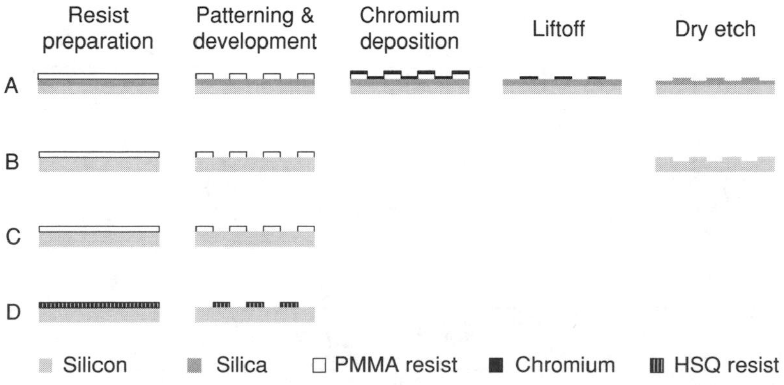

Figure 7.6 Electron beam lithography fabrication of DFB resonator gratings.

of a chromium layer and a liftoff process, the pattern is transferred into the substrate by dry etching.[41] Process B illustrates the possibility of skipping the process steps of *chromium layer deposition* and *liftoff* when the patterned resist is used as etch mask. In this way a dry etching process is applied to directly pattern a silicon wafer substrate, using the patterned resist as etch mask. A silicon wafer with structured PMMA resist is directly used as master in process chain (C).[42] As alternative to the high-resolution positive resist PMMA, the negative resist hydrogen silsesquioxane (HSQ) may be used; see process D. HSQ is spin-coated on a substrate, followed by a prebake step. The resist is then exposed with the electron beam, resulting in directly written inorganic structures with silica-like properties. The unexposed areas may be dissolved in the subsequent development step. The resolution of structures written by EBL is determined by the spot size of the electron beam and secondary effects, affecting the actual size of the nominally exposed and unexposed areas. The influence of this effect may be reduced by applying a higher acceleration voltage to the electrons (i.e., higher energy) and by reducing the resist thickness. PMMA resist may be patterned with minimum lateral structure sizes in the range of 10 nm. HSQ may also be patterned with minimum lateral structure sizes down to approximately 10 nm, depending on resist thickness and acceleration voltage.

7.3.2 Master Fabrication: Direct Laser Writing

Conventional direct laser writing is used as a cost saving alternative to electron beam lithography. Commercially available systems are able to pattern substrates as large as 400 × 400 mm^2 with feature sizes down to 0.6 μm. In the direct laser writing process a photoresist is patterned by locally exposing it with a focused laser beam. Common systems often use a HeCd gas laser source with an emission wavelength of 442 nm.

An extension to these systems can be made by using a femtosecond laser as light source.[43–46] Titanium-sapphire-based femtosecond lasers emit in the near infrared (NIR) spectrum at around 800 nm. Most photoresists are only sensitive in the UV spectral range and cannot be cured through NIR radiation directly. However, a reaction of the resist to the NIR radiation is possible with a two-photon absorption (TPA) process. This means that the simultaneous absorption of two low-energy photons causes a reaction in the material which normally can only happen with twice as much of the excitation energy in a single photon process. This non-linear phenomenon requires very high irradiation densities. To prevent thermal destruction of the resist, very short, high-intensity laser pulses are required.

A special feature of the TPA process is the distinct threshold characteristic for the exposure of photoresists. TPA requires a critical radiation dosage at which the process can start. This threshold behavior

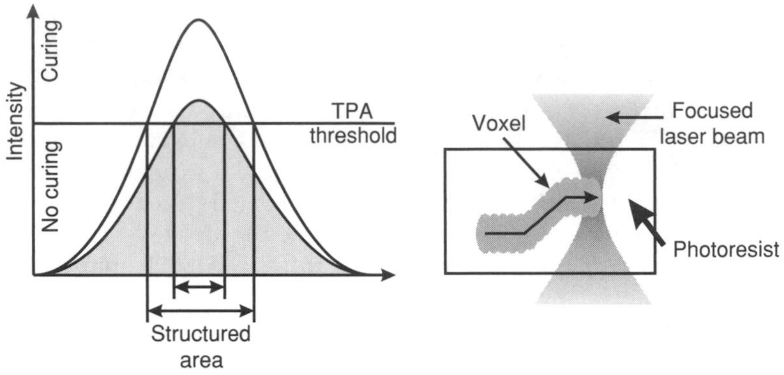

FIGURE 7.7 Left: Intensity profiles of two Gaussian beams. Depending on the TPA threshold, different-sized areas can be patterned. Right: By moving the beam focus relative to the sample, arbitrary structures can be written.

can be used to get below the resolution limit of traditional optical systems (see Fig. 7.7 left).

This characteristic also enables one to write three-dimensional patterns into the photoresist. As the chemical reaction to the irradiation only occurs in the focal volume, the so-called voxel, of the femtosecond laser beam, it is possible to pattern arbitrary structures by moving the sample (or the laser beam) (see Fig. 7.7 right). Depending on the numerical aperture of the focusing optics and the chemical properties of the photoresist, voxels with axial dimensions of 500 nm and lateral dimensions down to 100 nm can be realized.[47]

7.3.3 Master Fabrication: Laser Interference Lithography

Laser interference lithography (LIL) is one of the most important techniques for the fabrication of periodic structures. In contrast to electron beam lithography, it is possible to quickly pattern large areas with this technique. Although the basic principle behind the LIL is quite simple, the realization of large-area and high-quality structures is an elaborate task.[48]

A scheme of an LIL setup is shown in Fig. 7.8. A laser beam is split into two single beams which are directed onto a photosensitive resist. Typically the setup is symmetrical, leading to an interference of both beams, thus forming a linear light-dark pattern. This results in a locally varying exposure dosage, leaving a periodic structure after developing the resist.

The interference pattern grating constant Λ can be calculated by the formula $\Lambda = \lambda/(2\sin\alpha)$, with the angle α between the substrate normal and the laser beam. Therefore the lower limit of the grating constant is one-half of the laser wavelength λ. In case of the commonly used argon ion laser emitting at 364 nm, this means theoretically a

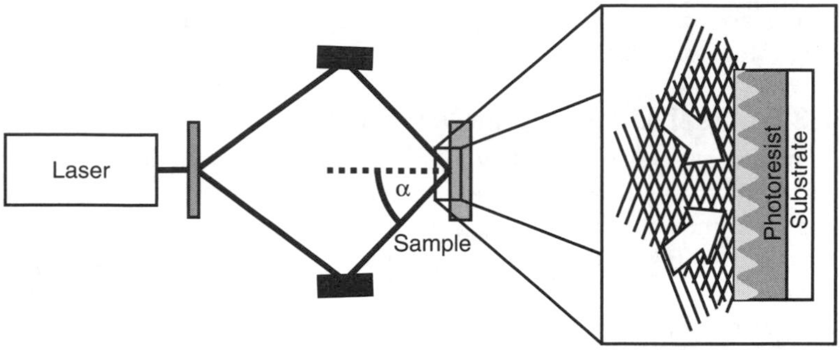

Figure 7.8 Schematic experimental setup for laser interference lithography. The inset shows the pattern created by the interfering beams. (*Reprinted wiht permission from Ref. 25, "Organic semiconductor lasers as integrated light sources for optical sensor systems," SPIE 2007.*)

minimal lattice period of 182 nm. A long coherence length of the laser is needed to realize a substrate-sized interference pattern. Additionally, a mechanically stable optical setup is obligatory. As a rough estimation the displacement of the interference pattern should be lower than one-tenth of the period during exposure.

After developing the structure, it can be either used directly as a laser resonator or transferred into another material. As the substrate can be rotated by 90° or 60°, square and triangular lattices can also be patterned with LIL using multiple exposures.

7.3.4 Master Fabrication: Laser Interference Ablation

Laser interference ablation (LIA) produces similar periodic structures as LIL by interfering two laser beams. The basic experimental setup is similar to LIL (see Fig. 7.8). However, there is no need for exposing and developing a resist. The structure is directly realized by partly ablating the surface instead. Grating period and thus the emission wavelength can be chosen on demand. The only requirement for the material is a good absorbance of the ablation laser light. Using UV lasers, many organic materials can be ablated.[49] Additionally the contamination of the surface through ablation products should be avoided. Pulsed laser systems are preferred for LIA, since they provide high pulse energies while reducing heating of the sample to a minimum.

When only a periodic surface grating is needed, LIA has the advantage that the laser resonator can be generated directly into the active organic layer, thus eliminating the need for a master and further processing. Essentially the production steps are reduced to active layer deposition and ablation. The pulse energy required for LIA on an Alq_3:DCM [aluminum tris(8-hydroxyquinoline) doped with the laser dye 4-dicyanomethylene-2-methyl-6-(p-dimethylaminostyryl)-4H-pyran] film is about 10 mJ/cm^2 using a frequency triple Nd:YAG laser emitting 150 ps pulses at 355 nm.[50]

7.3.5 Replication: Imprint Techniques

Imprinting plastics is one key to low-cost mass production. The first microstructures molded in polymer materials were fabricated with the well-known hot embossing process. The basic process steps are depicted in Fig. 7.9. The microstructure is replicated by pressing a master onto a substrate at temperatures above the substrate material's glass temperature.[51] With the introduction of vacuum embossing of plastics it became possible to imprint large areas uniformly since it avoids an air cushion between stamp and sample.[52] As the master may be damaged by the embossing step, the so-called LIGA process was adapted to fabricate a durable metal imprinting tool of the master microstructure.

LIGA is the German acronym for the main steps of the process, i.e., lithography, electroforming, and plastic molding.[53–55] The single process steps are shown in Fig. 7.10. To generate the metal tool by electroforming the insulating master structure, a conducting starting layer is deposited onto the master. This is usually done by evaporating a gold layer upon the master structure. By electroplating these structures, the negative pattern of the resonator structure is formed as a secondary metallic structure. Commonly used metals for plastic

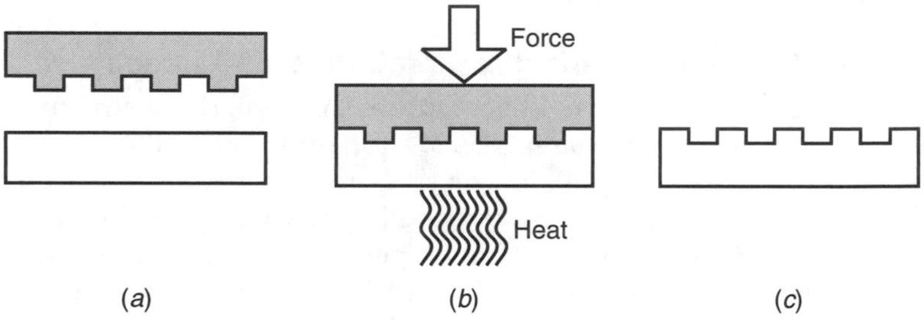

FIGURE 7.9 Processing steps for hot embossing.

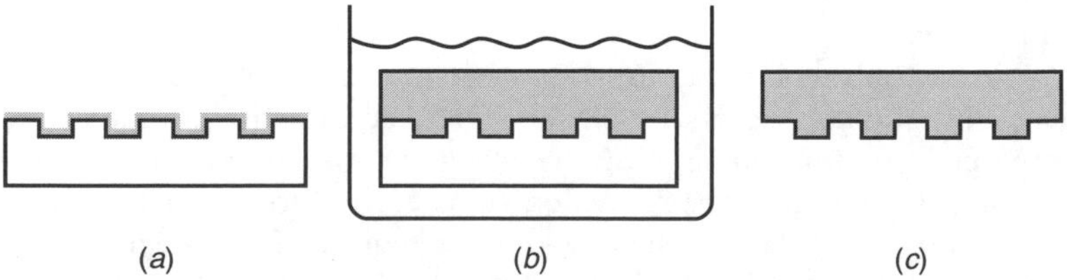

FIGURE 7.10 Fabrication scheme of a Ni shim following the LIGA process.

molding tools are nickel alloys, such as nickel-cobalt and nickel-iron. The tool, a so-called shim,[56, 57] is plated to a thickness of about 200 to 300 μm. As this process does not wear the master tool, basically an unlimited number of replicas can be fabricated. This process therefore enables the mass production of nano- and microcomponents at low cost.

A possible approach toward decreased process times by eliminating the heat-up and cooldown phase of the hot embossing is the UV nanoimprint lithography. Here, a transparent imprinting tool is pressed into a UV-curable liquid polymer or solgel material. The material is then cured by exposing it to UV radiation.[58, 59]

Another replication method used for laser fabrication is the microcontact printing (μCP) process.[60, 61]

7.3.6 Replication: Cast Molding and Photolithography

There is a major difference between imprinting and cast molding. Instead of applying pressure for the imprinting process, the master is cast with the material in the latter approach. The transition from imprinting to casting can be seamless. For example, the step-and-flash imprint lithography (SFIL) describes the same process steps as the UV nanoimprint lithography. The sole difference is the pressure applied to the substrate for the UV-NIL process. Depending on the material the master is made of, casting by itself can be separated into hard and soft casting. Therefore the mentioned SFIL is an instance of hard casting whereas an example of soft casting would be the use of the flexible PDMS [poly-(dimethylsiloxane)] as the master's material.[62]

Besides the imprinting approaches described above, conventional photolithography can be applied to fabricate laser resonators.[63] The required feature size can be structured without difficulties using 193 nm immersion lithography. However, these lithography systems are very complex and expensive compared to imprinting methods.[64, 65] Therefore a high packing density is required for economic reasons when such equipment is used. As this is given for high-price products such as microprocessors, it is not the case for lab-on-chip devices where most of the wafer space is filled with large-scale structures such as fluidic channels and optical waveguides.

7.3.7 Active Layer Deposition

For the manufacturing of an integrated organic laser it is necessary to selectively deposit the active layer. This typically means that layers with thicknesses between 130 and 400 nm and areal dimensions down to 500 × 500 μm^2 have to be fabricated. The two main deposition techniques for organic lasers are spin coating and thermal vapor deposition. Often only one deposition technique can be used for an organic material.

Obligatory for every thin-film deposition is a clean substrate. The possible cleaning techniques are often limited by the substrate's resistance to chemical treatment. If there are no limitations, a cleaning cycle with acetone, isopropanol, and deionized water rinsing will remove most residues. Treating the substrate subsequently with a plasma cleaning step (e.g., argon/oxygen plasma) removes remaining organic compounds on the surface and additionally enhances the wettability.

For the spin-coating process, the material has to be dissolved in a suitable solvent and applied to the substrate. The substrate is then rotated at up to 4000 rpm, causing the superfluous liquid to be spun out. After evaporation of the solvent, a thin solid film remains. Although being of impressive simplicity, this technique is problematic when patterning of the layer is required. Due to the good solubility of organic materials, traditional concepts used for inorganic materials such as masking and etching are not applicable.

In contrast to spin coating, the vapor deposition of the active material is ideal for producing patterned laser areas. To vapor-deposit organic materials, a controlled thermal evaporation source inside a vacuum chamber is needed. Typically the source consists of a crucible, a heating wire, and a thermocouple element. The crucible holds the organic material and is wrapped with the heating wire. The thermocouple element measures the temperature of the crucible according to which the current through the heating wire is regulated. Usually the deposition rate is determined by an oscillating crystal detector placed above the evaporation source. The best position for the detector is close to the substrate. However, this becomes problematic when two materials have to be evaporated simultaneously, e.g., for guest-host systems such as Alq_3:DCM. In this case it has to be ensured that the deposition rate of the dopant can be measured without being influenced by the host source.

For calibration and a check of the evaporation conditions, a surface roughness meter can be used to determine the layer thickness after the deposition. For measuring the dopant ratio of the film, the amplified spontaneous emission (ASE) wavelength gives feedback in the case of optical gain materials. This wavelength is independent of the film thickness given that the film forms a waveguide. Additionally, absorption and transmission spectra can be used.

For coevaporation the control should be automated to ensure reproducible evaporation conditions and deposition results. For an Alq_3:DCM laser with a typical doping concentration of around 3% DCM, the deposition rates for the two materials are on the order of 0.3 and 0.01 nm/s, respectively. The ratio has to be held constant for more than 15 min, assuming a desired layer thickness of about 300 nm.

To deposit only defined areas of the substrate, a mask with cutouts is fixed close to the substrate. This enables the coating of specific single-laser fields. The change of masks also allows the deposition of different materials on different areas. This is of particular interest for devices where several laser materials on one substrate with a wide range of different emission wavelengths are needed.

Different concepts can be used for tuning the laser emission wavelength within the gain spectrum of one single material.

The most common method is to produce one substrate exhibiting several resonators with varying grating periods. The active layer can then be deposited onto all gratings simultaneously. Due to the different feedback conditions the emission wavelength can be tuned. Figure 7.11 shows an example for Alq_3:DCM lasers where the grating constants cover the range between 390 and 440 nm. The corresponding wavelength range extends over more than 120 nm.[25]

Additional emission wavelength tuning can be achieved by altering the film thickness. Increasing the film thickness results in a bigger overlap of the guided mode into the active layer, increasing the effective refractive index n_{eff} and thus shifting the emission wavelength spectrally into the red (see Sec. 7.2.1). With this method a tuning range of 44 nm for Alq_3:DCM lasers could be demonstrated.[10]

Such a laser can be fabricated also in a single pass by using a shutter to cover the areas of lower thickness. Another way to alter the effective refractive index is via a buffer layer between substrate and active material with a different refractive index. For this approach it is required to add an extra deposition step to the process chain before

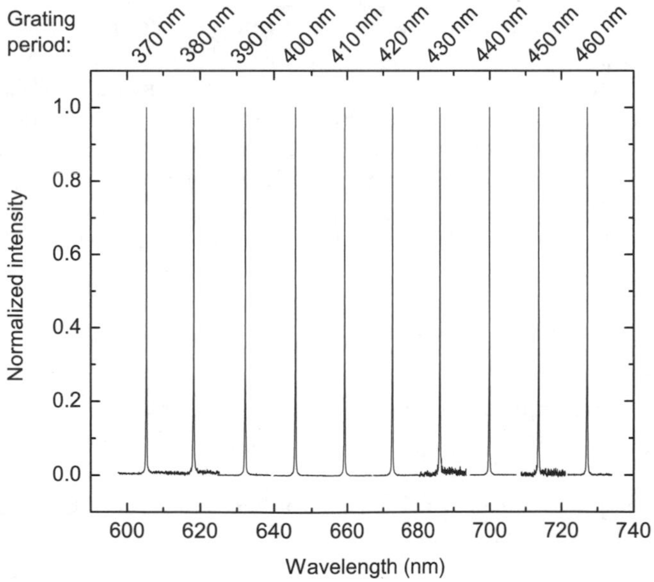

FIGURE 7.11 Spectra of Alq_3:DCM lasers based on resonator gratings with different periodicities. (*Reprinted with permission from Ref. 25, "Organic semiconductor lasers as integrated light sources for optical sensor systems," SPIE 2007.*)

the organic material deposition.[66] The achievable tuning range is primarily limited by the optical gain spectrum of the material. Other influences such as absorption of the substrate, deviations in the lattice form and the lattice period, the grating design (e.g., the duty cyle) itself, and the quality of the laser resonator may restrict the range further and increase the laser threshold to unacceptable levels.

7.4 Integrated Optical Sensor Systems

The examination of biological and chemical samples is essential for medicine, biology, and environmental monitoring. These analyses normally take place in central laboratories. In recent years, a clear trend toward miniaturized analytical devices, which are applicable at the "point of care," is recognizable.

This chapter deals with the integration of organic laser sources into such systems. The technology of miniaturized optical sensor systems will be introduced with respect to the possible use of organic lasers. Different sensing schemes that are relevant for laser-based analysis systems will be described including a short summary of waveguide and microfluidic based systems. The last part describes the work on the first waveguide coupled organic semiconductor lasers that are based on low-cost polymeric substrates.

7.4.1 Sensing Schemes

The miniaturization of bioanalytical techniques has an enormous relevance for the rapidly growing life sciences. Therefore, the realization of so-called lab-on-a-chip or micro total analysis system (μTAS) has become an important goal during the last years. This section discusses the major optical sensing principles which could be incorporated into such a system making use of an integrated laser source.

The determination of biological, physical, or chemical parameters[67] such as molecule type, binding behavior, or concentration can be realized by the optical investigation of

1. Absorbance and reflectance
2. Fluorescence
3. Interference phenomena
4. Resonance phenomena

The optical sensor consists preferably of a laser as a light source, the sensing setup, and a detector. See Fig. 7.12.

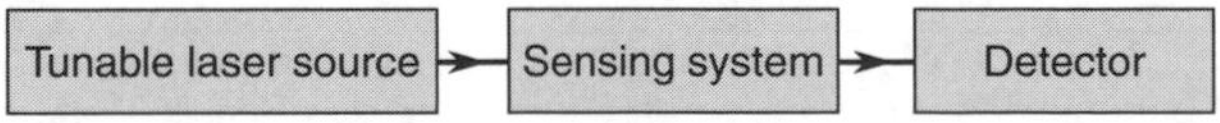

FIGURE 7.12 Sensing scheme of an optical sensor.

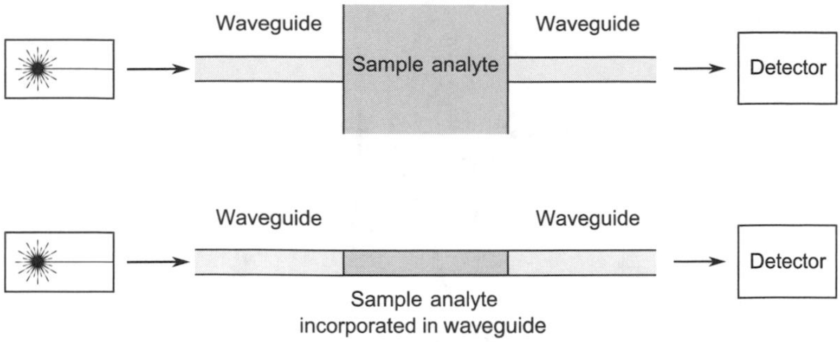

FIGURE 7.13 Scheme of a transmission measurement.

1. **Absorbance measurements**

The simplest type of optical measurement is the determination of the absorbance. Light of a certain frequency ω_0 travels a defined distance through the analyte. By detecting the fraction of transmitted light the extinction coefficient κ of the analyte for the frequency ω_0 can be determined. The extinction dependence of the frequency $\kappa(\omega)$ yields information about the chemical composition of the analyte. Sensor schemes are shown in Fig. 7.13.

For the extinction to be sufficiently high for detection, the light traveling distance *d* through the analyte has to be long enough. This can cause a problem, in particular if the analyte is a gas, where *d* can be more than 100 m. In this case, one could incorporate a photonic cavity into the sensing region which causes the light to be reflected back and forth multiple times, resulting in a much longer efficient traveling distance. See Fig. 7.14.

The light incident on the cavity (shown symbolically as a *Bragg* reflector here) has to be coherent, which means that a laser source is required. Additionally, the cavity has to be tuned according to the wavelength of the laser source.

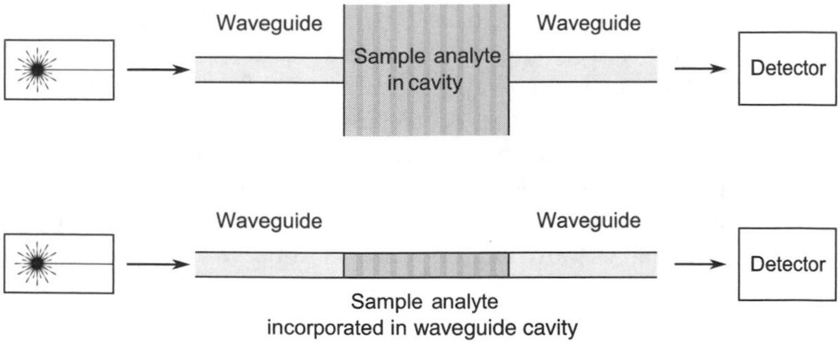

FIGURE 7.14 Scheme of a resonator-enhanced transmission measurement.

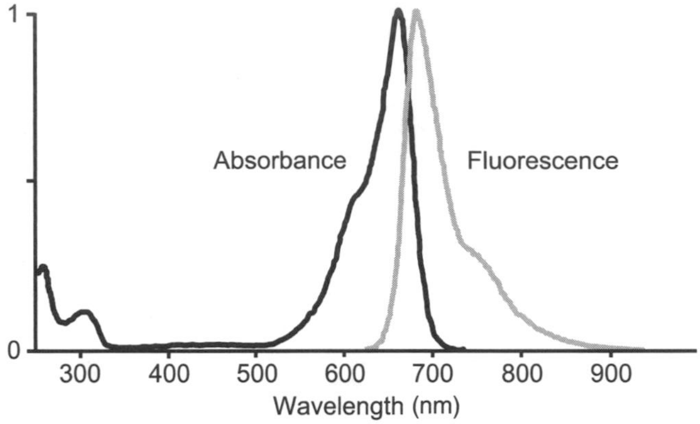

FIGURE 7.15 Schematic absorption/emission spectrum of a fluorescent dye.

2. **Fluorescence measurements**

   Nowadays the most frequent optical method applied for biological sensing purposes is the measurement of fluorescence. It is also a promising approach for integrated photonic sensing systems. In principle, a fluorescent analyte is excited at a specific wavelength and emits at a longer wavelength which is then detected. The excitation and emission wavelengths are specific to the fluorescent molecule. Thus, one uses this technique either to detect the fluorescent molecules directly or to label certain molecules, cells, etc., with a fluorescent marker which can be detected afterward. An example for the absorption and emission spectrum of a fluorescent dye is given in Fig. 7.15.

   Since the absorption and emission peaks are typically very close to each other, one needs a narrow-band light source to excite the dye without overlap with the emission spectrum. Therefore, a laser is used for excitation. See Fig. 7.16.

3. **Interference methods**

   Another sensing principle is the classical interferometer, where a coherent light beam is split up into two or more light beams which, after traveling different optical paths, are combined again, yielding interference. By changing one or several of the optical paths with an analyte, the change of the interference can be detected. The type most suitable for an integrated optical system is the Mach-Zehnder interferometer, which is shown in the schemes in Fig. 7.17.

   Here, a laser beam propagating through a waveguide is split up into two parts, the reference beam and the probing beam. The reference beam propagates through one waveguide. The probing beam travels through a certain distance of

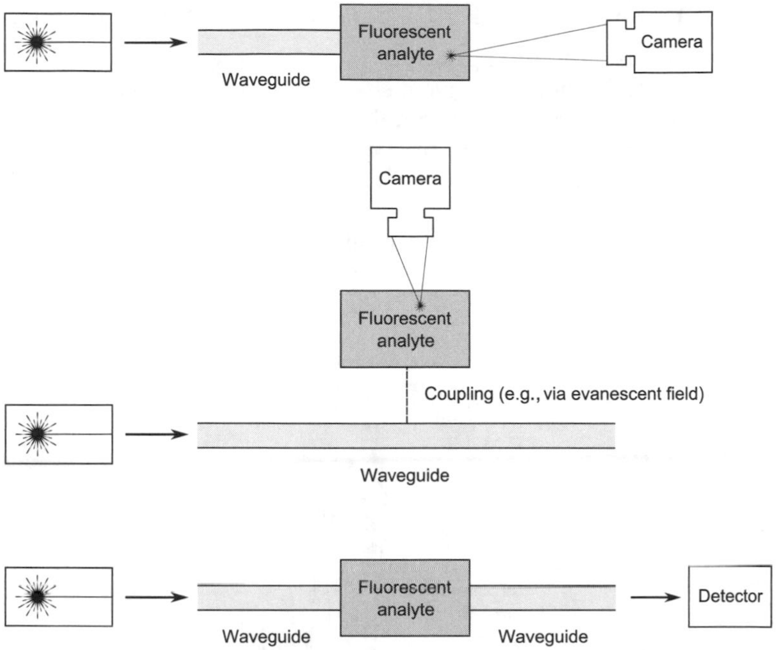

FIGURE 7.16 Scheme of different fluorescence detection methods.

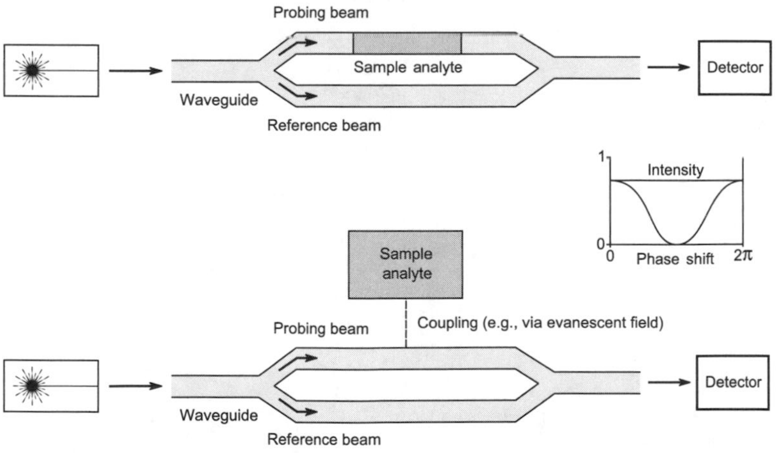

FIGURE 7.17 Schemes of Mach-Zehnder interferometers.

analyte (upper part of figure) or is otherwise coupled to an analyte (lower part of figure) in a second waveguide. After traveling the different paths, the two beams interfere. The difference in optical path length results in a characteristic change of intensity due to the interference of both beams.

4. **Resonant methods**

A very promising principle for an integrated optical sensor is the utilization of resonant effects in a photonic cavity coupled to a waveguide. A chemical modification of the cavity by an infiltrated analyte or a binding molecule results in a shift of the resonant frequency $\omega_0 \to \omega_1$. This shift can then be detected. See Fig. 7.18.

The photonic cavity can be of various types. The quality factor which determines the width of the resonance peak

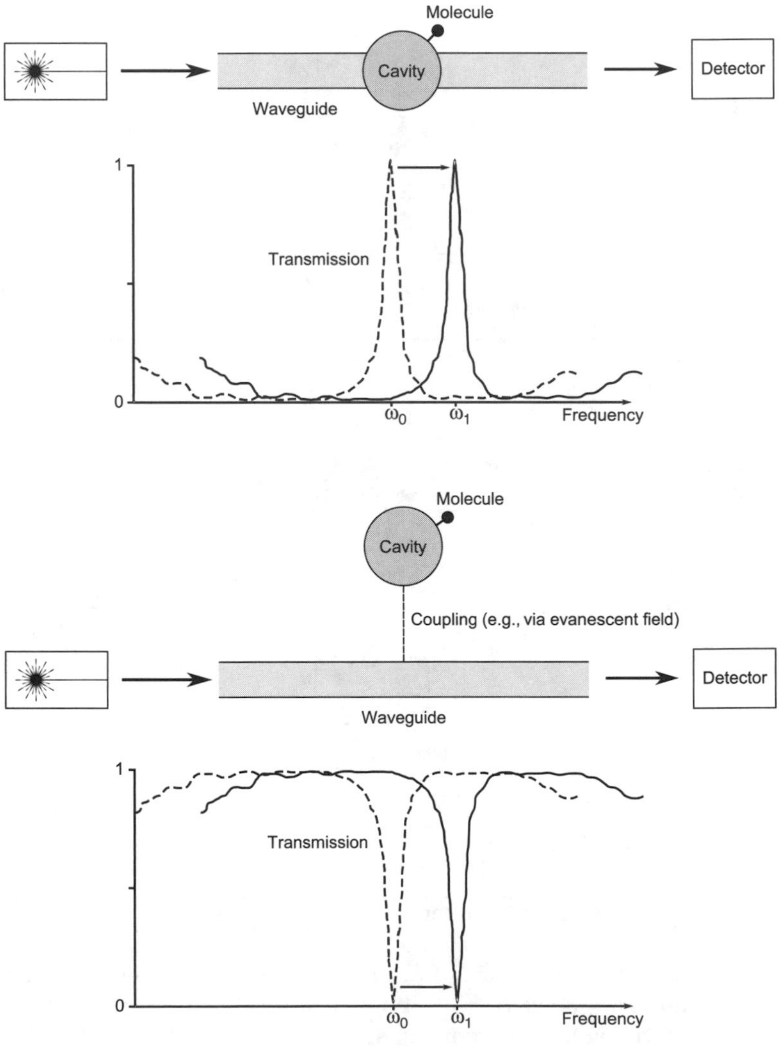

FIGURE 7.18 Scheme of the resonant detection method.

	Micropillar	Microdisc	Microtoroid	Microsphere	Photonic crystal
Scheme					
Quality factor Q	10^3	10^4	10^8	10^{10}	10^4
Modal volume $V_M = \frac{V}{\lambda^3}$	5	5	10^3	10^3	1

Figure 7.19 Overview of different microresonator types.

should be high enough (i.e., ≥ 1000). Additionally, the modal volume of the cavity mode should be small to tap the full potential of a lab-on-a-chip (LOC) using very little amounts of analyte. For an overview, the different types of resonators are listed[68] in Fig. 7.19 together with the corresponding quality factors and modal volumes:

Waveguide-Based Sensors

The use of waveguide-based sensor systems has numerous advantages over traditional concepts.[69–71] In a waveguide excitation scheme, light needed for the analysis can be guided efficiently to the detection zone. At the detection zone a well-defined radiation characteristic without the need for sophisticated alignment procedures is possible. Also, the integration of further functional components such as splitters or couplers requires no additional fabrication processes. Waveguide applications span from glass fibers to integrated stripe or rib waveguides. Waveguide-based sensors are operated nearly solely with a laser as light source. Main reasons are the typically high spectral intensity densities, efficient coupling, and monochromatic emission spectrum. Especially, the measuring method laser-induced fluorescence (LIF) depends on a laser light source. This procedure uses a laser to excite a sample substance and relies on the analysis of the resulting fluorescence signal (see Sec. 7.4.1). Normally specific marker dyes with a characteristic emission spectrum are used to distinguish different substances or DNA sequences.[72] Despite the expensive and often patented marker dyes, the LIF method is quite popular because it provides a high sensitivity with a very low limit of detection (LOD).

The integration of conventional laser systems on a chip-based analyzing system is not applicable. Studies on miniaturizing dye lasers may be an alternative.[73] These laser sources are optically pumped and can be integrated directly on the chip with the possibility of the integration with rib waveguides.[24] However, this approach still requires a costly pump laser source, and the achievable spectral emission range is limited.

Microfluidics

Many lab-on-a-chip designs make use of microfluidic systems. Basic components are microfluidic channels with widths ranging from 5 to 100 μm. The substance to analyze is handled in these channels by using pump systems or electrophoretic methods.[74–76] The use of microfluidics promises a fast and precise analysis with small amounts of needed substances. This is especially important in the field of DNA and high-throughput screening.[14, 77] By joining different system components for handling, mixing, and dosing, many tasks required for the preparation of the analyte can be done directly on the chip. With the combination of such systems with waveguide-based optical sensors, highly integrated analytic devices can be realized.[78]

7.4.2 Integration of Organic Lasers in Optical Sensor Systems

It has been shown that in principle OLEDs are possible light sources to excite a fluorophore. However, OLEDs have decisive drawbacks in comparison to organic lasers. First, their large spectral emission width is disadvantageous for fluorescence excitation, because the excitation light has to be eliminated from the measurement signal by complex methods. Second, even though coupling to waveguides is possible, the efficiency of the coupling is weak.

An organic laser is a monochromatic light source with a wide tuning range. Optical pumping of the laser allows operation without the need for electrical contacts. In comparison to dye-doped polymer lasers[24] organic semiconductor lasers have certain advantages. The use of *Förster* energy transfer systems allows organic lasers to be efficiently pumped with only one light source in the UV range and to emit laser light in the whole visible spectrum.

A scheme of the proposed integration of organic lasers as light sources in lab-on-a-chip systems is shown in Fig. 7.20. Light of the optically pumped organic lasers is coupled into a polymeric waveguide. The laser light is guided to the detection area, being a cross section of the waveguide and a microfluidic channel. The resulting optical signal, e.g., a change in the absorption pattern or a laser-induced fluorescent signal, is then detected by an integrated photodiode.

In conventional chemical analysis systems, often several solid state or gas lasers are needed to generate light at different wavelengths, rendering

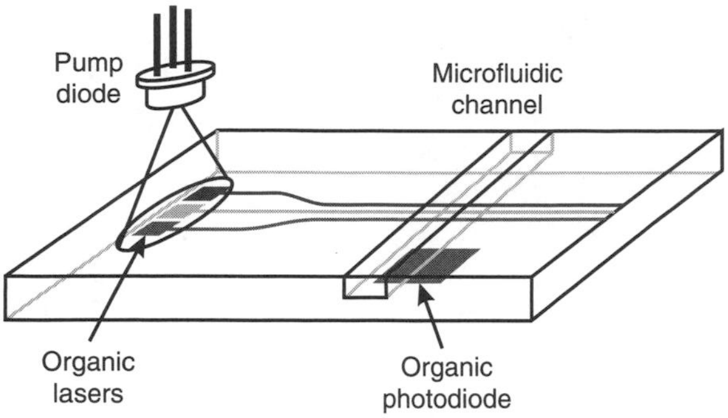

FIGURE 7.20 Scheme of a possible lab-on-a-chip design incorporating a microfluidic system for analyte preparation and handling, multiple laser sources, and photodetection. (See also color insert.)

these systems complex and expensive. In the concept discussed here, several organic lasers with different emission wavelengths will be integrated on a single chip.

This system is based on the polymer PMMA which is well suited for hot embossing nanostructures. Additionally PMMA is a low-cost biocompatible material.[79] Also it is quite simple to define stripe waveguides in PMMA.

PMMA-Based Waveguides

UV-Induced Refractive Index Modification Polymers are promising materials for the manufacturing of integrated optical elements, due to their low cost and excellent processing capabilities. Fabrication of polymer waveguides has been demonstrated using different approaches, such as reactive ion etching,[80, 81] photolocking,[82, 83] direct laser or ion beam writing,[84] as well as by replication techniques.[85, 86] A further way is the modification of the dielectric properties of methacrylate polymers by UV radiation, which will be discussed here.

Methacrylate polymers such as PMMA exhibit a significant change in the refractive index after exposure to ion radiation[87, 88] or UV light of a short wavelength (often referred to as deep UV, DUV).[89] This change of the dielectric properties of the material due to radiation was described by Tomlinson et al.[82] in 1970. The mechanism of the chemical reaction leading to this change depends on the type of radiation and the absorbed energy. Figure 7.21 illustrates the dominating reaction for deep UV radiation.[90] UV light with an energy of approx. 5 eV (wavelength of approx. 250 nm) leads to an excitation of the carbonyl group within the ester side chain. This can lead to a cleavage of the chemical bond. The dominating process induced by the radiation is the cleavage of the side chain, which can degrade into smaller, volatile fragments, such as CH_4, CH_3OH, CO_2, and CO.[91–93] The remaining

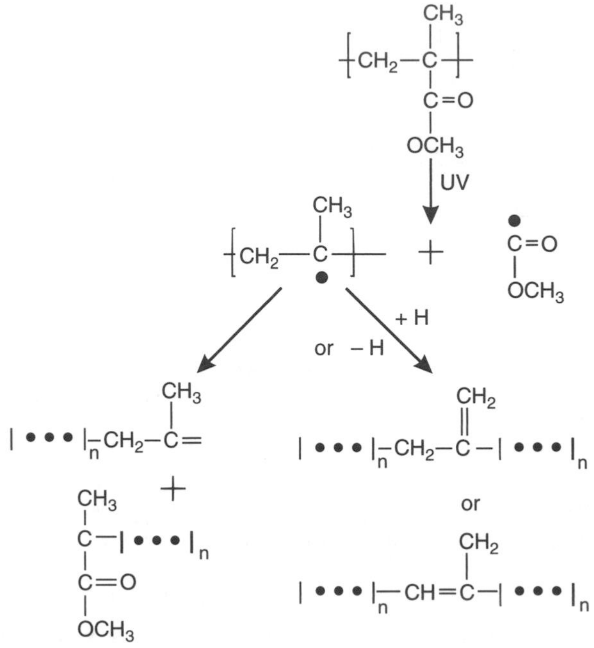

FIGURE 7.21 Photolysis reaction of PMMA.

alkyl radical can be dissipated either by hydrogen abstraction and formation of an unsaturated C=C double bond within the main chain of the polymer, or by a scission of the main chain. Hydrogen generated during the C=C formation recombines with the side chain fragments into methyl formate (HCO_2CH_3). Remaining radicals after the main-chain scission can react with other radicals, forming larger, more stable molecules.

Due to the reactions described, the ratio of C=C double bonds within the material and therefore its molar refraction, as well as its density, increase. The overall optical effect is an increase in the refractive index.[89]

Masking Process Masking of the polymer to define the wave guiding core can be done in two ways: either a photolithography step utilizing a chromium mask or self-masking of a prestructured substrate is employed.

The classical photolithography process leads to a local increase of the refractive index in the unmasked areas, defining integrated optical structures. Only a thin layer of the material surface is chemically altered, as the penetration depth of the radiation is limited by the absorption of the material. The dimensions of the waveguide core are therefore defined by the photomask and the dose of absorbed radiation. As substrates, either

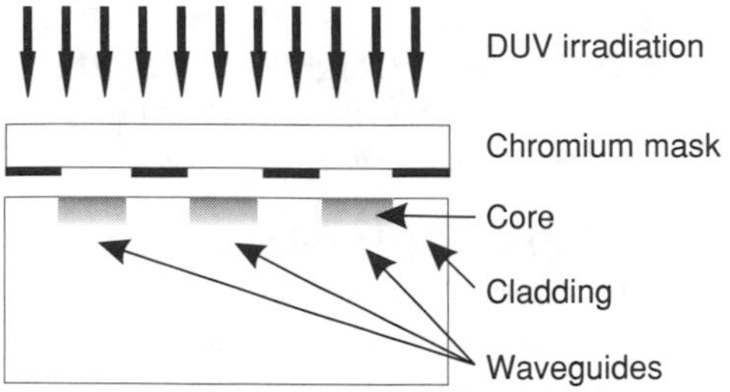

FIGURE 7.22 Fabrication of planar waveguide structures in PMMA using DUV irradiation. (*Reprinted with permission from Ref. 96. Copyright 2007 IEEE.*)

spin-coated PMMA layers on a carrier with a lower refractive index or single-material PMMA foils with several hundred micron thickness can be used. Figure 7.22 illustrates the process.

This technique has several advantages because only a single polymer layer is used, which serves as the substrate and waveguide as well. No further etching or development steps are required. Waveguides can be fabricated using a commercial mask aligner (EVG 620 from EV Group) at a dosage between 3 and 5 J/cm^2 at 240 nm, leading to an increase in the refractive index between 0.008 and 0.015. A waveguide loss between 0.7 and 0.8 dB/cm is obtained at a wavelength of 1550 nm, which is mainly attributed to material losses. For visible light of 635 nm wavelength, losses as low as 0.1 dB/cm were observed.

It has also been demonstrated that it is possible to fabricate passive optical components such as planar waveguides, splitters, and couplers using this approach.[94]

Certain structures and devices such as sharp bends or resonators cannot, however, be designed as planar structure, as they require strong guiding rib structures. To fabricate this type of structure, the combination of replication by hot embossing and refractive index modification is required. Figure 7.23 summarizes the processing steps for the replication of the molded part through (*a*) hot embossing and

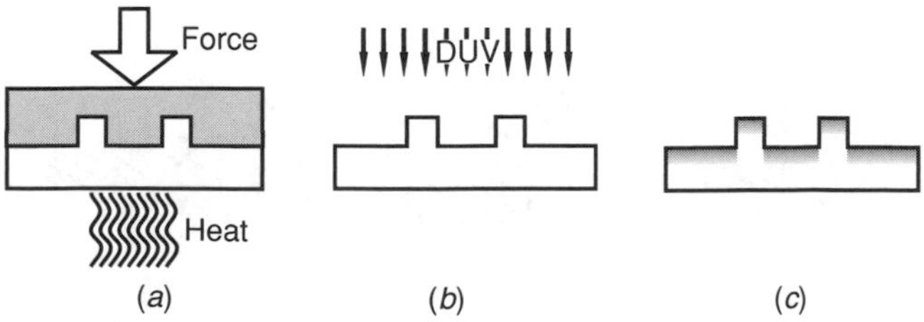

FIGURE 7.23 Process steps for hot embossing of ridge waveguide structures and deep UV flood exposure.

(*b*) the realization of a photochemically altered surface layer for increasing the refractive index. Straight single-mode waveguides and multimode interference couplers for the NIR spectrum fabricated by this approach have already been demonstrated.[95]

Waveguide-Coupled Organic Semiconductor Lasers

The above-mentioned fabrication techniques enable the fabrication of an integrated waveguide-coupled organic semiconductor laser. A first success was the coupling of amplified spontaneous emission generated in an organic semiconductor layer into an underlying deep UV patterned waveguide.[96] Recently, an integrated laser with strongly (20×) enhanced coupling efficiency has been demonstrated.[25] This waveguide-coupled laser was fabricated with three main process steps.

- Hot embossing of first-order DFB laser resonators
- Waveguide definition through deep UV exposure of PMMA
- Deposition of the active laser material Alq_3:DCM

By keeping the complete process chain compatible with 4 in wafers, the mass production potential was shown. The whole process includes several steps, which are schematically illustrated in Fig. 7.24.

With a possible target application being a high-volume production, a simple and fast tool replacement is necessary. Therefore the hot embossing process makes use of a nickel shim, which is based on a master fabricated by electron beam lithography and dry etching out of an oxidized silicon wafer (see Sec. 7.3). With the silicon wafer

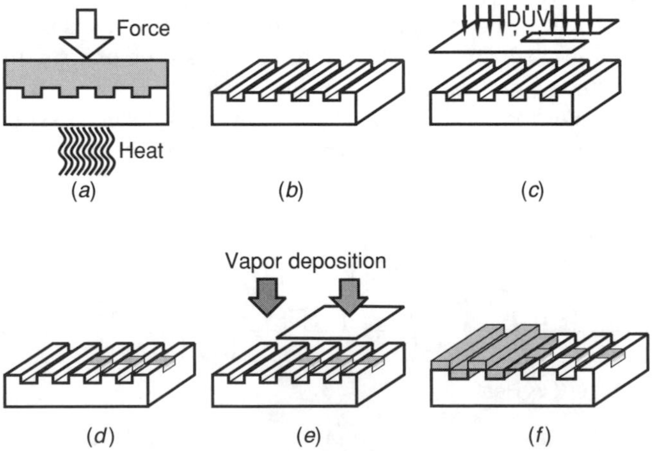

FIGURE 7.24 Schematic of the process chain for the fabrication of integrated waveguide-coupled organic solid-state lasers.

multiple shims can be electroplated which are good for a couple thousand embossing cycles. After hot embossing, waveguides are defined by deep UV exposure of the PMMA substrate. The PMMA substrate is split into sections by a wafer saw. This sawing process creates sample end facets of optical quality. However, the substrate is not completely cut to allow further processing on wafer scale. Subsequently, the active laser material Alq_3:DCM is evaporated through 1 mm^2 wide holes of a high-grade steel mask only onto resonators (thickness: 180 μm).[97]

The demonstrated waveguide-coupled organic semiconductor lasers uses first-order resonators with a period of 200 nm. The waveguides extend over the whole sample width of 25 mm and pass the resonator fields. The organic lasers are pumped with an elliptical excitation spot with a spot size of 500 × 50 μm^2.

A spectrum of one of these waveguide-coupled organic lasers is presented in Fig. 7.25. Laser light can be coupled into single-mode waveguides of 3 μm width as well as into multimode waveguides of 50 μm width. The spectrum shows single-mode lasing with a spectral width of less than 0.3 nm at a wavelength of 630 nm.

Photodetection

The detection of the sensor signal is of utmost importance for the overall sensor system. Here the combination of a suitable detector and signal processing will lead to a good sensitivity of the system.

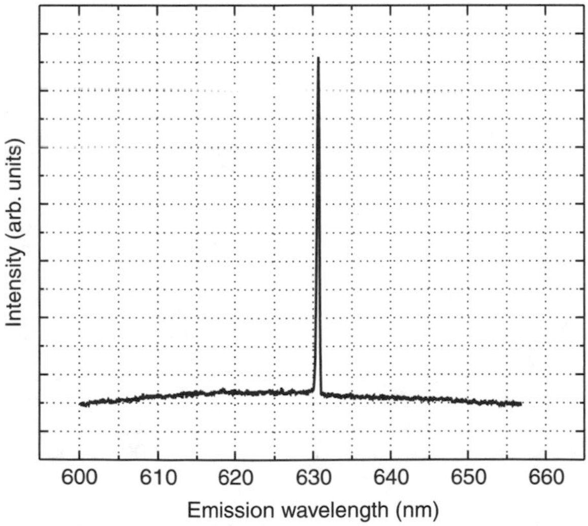

FIGURE 7.25 Emission spectrum of a waveguide-coupled organic DFB laser on PMMA (excitation parameters: wavelength 349 nm, repetition rate 1 kHz, spot size 500 x 50 μm^2).

Depending on signal level and wavelength and needed signal information, most often semiconductor-based detectors, photomultipliers, or spectrophotometers are employed. Lock-in amplifiers and boxcar integrators offer a high sensitivity sensing when combined with these detectors.

Organic semiconductor-based detectors would, of course, be an interesting alternative to these standard techniques. The possibilities offered by organic devices regarding integration and low-cost manufacturing render such devices very attractive for integration in optical sensor systems.

Research activities in the field of photodetectors are mainly driven by the strong interest in organic solar cells as a future cost-efficient way to generate electricity.[6, 5] The rapid progress in device efficiencies also led to investigations of the usage of organic materials in photodetectors.[3, 4] Optical sensor systems[98] as well as optical data transfer setups comprising organic photodiodes (OPDs) have been demonstrated.[99] Even a complete data transmission system using solely organic optoelectronic devices was presented.[23, 100] A high sensitivity to light pulses is one key factor in these applications.

For certain sensing schemes, a fast photoresponse of the detectors is important. The fastest photodetectors today are fabricated using small-molecule materials.[101, 102] They show response times in the nanosecond regime, but unfortunately their fabrication is relatively complicated due to the needed multilayer structures. Advances in the temporal response of photodiodes made of polymer materials show their prospects.[103, 104] Figure 7.26 shows the temporal behavior of the

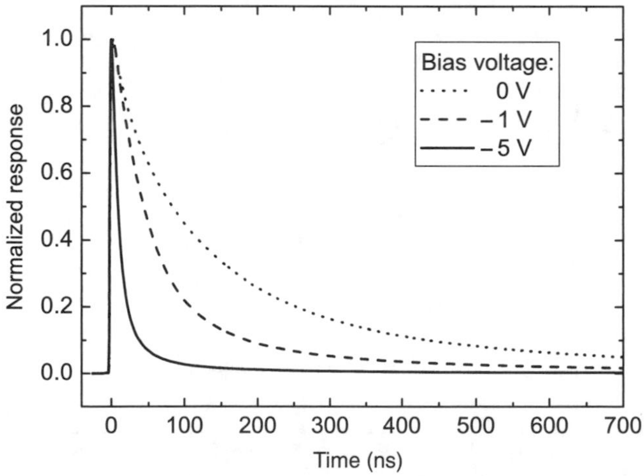

FIGURE 7.26 Normalized pulse response of a P3HT:PCBM photodiode following a 1.6 ns laser pulse with a wavelength of 532 nm. The FHWM of the −5 V biased device is 11 ns. (*Reprinted with permission from Ref. 104. Copyright 2007, American Institute of Physics.*)

photocurrent of a P3HT:PCBM [poly(3-hexylthiophene-2,5-diyl blended with the fullerene derivative [6,6]-phenyl C_{61}-butyric acid methyl ester] photodiode after excitation with a nanosecond laser pulse. With a reverse bias voltage of –5 V, a quantum efficiency comparable to the best Si photodiodes and a response time of about 11 ns are observed.

Organic photodetectors not only exhibit functionalities and specifications comparable to their inorganic counterparts, but also offer a range of other features which hardly can be realized with conventional devices. One example is the possibility to fabricate devices with customized spectral sensitivity. This can be realized by either choosing special organic materials with a certain spectral sensitivity or with the help of microcavity photodetectors.[105]

Organic photodiodes could also be integrated into organic laser-based sensor systems. The spectral response of such devices can ideally be tuned to the emission spectra of the many organic lasers and fluorescence markers. The temporal response is sufficient to detect even short laser pulses at high repetition rates. Furthermore, organic detectors can be fabricated on a wide range of substrates,[106, 107] thus offering the possibility to integrate them on the same substrate with the laser source.

The structuring options for organic photodetectors are another important issue. Depending on the sensor system layout, the detectors can be structured to have either a very large sensitive area (> 1 cm^2) or a specific and small active area. The structuring can be realized by simple photolithographic methods which define the active area of the photodetector.[108] In such a way the photodetectors could also be combined with waveguide structures.

7.5 Conclusions

In this chapter we discussed the opportunities given by organic lasers for biosensing applications. Due to their spectral tunability and the ease of processing such devices bear a huge potential for integrated analysis systems. We showed the possibilities for integrating such devices with optical waveguides for harvesting the laser radiation in microfluidic structures based on cost-effective replication techniques. In addition, we pointed out that the performance of organic photodiodes is already comparable to that of their inorganic counterparts. In combination with the techniques for integration, such devices might pave the way for future fully organic lab-on-a-chip structures.

References

1. Forrest, S. R.: "The path to ubiquitous and low-cost organic electronic appliances on plastic," *Nature* 428:911–918 (2004).
2. Bardsley, J. N.: "International OLED technology roadmap," *IEEE J. Sel. Top. Quantum Electron.* 10:3–9 (2004).
3. Peumans, P.; Yakimov, A.; and Forrest, S. R.: "Small molecular weight organic thin-film photodetectors and solar cells," *J. Appl. Phys.* 93(7):3693–3723 (2003).

4. Schilinsky, P.; Waldauf, C.; Hauch, J.; and Brabec, C. J.: "Polymer photovoltaic detectors: Progress and recent developments," *Thin Solid Films* 451–452: 105–108 (2004).
5. Hoppe, H., and Sariciftci N. S.: "Organic solar cells: An overview," *J. Mater. Res.* 19(7):1924–1944 (2004).
6. Brabec, C. J.; Sariciftci, N. S.; and Hummelen, J. C.: "Plastic solar cells," *Adv. Func. Mat.* 11(1):15–26 (2001).
7. Samuel, I. D. W., and Turnbull, G. A.: "Organic semiconductor lasers," *Chem. Rev.* 107:1272–1295 (2007).
8. Kozlov, V. G.; Bulovic, V.; Burrows, P. E.; Baldo, M.; Khalfin, V. B.; Parthasarathy, G.; Forrest, S. R., et al.: "Study of lasing action based on Förster energy transfer in optically pumped organic semiconductor thin films," *J. Appl. Phys.* 84(8): 4096–4108 (1998).
9. Scherf, U.; Riechel, S.; Lemmer, U.; and Mahrt, R. F.: "Conjugated polymers: Lasing and stimulated emission," *Pergamon Curr. Opin. Solid State Mater. Sci.* 5:143–154 (2001).
10. Riechel, S.; Lemmer, U.; Feldmann, J.; Berleb, S.; G. Mückl, A., Brütting, W.; Gombert, A., et al.: "Very compact tunable solid-state laser utilizing a thin-film organic semiconductor," *Opt. Lett.* 26(9):592–595 (2001).
11. Schneider, D.; Rabe, T.; Riedl, T.; Dobbertin, T.; Kröger, M.; Becker, E.; Johannes, H.-H., et al.: "Ultrawide tuning range in doped organic solid-state lasers," *Appl. Phys. Lett.* 85(11):1886–1888 (2004).
12. Gaal, M.; Gadermaier, C.; Plank, H.; Moderegger, E.; Pogantsch, A.; Leising, G.; and List, E. J. W.: "Imprinted conjugated polymer laser," *Adv. Mater.* 15: 1165–1167 (2003).
13. Ichikawa, M.; Tanaka, Y.; Suganuma, N.; Koyama, T.; and N. Taniguchi, Y.: "Low-threshold photopumped distributed feedback plastic laser made by replica molding," *Jpn. J. Appl. Phys., Part 1*, 42:5590–5593 (2003).
14. Burns, M. A.; Johnson, B. N.; Brahmasandra, S. N.; Handique, K.; Webster, J. R.; Krishnan, M.; Sammarco, T. S., et al.: "An integrated nanoliter DNA analysis device," *Science* 282:484–487 (1998).
15. Verpoorte, E.: "Chip vision—Optics for microchips," *Lab Chip* 3:42N–52N (2004).
16. Choudhury, B.; Shinar, R.; and Shinar, J.: "Glucose biosensors based on organic light-emitting devices structurally integrated with a luminescent sensing element," *J. Appl. Phys.* 96(5):2949–2954 (2004).
17. Hofmann, O.; Wang, X.; deMello, J. C.; Bradley, D. D.; and deMello, A. J.: "Towards microalbuminuria determination on a disposable diagnostic microchip with integrated fluorescence detection based on thin-film organic light emitting diodes," *Lab Chip* 5:863–868 (2005).
18. Yao, B.; Luo, G.; Wang, L.; Gao, Y.; Lei, G.; Ren, K.; Chen, L., et al.: "A microfluidic device using a green organic light emitting diode as an integrated excitation source," *Lab Chip* 5:1041–1047 (2005).
19. Ahn, C. H.; Choi, J.-W.; Beaucage, G.; Nevin, J. H.; Lee, J.-B.; Puntambekar, A.; and Lee, J. Y.: "Disposable smart lab on a chip for point-of-care clinical diagnostics," *Proc. IEEE* 92(1):154–173 (2004).
20. Gardeniers, J. G. E., and van den Berg, A.: "Lab-on-a-chip systems for biomedical and environmental monitoring," *Anal. Bioanal. Chem.* 378(7):1700–1703 (2004).
21. Psaltis, D.; Quake, S. R.; and Yang, C.: "Developing optofluidic technology through the fusion of microfluidics and optics," *Nature* 442:381–386 (2006).
22. Balslev, S.; Jorgensen, A. M.; Bilenberg, B.; Mogensen, K. B.; Snakenborg, D.; Geschke, O.; Kutter, J. P., et al.: "Lab-on-a-chip with integrated optical transducers," *Lab Chip* 6:213–217 (2006).
23. Punke, M.; Mozer, S.; Stroisch, M.; Bastian, G.; Gerken, M.; Lemmer, U.; Rabus, D.G., et al.: "Organic semiconductor devices for micro-optical applications," *Proc. SPIE* 6185:618505-1 to 618505–13 (2006).
24. Christiansen, M. B.; Schøler, M.; and Kristensen, A.: "Integration of active and passive polymer optics," *Opt. Exp.* 15:3931–3939 (2007).

25. Punke, M.; Woggon, T.; Stroisch, S.; Ebenhoch B.; Geyer U.; Karnutsch, C.; Gerken, M, et al.: "Organic semiconductor lasers as integrated light sources for optical sensor systems," *Proc. SPIE* 6659:665909 (2007).
26. Schneider, D.; Lemmer, U.; Riedl, T.; and Kowalsky, W.: "Low threshold organic semiconductor lasers," *Organic Light Emitting Devices*, K. Müllen and U. Scherf (eds.), Wiley-VCH, Weinheim, 2005.
27. Schneider, D.: "Organische Halbleiterlaser," Technical University Braunschweig, Ph.D. thesis, 2005.
28. Turnbull, G. A.; Andrew, P; Barnes, W. L.; and Samuel, I. D. W.: "Operating characteristics of a semiconducting polymer laser pumped by a microchip laser," *Appl. Phys. Lett.* 82:313 (2003).
29. Riedl, T.; Rabe, T.; Johannes, H. H.; Kowalsky, W; Wang, J.; Weimann, J.; Hinze, P., et al.: "Tunable organic thin-film laser pumped by an inorganic violet diode laser," *Appl. Phys. Lett.* 88:241116 (2006).
30. Karnutsch, C.; Haug, V.; Gärtner, C.; Lemmer, U.; Farrell, T.; Nehls, B.; Scherf, U., et al.: "Low threshold blue conjugated polymer DFB lasers," *CLEO*, CFJ3 (2006).
31. Vasdekis, A. E.; Tsiminis, G.; Ribierre, J. C.; O'Faolain, L.; Krauss, T. F.; Turnbull, G. A.; and Samuel, I. D. W.: "Diode pumped distributed Bragg reflector lasers based on a dye-to-polymer energy transfer blend," *Opt. Exp.* 14(20):9211–9216 (2006).
32. Yang, Y.; Turnbull, G. A.; and Samuel, I. D. W.: "Hybrid optoelectronics: A polymer laser pumped by a nitride light-emitting diode," *Appl. Phys. Lett.* 92:163306 (2008).
33. Campbell,V.; Smith, D.; Neef, C.; and Ferraris, J.: "Charge transport in polymer light-emitting diodes at high current density," *Appl. Phys. Lett.* 75(6): 841–843 (1999).
34. Yokoyama, W.; Sasabe, H.; and Adachi, C.: "Carrier injection and transport of steady-state high current density exceeding 1000 A/cm^2 in organic thin films," *Jpn. J. Appl. Phys.* 42:L1353–L1355 (2003).
35. Yamamoto, H.; Kasajima, H.; Yokoyama, W.; Sasabe, H.; and Adachi, C.: "Extremely-high-density carrier injection and transport over 12,000 A/cm^2 into organic thin films," *Appl. Phys. Lett.* 86:083502 (2005).
36. Gärtner, C.; Karnutsch, C.; Pflumm, C.; and Lemmer, U.: "Numerical device simulation of double heterostructure organic laser diodes including current induced absorption processes," *IEEE J. Quantum Electron.* 43(11):1006–1017 (2007).
37. Kavc, T.; Langer, G.; Kern, W.; Kranzelbinder, G.; Toussaere, E.; Turnbull, G. A.; Samuel, I. D. W., et al.: "Index and relief gratings in polymer films for organic distributed feedback lasers," *Chem. Mat.* 14:4178–4185 (2002).
38. Lawrence, J. R.; Andrew, P.; Barnes, W. L.; Buck, M; Turnbull, G. A.; and Samuel, I. D. W.: "Optical propertiers of light-emitting polymer directly patterned by soft lithography," *Appl. Phys. Lett.* 81:1955–1958 (2002).
39. Pisignano, D.; Persano, L.; Visconti, P; Cingolani, R.; and Gigli, G.: "Oligomer-based organic distributed feedback lasers by room-temperature nanoimprint lithography," *Appl. Phys. Lett.* 83:2545 (2003).
40. Forberich, K.: "Organische Photonische-Kristall-Laser," University of Freiburg, Ph.D. thesis, 2005.
41. Wang, J.; Weimann, T.; Hinze, P.; Ade, G.; Schneider, D.; Rabe, T.; Riedl, T., et al.: "A continuously tunable organic DFB laser," *Microelectronic Eng.* 78–79: 364–368 (2005).
42. Gadegaard, N., and McCloy D.: "Direct stamp fabrication for NIL and hot embossing using HSQ," *Microelectronic Eng.* 84 (12):2785–2789 (2007).
43. Sun, H. B.; Kawakami, T.; Xu, Y.; Ye, J.-Y.; Matuso, S.; Misawa, H.; Miwa, M., et al.: "Real three-dimensional microstructures fabricated by photopolymerization of resins through two-photon absorption," *Opt. Lett.* 25(5):1110–1112 (2000).
44. Sun, H. B., and Kawata, S.: "Two-photon laser precision microfabrication and its applications to micro–nano devices and systems," *J. Lightwave Techn.* 21(3): 624–633 (2003).

45. Deubel, M.; von Freymann, G.; Wegener, M.; Pereira, S.; Busch, K.; and Soukoulis, C. M.: "Direct laser writing of three-dimensional photonic-crystal templates for telecommunications," *Nature Mater.* 3(7):444–447 (2004).
46. Li, L., and Fourkas, J. T.: "Multiphoton polymerization," *Mater. Today* 10(6): 30–37 (2007).
47. Sun, H. B.; Maeda, M.; Takada, K.; Chon, J. W. M.; Gu, M.; and Kawata, S.: "Experimental investigation of single voxels for laser nanofabrication via two-photon photopolymerization," *Appl. Phys. Lett.* 83:819–821 (2003).
48. Gombert, A.; Bläsi, B.; Bühler, C.; Nitz, P.; Mick, J.; Hoßfeld, W.; and Niggemann, M.: "Some application cases and related manufacturing techniques for optically functional microstructures on large areas," *Opt. Eng.* 43: 2525–2533 (2004).
49. Srinivasan, R., and Braren, B: "Ultraviolet laser ablation of organic polymers," *Chem. Rev.*, 89:1303–1316 (1989).
50. Stroisch, M.; Woggon, T.; Lemmer, U.; Bastian, G.; Violakis, G.; and Pissadakis, S.: "Organic semiconductor distributed feedback laser fabricated by direct laser interference ablation," *Opt. Exp.* 15(7):3968–3973 (2007).
51. Chou, S. Y.; Krauss, P. R.; and Renstrom, P. J.: "Imprint lithography with 25-nanometer resolution," *Science* 272(5258):85 (1996).
52. Hanemann, T.; Heckele, M.; and Piotter, V.: "Current status of micromolding technology," *Polymer News* 25:224–229 (2000).
53. Bacher, W.; Menz, W.; and Mohr, J.: "The LIGA technique and its potential for microsystems—A survey," *Ind. Electroni., IEEE Trans.* 42(5):431–441 (1995).
54. Mappes, T.; Worgull, M.; Heckele, M.; and Mohr, J.: "Submicron polymer structures with X-ray lithography and hot embossing," *Microsyst. Technol.*, DOI:10.1007/s00542-007-0499-6, 14:1721–1725 (2008).
55. Becker, E. W.; Ehrfeld, W.; Hagmann, P.; Maner, A.; and Münchmeyer, D.: "Fabrication of microstructures with high aspect ratios and great structural heights by synchrotron radiation lithography, galvanoforming and plastic molding (LIGA process)," *Microelectron. Eng.* 4:35–56 (1986).
56. Kim, I., and Mentone, P. F.: "Electroformed nickel stamper for light guide panel in LCD back light unit," *Electrochim. Acta.* 52:1805–1809 (2006).
57. Gale, M. T.: "Replication techniques for diffractive optical elements," *Microelectron. Eng.* 34(3):321–339 (1997).
58. Meier, M.; Dodabalapur, A.; Rogers, J. A.; Slusher, R. E.; Mekis, A.; and Timko, A.: "Emission characteristics of two-dimensional organic photonic crystal lasers fabricated by replica molding," *J. Appl. Phys.* 86(7):3502–3507 (1999).
59. Berggren, M.; Dodalapur, A.; and Slusher, R. E.: "Organic solid-state lasers with imprinted gratings on plastic substrates," *Appl. Phys. Lett.* 72(4):410–411 (1998).
60. Rogers, J. A.; Meier, M.; and Dodabalapur, A.: "Using printing and molding techniques to produce distributed feedback and Bragg reflector resonators for plastic lasers," *Appl. Phys. Lett.* 73:1766 (1998).
61. Rogers, J. A.; Meier, M.; Dodabalapur, A.; Laskowski, E. J.; and Cappuzzo, M. A.: "Distributed feedback ridge waveguide lasers fabricated by nanoscale printing and molding on nonplanar substrates," *Appl. Phys. Lett.* 74(22): 3257–3259 (1999).
62. Bender, M.; Plachetka, U.; Ran, J.; Fuchs, A.; Vratzov, B.; Kurz, H.; Glinsner, T., et al.: "High resolution lithography with PDMS molds," *J. Vac. Sci. Technol., B* 22:3229–3232 (2004).
63. Nilsson, D.; Balslev, S.; Gregersen, M. M.; and Kristensen, A.: "Microfabricated solid-state dye lasers based on a photodefinable polymer," *Appl. Opt.* 44: 4965–4971 (2005).
64. Muzio, E.; Seidel, P.; Shelden, G.; and Canning, J.: "An overview of cost of ownership for optical lithography at the 100 nm and 70 nm generations," *Semicond. Fabtech* (11):191–194 (2000).
65. Sreenivasan, S. V.; Willson, C. G.; Schumaker, N. E.; and Resnick, D. J.: "Cost of ownership analysis for patterning using step and flash imprint lithography," *Proc. SPIE*, 4688 (2002).

66. Stroisch, M.: "Organische Halbleiterlaser auf Basis Photonischer-Kristalle," University Karlsruhe (TH), Ph.D. thesis, 2007.
67. Borisov, S. M., and Wolfbeis, O. S.: "Optical biosensors," *Chem. Rev.* 108: 423–461 (2008).
68. Vahala, K. J.: "Optical microcavities," *Nature* 424:839–846 (2003).
69. Potyrailo, R. A.; Hobbs, S. E.; and Hieftje, G. M.: "Optical waveguide sensors in analytical chemistry: Today's instrumentation, applications and trends for future development," *Fresenius J. Anal. Chem.* 362(4):349–373 (1998).
70. Lading, L.; Nielsen, L. B.; Sevel, T.; Center, S. T.; and Brondby, D.: "Comparing biosensors," *Proc. IEEE*, 229–232 (2002).
71. Mogensen, K. B.; El-Ali, J.; Wolff, A.; and Kutter, J. P.: "Integration of polymer waveguides for optical detection in microfabricated chemical analysis systems," *Appl. Opt.* 42(19):4072–4079 (2003).
72. Lakowicz, J. R.: *Principles of Fluorescence Spectroscopy*, Kluwer Academic, New York, 1999.
73. Nilsson, D.: "Polymer based miniaturized dye lasers for Lab-on-a-chip systems," Technical University of Denmark (DTU), Dep. Micro and Nanotechn., Ph.D. thesis, 2005.
74. Verpoorte, E.: "Microfluidic chips for clinical and forensic analysis," *Electrophoresis* 23(5):677 (2002).
75. Geschke, O.; Klank, H.; and Telleman, P.: *Microsystem Engineering of Lab-on-a-Chip Devices*, Wiley-VCH, Weinheim, 2004.
76. Bousse, L. J.; Kopf-Sill, A. R.; and Parce, J. W.: "Parallelism in integrated fluidic circuits," *Proc. SPIE*, 179–186 (2004).
77. Dittrich, P. S.; and Manz, A.: "Lab-on-a-chip: microfluidics in drug discovery," *Nature* 5:210–218 (2006).
78. Mogensen, K. B.; Klank, H.; and Kutter, J. P.: "Recent developments in detection for microfluidic systems," *Electrophoresis* 25(21–22):3498–3512 (2004).
79. Rabus, D. G.; Bruendel, M.; Ichihashi, Y.; Welle, A.; Seger, R. A.; and Isaacson, M.: "A bio-fluidic photonic platform based on deep UV modification of polymers," *IEEE J. Sel. Top. Quantum Electron.* 13(2):214–222 (2007).
80. Kobayashi, J.; Matsuura, T.; Sasaki, S.; and Maruno, T: "Single-mode optical waveguides fabricated from fluorinated polyimides," *Appl. Opt.* 37(6): 1032–1037 (1998).
81. Keil, N.; Yao, H. H.; Zawadzki, C.; Bauer, J.; Bauer, M.; Dreyer, C.; and Schneider, J.: "A thermal all-polymer arrayed-waveguide grating multiplexer," *Electron. Lett.* 37(9):579–580 (2001).
82. Tomlinson, W. J.; Kaminow, I. P.; Chandross, A.; Fork, R. L.; and Silvast, W. T.: "Photoinduced refractive index increase in poly(methylmethacrylate) and its applications," *Appl. Phys. Lett.* 16(12):486–489 (1970).
83. Keil, N.; Yao, H. H.; and Zawadzki, C.: "(2 × 2) Digital optical switch realised by low cost polymer waveguide technology," *Electron. Lett.* 32(16):1470–1471 (1996).
84. Keil, N.; Strebel, B. N.; Yao, H. H.; Zawadzki, C.; and Hwang, W. Y.: "Optical polymer waveguide devices and their applications to integrated optics and optical signal processing," *Proc. SPIE*, 1774:130–141 (1993).
85. Kragl, H.; Hohmann, R.; Marheine, C.; Pott, W.; and Pompe, G.: "Low cost monomode, integrated optics polymeric components with passive fibre-chip coupling," *Electron. Lett.* 33(24):2036–2037 (1997).
86. Bauer, H. D.; Ehrfeld, W.; Harder, M.; Paatzsch, T; Popp, M; and Smaglinski; I.: "Polymer waveguide devices with passive pigtailing: An application of LIGA technology," *Synth. Met.* 115(1–3):13–20 (2000).
87. Ruck, D. M.; Brunner, S.; Tinschert, K.; and Frank, W. F. X.: "Production of buried waveguides in PMMA by high energy ion implantation," *Nucl. Instrum. Methods Phys. Res., Sect. B* 106(1–4):447–451 (1995).
88. Hong, W.; Woo, H. J.; Choi, H. W.; Kim, Y. S.; and Kim, G. D.: "Optical property modification of PMMA by ion-beam implantation," *Appl. Surf. Sci.* 169: 428–432 (2001).

89. Schoesser, A.; Knoedler, B; Tschudi, T. T.; Frank, W. F.; Stelmasyzk, A.; Muschert, D.; Rueck, D. M., et al.: "Optical components in polymers," *Proc. SPIE*, 2540:110–117 (1995).
90. Choi, J. O.; Moore, J. A.; Corelli, J. C.; Silverman, J.; and Balkhru, H.: "Degradation of poly(methylmethacrylate) by deep ultraviolet, x-ray, electron beam, and proton beam irradiations," *J. Vac. Sci. Technol., B* 6(6):2286–2289 (1988).
91. Jellinek, H. H. G.: *Aspects of Degradation and Stabilization of Polymers*, Elsevier, Amsterdam, 1978.
92. Wochnowski, C.; Metev, S.; and Sepold, G.: "UV-laser-assisted modification of the optical properties of polymethylmethacrylate," *Appl. Surf. Sci.* 154–155: 706–711 (2000).
93. Schoesser, A.; Tschudi, T. T.; Frank, W. F.; and Pozzi, F.: "Spectroscopic study of surface effects in polymer waveguides generated by ionizing radiation related to guiding properties," *Proc. SPIE*, 2851:73–81 (1996).
94. Rabus, D.; Henzi, P.; and Mohr, J.: "Photonic integrated circuits by DUV-induced modification of polymers," *IEEE Phot. Technol. Lett.* 17(3):591–593 (2005).
95. Bruendel, M.; and Rabus, D. G.: "1 × 2 and 1 × 3 multimode interference couplers fabricated by hot embossing and DUV-induced modification of polymers," LEOS Annual Meeting, Montreal, 2006.
96. Punke, M.; Mozer, S.; Stroisch, M.; Heinrich, M. P.; Lemmer, U.; Henzi, P.; and Rabus, D. G.: "Coupling of organic semiconductor amplified spontaneous emission into polymeric single mode waveguides patterned by deep-UV irradiation," *IEEE Phot. Technol. Lett.* 19:61–63 (2007).
97. Punke, M.: "Organische Halbleiterbauelemente für mikrooptische Systeme," Universitätsverlag Karlruhe, Ph.D. thesis, 2008.
98. Hofmann, O.; Miller, P.; Sullivan, P.; Jones, T. S.; deMello, J. C.; Bradley, D. C.; and deMello, A. J.: "Thin-film organic photodiodes as integrated detectors for microscale chemiluminescence assays," *Sens. Actuat. B*, 106:878–884 (2005).
99. Ohmori, Y.; Kajii, H.; Kaneko, M.; Yoshino, K.; Ozaki, M.; Fujii, A.; Hikita, M., et al.: "Realization of polymer optical integrated devices utilizing organic light-emitting diodes and photodetectors fabricated on a polymer waveguide," *IEEE J. Sel. Top. Quantum Electron.* 10(1):70–78 (2004).
100. Punke, M.; Valouch, S., Kettlitz, S. W.; Gerken, M.; and Lemmer, U.: "Optical data link employing organic light-emitting diodes and organic photodiodes as optoelectronic components," *J. Lightwave Technol.* 26(7):816–823 (2008).
101. Peumans, P.; Bulovic, V.; and Forrest, S. R.: "Efficient, high-bandwidth organic multilayer photodetectors," *Appl. Phys. Lett.* 76(26):3855–3857 (2000).
102. Morimune, T.; Kajii, H.; and Ohmori, Y.: "High-speed organic photodetectors using heterostructure with phthalocyanine and perylene derivative," *Jpn. J. Appl. Phys.* 45(1B):546–549 (2006).
103. Komatsu, T.; Kaneko, S.; Miyanishi, S.; Sakanoue, K.; Fujita, K.; and Tsutsui, T.: "Photoresponse studies of bulk heterojunction organic photodiodes," *Jpn. J. Appl. Phys.* 43(11A):L1439–L1441 (2004).
104. Punke, M.; Valouch, S.; Kettlitz, S. W.; Christ, N.; Gärtner, C.; Gerken, M.; and Lemmer, U.: "Dynamic characterization of organic bulk heterojunction photodetectors," *Appl. Phys. Lett.* 91:071118 (2007).
105. Koeppe, R.; Müller, J. G.; Lupton, J. M.; Feldmann, J.; Scherf, U.; and Lemmer, U.: "One- and two-photon photocurrents from tunable organic microcavity photodiodes," *Appl. Phys. Lett.* 82:2601 (2003).
106. Brabec, C. J.; Padinger, F.; Hummelen, J. C. ; Janssen, R. A. J.; and Sariciftci, N. S.: "Realization of large area flexible fullerene-conjugated polymer photocells: A route to plastic solar cells," *Synth. Met.* 102:861 (1999).
107. Sariciftci, N. S.: "Polymer photovoltaic materials," *Curr. Opin. Solid-State Mater. Sci.* 4:373–378 (1999).
108. Peters, S.; Sui, Y.; Glöckler, F.; Lemmer, U.; and Gerken, M.: "Organic photo detectors for an integrated thin-film spectrometer," *Proc. SPIE* 6765:676503–1 (2007).

CHAPTER 8

Organic Electronics in Memories and Sensing Applications

Th. Birendra Singh*

Linz Institute of Organic Solar Cells (LIOS)
Institute of Physical Chemistry
Johannes Kepler University, Linz, Austria

Siegfried Bauer

Soft Matter Physics, Johannes Kepler University, Linz, Austria

Niyazi Serdar Sariciftci

Linz Institute of Organic Solar Cells (LIOS)
Institute of Physical Chemistry
Johannes Kepler University, Linz, Austria

Organic electronics is a rapidly developing field with a wide range of applications in flexible and conformable electronic devices. Functional materials allow for the implementation of new functionalities such as bistability to be employed in memory elements as well as physical mechanisms to be used in sensors for measuring various physical parameters, such as temperature or pressure changes. Thereby, new application markets may arise for organic electronic components. In this chapter first we provide a short synopsis of available organic semiconductors and dielectrics, useful for field-effect transistor devices and sensors. Then we discuss sensing principles for detecting temperature and pressure changes, a field which is still in its infancy. A tour d'horizon through selected

**Current affiliation*: CSIRO Molecular and Health Technology, Ian Work Laboratory, Clayton, Victoria, Australia.

applications in memory elements and sensor systems concludes the chapter.

8.1 Functional Organic Materials

8.1.1 Organic Semiconductors

The requirements for organic semiconductors in organic electronics are rigorous. A large number of equally important criteria determine the performance of these devices.

- Organic semiconductors need to exhibit good chemical stability when exposed to ambient conditions.
- They need to be stable under bias stress in typical operating conditions.
- They must be compatible with gate dielectrics to form interfaces with low interface trap density.
- Organic semiconductors must allow for efficient charge injection and low contact resistance, when in contact with metal electrodes.
- They need to exhibit charge carrier mobilities exceeding 0.1 or even 1 $cm^2/(V \cdot s)$ in practical applications to be comparable with inorganic competing technologies such as amorphous silicon.
- They should be sustainable to bending in flexible displays, circuits, and sensors.

Since charge transport in organic field-effect transistor (OFET) devices takes place at the interface between the organic semiconductor film and the gate dielectric, charge transport depends on achieving close intermolecular stacking throughout the length scale of the OFET channel. One of the key tasks is to achieve long-range efficient charge transport by self-organization schemes. This long-range order is often prevented by side groups that allow for producing solution processability in molecular semiconductors and polymers. Impurities and by-products from synthesis often hinder achieving high mobilities and result in low "off" currents.[1] These impurities are also a potential reason for the slow degradation of devices, as often observed in experimental investigations. Traps, e.g., due to impurities, may easily immobilize charge carriers and thereby decrease charge carrier mobility.

There is an intense ongoing research on high mobility organic semiconductors with all the aforementioned desirable properties. Indeed, significant recent advancement has been made in developing both p-type and n-type semiconductors with recent reviews describing in detail what has been achieved.[1–7] Among these reviews one can divide research efforts into synthetic aspects of organic semiconductors,[1] material challenges and applications,[2] n-type organic semiconductors[4] and oligomers,[6] and interface effects.[7] Depending on the requirements of device functionality, we put organic semiconductors into three categories: p-type semiconductors, n-type semiconductors, and ambipolar semiconductors.

p-Type Semiconductors

If holes can be easily injected into the valence band (HOMO level) of an organic semiconductor, i.e., can sustain stable radical cations, and these cations (positive polarons) can move throughout the solid phase, than we will call this material *p type*. To achieve high-performance, solution-processable organic semiconductors with high charge carrier mobilities, ordered structures are needed at the tertiary nanostructure of the organic thin films. Designing the material to exhibit microcrystallinity or liquid crystallinity[8] or self-organization or making use of specific interactions with a templating substrate is suggested for this route. The other approach aims to produce a completely amorphous micro/nanostructure to provide a uniform path for charge transport, with a minimum degree of site energy fluctuations.[9] Among the traditional molecular semiconductors, pentacene, thiophene oligomers, and metallophthalocyanines are well known as p type (see Fig. 8.1). Among polymers, polythiophenes, polyfluorenes, polyarylamines, and poly(benzo-bis-imidazobenzophenanthroline) are promising. Currently semiconducting polymers with considerable air stability and high charge carrier mobility are the subject of intense research interests. Among the polymeric semiconductors, significant efforts continue to be focused on derivatives of poly(3-hexylthiophene) (P3HT). One of the reasons for using P3HT thin films is the presence of microcrystalline and lamellar π stacking, which results in large charge carrier mobilities. The ionizaion potential (typically around 4.9 to 5.0 eV) is best suited to form ohmic contacts with many air-stable electrodes such as Au or conducting polymers such as polyethylenedioxy-thiopehene doped with polystyrene sulfonic acid (PEDOT/PSS). However, P3HT tends to exhibit large positive threshold voltage shifts V_T upon exposure to air, presumably due to slight doping of the polymer with oxygen.[10, 11] This can be improved by increasing the ionization potential of the polythiophene backbone either by adopting a fully planar conformation through the side chain substitution pattern[12] or by incorporating partially conjugated co-monomers into the main chain.[13] Field effect mobilities exceeding 0.15 $cm^2/(V \cdot s)$ have been reported from such materials in air.[13] Similarly, poly(2,5-bis(3-tetradecylthiophen-2-yl)thieno[3,2-b] thiophene (PBTTT) exhibits mobilities of 0.7 to 1 $cm^2/(V \cdot s)$.[14]

An alternative route to solution-processable polymeric materials is to use small-molecular semiconductors either processed from solution or evaporated (sublimated) from a heated source onto a target substrate. Such an example is pentacene, an aromatic compound with five

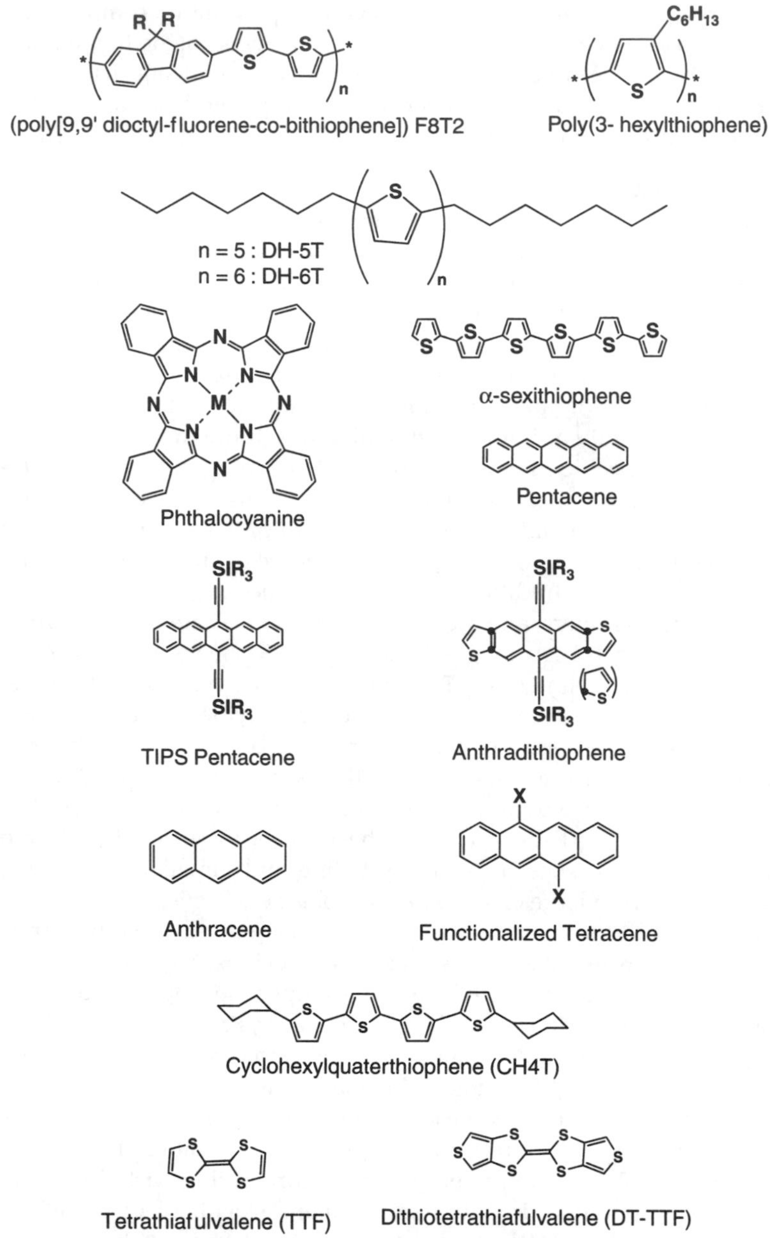

Figure 8.1 Commonly used p-type organic semiconductor: F8T2 (poly[9,9' dioctyl-fluorene-co-bithiophene]); P3HT: regioregular poly [3-hexylthiophene]; DH-5T: (α, ω-dihexylquinquethiophene); DH-6T: (α, ω-dihexylsexithiophene); phthalocyanine, (α-6T) α-sexithiophene, pentacene, TIPS pentacene, anthradithiophene, anthracene, functionalized tetracene, CH4T: cyclohexylquaterthiophene, TTF: tetrathiafulvalene, DT-TTF: dithiotetrathiafulvalene.

fused benzene rings. Pentacene has been widely studied as a p-type semiconductor for OFETs. The highest field-effect mobility obtained is 3 $cm^2/(V \cdot s)$ for pentacene thin films on polymer dielectrics[15] and 6 $cm^2/(V \cdot s)$ on chemically modified SiO_2/Si substrates.[16]

Substituted acenes based on silylethynyl,[17–19] such as 6,13-bis (triisopropylsilylethynyl) (TIPS) pentacene and triethylsilylethynyl anthradiophene (TES-ADT), not only are highly soluble but also exhibit high crystallinity with mobility as high as 1 $cm^2/(V \cdot s)$. TES-ADT as-deposited film is known to be amorphous and subsequent to annealing-induced controlled crystallization of the thin films.[19] This shows enormous progress in the direction of solution-processed organic semiconductors for OFETs.

n-Type Semiconductors

As far as n-type semiconductors are concerned, there are still only a few which show high charge mobilities as compared to p-type semiconductors. Among them naphthalene/perylene derivatives, copper perfluorophthalocyanine, and fluoroalkyl-substituted oligothiophene are promising (see Fig. 8.2). Among all n-type oligomers such as *N, N*-dioctyl-3,4,9,10-perylenetetracarboxylic diimide have shown the highest mobilities 0.6 to 1.7 $cm^2/(V \cdot s)$. Quinode oligomers such as terthiophene-based quinodimethane stabilized by dicyanomethylene groups at each end show large electron mobilities up to 0.2 $cm^2/(V \cdot s)$ with high on/off ratios. However, none of these materials have been applied so far in devices with the exception of devices for the measurement of charge carrier mobilities. Among the fullerenes, C_{60} is well known to have high electron mobilities up to 6 $cm^2/(V \cdot s)$ depending on the crystallinity of the film. Based on these high mobility C_{60} thin-film devices, seven-stage ring oscillators were demonstrated with an operating frequency as high as 50 kHz.[20] These devices have been operated only under inert conditions. Small-molecule organic semiconductors can also be rendered solution-processable by attachment of flexible side chains. Among the fullerene derivatives, phenyl C_{61}-butyric acid methyl ester (PCBM) have mobilities as high as 0.2 $cm^2/(V \cdot s)$ when the film is solution-cast on polymeric dielectrics.[21]

Design of molecules for electron transport with increased electron affinity has been achieved through substitution with electronegative elements to allow efficient electron injection into the lowest occupied molecular orbital (LUMO) level and to increase hydrophobicity. Fluorinated compounds are more air-stable, for example, $F_{16}CuPc$, perfluoropentacene. Interestingly there are plenty of n-type organic semiconductors with unpublished mobilities (see Fig. 8.3). Among polymeric n-type semiconductors, ladder-type polymers such as BBL have shown electron mobilities of 0.1 $cm^2/(V \cdot s)$. A polymer synthesized by Stille coupling of *N, N'*-dialkyl-1,7-dibromo-3,4,9,10-perylene diimide with a distannyl derivative of dithienothiophene has shown mobilities of 1.3×10^{-2} $cm^2/(V \cdot s)$ (see Fig. 8.2).

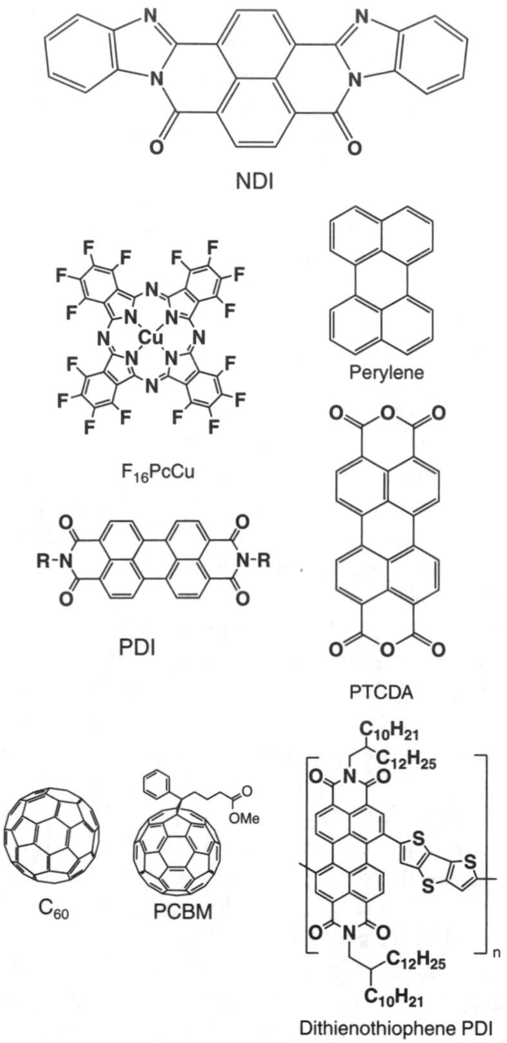

FIGURE 8.2 Commonly used n-type organic semiconductor: NDI: naphthalene diimide; F_{16}CuPc: perfluorocopper phthalocyanine; Perylene; PTCDA: 3,4,9,10-perylene-tetracarboxylic dianhydride and its derivaties; PDI: *N*, *N*'. dimethyl 3,4,9,10-perylene tetracarboxylicdiimide; C_{60}; and PCBM: methanofullerene [6,6]-phenyl C_{61}-butyric acid methyl ester and alternating dithienothiophene and perylenediimide units.

Ambipolar Semiconductors

The *ambipolar* charge transport in organic transistors is a highly desirable property because it enables the design of circuits with low-power dissipation and good noise margin similar to complementary metal-oxide semiconductor (CMOS) logic circuits. Thin films of pentacene

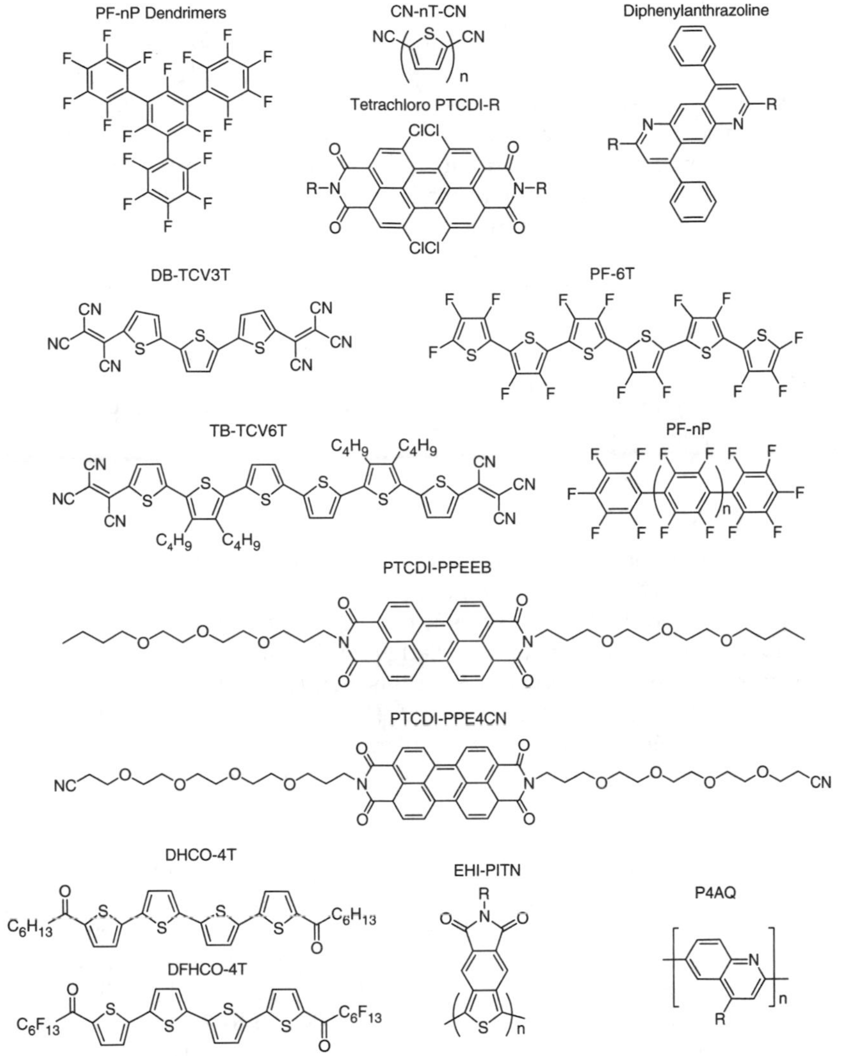

FIGURE 8.3 Structures of some other n-channel semiconductors. (*Reproduced with permission from Ref. 4. Copyright 2004, American Chemical Society.*)

have been shown to exhibit good electron and hole mobilities.[22–23] We have demonstrated ambipolar CMOS-like inverters employing ambipolar pentacene OFETs.[24] Fullerene derivatives such as PCBM and C_{70} have also shown ambipolar transport and circuits.[25–26] Among the phthalocyanines, copper phthalocyanine, CuPc,[27] and iron phthalocyanine, FePc, and among polymers, poly(3,9-di-*t*-butylindeno[1,2-b] fluorene) (PIF),[28] and bis[4-dimethylaminodithiobenzyl]-nickel have shown ambipolar transport properties in OFETs.[29] Ambipolar transport is also a necessary condition for light-emitting transistors,[30–31] and some of the highly luminescent polymers such as MDMO-PPV and F8BT have been shown to be not only ambipolar transport materials but also useful materials for light-emitting devices.[32–33] There is probably a large number of organic semiconductors with unknown ambipolar characteristics. One of the latest examples is the functionalized heptacene as reported by Chun et al.[34]

Self-assembled monolayer (SAM) modified gate dielectrics also influence the morphology and electronic properties, including ambipolar transport of all the aforementioned semiconductors.[35] SAMs bring an improvement in the charge injection by interposing an appropriately oriented dipole layer between the contact and the semiconductor in OFETs.[36] Some of the commonly used SAMs in OFETs are shown in Fig. 8.4.

8.1.2 DNA

For practical use of DNA as an electroactive material in organic electronics, natural DNA, fish waste, e.g., salmon sperm, which is normally a waste product of the salmon fishing industry, are attractive. Although there is a wealth of knowledge on the nature of the transport properties on synthetic DNA, in this chapter we focus only on DNA materials derived from salmon milt. The reader may also note that there is an ongoing debate on the insulating,[37–42] semiconducting, [43-44] highly conducting,[45] as well as superconducting nature[46] of transport in DNA molecules.

The DNA used for our research in optoelectronic devices was purified DNA provided by the Chitose Institute of Science and Technology (CIST).[47] The processing steps involved first the isolation of the DNA from frozen salmon milt and roe sacs through a homogenization process. It then went through an enzymatic treatment to degrade the proteins by protease. The resulting freeze-dried purified DNA has a molecular weight ranging from 500,000 to 8,000,000 Da with purity as high as 96% and protein content of 1 to 2%. The molecular weight of the DNA provided by CIST is on average greater than 8,000,000 Da. If necessary, the molecular weight of DNA supplied by CIST can be tailored and cut using an ultrasonic procedure[48] which gives rise to a lower molecular weight of 200,000 Da depending on the sonication energy, as shown in Fig. 8.5. It was found that purified DNA is soluble only in aqueous media; the resulting films are water-sensitive and have insufficient mechanical strength, so they are not compatible with typical fabrication processes used in polymer-based devices. It has also been observed that many particulates are present in the DNA films. Therefore, additional processing steps are performed to render DNA more suitable for organic device fabrication with better film quality. From the knowledge of stoichiometric

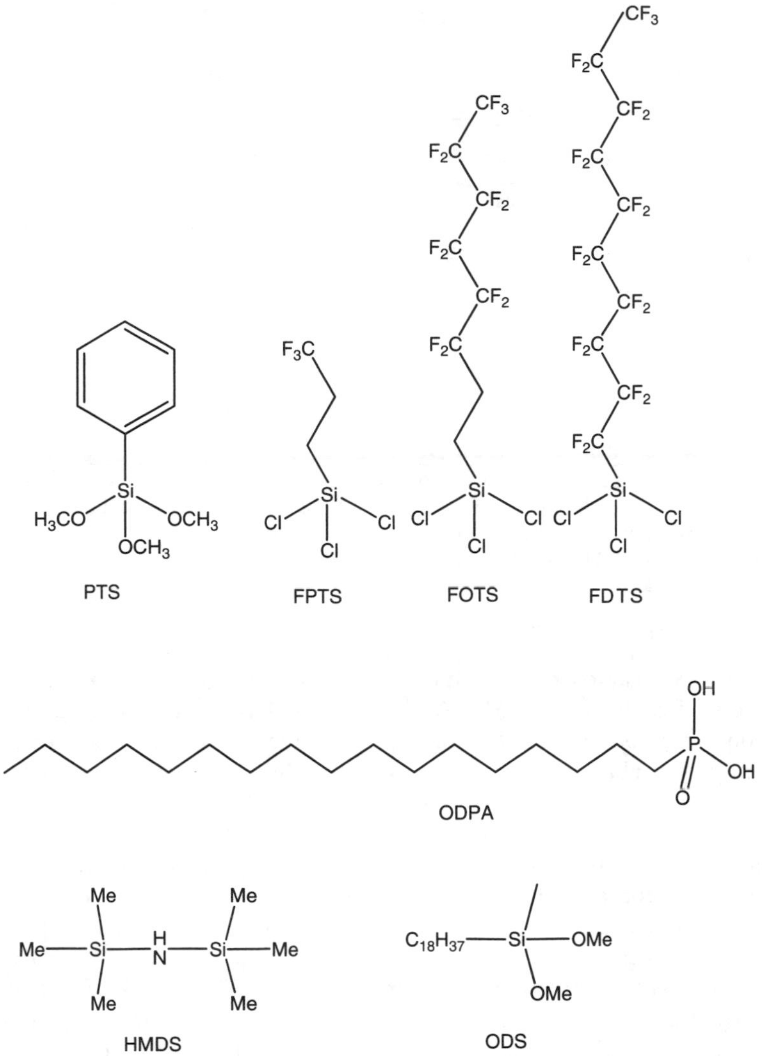

FIGURE 8.4 Molecular structures of the commonly used SAM in organic electronics: PTS, phenyltrimethoxysilane; FPTS, trichloro(3,3,3-trifluoropropyl) silane; FOTS, trichloro(1*H*,1*H*,2*H*,2*H*-perfluorooctyl)silane; FDTS, trichloro(1*H*,1*H*,2*H*,2*H*-perfluorodecyl)silane; ODPA, noctadecylphosphonic acid; HMDS, hexamethyldisilazane; ODS, octadecyltrimethoxysilane.

combination of an anionic polyelectrolyte with a cationic surfactant, it has been shown that DNA, which is an anionic polyelectrolyte, could be quantitatively precipitated with cationic surfactant in water,[49] using hexadecyltrimethylammmonium chloride (CTMA), by an ion-exchange reaction[50–51] (see Fig. 8.5). The resulting DNA-lipid

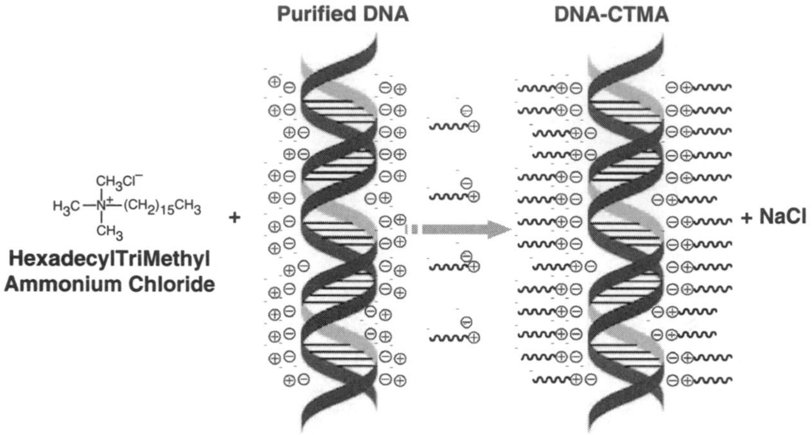

FIGURE 8.5 The purified DNA is initially soluble only in aqueous solutions and does not dissolve in any organic solvent. Purified DNA is modified through a cationic surfactant (hexadecyltrimethyl ammonium chloride—CTMA) cation exchange reaction to enhance solubility, processing, and stability. (See also color insert.)

complex became water-insoluble and more mechanically stable due to the alkyl chain of the CTMA. Adding the CTMA complex, DNA-CTMA could now be dissolved using organic solvents, such as chloroform, ethanol, methanol, butanol, or chloroform/alcohol blends. When dissolved in such organic solvents, the DNA-CTMA was passed through a 0.2 μm filter to remove large particulates. DNA-CTMA films can be cast by standard methods such as spin coating, doctor blading, dip coating, drop casting, etc. and exhibit excellent transmission over a wide wavelength range from 300 to 1600 nm. DNA-CTMA is also a very low-loss optical material applicable over a wide range of wavelengths with a refractive index ranging from 1.526 to 1.540. The electrical resistivity of DNA-CTMA films with molecular weights of 500,000 and 6,500,000 as a function of temperature is in the range of 10^9 to 10^{14} $\Omega \cdot$cm depending on the molecular weight. The dielectric constant of DNA-CTMA decreases from 7.8 to 6 in the frequency range between 1 and 1000 kHz. From thermogravimetric analysis (TGA) of the DNA-CTMA complex, thermal stability up to 230°C and a water uptake of 10% in air at room temperature are obtained.

Recently we demonstrated that thin films of DNA-CTMA can be employed as gate dielectric in low-voltage operating OFETs.[52–53] A smooth dielectric film is a prerequisite in order to allow for the deposition of smooth organic semiconductor films, thereby creating a better interface for charge transport. Another important feature is the large capacitive coupling enabled by the rather large dielectric constant of 7.8 for DNA-CTMA. A study on the thin-film morphology of DNA-CTMA reveals formation of self-organized structures in the thin films with a high molecular weight of 8,000,000 Da. These self-assembled DNA-surfactant complex materials, with good processability, may have applications in molecular optoelectronics. It has been proposed that such a self-organized structure arises because the alkyl chains are oriented perpendicular to the film plane, and chiral DNA helices were oriented in the direction parallel to the film plane.[46]

8.1.3 Electroactive Polymers

The role played by the polymeric gate dielectric is as important as the role of the organic semiconductor in organic electronic devices. In OFET sensors, the gate dielectric plays an even more important role than the organic semiconductor.

1. Gate dielectrics permit the creation of the gate field in order to establish a two-dimensional channel charge sheet.
2. The accumulated charge carriers transit from the source to the drain electrode in an area close to the dielectric/semiconductor interface. Hence, the chemical nature of the semiconductor/dielectric interface greatly affects how the accumulated charges move in the semiconductor.
3. Gate dielectrics are responsible for device stability.
4. They determine the operating voltage of devices according to the dielectric constant and thickness of the gate dielectric.
5. Gate dielectrics may display quasi-permanent charge storage or polarization used in nonvolatile memory elements.
6. They are essential in flexible displays and sensors since polymers are usually easily bendable.
7. When transparent, gate dielectrics enable the fabrication of photosensitive transistors. Hence the development of polymeric gate dielectric materials is of fundamental importance to the progress of organic electronic devices.

   Solution-processable polymeric dielectric materials are attractive, partly because films with excellent characteristics can often be formed by spin coating, casting, or printing at low process temperatures under ambient conditions. Moreover, the capability of easy film formation has practical advantages when coupled with low-cost patterning techniques for polymeric dielectrics as well as other materials needed in the fabrication of OFETs. From this point of view, polymeric dielectrics offer large potential as compared to their inorganic counterparts. Due to the limitations of space, no attempt is made to review the device physics of gate dielectrics in depth, nor will we give an overview of single crystal OFETs. For these important subjects we refer to other excellent recent reports and review articles.[54–56]

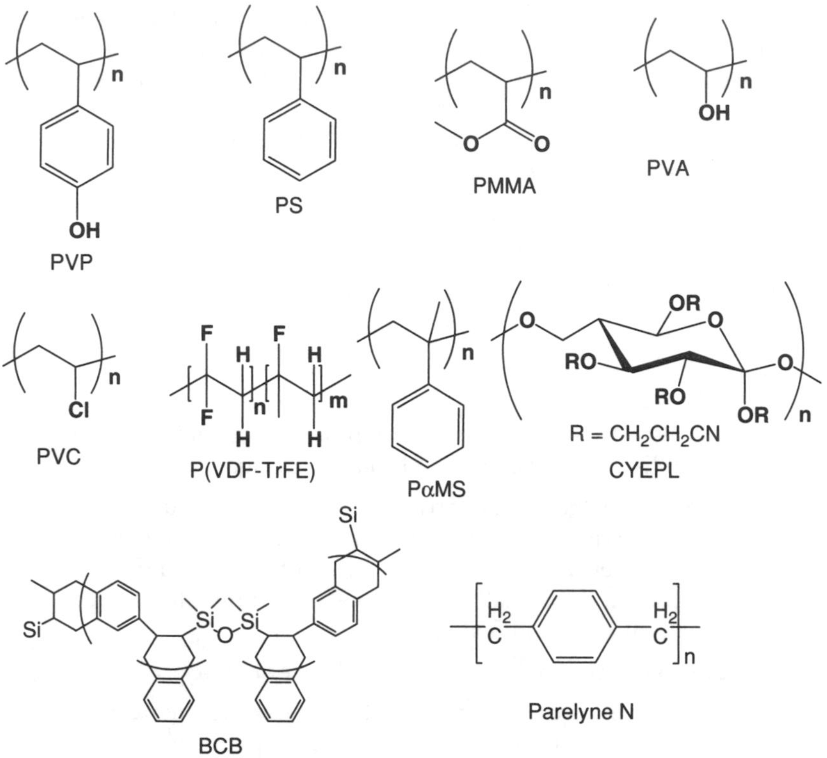

FIGURE 8.6 Chemical structure of some commonly used electroactive polymer dielectrics. PVP: poly(4-vinyl phenol); PS: polystyrene; PMMA: polymethylmethacrylate; PVA: polyvinyl alcohol; PVC: polyvinylchloride; P(VDF-TrFE): poly(vinylidene fluoride-trifluoro-ethylene); PαMS: poly[α-methylstyrene], CYEPL: cyano-ethylpullulan, BCB: divinyltetramethyldisiloxane-bis(benzocyclobutene) and parylene N.

A large variety of polymeric dielectrics have been investigated for their use in OFETs. Their chemical structures are shown in Fig. 8.6. For example, polyimides are commonly used as dielectric materials in flexible pressure sensors[57, 58] and circuitry.[59] *Divinyltetramethyldisiloxane-bis(benzocyclobutene)* (BCB), another type of spin-on dielectric broadly used in the microelectronics industry, is used for high-performance transistors and circuits.[20] Poly(vinylphenol) is used in the fabrication of all-polymer logic circuits.[60] Poly(vinyl alcohol) (PVA) is used as an electret in electret field-effects transistors (EFETs).[61, 62] Poly(dimethylsiloxane) (PDMS) is well known to form "PDMS stamps" which allow one to make conformal contact with organic semiconductors using deposition methods for dielectric layers that may cause degradation of the organic semiconductor.[63] Photocurable organic polymeric dielectrics such as thin films of poly(4-vinylphenol) (PVP) can be patternable and printable when crosslinked with a crosslinker and also meet requirements for fabrication of organic circuits.[64, 65] Polymer electrolytes are also used as high-capacitance dielectric layers to boost OFET currents and to enable low operating voltages.[66, 67] Very few polymers such as PVA support ambipolar transport which resulted in complimentary-like inverters.[68, 69] Poly(vinylidene fluoride-trifluoroethylene) (P(Vdf-TrFE) is a ferroelectric polymer used in the fabrication of ferroelectric field-effect transistors (FeFETs) and memory elements.[70] In its nonpolar form it facilitates the fabrication of hysteresis-free OFETs.[71] Poly(*m*-xylylene adipamide) (MXD6) is most likely a glassy dipolar polymer which has been used in nonvolatile memory elements, but not a ferroelectric polymer.[72] Poly(α-methylstyrene) (PαMS) has also been demonstrated to be useful as electret in a nonvolatile memory element.[73] However, criteria for memory elements are extremely tight (in terms of access times, retention time, and endurance to mention only a few), and so far none of the demonstrated polymer-based memories fulfill all these requirements.[74] The aforementioned applications suggest that polymeric dielectrics have to be very robust. For example, it should be possible to print them as large-area thin films without pinholes; they should adhere firmly to a variety of conducting substrates, compatible with p- and n-type organic semiconductors; and they should display low-leakage currents and high thermal stability. There is also debate on choosing high or low dielectric constant materials. Insulator layers with a high dielectric constant can negatively affect the mobility because they usually have randomly oriented dipole moments near the interface which increase the energetic disorder inside the semiconductors.[75, 76] On the other hand, high dielectric constant materials are employed for reduced operating voltage OFETs.

8.2 Single-Element Devices

8.2.1 Memory Elements

Nonvolatile flash memory, which uses silicon and its oxides, revolutionized consumer electronics: it is used to store information in mobile phones, pictures taken with digital cameras, data in memory sticks, and even as hard-disk replacement in cheap laptop computers. Bistability in such flash memories is achieved by introducing a second "floating gate" to a silicon transistor between the normal control gate (which regulates the flow of current through the transistor) and the semiconducting substrate, in order to define the spatial position of trapped charges (see Fig. 8.7*a*). However, the existing technology based on Si has not been currently employed in organic circuits. Hence, there are tremendous research efforts ongoing to develop an electrically accessible nonvolatile organic memory technology.[77] At present only two approaches to these challenges have been reported:

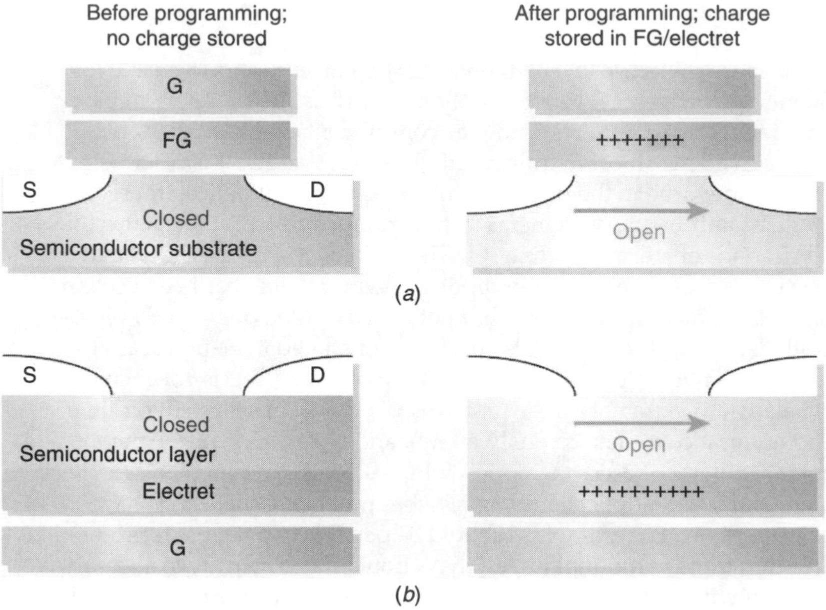

FIGURE 8.7 Methods against memory loss. The basic transistor is a device in which a small voltage applied at the control gate (G) modulates a much larger current flow from source (S) to drain (D) through a semiconductor substrate. (*a*) In flash memories, an amount of charge is trapped on a floating gate (FG) that modifies the control voltage required for current to flow from S to D. Whether current flows or not defines a boolean 1 or 0. The memory of this state persists as long as the charge remains trapped on the floating gate. (*b*) In Baeg and colleagues' organic device shown, the same principle is used, but the charge is trapped locally on a thin *electret* of chargeable polymer, rather than on an isolated floating gate.[73] (*Reproduced with permission from Ref. 77. Copyright 2004, Nature Publishing Group.*) (See also color insert.)

EFETs (which can be seen as crude first steps toward organic flash elements) and FeFETs.

EFETs

EFETs are field-effect transistors with a space charge electret as gate dielectric. The external field of the charged electret alters the conductance of the semiconductor channel between the source and the drain electrode, enabling applications in nonvolatile memories and sensors. Such an electret-based memory element developed by Baeg et al.[73] employs a thin layer of electret instead of a floating gate in between a gate dielectric and semiconductor, as depicted in Fig. 8.7*b*, and comes quite close to architectures used in silicon flash memories. However, operation voltages are still much too high to be of practical use.

First attempts to use polarizable gate insulators in combination with organic semiconductors were reported by Katz et al. in a most important seminal publication.[78] The FETs showed floating gate effects,

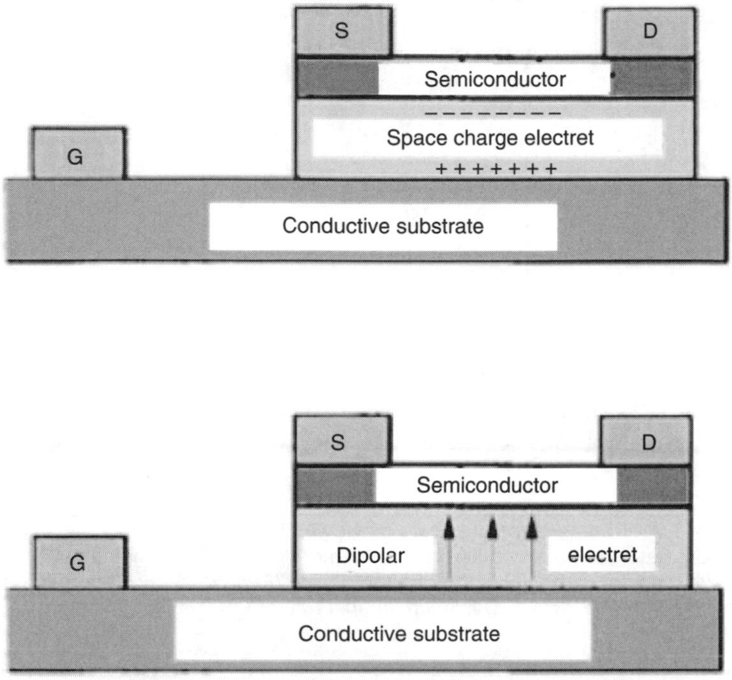

FIGURE 8.8 Organic field-effect transistor configurations with "polarizable" gate dielectrics. Upper panel: space charge electrets; lower panel: dipole electrets with frozen or ferroelectric polarization.

but the potential for organic memories was not fully exploited. A schematic view of devices with space charge or dipolar electrets is shown in Fig. 8.8. Due to charge separation and trapping in space charge electrets and frozen metastable or permanent ferroelectric polarization, the charge density at the interface between the semiconductor and the gate electret is altered. This results in significant effects in the transfer characteristics of the transistor as outlined in detail below.

As revealed in Fig. 8.9, the memory effect of the EFET is demonstrated by showing the drain-source current I_{ds} versus the gate-source voltage V_{gs} at a constant drain-source voltage $V_{ds} = 80$ V. The magnitude of the source-drain current I_{ds} increases with an amplification of up to 10^4 at $V_g \approx 50$ V with respect to the initial "off" state with $V_g = 0$ V. The saturated I_{ds} remains at a high value even when V_g is reduced back to $V_g = 0$ V (hysteresis or bistability). To completely deplete I_{ds}, one needs to apply a reverse voltage of $V_g \approx -30$ V. A large shift in threshold voltage, V_t by 14 V, is observed when measured the second time in comparison to the initial cycle. After that, there is practically no more shift in V_t. The 10th cycle showed no significant shift in V_t in comparison to the 2nd cycle. Each measurement was performed with a long integration time of 1 s.

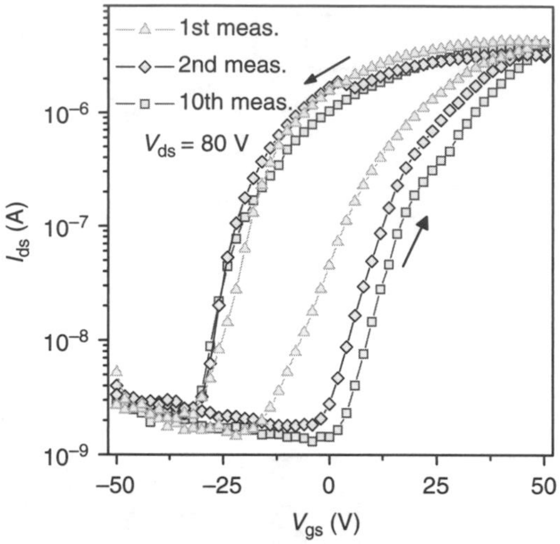

FIGURE 8.9 Transfer characteristics of the OFET with V_{ds} = 80 V demonstrating the nonvolatile organic memory device. Each measurement was carried out with an integration time of 1 s. *(Reproduced with permission from Ref. 61. Copyright 2004, American Institute of Physics.)*

A similar memory element was realized with a gate dielectric composed of marine-based DNA, as shown in Fig. 8.10. A I_{ds} on the order of 10^{-10} A in the off state is modulated by the gate field up to 4 orders of magnitude, reaching a saturated I_{ds} of 10^{-6} A with an applied $V_g < 10$ V. One can also clearly observe a large hysteresis with a shift of $V_t \approx 7$ V. Figure 8.10*a* shows that $I_{drain,\,sat}$ is bistable around $V_{gs} = 0$ V. In the case of PVA gate dielectrics, ionic residuals from the polymerization process were identified as source of the current bistability.[79]

Memory elements are characterized by their retention time, the time when the stored charges decrease to 50% of the initial value. To estimate the retention time of the stored charges in the memory, time resolved measurements were performed as depicted in Fig. 8.10*a*. In Fig. 8.10*a*, I_{ds} is measured for a gate voltage pulse V_{gs} applied for 200 s. After switching off the voltage pulse, the current decays, after 800 s the current is still more than one order of magnitude larger than the off current. It is also evident in Fig. 8.10*a* that the relaxation of the current slows down, so one might expect a sizable memory even after much longer times.

A modified EFET developed by Baeg et al. makes use of two-layer gate dielectrics: SiO_2 as gate dielectric and a thin layer of the charged electret PαMS in between the SiO_2 and the pentacene semiconductor layer. When a high gate voltage is applied to the device, the electret layer is charged. When a reverse voltage is applied, the electret layer is discharged and thus the initial state is restored. However, a large amount of trapped charges in the electret impose an added voltage

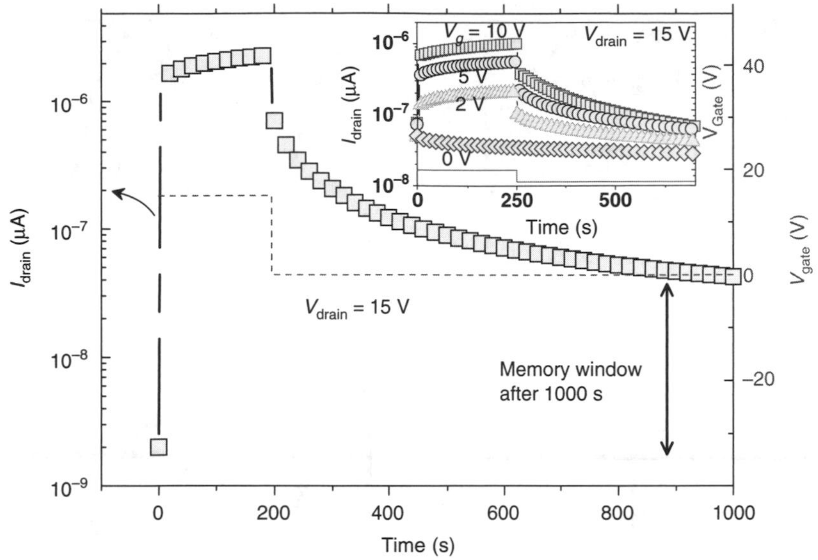

(*a*)

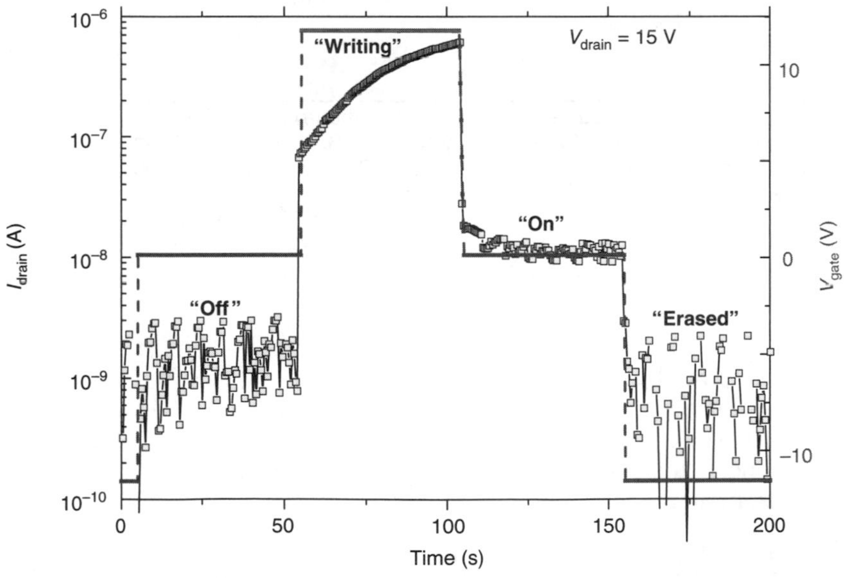

(*b*)

FIGURE 8.10 Transient response characteristics of BiOFETs (*a*) indicating a long retention time (inset: transient decay characteristics of BiOFET with different gate bias conditions) and (*b*) as a memory element with an applied gate voltage pulse showing memory off, on, write, and erase. *Note*: Gate voltage pulse height is shown in right-hand scale of each graph. (*Reproduced with permission from Ref. 53. Copyright 2007, Elsevier.*)

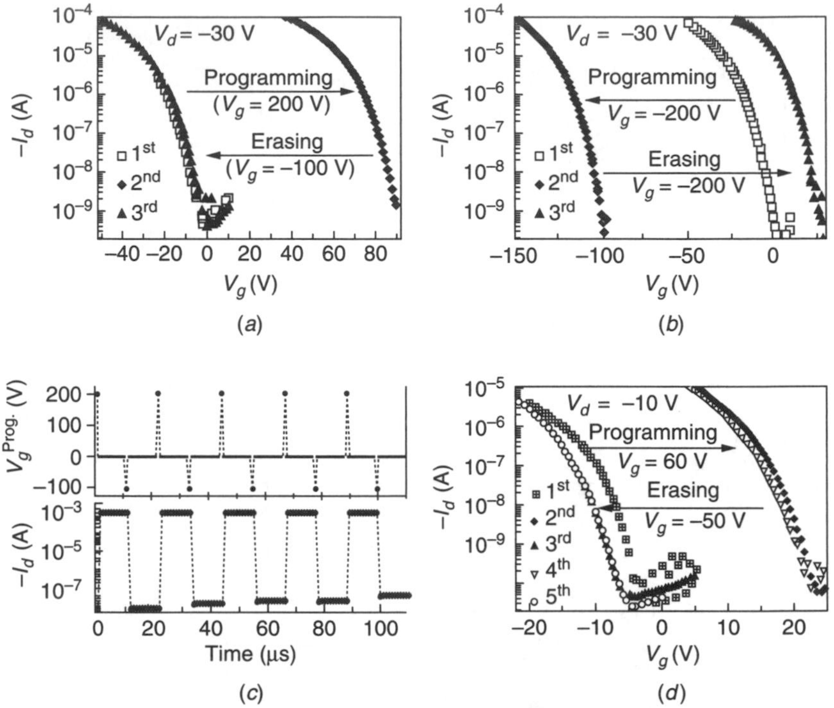

FIGURE 8.11 Shifts in transfer curves at V_{ds} = −30 V in the (*a*) positive and (*b*) negative directions for an OFET memory device with a 300 nm thick SiO_2 layer. (*c*) Reversible switching for on and off current states. (Programming: V_{gs} = 200 V and V_{ds} = 0 V, reading: V_{gs} = 0 V and V_{ds} = −30 V, and erasing: V_{gs} = −100 V and V_{ds} = 0 V). (*d*) Shifts in transfer curves at V_{ds} = −10 V for an OFET memory device with a 100 nm thick SiO_2 layer, where V_{gs} = 60 V and V_{gs} = −50 V were applied for 1 μs for programming and erasing, respectively. (*Reproduced with permission from Ref. 73. Copyright 2006, Wiley VCH Verlag GmbH & Co.*)

on the threshold gate voltage analogous to a floating gate in a flash memory. Such a bistable state is demonstrated in Fig. 8.11. Resulting devices have switching speeds on the order of 1 μs, which is the fastest among all the devices reported, but at voltage levels still too high to be useful in practical applications.

FeFETs

FeFETs are field-effect transistors with ferroelectric gate insulators. These devices make use of the dipole polarization of ferroelectric capacitors which shows butterflylike hysteresis loops. Memory functionality is obtained by the bistable polarization of the ferroelectric gate dielectric, which remanently attenuates the charge density in the semiconductor channel.

For this class of devices, ferroelectric copolymers such as P(VDF-TrFE)[80,81] and glassy dipolar polymers such as MXD6[82] are used. Because

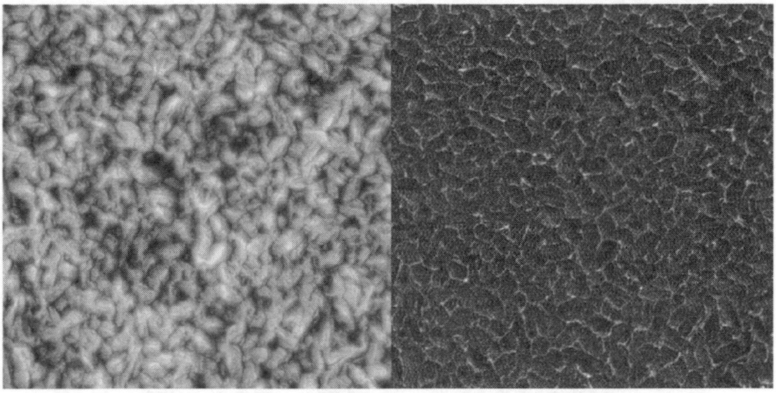

FIGURE 8.12 AFM measurement on an annealed P(VDF-TrFE) layer with a thickness of 200 nm. The area size is 2 × 2 μm² with, on the left, the topography (20 nm gray scale range) and, on the right, the phase response of the same area. (*Reproduced with permission from Ref. 81. Copyright 2005, American Institute of Physics.*)

of the fact that ferroelectric switching takes place only when the applied field exceeds the coercive field, high-quality thin films of P(VDF-TrFE) were prepared by using specific solvents (cyclohexane) (see Fig. 8.12). Typical coercive fields are on the order of 50 MV/m. The important findings on ferroelectric switching in P(VDF-TrFE) suggest that the coercive field does not saturate with decreasing film thickness.[83] An overview of remanent polarization Pr versus ferroelectric layer thickness is shown in Fig. 8.13. For the ferroelectric characterization, the Sawyer-Tower

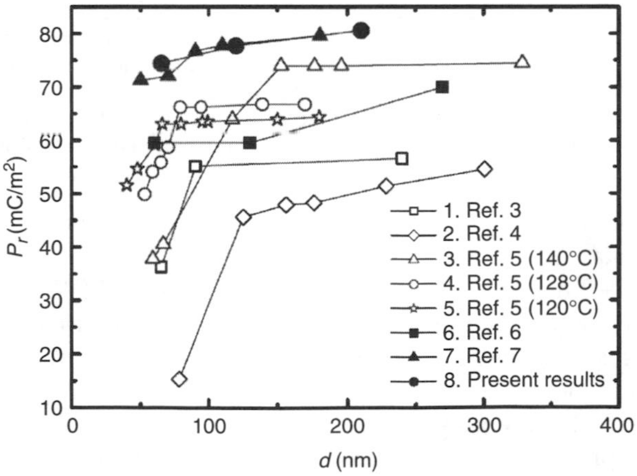

FIGURE 8.13 Summary of the remanent polarization of spin-cast P(VDF-TrFE) capacitors as a function of the ferroelectric layer thickness. The graph includes reported as well as present results. The lines are drawn as a guide to the eye. (*Reproduced partly with permission from Ref. 81. Copyright 2004, American Institute of Physics.*)

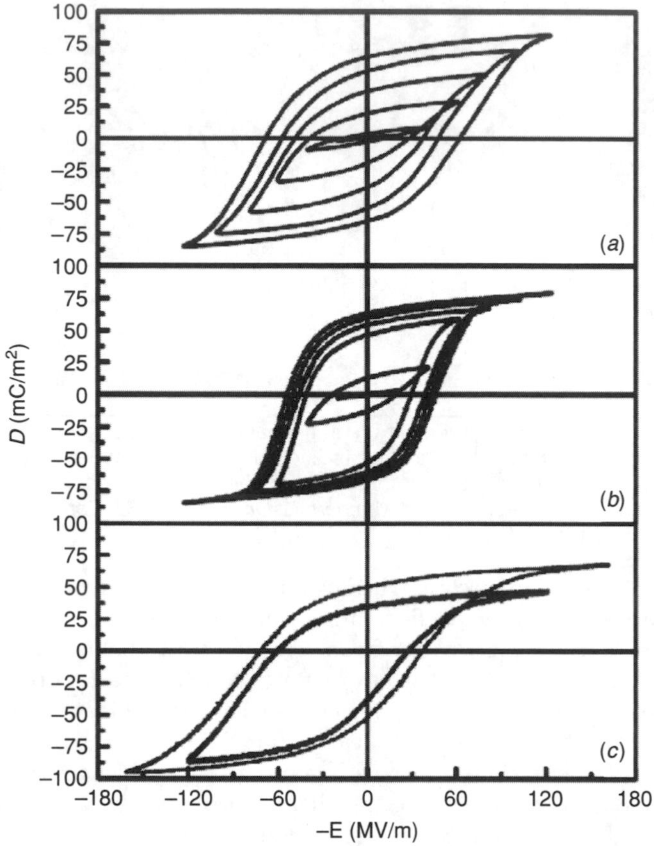

FIGURE 8.14 Displacement charge *D* vs. applied field *E* hysteresis loop measurements. (*a*) Obtained with a 190 nm ferroelectric layer thickness and 100 Hz frequency; (*b*) the same as (*a*) but at 1 Hz; (*c*) obtained with a ferroelectric layer thickness of 60 nm and a frequency of 1 Hz. (*Reproduced with permission from Ref. 81. Copyright 2004, American Institute of Physics.*)

circuit is used, where the displacement is measured versus the applied field to obtain *D–E* hysteresis loops.

The *D–E* hysteresis loops for ITO/PEDOT:PSS/P(VDF-TrFE)/Au capacitors, which demonstrate square and symmetrical hysteresis loops are shown in Fig. 8.14. At high fields, polarization saturates with a remanent polaraization of 75 mC/m^2 and a coercive field of 55 MV/m. Low-voltage (20 V) operating FeFETs with poly(3-hexylthiophene) as solution-processed semiconductor were demonstrated[81] as shown in Fig. 8.15. Retention times up to 1 week have been measured under floating gate conditions. One of the drawbacks of the floating gate operating mode is the slowing down of the retention loss because charges in the gate electrode are not free to exit the device. Unni et al. performed experiments under nonfloating conditions

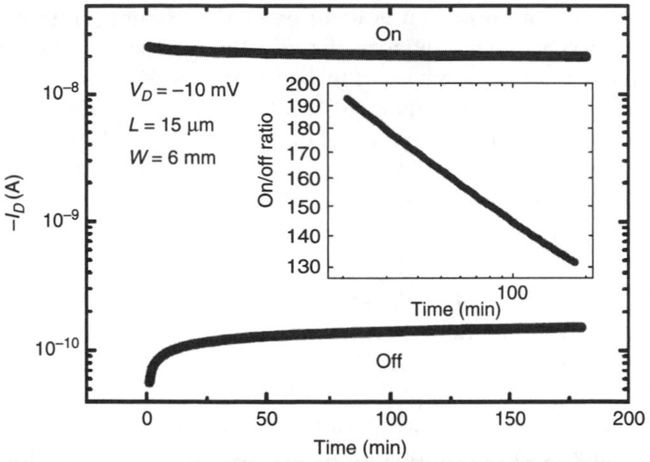

Figure 8.15 Data retention measurement of the on- and off-state drain current of an FeFET with a gate insulator layer similar to that in Fig. 8.14*a*, obtained with a continuous drain voltage and with the gate and source electrodes connected to 0 V. The inset shows the drain current on/off ratio on a double logarithmic scale. (*Reproduced with permission from Ref. 81. Copyright 2005, American Institute of Physics.*)

which does not seem to be a nondestructive mode readout operation of FeFET.[80] Stadlober et al. demonstrated hysteresis-free ferroelectric polymer transistors, using films in the nonpolar a-phase.[71]

In a completely different approach, without using any electret or ferroelectricity, bilayers of ZnO/pentacene OFETs also give rise to floating gate mode operating memory elements.[84] However, the retention curves as shown in Fig. 8.16 are obtained with a writing gate voltage of –100 V

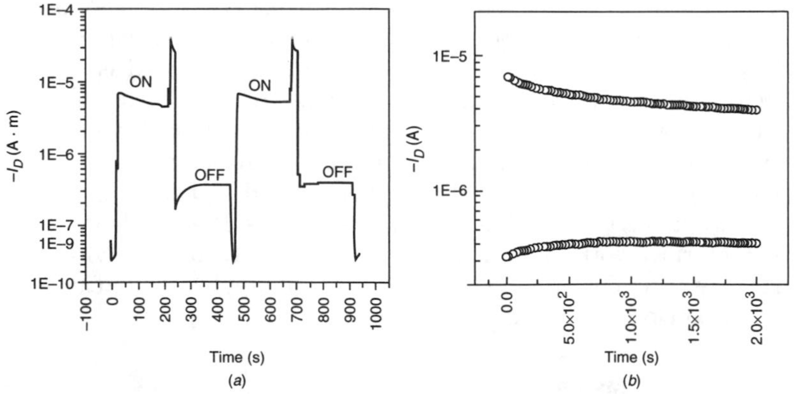

Figure 8.16 Retention of the on and off state for device n-ZnO/p-pentacene bilayer FET (*a*) in a cyclic way and (*b*) for long time experiment. (*Reproduced with permission from Ref. 85. Copyright 2008, Wiley-VCH Verlag GmbH & Co. KGaA, Weinhem.*)

for 1 s and a constant drain voltage of –100 V. Retention times up to 200 s at zero gate voltage were achieved although the on/off ratio is on the order of 10. As a summary of this section, it is obvious that the field of organic memory elements is in its infancy and requires additional extensive research to reach practical interest in real-world applications.

As another example with potential practical applications, we next describe briefly organic electronics-based sensors for monitoring physical parameters and for photodetection.

8.2.2 Single-Element Temperature and Pressure Sensors

There is a large amount of literature on chemical sensing with OFETs, which we will not review here. Rather we limit our discussion to sensors recording changes in physical environmental parameters such as changes in temperature and pressure. Such sensors may be useful in a wide range of applications and may also form the basis for large area electronic skin.

The field of temperature and pressure sensing with OFETs is also in its infancy, and only a limited number of publications are available, despite the large prospect for applications. Darlinski et al.[85] have outlined a pressure sensitivity in polyvinylphenol-based OFETs, which they ascribe to changes in carrier mobility, threshold voltage, and contact resistances. The origin of the effects measured has not been completely clear yet, so there is still research required to clarify the inherent pressure dependence of OFETs. Trapped charges may play an important role, but more experiments are needed to elucidate this suggestion. Jung et al. exploited thermal transport in the subthreshold regime of organic thin-film transistors to demonstrate temperature sensing with pentacene-based OFETs with silicon dioxide gate dielectrics on silicon substrates.[86] Maccioni et al. have suggested the use of OFETs as sensors for environmental properties in smart textiles.[87] Graz et al. and Zirkl et al. employed functional polymers such as ferroelectrets[88] or ferroelectric copolymers[89] for pressure and temperature sensing. In their approach, the functional polymer is not directly employed as gate dielectric, but is connected to the gate. Temperature and pressure sensitivity is obtained by means of the pyro- and piezoelectric effect of ferroelectrets or ferroelectric polymers. Thereby piezoelectric switches, pressure sensors, and paper-thin microphones could be demonstrated, as well as optothermal switches and infrared sensors. There is plenty of room for further work on sensing principles with OFET-based devices, which may be used in applications such as mobile appliances, but also sensitive skin in robotics, etc.

8.2.3 Light Sensors

Light response in OFETs is of interest from both a fundamental science and an application point of view.[90–109] A scheme of a light-responsive OFET is shown in Fig. 8.17. From the fundamental point of view, so

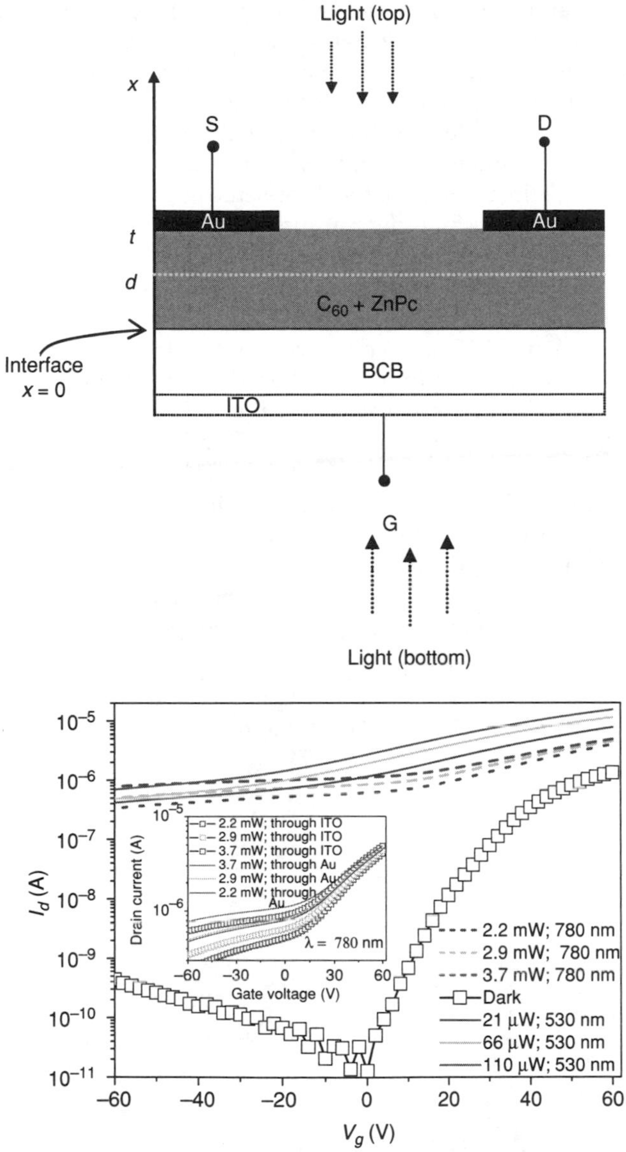

FIGURE 8.17 Top: Device scheme of the C_{60} and ZnPc-based phototransistors. Bottom: Photoresponse of the above device fabricated in the author's laboratory. (See also color insert.)

far very little is known about the origin of the nature of charge transport under illumination and whether it is unipolar or ambipolar.[90] Ambipolar transport is sensitive to the interface between the organic semiconductor and the gate dielectric[23] as well as the nature of the traps.[110] Likewise, in the studies of photoresponse, an observation of high responsivity R, which is defined as J_{ph}/P, where J_{ph} is the photocurrent density (total drain current I_d upon illumination minus the dark current per unit area) and P is the illumination power density,[101] can be affected by electron trapping at the interface near the gate dielectric layer[96] and the electrode work function.[98] From the application side, a distinct feature of the light-sensing properties of OFETs is that R can be tuned by orders of magnitude by an applied V_g. The largest measured R values are on the order of 1 to 100 A/W among all organic devices prepared by Narayan and Kumar on single-layer poly P3HT OFETs[90] and by Noh et al. on single-layer 2,5-dibromothieno [3,2-b]thiophene (BPTT).[101] A photovoltaic effect[111] and high R[112, 113] are reported on ambipolar transistors based on a bulk heterojunction concept. Such a highly photoresponsive OFET based on photoinduced charge transfer layer of C_{60} and ZnPc mixed layer as a photoactive layer is shown in Fig. 8.17. In this experiment, using a mask to illuminate only the active channel of the OFET with minimized artifacts, a large R of 10^{-3} to 10^{1} A/W depending on the intensity of illumination and on applied V_g could be obtained.[114]

Under illumination I_d in a transistor can be simply written as the sum of dark current I_{dark} and wavelength-dependent photocurrent I_{ph} (λ):

$$I_d(\lambda) = I_{dark} + I_{ph}(\lambda) \tag{8.1}$$

$$I_d(\lambda) = \overbrace{\frac{\mu_e C_{ins} W}{L}\left[(V_G - V_T)V_D - \frac{V_D^2}{2}\right]}^{I_{dark}} + \overbrace{\left[\frac{N(\lambda)e\mu_e Wt}{L}\right]}^{I_{ph}} V_D \qquad V_D \leq V_G \tag{8.2}$$

$$I_d(\lambda) = \overbrace{\frac{\mu_e C_{ins} W}{L}\left[\frac{(V_G - V_T)^2}{2}\right]}^{I_{dark}} + \overbrace{\left[\frac{N(\lambda)e\mu_e Wt}{L}\right]}^{I_{ph}} V_D \qquad V_D \geq V_G \tag{8.3}$$

where C_{ins} = capacitance per unit area
W, L, and t = channel width, length, and the active layer thickness, respectively
N = charge carrier density in the bulk

Charge carrier density N depends on the number of photogenerated charge carriers (PCCs), denoted by n_{ph}.

$$n_{ph} = \eta G \tau \tag{8.4}$$

which depends on the quantum efficiency for charge carrier generation η, the generation rate G, and the average lifetime of the PCC τ.

The generation rate G is given by the product of absorption coefficient $\alpha(\lambda)$ and the photon flux density ϕ per unit area per unit time. If we assume that charge generation takes place within the active region, i.e., $\eta \neq 0$ for $0 \leq x \leq d$ and $\eta = 0$ for $d \leq x \leq t$ and $I_{ph}(\lambda)$ is primarily controlled by the generation process in the regime above the absorption edge, then $I_{ph}(\lambda)$ can be written in terms of

$$I_{ph}(\lambda) \sim n_{ph}(\lambda) = \begin{cases} K_\lambda (e^{-\alpha(t-d)} - e^{-\alpha_\lambda t}) & \text{for light from top} \\ K_\lambda (1 - e^{-\alpha_\lambda d}) & \text{for light from bottom} \end{cases} \tag{8.5}$$

Where $K_\lambda = (\alpha_\lambda \eta \tau \lambda P_0)/hc$.

Equation (8.5) can be used to estimate the depletion zone d in a phototransistor.[103]

The photocurrent spectral response of photoresponsive P3HT OFETs depending on the gate voltage is shown in Fig. 8.18. The aforementioned dependencies are well studied in multilayer thin-film diode structures; however, studies of the photogeneration processes in OFETs are scarce.

8.3 Large-Area Pressure and Temperature Sensors

Most advanced large-area, flexible sensor skins have been reported by Someya et al.[57] By using pressure-sensitive rubber and temperature-sensitive organic diodes, large-area pressure and temperature sensor arrays capable of capturing images of pressure and temperature distributions have been demonstrated. As an example, images of pressure distributions from a kiss mark are shown in Fig. 8.19. This example outlines the huge potential of organic electronics in large-area electronic surfaces.

8.4 Summary

In this chapter we provided a brief overview of the state of the art in organic semiconductors, in polymeric gate dielectrics, in functional gate dielectrics, and in OFETs derived from such materials. We have outlined practical applications of OFETs in nonvolatile memories, as well as in sensors for recording changes in ambient conditions, such as changes in temperature and pressure. Although there has been tremendous progress in these fields of organic electronics, there is still a vast amount of research necessary to bring the technology to maturity. Hence there is still plenty of room for exciting new developments in memories and sensor applications, and organic electronics has a bright future in research and development.

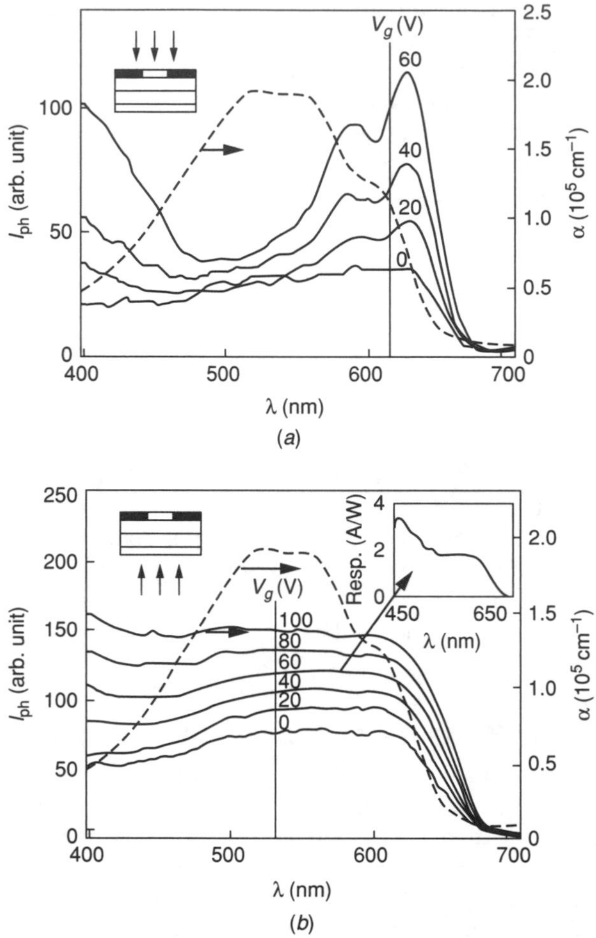

FIGURE 8.18 (a) I_{ph} (λ) as a function of V_g for the light incident from the top for a device with $t \sim$ 150 nm. The right ordinate displays for the absorption spectra (dashed line). The inset is the schematic representation of the illumination from the top. (*b*) I_{ph}(λ) as a function of V_g with the absorption spectra (dashed line). The left inset shows the schematic representation of bottom illumination; the right inset is the responsivity curve corresponding to V_g = 60 V, V_{ds} =−60 V. (*Reproduced with permission from Ref. 103. Copyright 2005, American Institute of Physics.*)

Acknowledgments

The authors thank Dr. James G. Grote, Prof. Gilles Horowitz, DI. Christoph Lackner, and Dr. Reinhard Schwödiauer for fruitful discussions and/or suggestions during experimental design. This work has been financially supported by the Austrian Science Foundation "FWF" within the National Research Network NFN on Organic Devices (S09712-N08, S097-6000, S9711-N08 and P20724-N20).

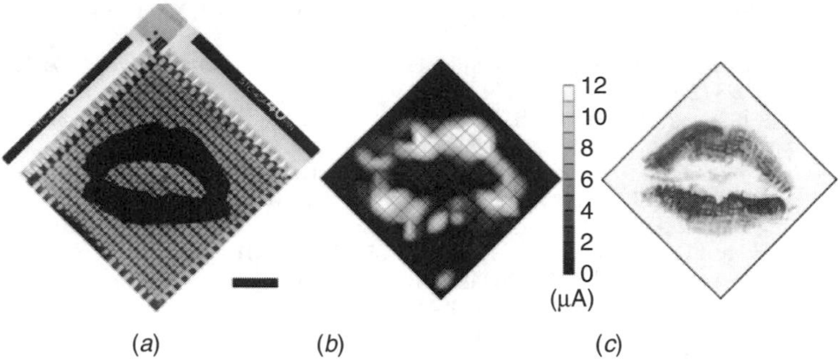

Figure 8.19 A pressure image of a kiss mark taken by using sensors consisting of 16 × 16 sensor cells. (*a*) The device is pressed with a lip-shaped rubber replica, (*b*) and the pressure image is compared with (*c*) the print on paper. The two bright spots at the bottom of (*b*) are due to a failure of sensors around those two spots (low local resistance of the pressure-sensitive rubber). (Scale bar = 1 cm.) (*Reproduced with permission from Ref. 57. Copyright 2004, PNAS.*)

References

1. J. E. Anthony, M. Heeny, and B. S. Ong, *Mater. Res. Bull.* 33:698 (2008); J. E. Anthony, *Angew. Chem.* 47:452 (2008).
2. H. Sirringhaus and M. Ando, *Mater. Res. Bull.* 33:676 (2008).
3. S. Allard, M. Forster, B. Souharce, H. Thiem, and U. Scherf, *Angew. Chem.* 47:4070 (2008).
4. C. R. Newman, C. D. Frisbie, D. A. da Silva Filho, J-L. Brédas, P. C. Ewbank, and K. R. Mann, *Chem. Mater.* 4436:16 (2004).
5. H. Yamada, T. Okujima, and N. Ono, *ChemComm.* 26:2957 (2008).
6. A. R. Murphy and J. M. J. Fréchet, *Chem. Rev.* 107:1066 (2007).
7. Th. B. Singh and N. S. Sariciftci, *Ann. Rev. Mater. Res.* 36:199 (2006).
8. H. Sirringhaus, P. J. Brown, R. H. Friend, M. M. Nielsen, K. Bechgaard, et al., *Nature* 401:685 (1999); H. Sirringhaus, R. J. Wilson, R. H. Friend, M. Inbasekaran, and W. Wu, *Appl. Phys. Lett.* 77:406 (2000).
9. H. Koezuka, A. Tsumara, and T. Ando, *Synth. Met.* 18:699 (1987).
10. H. Sirringhaus, N. Tessler, D. S. Thomas, P. J. Brown, and R. H. Friend, in *Advances in Solid-State Physics*, vol. 39 (Ed. B. Kramer), Vieweg, Wiesbaden, Germany, 1999, pp. 101–110
11. M. S. A. Abdou, F. P. Orfino, Y. Son, and S. Holdcroft, *J. Am. Chem. Soc.* 119:4518 (1997).
12. B. S. Ong, Y. L. Wu, P. Liu, and S. Gardner, *J. Am. Chem. Soc.* 126:3378 (2004).
13. M. Heeney, C. Bailey, K. Genevicius, M. Shkunov, D. Sparrow, et al., *J. Am. Chem. Soc.* 127:1078 (2005).
14. Mcculloch, M. Heeney, C. Bailey, K. Genevicius, I. Macdonald, M. Shkunov, D. Sparrowe, et al., *Nature Mater.*, 5:328 (2006).
15. F. Eder, H. Klauk, M. Halik, U. Zschieschag, G. Schmid, and C. Dehm., *Appl. Phys. Lett.* 84:2673 (2004).
16. T. W. Kelley, D. V. Muyres, P. F. Baude, T. P. Smith, and T. D. Jones, *Mater. Res. Soc. Symp. Proc.* 771 (Warrendale, PA):L6.5.1 (2003).
17. J. E. Anthony, *Chem. Rev.* 106:5028 (2006).
18. M. M. Payne, S. R. Parkin, J. E. Anthony, C. C. Kuo, and T. N. Jackson, *J. Am. Chem. Soc.* 127:4986 (2005).
19. K. C. Dickey, J. E. Anthony, and Y. L. Loo, *Adv. Mater.* 18:1721 (2006).
20. T. Anthopoulos, Th. B. Singh, N. Marjanovic, N. S. Sariciftci, A. M. Ramil, H. Sitter, M. Cölle, et al., *Appl. Phys. Lett.* 89:213504 (2006).

21. Th. Singh. Th, N. Marjanović, P. Stadler, M. Auinger, G. J. Matt, et al., *J. Appl. Phys.* 97:083714 (2005).
22. T. Yasuda, Takeshi Goto, Katsuhiko Fujita, and Tetsuo Tsutsui, *Appl. Phys. Lett.* 85:2098 (2004).
23. Th. B. Singh, F. Meghdadi, S. Gunes, N. Marjanovic, F. Lang, G. Horowitz, S. Bauer, et al., *Adv. Mater.* 17:2315 (2005).
24. Th. B. Singh, P. Senkarabacak, N. S. Sariciftci, A. Tanda, C. Lackner, R. Hagelauer, and G. Horowitz, *Appl. Phys. Lett.* 89:033512 (2006).
25. T. D. Anthopoulos, D. M. de Leeuw, E. Cantatore, C. Tanase, J. C. Hummelen, and Paul W. M. Blom, *Appl. Phys. Lett.* 85:4205 (2004).
26. T. D. Anthopoulos, D. M. de Leew, E. Cantatore, P. van't Hof, J. Alma, and J. C. Hummelen, *J. Appl. Phys.* 98:054503 (2005).
27. J. Locklin, K. Shinbo, K. Onishi, F. Kaneko, Z. Bao, and R. C. Advincula, *Chem. Mater.* 15:1404 (2003).
28. E. J. Meijer, D. M. de Leeuw, S. Setayesh, E. Van Veenendaal, B.-H. Huisman, P. W. M. Blom, J. C. Hummelen, et al., *Nature Mater.* 2:678 (2003).
29. Edsger C. P. Smits, Thomas D. Anthopoulos, Sepas Setayesh, Erik van Veenendaal, Reinder Coehoorn, Paul W. M. Blom, Bert de Boer, et al., *Phys. Rev. B* 73:205316 (2006).
30. C. Santato, R. Capelli, M. A. Loi, M. Murgia, and F. Cicoira, *Synth. Met.* 146:329 (2004).
31. C. Rost, S. Karg, W. Reiss, M. A. Loi, M. Murgia, and M. Muccini. *Appl. Phys. Lett.* 85:1613 (2004).
32. J. Zaumseil, R. H. Friend, H. Sirringhaus, *Nature Mater.* 5:69 (2006).
33. J. Zaumseil, C. L. Donley, J-S. Kim, R. H. Friend, and H. Sirringhaus, *Adv. Mater.* 18:2708 (2006).
34. D. Chun, Y. Cheng, and F. Wudl, *Angew. Chem.* 120:1 (2008).
35. P. Marmont, N. Battaglini, P. Lang, G. Horowitz, J. Hwang, A. Kahn, C. Amato, et al., *Org. Ele.* 9:419 (2008).
36. C. Huang, J. E. West, and H. E. Katz, *Adv. Mater.* 17:142 (2007).
37. E. Braun, Y. Eichen, U. Sivan, and G. B. Yoseph, *Nature* 775:291 (1998).
38. P. J. de Pablo, F. Moreno-Herrero, J. Colchero, J. Gómez Herrero, P. Herrero, A. M. Baró, Pablo Ordejón, et al., *Phys. Rev. Lett.* 4992:85 (2000).
39. L. Cai, H. Tabata, and T. Kawai, *Appl. Phys. Lett.* 3105:77 (2000).
40. A. J. Storm, J. van Noort, S. De Vries, and C. Dekker, *Appl. Phys. Lett.* 3881:79 (2001).
41. J. S. Hwang, K. J. Kong, D. Ahn, G. S. Lee, D. J. Ahn, and S. W. Hwang, *Appl. Phys. Lett.* 1134:81 (2002).
42. Y. Zhang, R. H. Austin, J. Kraeft, E. C. Cox, and N. P. Ong, *Phys. Rev. Lett.* 89: 198102 (2002).
43. D. Porah, A. Bezrydin, S. de Vries, and C. Dekker, *Nature* 635:403 (2000).
44. Y. Yang, P. Yin, X. Li, and Y. Yan, *Appl. Phys. Lett.* 203901 (2005).
45. H. Fink and C. Schönenberger, *Nature* 407:398 (1999).
46. A. Yu. Kasumov, M. Kociak, S. Guéron, B. Reulet, V. T. Volkov, D. V. Klinov, and H. Bouchiat, *Science* 291:5502 (2001).
47. L. Wang, J. Yoshida, N. Ogata, S. Sasaki, and T. Kajiyama, *Chem. Mater.* 13(4):1273 (2001).
48. Emily M. Heckman, Joshua A. Hagen, Perry P. Yaney, James G. Grote, and F. Kenneth Hopkins, *Appl. Phys. Lett.* 87:211115 (2005).
49. R. Ghirlando, E. J. Wachtel, T. Arad, and A. Minsky, *Biochem.* 7110:31 (1992).
50. H. Kimura, S. Machida, K. Horie, and Y. Okahata, *Polymer J.* 30:708 (1998).
51. J. Grote, N. Ogata, J. Hagen, E. Heckman, M. Curley, P. Yaney, M. Stone, et al., *SPIE Proc.—Nonlinear Optical Transmission and Multiphoton Processes in Organics*, A. Yates, K. Belfield, F. Kajzar, and C. Lawson (eds.), 5221:53 (2003).
52. Th. B. Singh, N. S. Sariciftci, J. Grote, and F. Hopkins, *J. Appl. Phys.* 100:024514 (2006).
53. P. Stadler, K. Oppelt, B. Singh, J. Grote, R. Schwödiauer, S. Bauer, H. Piglmayer-Brezina, et al., *Org. Electron.* 8:648 (2007).

54. A. Facchetti, M-H. Yoon, and T. J. Marks, *Adv. Mater.* 17:1705 (2005).
55. J. Veres, S. Ogier, and G. Lloyd, *Chem. Mater.* 16:4544 (2004).
56. C. Kim, Z. Wang, H.-J. Choi, Y-G. Ha, A. Facchetti, and T. J. Marks, *J. Am. Chem. Soc.* 130:6867 (2008).
57. T. Someya, T. Sekitani, S. Iba, Y. Kato, H. Kawaguchi, and T. Sakurai, *Proc. Natl. Acad. Sci.* 101:9966 (2004).
58. T. Someya, Y. Kato, T. Sekitani, S. Iba, Y. Noguchi, Y. Murase, H. Kawaguchi, et al., *Proc. Natl. Acad. Sci.* 102:12321 (2005).
59. D-H. Kim, J-H. Ahn, W. M. Choi, H-S. Kim, T-H. Kim, J. Song, Y. Y. Huang, et al., *Science* 320:507 (2008).
60. C. J. Drury, C. M. J. Mutsaers, C. M. Hart, M. Matters, and D. M. de Leeuw, *Appl. Phys. Lett.* 73:108 (1998).
61. Th. B. Singh, N. Marjanović, G. J. Matt, N. S. Sariciftci, R. Schwödiauer, and S. Bauer, *Appl. Phys. Lett.* 85:5409 (2004)
62. Th. B. Singh, N. Marjanović, N. S. Sariciftci, R. Schwödiauer, and S. Bauer, *IEEE Trans. Dielectrics & Electrical Insul.* 13:1082 (2006).
63. V. C. Sundar, J. Zaumseil, V. Podzorov, E. Menard, R. L. Willett, T. Someya, M. E. Gershenson, et al., *Science* 303:1644 (2004).
64. T-W. Lee, J. H. Shin, I-N. Kang, and S. Y. Lee, *Adv. Mater.* 19:2702 (2007).
65. Choongik Kim, Zhiming Wang, Hyuk-Jin Choi, Young-Geun Ha, Antonio Facchetti, and Tobin J. Marks, *J. Am. Chem. Soc.* 130:6867 (2008).
66. M. J. Panzer and C. D. Frisbie, *J. Am. Chem. Soc.* 127:6960 (2005).
67. J. H. Cho, J. Lee, Y. He, B. Kim, T. P. Lodge, and C. D. Frisbie, *Adv. Mater.* 20:686 (2008).
68. Th. B. Singh, P. Senkarabacak, N. S. Sariciftci, A. Tanda, C. Lackner, R. Hagelauer, and G. Horowitz, *Appl. Phys. Lett.* 89:033512 (2006).
69. T-F. Guo, Z-J. Tsai, S-Y. Chen, T-C. Wen, and C-T. Chung, *Appl. Phys. Lett.* 101:124505 (2007).
70. R. C. G. Naber, P. W. M. Blom, A. W. Marsman, and D. M. de Leeuw, *Appl. Phys. Lett.* 85:2032 (2004). R. C. G. Naber, C. Tanase, P. W. M. Blom, G. H. Gelinck, A. W. Marsman, F. J. Touwslager, S. Setayesh, et al., *Nature Mater.* 4:243 (2005).
71. B. Stadlober, M. Zirkl, M. Beutl, G. Leising, S. Bauer-Gogonea, and S. Bauer, *Appl. Phys. Lett.* 86:242902 (2005).
72. R. Schroeder, L. A. Majewski, M. Voigt, and M. Grell, *IEEE Electron. Dev. Lett.* 26:69 (2005).
73. K-J. Baeg, Y-Y. Noh, J. Ghim, S-J. Kang, H. Lee, and D-Y. Kim, *Adv. Mater.* 18:31–79 (2006).
74. J. C. Scott and L. D. Bozano, *Adv. Mater.* 19:1452 (2007).
75. J. Veres, S. D. Ogier, S. W. Leeming, D. C. Cupertino, and S. M. Khaffaf, *Adv. Funct. Mater.* 13:199–204 (2003).
76. I. N. Hulea, S. Fratini, H. Xie, C. L. Mulder, N. N. Iossad, G. Rastelli, S. Cluchi, and A. F. Morpurgo, *Nature Mater.* 5:982 (2006).
77. G. Gelinck, *Nature* 445:268 (2007).
78. H. E. Katz, X. M. Hong, A. Dodabalapur, and R. Sarpeshkar, *J. Appl. Phys.* 91:1572 (2002).
79. M. Egginger, M. Irimia-Vladu, R. Schwö diauer, A. Tanda, I. Frischauf, S. Bauer, and N. S. Sariciftci, *Adv. Mater.* 20:1018 (2008).
80. K. N. N. Unni, R. de Bettignies, S. Dabos-Seignon, and J.-M. Nunzi, *Appl. Phys. Lett.* 85:1823 (2004).
81. R. C. G. Naber, B. de Boer, P. W. M. Blom, and D. M. de Leeuw, *Appl. Phys. Lett.*, 87:203509 (2005).
82. R. Schroeder, L. A. Majewski, M. Voigt, and M. Grell, *IEEE Electron. Dev. Lett.* 26:69 (2005).
83. H. Kliem and R. Tadros-Morgane, *J. Phys. D: Appl. Phys.* 38:1860 (2005).
84. B. N. Pal, P. Trottman, J. Sun, and H. E. Katz, *Adv. Funct. Mater.* 18:1832 (2008).
85. G. Darlinski, U. Böttger, R. Waser, Hagen Klauk, Marcus Halik, U. Zschieschang, G. Schmid, et al., *J. Appl. Phys.* 97:093708 (2005).
86. S. Jung, T. Ji, and V. K. Varadan, *Appl. Phys. Lett.* 90:062105 (2007).

87. M. Maccioni, E. Orgiu, P. Cosseddu, S. Locci, and A. Bonfiglio, *Appl. Phys. Lett.* 89:143515 (2006).
88. I. Graz, M. Kaltenbrunner, C. Keplinger, R. Schwödiauer, S. Bauer, S. P. Lacour, and S. Wagner, *Appl. Phys. Lett.* 89:073501 (2006).
89. M. Zirkl, A. Haase, A. Fian, H. Schön, C. Sommer, G. Jakopic, G. Leising, et al., *Adv. Mater.* 19:2241 (2007).
90. K. S. Narayan and N. Kumar, *Appl. Phys. Lett.* 79:1891 (2001).
91. S. Man Mok, F. Yan, and H. L. W. Chan, *Appl. Phys. Lett.* 93:023310 (2008).
92. D. Knipp, D. K. Murti, B. Krusor, R. Apte, L. Jiang, J. P. Lu, B. S. Ong, et al., *Mater. Res. Soc. Symp. Proc.* 665:C5.44.1 (2001).
93. J-M. Choi, J. Lee, D. K. Hwang; J. H. Kim, S. Im, and E. Kim, *Appl. Phys. Lett.* 88:043508 (2006).
94. M. Breban, D. B. Romero, S. Mezhenny, V. W. Ballarotto, and E. D. Williams, *Appl. Phys. Lett.* 87:203503 (2005).
95. M. Debucquoy, S. Verlaak, S. Steudel, K. Myny, J. Genoe, and P. Heremans, *Appl. Phys. Lett.* 91:103508 (2007).
96. Y. Hu, G. Dong, C. Liu, L. Wang, and Y. Qiu, *Appl. Phys. Lett.* 89:072108 (2006).
97. S. M. Cho, S. H. Han, J. H. Kim, J. Jang, and M. H. Oh, *Appl. Phys. Lett.* 88:071106 (2006).
98. J-M. Choi, K. Lee, D. K. Hwang, J. H. Kim, S. Im, J. H. Park, and E. Kim, *J. Appl. Phys.* 100:116102 (2006).
99. J-M. Choi, K. Lee, D. K. Hwang, J. H. Kim, S. Im, and E. Kim, *Appl. Phys. Lett.* 88:043508 (2006).
100. N. Mathews, D. Fichou, E. Menard, V. Podzorov, and S. G. Mhaisalkar, *Appl. Phys. Lett.* 91:212108 (2007).
101. Y-Y. Noh, D-Y. Kim, Y. Yoshida, K. Yase, B. Jung, E. Lim, and H-K. Shim, *Appl. Phys. Lett.* 86:043501 (2005).
102. Y-Y. Noh, D-Y. Kim, and K. Yase, *J. Appl. Phys.* 98:074505 (2005).
103. S. Dutta and K. S. Narayan, *Appl. Phys. Lett.* 87:193505 (2005).
104. S. Dutta and K. S. Narayan, *Synth. Met.* 146:321 (2004).
105. M. C. Hamilton and J. Kanicki, *IEEE J. Sel. Top. Quant. Electr.* 10:1077 (2004).
106. M. C. Hamilton, S. Martin, and J. Kanicki, *IEEE Trans. Electr. Dev.* 51:876 (2004).
107. T. P. I Saragi, R. Pudzich, T. Fuhrmann, and J. Salbeck, *Appl. Phys. Lett.* 84:2334 (2004).
108. T. P. I. Saragi, T. Spehr, A. Siebert, T. Fuhrmann-Lieker, and J. Salbeck, *Chem. Rev.* 107:1011 (2007).
109. Y. Xu, P. R. Berger, J. N. Wilson, and U. H. F. Bunz, *Appl. Phys. Lett.* 85:4219 (2004).
110. L. L. Chua, J. Zaumseil, J. Chang, E. C. W. Ou, P. K. H. Ho, H. Sirringhaus, and R. H. Friend, *Nature* 434:194 (2005).
111. S. Cho, J. Yuen, J. Y. Kim, K. Lee, and A. J. Heeger, *Appl. Phys. Lett.* 90:063511 (2007).
112. N. Marjanovic, Th. B. Singh, G. Dennler, S. Guenes, H. Neugebauer, N. S. Sariciftci, R. Schwödiauer, et al., *Org. Ele.* 7:188 (2006).
113. T. D. Anthopoulos, *Appl. Phys. Lett.* 91:113513 (2007).
114. Th. B. Singh, R. Koeppe, N. S. Sariciftci, M. Morana, and C. J. Brabec, *Adv. Funct. Mater*. 19:789 (2009).

CHAPTER 9

Luminescent Conjugated Polymers for Staining and Characterization of Amyloid Deposits

K. P. R. Nilsson

Department of Chemistry, Linköping University, Linköping, Sweden

9.1 Introduction

The evolution of living organisms has led to development of excellent biosensors, and when developing new sensory technologies, one should look to nature for the most favorable solution. The amazing sensory performance of biological systems is derived from a collective system response mostly involving analyte-triggered biochemical cascades. In addition, molecules with alternating single and double carbon bonds, i.e., conjugated molecules, having specific optical properties are frequently utilized in biological systems. The fascinating light harvesting complex taking part in the photosynthesis in green plants and the conformational transition of the retinal molecule, which is covalently attached to the protein rhodopsin in the retina of the eye, are examples of how nature makes use of conjugated molecules. Likewise, optical biosensors utilizing luminescent conjugated polymers (LCPs) are taking advantage of similar phenomena. The detection schemes of these sensors are mainly employing the efficient light harvesting properties or the conformation-sensitive optical properties of the conjugated polymers. LCPs offer a diverse sensor platform and can be used in a wide range of biomolecular recognition schemes to obtain sensory responses. Biosensors based on conjugated polymers are sensitive to very minor perturbations, due to amplification by a collective system response, and offer a key advantage compared to small-molecules based sensors.

Conjugated polymers are made of several repeating units, mers, and a wide range of biological active polymers can also be found in nature. For instance, the molecule carrying all the genetic information, DNA, has a repetitive helical structure made up from four nucleotides, whereas 20 amino acids are used to create a diversity of polypeptide chains that are folded into functional proteins. This molecular similarity between conjugated polymers and biological polymers offers a great possibility to create simple versatile biosensors, as these two classes of molecules are able to form strong complexes with each other due to multiple noncovalent interactions. The ability of conjugated polymers to noncovalently interact with individual biomolecules, such as proteins, and afford an optical fingerprint corresponding to a distinct conformational state of this biomolecule sets these molecules apart from conventional dyes and other sensor technologies, potentially enabling novel technologies for studying biological processes in a more refined manner. Most conventional techniques are limited by their reliance on detecting a certain biomolecule, whereas the LCPs are identifying a specific structural motif or a distinct conformational state of a biomolecule. Hence, the LCPs offer a possibility to monitor the biochemical activity of biological events on the basis of a structure-function relationship rather than on a molecular basis.

The unique conformational-sensitive optical properties of LCPs have proved to be a great asset for studying protein misfolding and aggregation. As the aggregation of proteins is associated with a wide range of serious diseases, the LCP technique can also be used to gain increasing knowledge regarding the pathological events of such diseases. In this chapter, the molecular structure and the optical properties of LCPs, as well as the use of LCPs as optical sensors for biological events, especially protein aggregation, will be discussed.

9.2 Luminescent Conjugated Polymers

9.2.1 Definition and Examples

In the unsubstituted form, conjugated polymers are insoluble, but with proper chemical modifications of the polymer backbone they can be dissolved in organic solvents. However, the use of conjugated polymers as detecting elements for biological molecules requires that

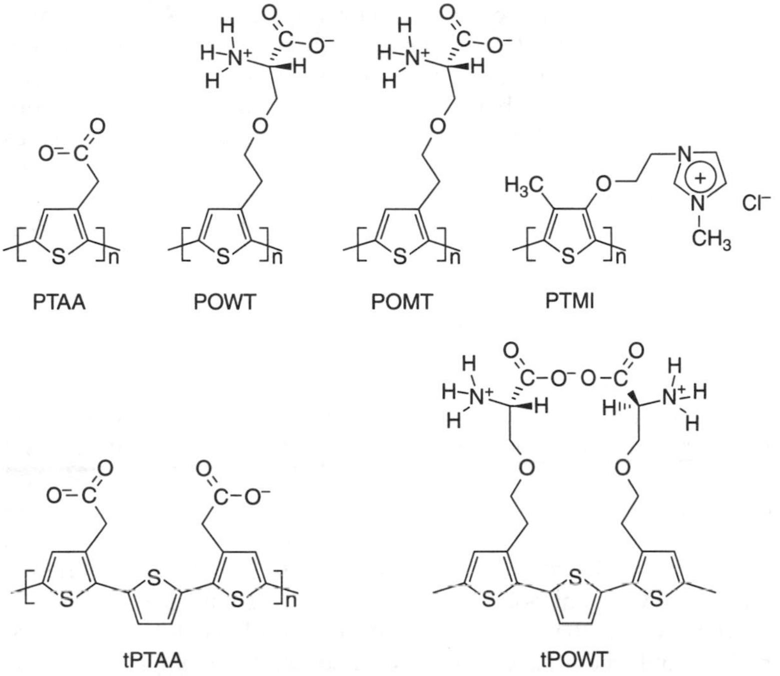

Figure 9.1 Chemical structures of some anionic, cationic, and zwitter-ionic luminescent conjugated polymers that have been used for staining and characterization of amyloid fibrils.

the polymers be compatible with an aqueous environment. This can be achieved by adding ionic substituents on the polymer backbone. To achieve excellent LCPs, e.g., polymers exhibiting fluorescence with high quantum efficiency, a polymer backbone consisting of substituted thiophene rings or fluorene building blocks is preferable, as photoluminescence in conjugated polymers requires a non degenerated ground state. A wide range of water-soluble LCPs have been reported,[1–14] and some examples are shown in Fig. 9.1. The functional groups of the conjugated polymers, being anionic or cationic at different pH values, make these polymer derivatives suitable for forming strong polymer complexes with negatively or positively charged biomolecules, such as DNA or proteins. In addition, the ionic groups are able to create versatile hydrogen bonding patterns with different molecules, which might be necessary to achieve specific interactions with different biomolecules.

9.2.2 Optical Properties

The optical processes in conjugated polymers are highly influenced by the conformation of the polymer backbone and the separation and aggregation of polymer chains. The optical transitions of conjugated polymers are believed to occur on different parts of the same polymer chain, intrachain events, or between adjacent polymer chains, interchain events. The intrachain events are mainly dependent on the conformation of the polymer backbone, and the interchain processes occur as nearby polymer chains come in contact with each other, leading to stacking of the aromatic ring systems, such as the thiophene rings. Polythiophenes have been shown to exhibit a variety of optical transitions upon external stimuli such as solvents,[15] heat,[15–18] ions,[3, 19, 20] or proteins,[21–24] leading to the design of a variety of sensory devices. Such effects have been termed *solvatochromism*, *thermochromism*, *ionochromism*, or *biochromism*, respectively. As there is a strong correlation between the electronic structure and the backbone conformation in conjugated polymers, any change in the main chain conformation will lead to an alteration of the effective conjugation length, coupled with a shift of the absorption in the UV-visible range.[25–28] A coil-to-rod (nonplanar-to-planar) conformation transition of the polymer backbone is observed as a red shift of the absorption maximum, due to an increased effective π conjugation length of the polymer backbone.[15–18] Studies on different polythiophene derivatives have shown that the conformational changes of the main chain, planar to twisted nonplanar, could be affected by order-disorder transitions of the side chains.[16–18, 29–31] This phenomenon has also been confirmed by theoretical investigations.[32, 33]

The highly conjugated planar form of the polymer backbone is also assumed to be coupled with intermolecular aggregates.[33] Hence, a red shift in absorption might also be associated with π aggregation between polymer chains which is seen as a distinct shoulder at longer wavelength in the UV-vis spectra.[34–36] This phenomenon might occur independently of the planarization of the backbone and should be distinguished from the red shift owing to planarization alone.[34, 35] However, studies performed on well-defined oligothiophene model compounds have shown that the structurally induced chromic effects are mainly due to conformational changes of the backbone instead of interchain interactions.[37–39] To visualize the optical transitions, the absorption spectra of a zwitter-ionic polythiophene derivative, POWT (Fig. 9.1), in different buffer systems are shown in Fig. 9.2*a*.[36] At pH 5, where most of the side chains are zwitter-ionic (both negatively and positively charged), the polymer backbone is in a nonplanar helical conformation, associated with a blue-shifted absorption spectrum. When the pH is increased or decreased, the backbone of the conjugated polymer adopts a more planar conformation, seen as a red shift of the absorption maximum. A closer look at the spectra for POWT in pH 2 and pH 8 buffer solutions (Fig. 9.2*a*) reveals an interesting observation. The absorption maximum is similar, but the spectrum recorded for POWT in the pH 8 buffer solution has a shoulder in the region at

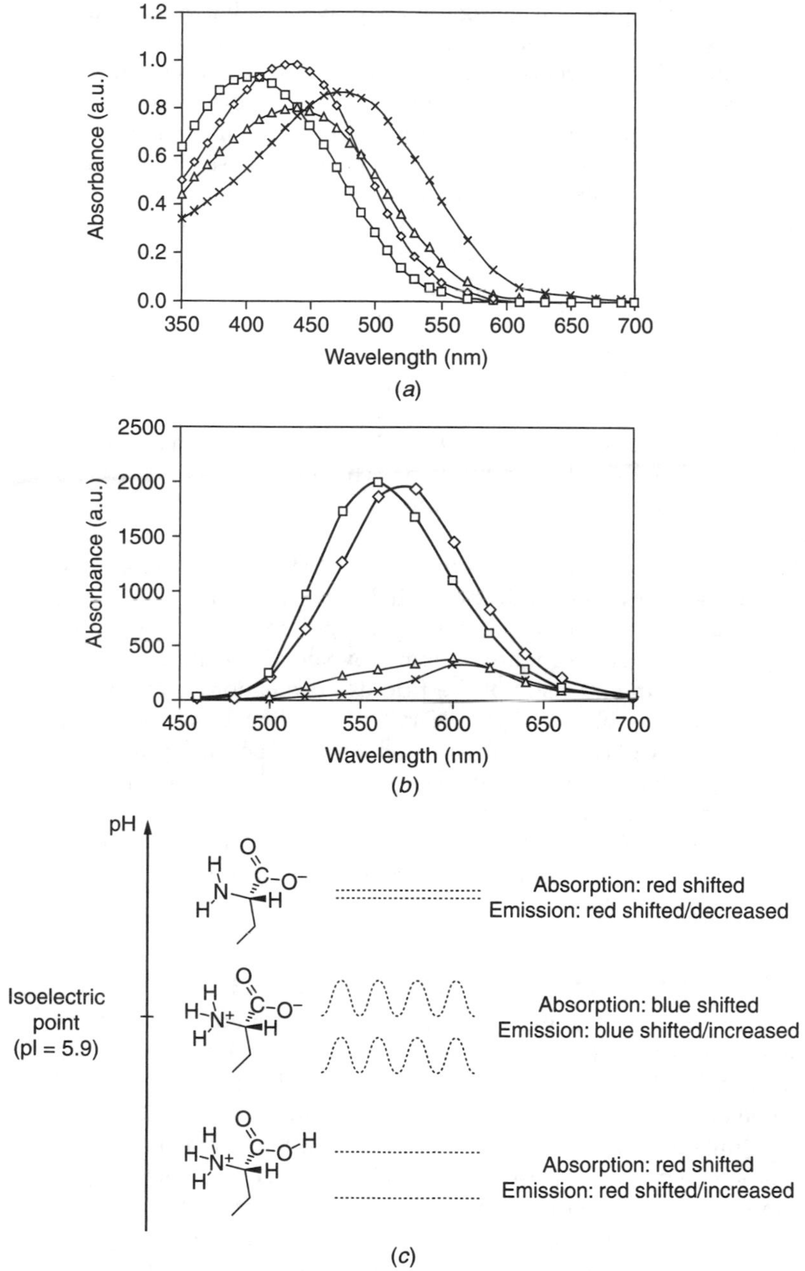

FIGURE 9.2 (*a*) Absorption and (*b*) emission spectra of POWT in different buffer solutions: pH 2 (◊), pH 5 (□), pH 8 (Δ), and pH 11 (×). (*c*) The charge of the zwitterionic side chain, schematic drawing of the proposed backbone conformations, and optical properties of POWT at different pH values. (*From Ref. 36. Represented with permission from the Institute of physics. Copyright 2002.*)

570 nm, indicating that an aggregation of polyelectrolyte chains takes place at alkaline pH but not at acidic pH.

The photoluminescence (fluorescence) efficiency of conjugated polymers is also dependent on the geometry of the polymer backbone, especially the separation or aggregation of polymer chains. The intensity of the fluorescence of the aggregated phase of polythiophene derivatives compared with the fluorescence of the single-chain state has been shown to be weaker by approximately one order of magnitude.[40–42] Similarly, studies of thin films of POWT (Fig. 9.1) have shown an analogous trend in the photoluminescence upon separation and aggregation of the polymer chains.[43, 44] As the polymer chains were separated, the photoluminescence maximum was also blue-shifted by approximately 105 nm, compared to the dense packing of the polymer chains. Similar changes also occur when placing POWT in different buffer solutions (Fig. 9.2*b*).[36] At pH 5, when the polymer side chains largely have a neutral net charge, the polymer chains adopt a nonplanar conformation, and the chains are separated, seen as a blue-shifted emission maximum and an increase of the intensity of the emitted light. In more alkaline pH (pH 8) the POWT peak emission is at a longer wavelength and with decreased intensity, related to a more planar backbone and aggregation of polymer chains. At acidic pH (pH 2), light with a slightly longer wavelength (relative to pH 5) is emitted, but the intensity of the fluorescence is not decreasing in the same way as observed for POWT in alkaline buffer solution. Hence, an acidic pH seems to favor a more rod shape conformation of the polymer chains, but aggregation of the polyelectrolyte chains is presumably absent. A schematic drawing of the polymer chain conformations for POWT in different buffer solutions and the conformational induced optical transitions relating to these geometric changes are shown in Fig. 9.2*c*.[36]

9.2.3 Conjugated Polymers as Optical Sensors

The application of conjugated polymers for colorimetric detection of biological targets (biochromism) was first described by Charych and coworkers[45] in 1993. The technique is utilizing a ligand-functionalized conjugated polymer, which undergoes a colorimetric transition (coil-to-rod transition of the conjugated backbone) upon interaction with a receptor molecule of interest (Fig. 9.3). The specificity in this first generation of conjugated polymer-based biosensors is due to the covalent integration of distinct ligands on the side chains of the conjugated polymers. Ligand-functionalized versions of polydiacetylenes have been used extensively for colorimetric detection of molecular interactions,[45–49] and polythiophene derivatives that display biotin[21–23] and different carbohydrates[24] have been synthesized and shown to undergo colorimetric transitions in response to binding of streptavidin and different types of bacteria and viruses, respectively.

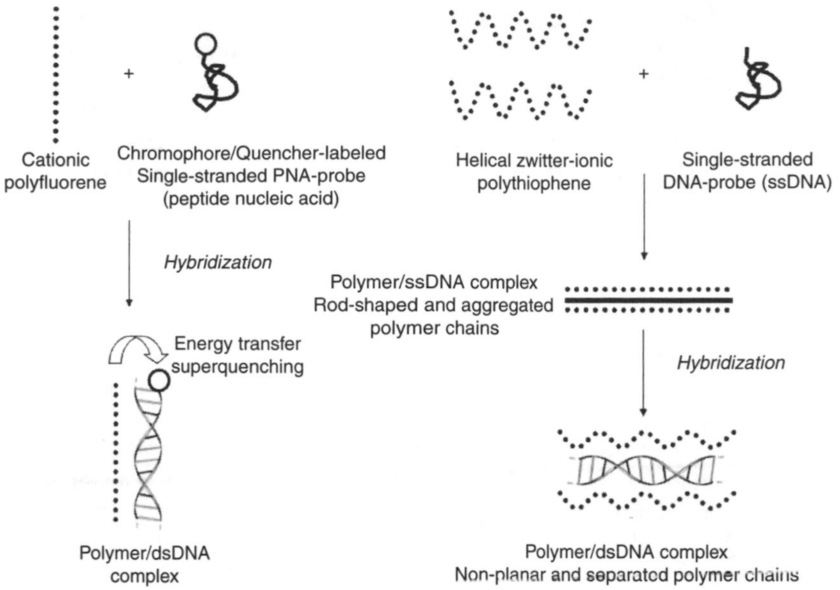

FIGURE 9.3 Schematic drawing of the detection of DNA hybridization by luminescent conjugated polymers. The technique using FRET or superquenching from the polymer chain is shown at left, and the technique using the geometric changes of the polymer chains to detect the hybridization event is shown at right.

However, in all cases, the detection and recognition events are due to the covalently attached side chains, and these chemical modifications require advanced synthesis and extensive purification of numerous monomeric and polymeric derivatives.Also the first-generation sensors were utilizing optical absorption as the source for detection, and the sensitivity of these sensors was much lower compared with other sensing systems for biological processes.

To avoid covalent attachment of the receptor to the polymer side chain and to increase the sensitivity of the biosensors, LCPs with repetitive ionic side chains have been utilized. These systems take advantage of the polymeric nature of the LCPs, and multivalent noncovalent interactions between a synthetic polymer, the LCP, and a natural polymer, i.e., the biomolecule, occur.[7, 13, 50–53] This is something quite different from what has been accomplished with many fluorescent detector dyes over the years. In addition to the covalent attachment of point-like fluorophores by covalent chemistry to biomolecules, noncovalent environment-sensitive dyes have been frequently employed for biomolecular recognition and can provide information on, e.g., local hydrophobicity and proximity within a complex. In contrast to stiff small-molecular dye binding, complexation between a flexible polymer, the LCP, and a biological polymer causes changes in the geometry of the LCP. The most sensitive biosensors based on conjugated polymers reported in the literature are utilizing changes in the absorption or emission properties from the conjugated polymers. Normally, fluorescence is the preferable method of choice for detection, as it is a widely used and rapidly expanding method in chemical sensing. Aside from inherent sensitivity, this method offers diverse transduction schemes based upon changes in intensity, energy transfer, wavelength (excitation and emission), polarization and lifetime. Optical sensors, based on LCPs, can be divided mainly into two different types, depending on which detection scheme is used. Schematic drawings of the two detection schemes for the detection of DNA hybridization are shown in Fig. 9.3.

In the first approach, superquenching of the fluorescence from the conjugated polymer chain is used, where a single site of quenching causes loss of fluorescence from the complete chain.[8, 50, 54–63] The quenching may be due to fluorescence resonance energy transfer (FRET) or excitation quenching. If a biomolecule labeled with a quencher is coordinated in close vicinity to the polymer chain by multiple noncovalent interactions (electrostatic or hydrophobic interactions), it is possible to detect the presence of a certain biomolecule in a sample by the quenching of the emitted light from the LCP. A wide range of biosensors, including sensors for DNA hybridization as well as ligand-receptor interactions and enzymatic activity, utilizing the impact of biomolecules on these conditions for FRET or excitation transfer have been reported.[64–79]

The second type of biosensors is based on detection of biological processes through their impact on the conformation and the geometry of the conjugated polymer chains.[7, 13, 52, 53, 80–85] Similar to the first technique described above, a complex between the conjugated polymer and a certain biomolecule is being formed due to noncovalent interactions. The complex being formed can then be studied in situ as the conformational flexibility of LCPs allows direct correlation between the geometry of chains and the resulting electronic structure and optical processes such as absorption and emission. If conformational changes of the biomolecule or other biomolecular events can lead to different conformations of the associated polymer backbone, an alteration of the absorption and emission properties of the polymer will be observed. Hence, this phenomenon can be used as a sensing element for a wide range of biological events, appropriate for making novel biosensors. Similar to the quenching method described above, this second technique has been used to detect DNA hybridization and ligand-receptor interactions.[7, 53, 81, 83-86] However, utilizing the structurally induced optical changes of the conjugated polymer backbone also allows the tantalizing possibility to detect conformational changes of biomolecules. This has been verified by using LCPs to detect conformational changes in synthetic peptides,[87, 88] and calcium-induced conformational changes in calmodulin.[86] The detection of these biological processes is carried out by using the conformational changes of cationic, anionic, and zwitter-ionic polythiophene derivatives, which are noncovalently attached to the biomolecule of interest.

The LCP, POWT (Fig. 9.1), which has a zwitter-ionic side chain functionality capable of forming strong electrostatic interactions and strong hydrogen bonding with biomolecules, was mixed with synthetic peptides.[87] These peptides, one cationic and one anionic, were designed to adopt random-coil formations by themselves, and when the two peptides were mixed, heterodimers with a four-helix bundle conformation were formed (Fig. 9.4*a*). The addition of a positively

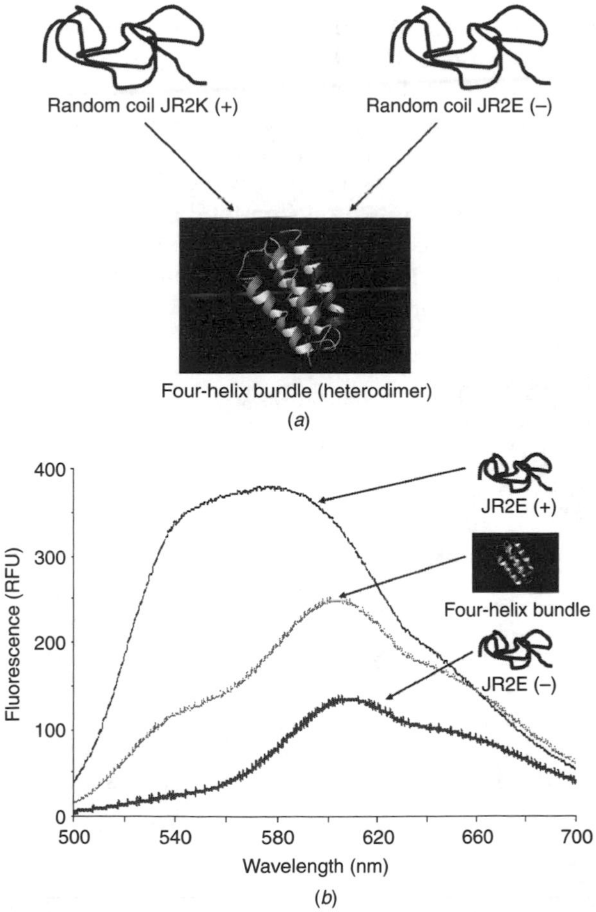

Figure 9.4 (*a*) Schematic drawing of the conformational changes of the synthetic peptides JR2E (negatively charged), JR2K (positively charged), and the JR2K-JR2E heterodimer. (*b*) Fluorescence spectra of POWT being bound to JR2E, JR2K, or the JR2K-JR2E heterodimer. (*Represented with permission from Ref. 87. Copyright 2003, National Academy of Sciences, USA.*)

charged peptide with a random-coil conformation, JR2K, forced the polyelectrolyte to adopt a nonplanar conformation with separated polyelectrolyte chains, observed as a blue shift and an increased intensity of the emitted light. In contrast, upon exposure to a negatively charged peptide with a random-coil conformation, JR2E, the backbone adopts a planar conformation, and aggregation of the polyelectrolyte chains occurs, seen as a red shift and a decreased intensity of the emitted light. Finally, by adding JR2K to the POWT-JR2E complex, the intensity of the emitted light is increased and blue-shifted, which is associated with separation of the polyelectrolyte chains. This geometric alteration of the polyelectrolyte chains is due to the conformational changes of the peptides upon formation of the four-helix-bundle motif. Hence, different emission spectra of POWT could be assigned to the charge distribution and/or the conformational state of the synthetic peptides (Fig. 9.4*b*).

Similar to the study described above, POWT was also reported to detect conformational changes in calmodulin (CaM), a calcium-binding protein important for intracellular cell signaling. The overall structure of CaM (Fig. 9.5*a*) consists of two globular calcium-binding domains, each containing two calcium-binding regions with characteristic motifs,[89–91] connected by a linker. Upon binding of calcium the relative orientation of the two α helices that define the EF-hand

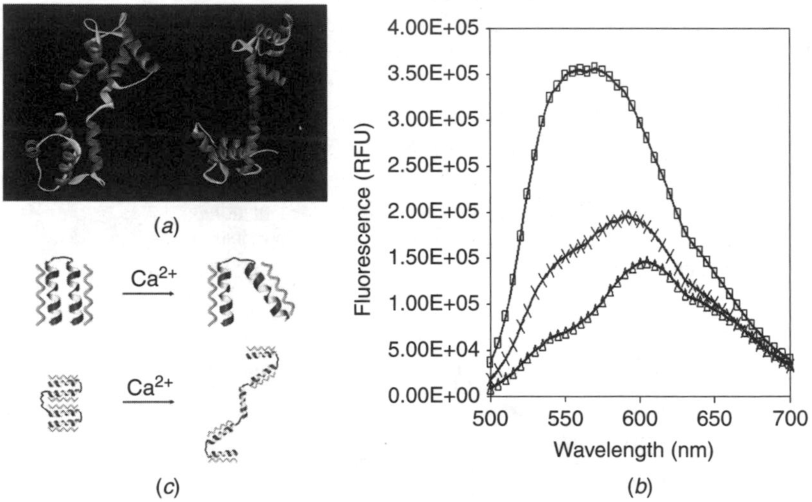

FIGURE 9.5 (*a*) Structure of calmodulin (CaM) without Ca^{2+} (left) and with Ca^{2+} (right). (*b*) Emission spectra of POWT (Δ), POWT-CaM (x), and POWT-CaM-Ca^{2+} (□) in 20 mM Tris-HCl pH 7.5. (*c*) Schematic drawing of the different conformational changes of the CaM molecule (striped helices) upon exposure to Ca^{2+} and the suggested geometries of the POWT chains (gray helices) (*Represented with permission from Ref. 86. Copyright 2004, American Chemical Society*).

changes considerably, resulting in a rearrangement from a closed to an open conformation of the protein motif. Structural studies also suggest that calcium activation of the protein is accompanied by a global conformational change, whereby the compact calcium-free form of CaM is converted to a more extended dumbbell-shaped molecule upon binding of calcium.[90, 91] The extended form of the protein consists of two lobes separated by a central α helix, and this central helix is flexible and allows considerable movements of the two lobes with respect to each other.

Upon formation of a complex between POWT and CaM, the emission maximum of POWT is red-shifted and the intensity of the emitted light is decreased (Fig. 9.5*b*) (relative to POWT alone in the same buffer), indicating that the POWT backbone becomes more planar and that aggregation of the POWT chains occurs. An addition of 10 m*M* Ca^{2+} to this complex will blue shift the emission maximum (594 nm), and the shoulder around 540 nm is increasing (Fig. 9.5*b*), suggesting that the polymer backbone becomes more nonplanar and that a separation of the polymer chains occurs. The ratio of the intensity of the emitted light at 540/670 nm is increased, showing that the conformational change of the CaM molecule upon exposure to Ca^{2+} is governing the geometry of the POWT chains. The increased emission at 540 nm, associated with separation of the polymer chains, is probably a result of the conformational changes of CaM that occur upon binding of calcium. A schematic presentation of the different conformational alterations of the CaM molecule upon exposure to calcium and the suggested POWT chain geometries interpreted from the spectral changes seen for the different POWT/CaM solutions is shown in Fig. 9.5*c*.[86]

As discussed above, it is rather evident that the conformation-sensitive optical properties of LCPs can be used as an optical fingerprint for distinct protein conformations. Hence, LCPs can be applied as a novel tool within the research field of protein folding and protein aggregation diseases. The underlying mechanism of protein aggregation, e.g., the formation of amyloid fibrils, and the diseases believed to be associated with this event are discussed in greater detail next.

9.3 Amyloid Fibrils and Protein Aggregation Diseases

9.3.1 Formation of Amyloid Fibrils

Proteins frequently alter their conformation due to different external stimuli, and many diseases are associated with misfolded proteins.[92, 93] Especially under conditions that destabilize the native state, proteins can self-assemble into aggregated β-sheet rich fibrillar assemblies, known as amyloid fibrils, which are around 10 nm wide and unusually stable biological materials (Fig. 9.6). However, the process where by a native protein is converted to amyloid fibrils is quite complex, and

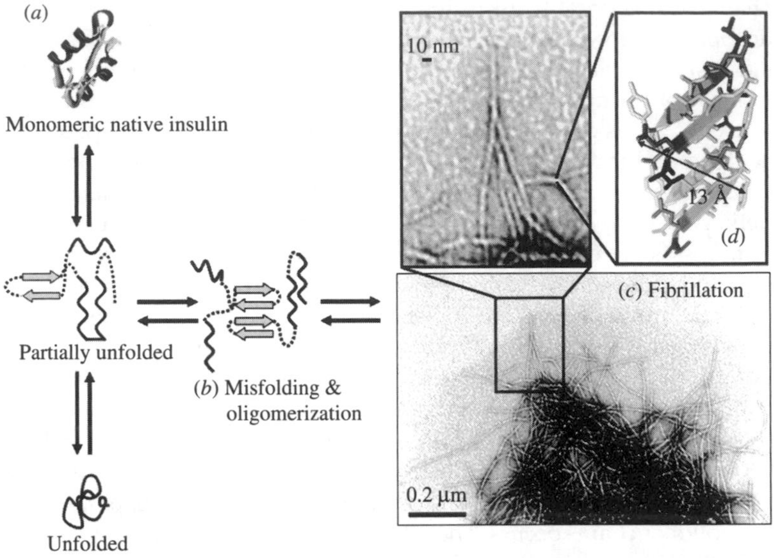

Figure 9.6 Schematic drawing of the formation of amyloid fibrils. (*a*) Monomeric insulin having an α-helical conformation. (*b*) β-sheet (arrows) rich oligomers are being formed. (c) Amyloid fibrils having a diameter around 10 nm are being formed. (*d*) Higher magnification of the intrinsic repetitive β-pleated sheet structure of the amyloid fibril. The pictures were taken by transmission electron microscopy (TEM).

there are many fundamental questions regarding this aggregation process that remain unanswered.[93] The full elucidation of the aggregation process requires the identification of all the conformational states and oligomeric structures adopted by the polypeptide chain during the process. It also entails characterizing each of the transitions in molecular detail and identifying the residues or regions of the sequence that are involved in and promote the various aggregation steps. The identification and characterization of prefibrillar states, such as oligomers, preceding the formation of well-defined fibrils are of particular interest because of an increasing awareness that these species are likely to play a critical role in the pathogenesis of protein deposition diseases (see Sec. 9.3.2).

Amyloid fibril formation has many characteristics of a "nucleated growth" mechanism, and the time course of the conversion of a peptide or protein into its fibrillar form typically includes a lag phase that is followed by a rapid exponential growth phase and a plateau phase (Fig. 9.7, Sec. 9.4.1). The lag phase is assumed to be the time required for "nuclei" to form, and once a nucleus is formed, fibril growth is thought to proceed rapidly by further association of either monomers or oligomers with the nucleus. As with many other processes dependent on a nucleation step, addition of preformed fibrillar species to a sample of a protein under aggregation conditions causes the lag phase to be shortened and finally abolished when the rate of the aggregation process is no longer limited by the need for nucleation.[94] Furthermore, changes in experimental conditions can also reduce or eliminate the length of the lag phase, again assumed to result from a situation wherein nucleation is no longer rate-limiting. Therefore, the absence of a lag phase does not inevitably imply the absence of a nucleated growth mechanism, but it may simply be that the time required for fibril growth is sufficiently slow relative to the nucleation process and that the latter is no longer the slowest step in the conversion of the monomeric protein into amyloid fibrils.

It is clear that the lag phase in fibril formation is an important event in which a variety of prefibrillar species are formed, including β-sheet rich oligomers and protofibrils clustered as spherical beads (2 to 5 nm in diameter) with β-sheet structure.[93] The past decade has seen very substantial efforts directed toward identifying, isolating, and characterizing these prefibrillar species that are present in solution prior to the appearance of fibrils, both because of their likely role in the mechanism of fibril formation and because of their implication as the toxic species involved in neurodegenerative disorders (see Sec. 9.3.2). Furthermore, the differing features of the aggregation processes, described in the previous paragraphs, reveal that polypeptide chains can adopt a multitude of conformational states. Therefore, it is not surprising that both the prefibrillar species and the fibrillar end products of amyloid fibril formation are characterized by morphological and structural diversity. Hence, techniques for detecting of variety of protein aggregates and methods for studying the molecular details of these aggregates are of great interest. This statement will become even more evident when the pathological events underlying the diseases associated with amyloid fibril formation are being discussed.

9.3.2 Protein Aggregation Diseases

A broad range of human diseases arise from the failure of a specific peptide or protein to adopt, or remain in, its native functional conformation. These pathological conditions are generally referred to as protein misfolding diseases and include pathological states in which an impairment in the folding efficiency of a given protein results in a reduction of the effective concentration of a functional protein that is available to play its normal role, as seen in cystic fibrosis. However, the largest group of misfolding diseases is associated with the conversion of peptides or proteins to amyloid fibrils. The amyloid fibrils are further assembled into higher-order structures that pathologically are termed *amyloid plaques* when they accumulate extracellularly, whereas the term *intracellular inclusions* has been suggested as more appropriate when fibrils morphologically and structurally related to extracellular amyloid form inside the cell.[95]

A list of known diseases that are associated with the formation of extracellular amyloid plaques or intracellular inclusions is given in Table 9.1, along with the specific proteins that in each case are the predominant components of the deposits. The diseases can be generally grouped into neurodegenerative conditions, in which aggregation occurs in the brain; peripheral localized amyloidoses, in which aggregation occurs in a single type of tissue other than the brain, and systemic amyloidoses, in which aggregation occurs in multiple tissues (Table 9.1). Some of these conditions are predominantly sporadic, although hereditary forms from specific mutations are also quite common. In addition, spongiform encephalopathies (prion diseases) can be transmissible in humans as well as in other mammals. According to the protein-only hypothesis, the transmissible agent, denoted prion, is composed solely of PrP^{Sc}, an aggregated form of the

Disease	Peptide/Protein
Alzheimer's disease	Amyloid β peptide
Transmissible spongiform encephalopathies (Prion diseases)	Prion protein or fragments thereof
Parkinson's disease	α-Synuclein
Amyotrophic lateral sclerosis (ALS)	Superoxide dismutase I
Huntington's disease	Huntingtin
Familial British dementia	ABri
Familial Danish dementia	ADan
AL amyloidosis	Fragments of immunoglobulin light chains
AA amyloidosis	Fragments of serum amyloid A protein
Familial Mediterranean fever	Fragments of serum amyloid A protein
Senile systemic amyloidosis	Wild-type transthyretin
Familial amyloidotic polyneuropathy	Mutants of transthyretin
Type II diabetes	Amylin (islet amyloid polypeptide, IAPP)
Inclusion body myositis	Amyloid β peptide

TABLE 9.1 Human Diseases Associated with Formation of Extracellular Amyloid Deposits or Intracellular Inclusions with Amyloidlike Characteristics

normal soluble prion protein PrP^{C}.[96] It has also been found that intravenous injection or oral administration of preformed fibrils from different sources can result in accelerated amyloidosis through a prionlike mechanism.[97, 98] Hence, an environment enriched with fibrillar material could act as a risk factor for amyloid diseases. In addition, prions can occur as multiple strains, giving rise to different symptoms and incubation periods.[96] The prion strain phenomenon is believed to be associated with multiple morphologies of the prion aggregates, and the specific properties of a prion strain are encoded in the tertiary or quaternary structure of the aggregates.

The presence of highly organized and stable fibrillar deposits, amyloid plaques, in the organs of patients suffering from protein deposition diseases led initially to the reasonable postulate that this material is the causative agent of the various disorders. However, as mentioned in Sec. 9.3.2, more recent findings have raised the possibility that precursors to amyloid fibrils, such as low-molecular-weight oligomers and/or structured protofibrils, are the real pathogenic species, at least in neurodegenerative diseases. For instance, the severity of cognitive impairment in Alzheimer's disease (AD) correlates with the levels of low-molecular-weight species of aggregated A-beta peptide (Aβ), including small oligomers, rather than with the amyloid burden.[99] Genetic evidence also supports the theory that the precursor aggregates are the pathogenic species. As seen from in vitro experiments, the introduction of an aggressive mutation in the protein favors the formation of prefibrillar states, and this mutation is associated with a heritable early onset of Alzheimer's disease.[100] In addition, the most highly infective form of the mammalian prion protein has been identified as an oligomer of about 20 molecules, indicating that such small aggregates are the most effective initiators of transmissible spongiform encephalopathies.[101]

9.3.3 Methods for Detection and Structural Characterization of Amyloid Fibrils

The formation and presence of amyloid fibrils or amyloid plaques can be visualized by small amyloid ligands dyes, such as derivatives of Congo red and thioflavins.[102] These dyes bind selectively to protein aggregates having an extensive cross β-pleated sheet conformation and sufficient structural regularity. Hence, the presence of mature amyloid fibrils is easily detected by these dyes, seen as an enhanced fluorescence (thioflavins) or green-yellow birefringence under cross-polarized light (Congo red) from the dye. However, these dyes are not able to recognize prefibrillar species, and amyloid fibrils of diverse morphological origin, as seen for prion strains, cannot be separated. Furthermore, amyloid ligands are not selective for a distinct protein and cannot differentiate between amyloid subtypes. The classification of amyloid subtypes is made by immunohistochemical stains utilizing antibodies that are selective for the peptide/protein associated with a distinct amyloid subtype. However, immunohistochemistry is fraught with specific problems. For one thing, most antibodies penetrate only poorly the compact beta-sheet accumulations of amyloid. Also, many amyloids incorporate nonspecifically other proteins, which can give rise to false-positive stains with diagnostic antibodies and—in worst-case scenarios—may lead to misdiagnoses.

For many years the only structural information about amyloid fibrils came from imaging techniques such as TEM (transmission electron microscopy), AFM (atomic force microscopy) and X-ray fiber diffraction.[103, 104] These experiments revealed that the fibrils predominantly consist of a number (typically two to six) of protofilaments that twist together to form the typical amyloid fibrils. X-ray fiber diffraction data have shown that in each individual protofilament, the protein or peptide molecules are arranged so that the polypeptide chain forms β-strands that run perpendicular to the long axis of the fibril. Today novel structural insight and molecular details have been provided by solid-state NMR spectroscopy,[105, 106] and by single crystal X-ray diffraction analysis of small amyloid-like peptide fragments.[107] The latter has allowed both the structure of the peptides and the way the molecules could be packed together to be determined with unprecedented resolution. A particularly significant aspect of the structures determined with the different techniques is that they are strikingly similar even for polypeptides having no sequence homology, suggesting that many amyloid fibrils could share similar core structures. However, the specific nature of the side chain packing, including characteristics as the alignment of adjacent strands and the separation of the sheets, provides an explanation for the occurrence of variations in the details of the structures for specific types of fibril.

Even though a wide range of techniques have been applied for studying the amyloid fibrillation event and the pathological mechanism underlying protein aggregation disease, there are still many questions that remain to be answered regarding these events. Hence, there is a need for novel tools that provide greater insight into these events, and next we discuss the utilization of LCPs for monitoring these events.

9.4 Luminescent Conjugated Polymers as Amyloid Specific Dyes

9.4.1 Detection of Amyloid Fibrils in Solution

As discussed earlier, novel tools that detect the conformational changes in proteins, especially the formation of amyloid fibrils, are of great importance, as this is an extremely complex event and many diseases are associated with conformational changes in proteins. Methods for the detection and quantification of diverse aggregated states of proteins are also of great importance with respect to the long-term stability and production of peptide pharmaceuticals in commercial pharmaceutical formulations used for the treatment of various diseases, e.g., insulin for diabetes. Although, it is rare to observe iatrogenic (i.e., treatment-induced) protein aggregates due to administration of peptide pharmaceuticals, the therapeutic effect of the peptide drug becomes limited if the peptide has been converted into amyloid fibrils.

LCPs, with their unique structural related optical properties, have proved to be an exceptionally powerful tool to study the amyloid fibrillation event. Novel conformation-sensitive optical methods for the detection of formation of amyloid fibrils in bovine insulin and chicken lysozyme based on conformational changes of the anionic polythiophene derivative, PTAA, were recently reported.[108] The technique is based on noncovalent assembly of the LCP and the different forms of the proteins (Fig. 9.7*a*). Depending on the conformation of the protein, different emission spectra from PTAA are observed (Fig. 9.7*b*). Upon binding to the native monomeric form of insulin, PTAA emits light with emission maximum of 550 nm, whereas PTAA bound to amyloid fibrils of insulin is emitting light with lower intensity and the emission maximum is red-shifted to 580 nm. The red-shift of the emission maximum and the decrease of the emission intensity from PTAA are associated with a planarization of the polymer backbone and an aggregation of adjacent polymer chains. Hence, specific optical fingerprint is achieved for the β-sheet containing amyloid fibrils. The detection of insulin fibrils can also be observed by absorption and visual inspection, and this can be useful for the development of simple screening methods for the detection of amyloid fibrils.[108] Furthermore, when plotting the ratio of emitted light at 550 nm and 580 nm, ratio 550/580 nm, the formation of insulin fibrils can be monitored (Fig. 9.7*c*). The kinetic plot is showing a lag phase, followed by a growth phase and a plateau phase, which are characteristic for the formation of amyloid fibrils.

PTAA binds both the native form of insulin and the fibrillar form of insulin. Nevertheless, these forms can be easily distinguished, due to the conformational changes of the polymer backbone upon binding to the different forms of the proteins, as minor perturbations of the geometry of the polymer backbone can be reflected as alterations of the electronic structure of the conjugated backbone. Thus, binding of the polymer to different forms of proteins will give rise to different optical features for the LCP. This is an improvement over small amyloid ligands, such as Congo red or thioflavins, as these probes only change in optical feature whether they are free in solution or binding to pockets in the protein or to the surface of the protein. So far, it has been shown that LCPs can be used to distinguish between the native

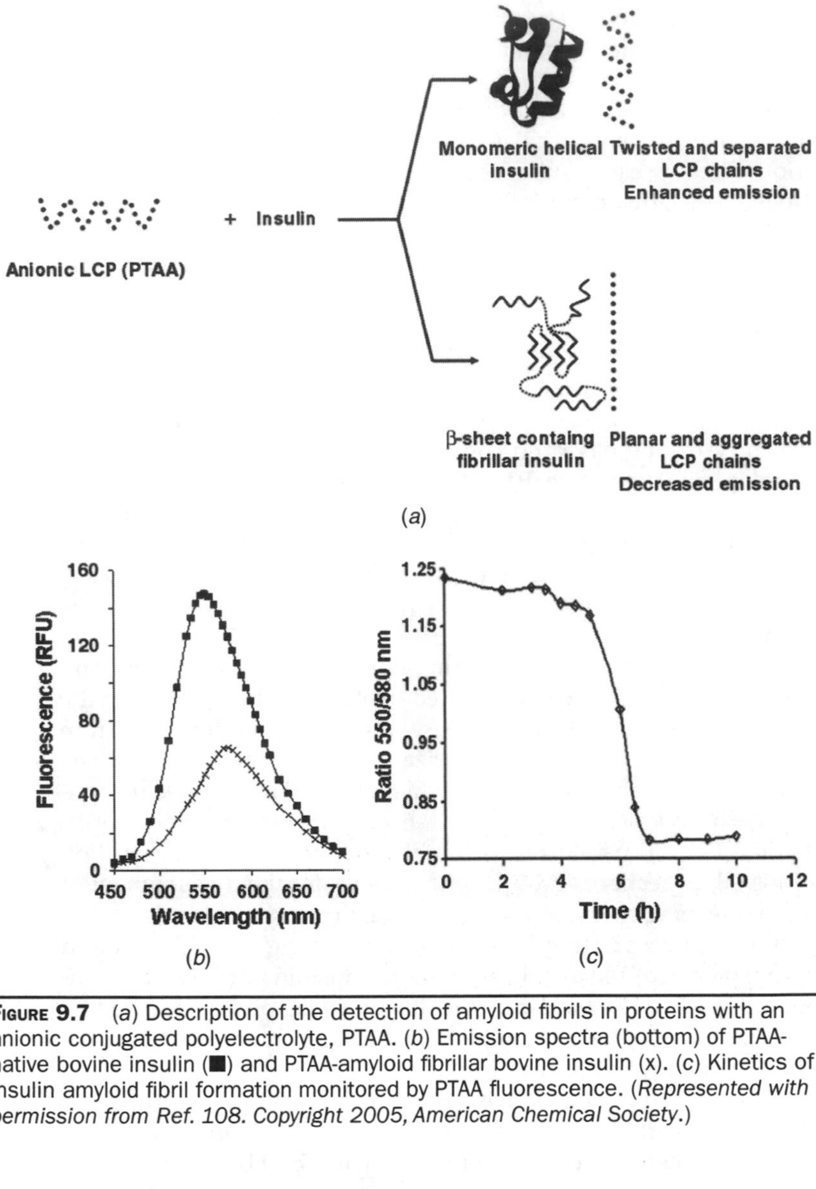

FIGURE 9.7 (*a*) Description of the detection of amyloid fibrils in proteins with an anionic conjugated polyelectrolyte, PTAA. (*b*) Emission spectra (bottom) of PTAA-native bovine insulin (■) and PTAA-amyloid fibrillar bovine insulin (x). (c) Kinetics of insulin amyloid fibril formation monitored by PTAA fluorescence. (*Represented with permission from Ref. 108. Copyright 2005, American Chemical Society.*)

form of proteins and the amyloid fibrillar form of proteins. However, the chemical design of novel LCPs might offer a novel approach to discriminate between different conformational structures observed during the amyloid formation processes.

A second generation of LCPs, containing a repetitive trimer block, was recently presented.[13, 14, 109] Except for just having ionic side chain functionalities, these molecules also include unsubstituted thiophene rings which give rise to a greater conformational freedom of the polymer backbone. Additionally, the chain length distribution of these materials was shown to be rather narrow, and around 90% of the material had a well-defined chain length of 9 or 12 monomers, although they were synthesized by random polymerization.[14] Both of these properties—the well defined chain length and the enhancement of the conformational freedom—were shown to improve the specificity for amyloid fibrils and the spectral assignment of distinct protein conformations.

As seen in Fig. 9.8*a*, the zwitter-ionic LCP, PONT, also known as tPOWT (Fig. 9.1), showed a huge increase of the emission intensity

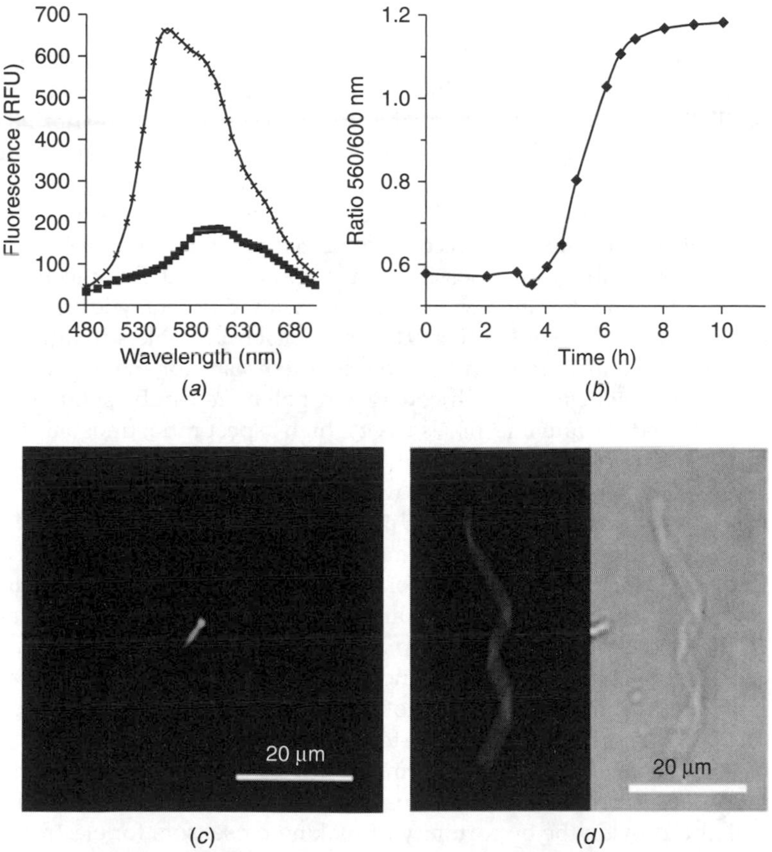

FIGURE 9.8 (*a*) Emission spectra (bottom) of PONT-native bovine insulin (■) and PONT-amyloid fibrillar bovine insulin (x). (*b*) Kinetics of insulin amyloid fibril formation monitored by PONT fluorescence. (c), (d) Images of self-assembled electroactive nanowires of PONT and insulin amyloid fibrils that are being formed when PONT is present during the fibrillation event. (*Parts (a) and (b) are reproduced with permission from Ref. 13. Copyright 2005, American Chemical Society. Parts (c) and (d) are from Ref. 109. Copyright 2005, Wiley-VCH Verlag GmbH & Co. KGaA.*) (See also color insert.)

and a blue shift of the emission maximum to 560 nm upon binding to insulin amyloid fibrils.[13] The spectrum for PONT mixed with native monomeric native insulin shows a weaker intensity with a maximum at 600 nm and resembles the spectrum for PONT free in solution, indicating that the interaction between PONT and native insulin is absent. In contrast to PTAA, the spectral observations indicate that the interaction with amyloid fibrils is leading to a twisted polymer backbone and separation of adjacent PONT chain, whereas the PONT chains are planar and aggregated when mixed with native insulin. The kinetics of the amyloid fibrillation was also followed by plotting the ratio of the intensity of the emitted light at 560 and 600 nm. Similar to the observations with PTAA, the characteristic three different phases of amyloid fibril formation are seen (Fig. 9.8*b*).

Another advantage of PONT, compared to the previously described PTAA, is that this LCP is stable under the acidic conditions used for fibrillation of insulin. Hence, PONT can be present during the fibrillation event, allowing the formation of a self-assembled hybrid material comprised of a natural nanostructure, the amyloid fibril, and an incorporated semiconductor, the LCP. Such electroactive luminescent self-assembled nanowires were recently reported by Herland et al.[109] This study showed that it was possible to incorporate the LCP in the amyloid fibrils, and luminescent nanowires with different widths and lengths could be obtained (Fig. 9.8*c* and *d*). Hence, stable structurally defined amyloid fibrils can be used as a template for making well-defined nanowires on which conjugated polymers can be symmetrically aligned. As amyloid fibrils have a high aspect ratio, unusual stability, and are rather symmetric, these findings might open up a wide range of tantalizing possibilities within the research field of bioelectronics and become an attractive avenue for functional nanodevices. The lack of control over the alignment of the materials appears to be a factor affecting the performance of electronic devices based on conjugated polymers, and the use of amyloid fibrils as a nano-structural motif might provides a solution to this problem. However, more work is needed to better characterize and understand the operating mechanism and the device architecture of such hybrid devices, as a complex set of processes is affecting the performance of electronic devices based on conjugated polymers. Furthermore, LCPs showing a high conductivity need to be developed to be used in electronic devices. Such LCPs will also provide the opportunity of making biosensors for electronic detection of amyloid fibrils.

9.4.2 Histological Staining of Amyloid Deposits in Tissue Samples

As described earlier, the LCPs are excellent tools for studying the amyloid fibril formation in vitro. However, those in vitro systems only contain the desired molecules. So the question remains: Can the sensory performance and the amyloid specificity of LCPs be utilized in more complex systems such as tissue sections? A couple of studies have shown that LCPs can be used as amyloid-specific dyes for histological staining of tissue sections and that LCPs provide additional information regarding the pathological events of protein aggregation diseases compared to conventional techniques.[110–112]

The proof of the concept of using the LCPs as amyloid-specific dyes in tissue samples was first shown by Nilsson et al.[110] Under certain conditions PTAA, POMT, and PONT (Fig. 9.1) were shown to selectively stain a plethora of amyloid plaques in formalin-fixed tissue sections (Fig. 9.9). The negatively charged PTAA was amyloid-specific under alkaline (pH 10) staining conditions, whereas the staining with the cationic and the zwitter-ionic LCPs was performed at acidic

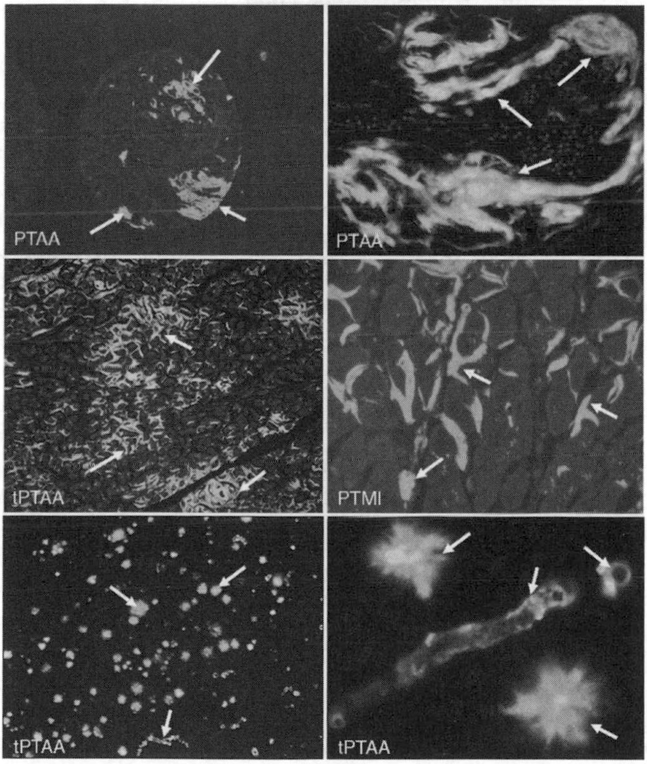

Figure 9.9 A plethora of amyloid deposits in tissue samples stained by different LCPs as indicated in the figure. Some typical amyloid deposits being stained by the LCPs are highlighted by arrows. Notably, PTAA and tPTAA bound to diverse amyloid deposits in the same tissue emit light with different colors, indicating that there is a heterogenic population of amyloid deposits in these samples (top right, bottom left and right) (See also color insert).

pH (pH 2). All the LCPs stained amyloid plaques associated with systemic amyloidoses and local amyloidoses, such as type 2 diabetes and Alzheimer's disease (AD). Similar to results obtained on in vitro formed amyloid fibrils, PTAA shows a red-shifted spectrum with a maximum around 580 nm upon binding to amyloid plaques, whereas amyloid deposits stained by PONT emit light with a more green-yellowish hue. Hence, upon binding to amyloid plaques in tissue samples, the rotational freedom of the thiophene rings and the geometry of the backbone are restricted, leading to a specific emission profile from the LCP similar to what was observed on pure amyloid fibrils in solution. Furthermore, some results were suggesting that PTAA emits light of different colors upon binding to different amyloid subtypes.[110]

The conformation-induced change of the fluorescence is a unique property seen for LCPs that cannot be achieved by sterically rigid conventional amyloid ligands dyes such as thioflavin T (ThT) and Congo red. Thus, LCPs offer the possibility to obtain a specific spectroscopic signature for individual protein aggregates. As mentioned earlier, there are many fundamental questions regarding this aggregation process that remain unanswered, and the underlying mechanism of amyloid or protein aggregate formation is poorly understood. [93] In this regard, the technique using LCPs, which provides a direct link between spectral signal and protein conformation, might provide an opportunity to gain more information concerning the morphology of the protein deposits and might facilitate a greater understanding of the conformational phenotype encoded in the protein aggregates. Instead of looking at the total amount of protein aggregates, heterogenic population of specific protein aggregates could be observed, and novel findings regarding toxic species and the molecular mechanism of these diseases could be obtained with the LCP technique. These assumptions have also been verified with experimental data using transgenic mouse model having AD pathology and transgenic mice infected with distinct prion strains.[111–112]

Upon application of LCPs to transgenic mouse models having AD pathology, a striking heterogenicity in the characteristic plaques composed of the A-beta peptide was identified.[111] LCP staining of brain tissue slices revealed different subpopulations of plaques, seen as plaques with different colors (Fig. 9.9). The spectral features of LCPs enable an indirect mapping of the plaque architecture, as the different colors of the LCPs are associated with different conformations of the polymer backbone. These findings can lead to novel ways of diagnosing AD and also provide a new method for studying the pathology of the disease in a more refined manner. Especially, the LCP technique might be valuable for identifying distinct toxic entities giving rise to cell death and loss of neurons, or for establishing a correlation between the type of plaque and the severity of AD. However, further studies of complexes between in vitro produced protein aggregates with defined conformations and LCPs with distinct ionic side chain functionalities or different chain lengths will likely be necessary to understand the correlation of the spectroscopic readout from the LCP and the molecular structure of the protein aggregate. Although the achievement of obtaining certain spectroscopic LCP signatures from heterogenic populations of protein aggregates is beneficial compared to conventional amyloid specific dyes, correlating this spectroscopic signature to a specific form of the aggregated protein is still necessary to gain novel insight into the pathological process of the disease. Nevertheless, the LCPs can be useful for comparison of heterogenic protein aggregates in well-defined experimental systems.

Heterogenic protein aggregates can also be found in other protein aggregation disorders, such as the infectious prion diseases. As mentioned previously, prion disease is caused by a proteinaceous agent called PrP^{Sc}, a misfolded and aggregated version of the normal prion protein. In addition, prions can occur as different strains, and the prion strain phenomenon is most likely encoded in the tertiary or quaternary structure of the prion aggregates. This belief was also verified when protein aggregates in brain sections from mice infected with distinct prion strains were being stained by LCPs.[112] The LCPs bound specifically to the prion deposits and different prion strains can be separated due to alternative staining patterns of LCPs with distinct ionic side chains. Furthermore, the anionic LCP, PTAA, emits light of different wavelengths when bound to distinct protein deposits associated with a specific prion strain (Fig 9.10*a*). As the emission profiles of LCPs are associated with geometric changes of the polymer backbone,[36, 43, 44] ratios of the intensity of the emitted light at certain wavelengths can be used as an indicator of the geometry of the polymer chains.[36, 86] Nonplanar and separated LCP chains emit light around 530 to 540 nm, whereas a planarization of the thiophene backbone will shift the emission maximum E_{max} toward longer wavelengths. A planar backbone might also give rise to an aggregation of LCP chains, seen as an increase of the intrinsic emission around 640 nm. When plotting the ratio $532/E_{max}$ and the ratio 532/639 nm in a correlation diagram, prion aggregates associated with distinct prion strains, chronic wasting disease (CWD) and sheep scrapie, were easily distinguished from each other, verifying the usefulness of spectral properties of LCPs for classification of protein deposits (Fig. 9.10*b*). These conformation dependent spectral characteristics can only be afforded by LCPs and provide the opportunity to get an optical fingerprint for protein aggregates correlating to a distinct prion strain.

Although it was shown that the emission profile of LCPs could be used to characterize protein deposits, further evidence was necessary to enable relating the geometric alterations of the LCPs to a structural

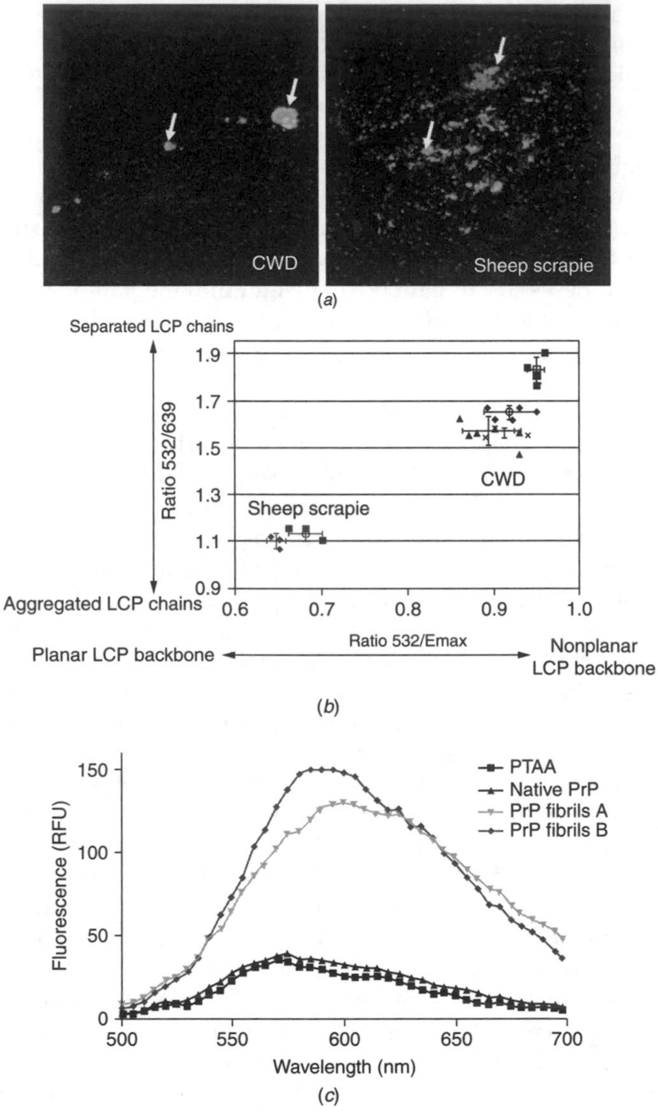

FIGURE 9.10 (*a*) Fluorescent images of prion deposits associated with distinct prion strains, chronic wasting disease (CWD, left) and sheep scrapie (right), which have been stained by PTAA. (*b*) Correlation diagram of the ratios, R532/639 and R532/E_{max}, of the intensity of the emitted light from PTAA bound to prion deposits originating from individual mice infected with CWD (four generations denoted with black symbols, ■, ♦, ▲, x) or sheep scrapie (two generations denoted with purple symbols, squares and diamonds). (*c*) Emission spectra of PTAA in buffer (black), PTAA bound to native (blue) or two different forms of fibrillar recombinant mouse PrP (green and purple). [*Part* (c) *reproduced with permission from Ref. 112. Copyright 2007, Nature Publishing Group.*] (See also color insert.)

variance of the protein deposits associated with the distinct prion strains. By taking recombinant mouse prion protein (mPrP) and converting it to two different types of amyloid fibrils by using varying conditions for fibrillation, Sigurdson et al. were able to show that the emission profile of PTAA could be used to distinguish the two fibril preparations (Fig. 9.10*c*).[112] As these two preparations of fibrils were chemically identical, having the same protein (mPrP) and being dialyzed against the same buffer, the spectral differences seen for PTAA were most likely due to structural differences between the fibrils. Hence, LCPs provide structural insights regarding the morphology of individual protein deposits and can be used as a complementary technique to conventional staining protocols for the characterization of protein deposits associated with individual prion strains. These findings might be of great value, as phenomena similar to those occurring in prion strains may be much more frequent than is now appreciated, and may extend to additional protein misfolding and aggregation disorders. As mentioned previously, strainlike conformational variants have been described for A-beta aggregation, which underlies AD.[98] LCPs might therefore aid in the fundamental understanding of conformational protein variants in a wide range of protein misfolding disorders. In addition, LCPs could improve the precision of diagnoses of protein aggregation diseases and facilitate analysis of amyloid maturity and origin.

9.4.3 Toward in Vivo Staining of Amyloid Deposits

Clearly, there is much to be gained through a multidisciplinary scientific approach; the integration of information can provide new insight and discoveries within diverse fields of research. Unexpectedly, conjugated polymer materials that originate from electronics and solar cells have provided novel insights into the biology and pathology of protein aggregation diseases. Apparently, the spectral information from LCPs will be useful to gain more information regarding the molecular details of the protein aggregation process and the pathological events underlying protein aggregation diseases. However, there is still a great extent of basic research that can be performed to take full advantage of the technique.

One tantalizing possibility is to develop LCPs that can be used for in vivo imaging of protein aggregates. In this regard, smaller, appropriately functionalized, but still selective LCPs that are able to cross the blood-brain barrier (BBB) need to be synthesized. These dyes can be utilized in powerful multiphoton imaging applications for in vivo imaging of protein aggregates, since previously reported LCPs have been shown to have an excellent cross-sectional area compared to small fluorescent dyes.[110, 113] However, these LCPs having a molecular weight between 1000 and 5000 Da will not cross the BBB, so chemical design of smaller, well-defined oligomeric thiophene molecules is necessary. Such molecules will be excellent for monitoring AD pathology and disease progress upon treatment, especially in transgenic mouse models. On the other hand, multiphoton imaging might not be suitable as a clinical diagnostic tool for protein aggregation diseases. In this regard, radiolabeled or fluorinated versions of the oligomeric thiophenes usable for positron emission tomography (PET) or single-photon emission computerized tomography (SPECT) will be advantageous.

Future chemical design of novel well-defined oligomeric thiophenes will certainly utilize combinatorial approaches for optimizing the thiophene core structure, which may provide more effective binders for different classes of protein aggregates. As mentioned in earlier sections, there is a great need for techniques identifying different molecular species that are present on the pathway to the formation of amyloid fibrils, and it may even be possible to develop LCPs that selectively stain oligomeric or fibrillar species. Recently, a successful approach for the synthesis of more well-defined oligomeric thiophenes was reported, and these molecules showed a higher selectivity for amyloid fibrils than their polymeric polydispersed counterparts.[14] LCPs having different and well-defined chain lengths would also be of interest to establish and optimize the maximum effective conjugation length necessary for selective fluorescence from LCPs being bound to a wide range of heterogenic protein aggregates. A focus must also be turned to the fundamental underlying photophysical processes of LCPs and the molecular details regarding the selective binding site of LCPs to specific protein aggregates. Therefore, a general understanding of photophysical processes of LCPs and design rules in the synthesis of LCPs will be important for continued progress in understanding protein aggregation disease.

As described in this chapter, LCPs have been implemented into a new area of science, and scientists have just begun to explore the use of this material within the field of biology and pathology. So the question remains, can the sensory performance of LCPs be utilized to gain novel insight into the mysterious pathogenic events of protein aggregation diseases? Hopefully, a multidisciplinary scientific approach will give the answer to this question and also lead to development of LCP-based sensors for studying other diseases.

References

1. Andersson, M.; Ekeblad, P. O.; Hjertberg, T.; Wennerström, O.; and Inganäs, O. Polythiophene with a free amino acid side chain. *Polymer Commun.* 32:546–548 (1991).
2. McCullough, R. D.; Ewbank, P. C.; and Loewe, R. S. Self-assembly and disassembly of regioregular, water soluble polythiophenes: Chemoselective ionchromatic sensing in water. *J. Am. Chem. Soc.* 119:633–634 (1997).
3. Kim, B.; Chen, L.; Gong, J.; and Osada, Y. Titration behavior and spectral transitions of water-soluble polythiophene carboxylic acid. *Macromolecules* 32:3964–3969 (1999).

4. Chayer, M.; Faid, K.; and Leclerc, M. Highly conducting water-soluble polythiophene derivatives. *Chem. Mater.* 9:2902–2905 (1997).
5. Ewbank, P. C.; Nuding, G.; Suenaga, H.; McCullough, R. D.; and Shinkai, S. Amine functionalized polythiophenes: Synthesis and formation of chiral ordered structures on DNA substrates. *Tetrahedron Lett.* 42:155–157 (2001).
6. Lukkari, J.; Salomäki, M.; Viinikanoja, A.; Ääritalo, T.; Paukkunen, J.; Kocharova, N.; and Kankare, J. Polyelectrolyte multilayers prepared from water-soluble poly(alkoxythiophene) derivatives. *J. Am. Chem. Soc.* 123: 6083–6091 (2001).
7. Ho, H-A.; Boissinot, M.; Bergeron, M. G.; Corbeil, G.; Dore, K.; Boudreau, D.; and Leclerc, M. Colorimetric and fluorometric detection of nucleic acids using cationic polythiophene derivatives. *Angew. Chem. Int. Ed.* 41:1548–1551 (2002).
8. Stork, M. S.; Gaylord, B. S.; Heeger, A. J.; and Bazan, G. C. Energy transfer in mixtures of water-soluble oligomers: Effect of charge, aggregation, and surfactant complexation. *Adv. Mater.* 14:361–366 (2002).
9. Tan, C.; Pinto, M. R.; and Schanze, K. S. Photophysics, aggregation and amplified quenching of a water-soluble poly(phenylene ethynylene). *Chem. Commun.* 446–447 (2002).
10. Kushon, S. A.; Ley, K. D.; Bradford, K.; Jones, R. M.; McBranch, D.; and Whitten, D. Detection of DNA hybridization via fluorescent polymer superquenching. *Langmuir* 18:7245–7249 (2002).
11. Bin, L.; Wang, S.; Bazan, G. C.; and Mikhailovsky, A. Shape-adaptable water-soluble conjugated polymers. *J. Am. Chem. Soc.* 125:13306–13307 (2003).
12. Nilsson, K. P. R.; Olsson, J. D. M.; Konradsson, P.; and Inganäs, O. Enantiomeric substituents determine the chirality of luminescent conjugated polythiophenes. *Macromolecules* 37:6316–6321 (2004).
13. Herland, A.; Nilsson, K. P. R.; Olsson, J. M. D.; Hammarström, P.; Konradsson, P.; and Inganäs, O. Synthesis of a regioregular zwitterionic conjugated oligoelectrolyte, usable as an optical probe for detection of amyloid fibrillation at acidic pH. *J. Am. Chem. Soc.* 127:2317–2323 (2005).
14. Åslund, A.; Herland, A.; Hammarström, P.; Nilsson, K. P. R.; Jonsson, B-H.; Inganäs, O.; and Konradsson, P. Studies of luminescent conjugated polythiophene derivatives: Enhanced spectral discrimination of protein conformational states. *Bioconjug. Chem.* 18:1860–1868 (2007).
15. Inganäs, O.; Salaneck, W. R.; Österholm, J.-E.; and Laakso, J. Thermochromic and solvatochromic effects in poly(3-hexylthiophene). *Synth. Met.* 22:395–406 (1988).
16. Roux, C., and Leclerc, M. Rod-to-coil transition in aloxy-substituted polythiophenes. *Macromolecules* 25:2141–2144 (1992).
17. Roux, C.; Faïd K.; and Leclerc, M. Polythiophene derivatives: Smart materials. *Polymer News* 19:6–10 (1994).
18. Faïd, K.; Fréchette, M.; Ranger, M.; Mazerolle, L.; Lévesque, I.; and Leclerc, M. Chromic phenomena in regioregular and nonregioregular polythiophene derivatives. *Chem. Mater.* 7:1390–1396 (1995).
19. Marsella, M. J.; and Swager, T. M. Designing conducting polymer-based sensors: Selective ionochromic response in crown ether-containing polythiophenes. *J. Am. Chem. Soc.* 115:12214–12215 (1993).
20. Crawford, K. B.; Goldfinger, M. B; and Swager, T. M. Na^+ specific emission changes in an ionophoric conjugated polymer. *J. Am. Chem. Soc.* 120:5187–5192 (1998).
21. Faid, K., and Leclerc, M. Functionalized regioregular polythiophenes: Towards the development of biochromic sensors. *Chem. Commun.*, no. 24:2761–2762 (1996).
22. Samuelson, L. A.; Kaplan, D. L.; Lim, J. O.; Kamath, M.; Marx, K. A.; and Tripathy, S. K. Molecular recognition between a biotinylated polythiophene copolymer and phycoerythrin utilizing the biotin-streptavidin interaction. *Thin Solid Films* 242:50–55 (1994).
23. Pande, R.; Kamtekar, S.; Ayyagari, M. S.; Kamath, M.; Marx, K. A.; Kumar, J.; Tripathy, S. K., et al. A biotinylated undecylthiophene copolymer bioconjugate

for surface immobilization: Creating an alkaline phosphatase chemiluminescence-based sensor. *Bioconjugate Chem.* 7:159–164 (1996).

24. Baek, M-G.; Stevens, R. C.; and Charych, D. H. Design and synthesis of novel glycopolythiophene assemblies for colorimetric detection of influenza virus and *E. coli*. *Bioconjugate Chem.* 11:777–788 (2000).
25. Bredas, J. L.; Street, G. B.; Themans, B.; and Andre, J. M. Organic polymers based on aromatic rings (polyparaphenylene, polypyrrole, polythiophene): Evolution of the electronic properties as a function of the torsion angle between adjacent rings. *J. Chem. Phys.* 83:1323–1329 (1985).
26. Orchard, B. J.; and Tripathy, S. K. Molecular-structure and electronic property modification of poly(diacetylenes). *Macromolecules* 19:1844–1850 (1986).
27. Dobrosavljevic, V., and Stratt, R. M. Role of conformational disorder in the electronic structure of conjugated polymers—substituted polydiacetylenes. *Phys. Rev. B* 35:2781–2794 (1987).
28. Tanaka, H.; Gomez, M. A.; Tonelli, A. E.; and Thakur, M. Thermochromic phase transitions of a polydiacetylene, poly(ETCD), studied by high-resolution solid-state ^{13}C NMR. *Macromolecules* 22:1208–1215 (1989).
29. Zerbi, G.; Chierichetti, B.; and Inganäs, O. Thermochromism in polyalkylthiophenes—molecular aspects from vibrational spectroscopy. *J. Chem. Phys.* 94:4646–4658 (1991).
30. Leclerc, M.; and Faïd, K. Electrical and optical properties of processable polythiophene derivatives: Structure-property relationship. *Adv. Mater.* 9:1087–1094 (1997).
31. Garreau, S.; and Leclerc, M. Planar-to-nonplanar conformational transitions in thermocromic polythiophenes: A spectroscopic study. *Macromolecules* 36:692–697 (2003).
32. DiCesare, N.; Belletete, M.; Leclerc, M.; and Durocher, G. Towards a theoretical design of thermochromic polythiophenes. *Chem. Phys. Lett.* 275:533–539 (1997).
33. DiCésare, N.; Belletete, M.; Marrano, C.; Leclerc, M.; and Durocher, G. Conformational analysis (ab ignition HF/3-21G*) and optical properties of symmetrically disubstituted terthiophenes. *J. Phys. Chem. A* 102:5142–5149 (1998).
34. Miteva, T.; Palmer, L.; Kloppenburg, L.; Neher, D.; and Bunz, U. H. F. Interplay of thermochromicity and liquid crystalline behavior in poly (*p*-phenyleneethynylene)s; π-π interactions or planarization of the conjugated backbone? *Macromolecules* 33:652–654 (2000).
35. Kim, J.; and Swager, T. M. Control of conformational and interpolymer effects in conjugated polymers. *Nature* 411:1030–1034 (2001).
36. Nilsson, K. P. R.; Andersson, M. R.; and Inganäs, O. Conformational transitions of a free amino-acid-functionalized polythiophene induced by different buffer systems. *J. Phys.: Condensed Matter* 14:10011–10020 (2002).
37. DiCésare, N.; Belletet, M.; Marrano, C.; Leclerc, M.; and Durocher, G. Intermolecular interactions in conjugated oligothiophenes. 1. Optical spectra of terthiophene and substituted terthiophenes recorded in various enviroments. *J. Phys. Chem. A* 103:795–802 (1999).
38. DiCésare, N.; Belletet, M.; Leclerc, M.; and Durocher, G. Intermolecular interactions in conjugated oligothiophenes. 2. Quantum chemical calculations performed on crystalline structures of terthiophene and substituted terthiophenes. *J. Phys. Chem. A* 103:803–811 (1999).
39. DiCésare, N.; Belletet, M.; Garcia, E. R.; Leclerc, M.; and Durocher, G. Intermolecular interactions in conjugated oligothiophenes. 3. Optical and photophysical properties of quarterthiophene and substituted quarterthiophenes in various enviroments. *J. Phys. Chem. A* 103:3864–3875 (1999).
40. Langeveld-Voss, B. M. W.; Janssen, R. A. J.; Christiaans, M. P. T.; Meskers, S. C. J.; Dekkers, H. P. J. M.; and Meijer, E. W. Circular dichroism and circular polarization of photoluminescence of highly ordered poly{3,4-di[(S)-2-methylbutoxy]thiophene}. *J. Am. Chem. Soc.* 118:4908–4909 (1996).

41. Langeveld-Voss, B. M. W.; Christiaans, M. P. T.; Janssen, R. A. J.; and Meijer, E. W. Inversion of optical activity of chiral polythiophene aggregates by a change of solvent. *Macromolecules* 31:6702–6704 (1998).
42. Langeveld-Voss, B. M. W.; Janssen, R. A. J.; and Meijer, E. W. On the origin of optical activity in polythiophenes. *J. Mol. Structure* 521:285–301 (2000).
43. Andersson, M. R.; Berggren, M.; Olinga, T.; Hjertberg, T.; Inganäs, O.; and Wennerström, O. Improved photoluminescence efficiency of films from conjugated polymers. *Synth. Met.* 85:1383–1384 (1997).
44. Berggren, M.; Bergman, P.; Fagerström, J.; Inganäs, O.; Andersson, M.; Weman, H.; Granström, M., et al. Controlling inter chain and intra-chain excitations of a poly(thiophene) derivative in thin films. *Chem. Phys. Lett.* 304:84–90 (1999).
45. Charych, D. H.; Nagy, J. O.; Spevak, W.; and Bednarski, M. D. Direct colorimetric detection of receptor-ligand interaction by a polymerized bilayer assembly. *Science* 261:585–588 (1993).
46. Reichert, A.; Nagy, J. O.; Spevak, W.; and Charych, D. Polydiacetylene liposomes functionalized with sialic acid bind and colorimetrically detect influenza virus. *J. Am. Chem. Soc.* 117:829–830 (1995).
47. Charych, D.; Cheng, Q.; Reichert, A.; Kuziemko, G.; Stroh, M.; Nagy, J. O.; Spevak, W., et al. A litmus test for molecular recognition using artificial membranes. *Chem. Biol.* 3:113–120 (1996).
48. Pan, J. J.; and Charych, D. Molecular recognition and colorimetric detection of cholera toxin by poly(diacetylene) liposomes incorporating G_{m1} ganglioside. *Langmuir* 13:1365–1367 (1997).
49. Okada, S. Y.; Jelinek, R.; and Charych, D. Induced color change of conjugated polymeric vesicles by interfacial catalysis of phospholipase A_2. *Angew. Chem. Int. Ed.* 38:655–659 (1999).
50. Chen, L.; McBranch, D. W.; Wang, H-L.; Helgeson, R.; Wudl, F.; and Whitten, D. G. Highly sensitive biological and chemical sensors based on reversible fluorescence quenching in a conjugated polymer. *Proc. Natl. Acad. Sci. USA* 96:12287–12292 (1999).
51. Gaylord, B. S.; Heeger, A. J.; and Bazan, G. C. DNA detection using water-soluble conjugated polymers and peptide nucleic acid probes. *Proc. Natl. Acad. Sci. USA* 99:10954–10957 (2002).
52. Ho, H-A.; Bera-Aberem, M.; and Leclerc, M. Optical sensors based on hybrid DNA/conjugated polymer complexes. *Chem. Eur. J.* 11:1718–1724 (2005).
53. Nilsson, K. P. R.; and Inganäs, O. Chip and solution detection of DNA hybridization using a luminescent zwitterionic polythiophene derivative. *Nature Mater.* 2:419–424 (2003).
54. Zhou, Q.; and Swager, T. M. Methodology for enhancing the sensitivity of fluorescent chemosensors—Energy migration in conjugated polymers. *J. Am. Chem. Soc.* 117:7017–7018 (1995).
55. Zhou, Q.; and Swager, T. M. Fluorescent chemosensors based on energy migration in conjugated polymers: The molecular wire approach to increased sensitivity. *J. Am. Chem. Soc.* 117:12593–12602 (1995).
56. Swager, T. M. The molecular wire approach to sensory signal amplification. *Acc. Chem. Res.* 31:201–207 (1998).
57. McQuade, D. T.; Pullen, A. E.; and Swager, T. M. Conjugated polymer-based chemical sensors. *Chem. Rev.* 100:2537–2574 (2000).
58. Wang, J.; Wang, D.; Miller, E. K.; Moses, D.; Bazan, G. C.; and Heeger, A. J. Photoluminescence of water-soluble conjugated polymers: Origin of enhanced quenching by charge transfer. *Macromolecules* 33:5153–5158 (2000).
59. Wosnick, J. H.; and Swager, T. M. Molecular photonics and electronic circuitry for ultra-sensitive chemical sensors. *Curr. Opin. Chem. Biol.* 4:715–720 (2000).
60. Harrison, B. S.; Ramey, M. B.; Reynolds, J. R.; and Schanze, K. S. Amplified fluorescence quenching in a poly(*p*-phenylene)-based cationic polyelectrolyte. *J. Am. Chem. Soc.* 122:8561–8562 (2000).

61. Wang, D.; Wang, J.; Moses, D.; Bazan, G. C.; and Heeger, A. J. Photoluminescence quenching of conjugated macromolecules by bipyridinium derivatives in aqueous media: Charge dependence. *Langmuir* 17:1262–1266 (2001).
62. Jones, R. M.; Bergstedt, T. S.; McBranch, D. W.; and Whitten, D. G. Tuning of superquenching in layered and mixed fluorescent polyelectrolytes. *J. Am. Chem. Soc.* 123:6726–6727 (2001).
63. Fan, C.; Wang, S.; Hong, J. W.; Bazan, G. C.; Plaxco, K. W.; and Heeger, A. J. Beyond superquenching: Hyper-efficient energy transfer from conjugated polymers to gold nanoparticles. *Proc. Natl. Acad. Sci. USA* 100:6297–6301 (2003).
64. Heeger, P. S.; and Heeger, A. J. Making sense of polymer-based biosensors. *Proc. Natl. Acad. Sci. USA* 96:12219–12221 (1999).
65. Kushon, S. A.; Bradford, K.; Marin, V.; Suhrada, C.; Armitage, B. A.; McBranch, D.; and Whitten, D. Detection of single nucleotide mismatches via fluorescent polymer superquenching. *Langmuir* 19:6456–6464 (2003).
66. Gaylord, B. S.; Heeger, A. J.; and Bazan, G. C. DNA hybridization detection with water-soluble conjugated polymers and chromophore-labeled single stranded DNA. *J. Am. Chem. Soc.* 125:896–900 (2003).
67. Liu, B.; and Bazan, G. C. Interpolyelectrolyte complexes of conjugated copolymers and DNA: Platforms for multicolor biosensors. *J. Am. Chem. Soc.* 126:1942–1943 (2004).
68. Xu, Q-H.; Gaylord, B. S.; Wang, S.; Bazan, G. C.; Moses, D.; and Heeger, A. J. Time-resolved energy transfer in DNA sequence detection using water-soluble conjugated polymers: The role of electrostatic and hydrophobic interactions. *Proc. Natl. Acad. Sci. USA* 101:11634–11639 (2004).
69. Wang, S.; Gaylord, B. S.; and Bazan, G. C. Fluorescein provides a resonance gate for FRET from conjugated polymers to DNA intercalated dyes. *J. Am. Chem. Soc.* 126:5446–5451 (2004).
70. Liu, B.; Baudrey, S.; Jaeger, L.; and Bazan, G. C. Characterization of tectoRNA assembly with cationic conjugated polymers. *J. Am. Chem. Soc.* 126:4076–4077 (2004).
71. Gaylord, B. S.; Massie, M. R.; Feinstein, S. C.; and Bazan, G. C. SNP detection using peptide nucleic acid probes and conjugated polymers: Applications in neurodegenerative disease identification. *Proc. Natl. Acad. Sci. USA* 102:34–39 (2005).
72. Liu, B.; and Bazan, G. C. Methods for strand-specific DNA detection with cationic conjugated polymers suitable for incorporation into DNA chips and microarrays. *Proc. Natl. Acad. Sci. USA* 102:589–593 (2005).
73. Wang, D.; Gong, X.; Heeger, P. S.; Rininsland, F.; Bazan, G. C.; and Heeger, A. J. Biosensors from conjugated polyelectrolyte complexes. *Proc. Natl. Acad. Sci. USA* 99:49–53 (2002).
74. Fan, C.; Plaxco, K. W.; and Heeger, A. J. High efficiency fluorescence quenching of conjugated polymers by proteins. *J. Am. Chem. Soc.* 124:5642–5643 (2002).
75. Pinto, M. R.; and Schanze, K. S. Amplified fluorescence sensing of protease activity with conjugated polyelectrolytes. *Proc. Natl. Acad. Sci. USA* 101:7505–7510 (2004).
76. Kumaraswamy, S.; Bergstedt, T.; Shi, X.; Rininsland, F.; Kushon, S.; Xia, W.; Ley, K., et al. Fluorescent-conjugated polymer superquenching facilitates highly sensitive detection of proteases. *Proc. Natl. Acad. Sci. USA* 101:7511–7515 (2004).
77. Dwight, S. J.; Gaylord, B. S.; Hong, J. W.; and Bazan, G. C. Perturbation of fluorescence by nonspecific interactions between anionic poly(phenylenevinylene)s and proteins: Implications for biosensors. *J. Am. Chem. Soc.* 126:16850–16859 (2004).
78. Rininsland, F.; Xia, W.; Wittenburg, S.; Shi, X.; Stankewicz, C.; Achyuthan, K.; McBranch, D., et al. Metal ion-mediated polymer superquenching for highly sensitive detection of kinase and phosphatase activities. *Proc. Natl. Acad. Sci. USA* 101:15295–15300 (2004).

79. Wosnick, J. H.; Mello, C. M.; and Swager, T. M. Synthesis and application of poly(phenylene ethynylene)s for bioconjugation: A conjugated polymer-based fluorogenic probe for proteases. *J. Am. Chem. Soc.* 127:3400–3405 (2005).
80. Faïd, K.; and Leclerc, M. Responsive supramolecular polythiophene assemblies. *J. Am. Chem. Soc.* 120:5274–5278 (1998).
81. Leclerc, M. Optical and electrochemical transducers based on functionalized conjugated polymers. *Adv. Mater.* 11:1491–1498 (1999).
82. Kumpumbu-Kalemba, L.; and Leclerc, M. Electrochemical characterization of monolayers of a biotinylated polythiophene: Towards the development of polymeric sensors. *Chem. Comm.*, no. 19:1847–1848 (2000).
83. Ho, H-A.; and Leclerc, M. Optical sensors based on hybrid aptamer/conjugated polymer complexes. *J. Am. Chem. Soc.* 126:1384–1387 (2004).
84. Dore, K.; Dubus, S.; Ho, H-A.; Levesque, L.; Brunette, M.; Corbeil, G.; Boissinot, M., et al. Fluorescent polymeric transducer for the rapid, simple, and specific detection of nucleic acids at the zeptomole level. *J. Am. Chem. Soc.* 126:4240–4244 (2004).
85. Bera-Aberem, M.; Ho, H-A.; and Leclerc, M. Functional polythiophenes as optical chemo- and biosensors. *Tetrahedron* 60:11169–11173 (2004).
86. Nilsson, K. P. R.; and Inganäs, O. Optical emission of a conjugated polyelectrolyte report calcium induced conformational changes in calmodulin and calmodulin-calcineurin interactions. *Macromolecules* 37:419–424 (2004).
87. Nilsson, K. P. R.; Rydberg, J.; Baltzer, L.; and Inganäs, O. Self-assembly of synthetic peptides control conformation and optical properties of a zwitterionic polythiophene derivative. *Proc. Natl. Acad. Sci. USA* 100:10170–10174 (2003).
88. Nilsson, K. P. R.; Rydberg, J.; Baltzer, L.; and Inganäs, O. Twisting macromolecular chains: Self-assembly of a chiral supermolecule from nonchiral polythiophene polyanions and random-coil synthetic peptides. *Proc. Natl. Acad. Sci. USA* 101:11197–11202 (2004).
89. Kretsinger, R. H. Structure and evolution of calcium-modulated proteins. *CRC Crit. Rev. Biochem.* 8:119–174 (1980).
90. Zhang, M.; Tanaka, T.; and Ikura, M. Calcium-induced conformational transition revealed by the solution structure of apo calmodulin. *Nature Struct. Biol.* 2:758–767 (1995).
91. Kuboniwa, H.; Tjandra, N.; Grzesiek, S.; Ren, H.; Klee, C. B.; and Bax, A. Solution structure of calcium-free calmodulin. *Nature Struct. Biol.* 2:768–776 (1995).
92. Carrell, R. W.; and Lomas, D. A. Conformational disease. *Lancet* 350:134–138 (1997).
93. Chiti, F., and Dobson, C. M. Protein misfolding, functional amyloid, and human disease. *Annu. Rev. Biochem.* 75:333–366 (2006).
94. Serio, T. R.; Cashikar, A. G.; Kowal, A. S.; Sawicki, G. J.; Moslehi, J. J.; Serpell, L. L.; Arnsdorf, M. F., et al. Nucleated conformational conversion and the replication of conformational information by a prion determinant. *Science* 289:1317–1321 (2000).
95. Westermark, P.; Benson, M. D.; Buxbaum, J. N.; Cohen, A. S.; Frangione, B.; Ikeda, S.; Masters, C. L., et al. Amyloid: Toward terminology clarification. Report from the Nomenclature Committee of the International Society of Amyloidosis. *Amyloid* 12:1–4 (2005).
96. Prusiner, S. B. Novel proteinaceous infectious particles cause scrapie. *Science* 216:36–144 (1982).
97. Lundmark, K.; Westermark, G. T.; Nystrom, S.; Murphy, C. L.; Solomon, A.; and Westermark, P. Transmissibility of systemic amyloidosis by a prion-like mechanism. *Proc. Natl. Acad. Sci. USA* 99:6979–6984 (2002).
98. Meyer-Luehmann, M.; Coomaraswamy, J.; Bolmont, T.; Kaeser, S.; Schaefer, C.; Kilger, E.; Neuenschwander, A., et al. Exogenous induction of cerebral beta-amyloidogenesis is governed by agent and host. *Science* 313:1781–1784 (2006).
99. Lue, L. F.; Kuo, Y. M.; Roher, A. E.; Brachova, L.; Shen, Y.; Sue, L.; Beach, T., et al. Soluble amyloid beta peptide concentration as a predictor of synaptic change in Alzheimer's disease. *Am. J. Pathol.* 155:853–862 (1999).

100. Nilsberth, C.; Westlind-Danielsson, A.; Eckman, C. B.; Condron, M. M.; Axelman, K.; Forsell, C.; Stenh, C., et al. The 'Arctic' APP mutation (E693G) causes Alzheimer's disease by enhanced Abeta protofibril formation. *Nat. Neurosci.* 4:887–893 (2001).
101. Silveira, J. R.; Raymond, G. J.; Hughson, A. G.; Race, R. E.; Sim, V. L.; Hayes, S. F.; and Caughey, B. The most infectious prion protein particles. *Nature* 437:257–261 (2005).
102. Nilsson, M. R. Techniques to study amyloid fibril formation in vitro. *Methods* 34:151–160 (2004).
103. Serpell, L. C.; Sunde, M.; and Blake, C. C. The molecular basis of amyloidosis. *Cell. Mol. Life. Sci.* 53:871–887 (1997).
104. Harper, J. D.; Lieber, C. M.; and Lansbury, P. T. Atomic force microscopic imaging of seeded fibril formation and fibril branching by the Alzheimer's disease amyloid-beta protein. *Chem. Biol.* 4:951–959 (1997).
105. Petkova, A. T.; Ishii, Y.; Balbach, J. J.; Antzutkin, O. N.; Leapman, R. D.; Delaglio, F.; and Tycko, R. A structural model for Alzheimer's beta-amyloid fibrils based on experimental constraints from solid state NMR. *Proc. Natl. Acad. Sci. USA* 99:16742–16747 (2002).
106. Ritter, C.; Maddelein, M. L.; Siemer, A. B.; Lührs, T.; Ernst, M.; Meier, B. H.; Saupe, S. J., et al. Correlation of structural elements and infectivity of the HET-s prion. *Nature* 435:844–848 (2005).
107. Makin, O. S.; Atkins, E.; Sikorski, P.; Johansson, J.; and Serpell, L. C. Molecular basis for amyloid fibril formation and stability. *Proc. Natl. Acad. Sci. USA* 102:315–320 (2005).
108. Nilsson, K. P. R.; Herland, A.; Hammarström, P.; and Inganäs, O. Conjugated polyelectrolytes—conformation sensitive optical probes for detection of amyloid fibril formation. *Biochemistry* 44:3718–3724 (2005).
109. Herland, A.; Björk, P.; Nilsson, K. P. R.; Olsson, J. D. M.; Åsberg, P.; Konradsson, P.; Hammarström, P., et al. Electroactive luminescent self-assembled bio-organic nanowires: Integration of semiconducting oligoelectrolytes within amyloidogenic proteins. *Adv. Mat.* 17:1466–1471 (2005).
110. Nilsson, K. P. R.; Hammarström, P.; Ahlgren, F.; Herland, A.; Schnell, E. A.; Lindgren, M.; Westermark, G. T., et al. Conjugated polyelectrolytes–conformation-sensitive optical probes for staining and characterization of amyloid deposits. *Chembiochem.* 7:1096–1104 (2006).
111. Nilsson, K. P. R.; Åslund, A.; Berg, I.; Nyström, S.; Konradsson, P.; Herland, A.; Inganäs, O., et al. Imaging distinct conformational states of amyloid-beta fibrils in Alzheimer's disease using novel luminescent probes. *ACS Chem. Biol.* 2:553–560 (2007).
112. Sigurdson, C. J.; Nilsson, K. P. R.; Hornemann, S.; Manco, G.; Polymenidou, M.; Schwarz, P.; Leclerc, M., et al. Prion strain discrimination using luminescent conjugated polymers. *Nat. Methods.* 4:1023–1030 (2007).
113. Stabo-Eeg, F.; Lindgren, M.; Nilsson, K. P. R.; Inganäs, O.; and Hammarström, P. Quantum efficiency and two-photon absorption cross-section of conjugated polyelectrolytes used for protein conformation measurements with applications on amyloid structures. *Chem. Phys.* 336:121–126 (2007).

CHAPTER 10

Electrophoretically Deposited Polymers for Organic Electronics

Chetna Dhand and B. D. Malhotra

Department of Science and Technology Centre on Biomolecular Electronics National Physical Laboratory, New Delhi, India

10.1 Introduction

The phenomenon of electrophoretic deposition (EPD) has been known since 1808 when an electric field-induced movement of clay particles in water was observed by the Russian scientist Ruess. However, the first practical use of the electrophoretic technique occurred in 1933 when thoria particles deposited on a platinum cathode were used as an emitter for electron tube application. This process is being industrially used for applying coatings to metal fabricated products.[1] EPD is a colloidal process wherein the materials are shaped directly from a colloidal suspension upon application of a DC electric field. All colloid particles that can carry a charge can be used in EPD. This includes materials such as polymers, pigments, dyes, ceramics, and metals. With regard to the technological application, the potential of EPD as a material processing technique is increasingly being recognized by scientists and technologists. In addition to its conventional applications in the fabrication of wear-resistant and antioxidant ceramic coatings, it is used for fabrication of films for advanced microelectronic devices and solid oxide fuel cells, as well as in the development of novel composites or bioactive coatings for medical implants. There is increased interest for application of these materials in nanoscale assembly for advanced functional materials.[2]

Organic electronics[3] is a branch of electronics that deals with (semi) conductive polymers, plastics, or small molecules. The field of organic electronic devices is characterized by fast-paced progress in both efficiency and device function. In addition to the prototypical devices, such as organic light-emitting diodes (OLEDs), organic thin-film transistors (OTFTs), and photovoltaic cells (OPVCs), many other applications, e.g., sensors, memory cells, and light-emitting transistors, have been demonstrated. The most intriguing benefits of using organic materials include mechanical flexibility and light weight, which makes this type of device most attractive for mobile applications or "smart clothing." Moreover, entirely new design concepts for consumer electronics have emerged, as organic electronic devices can be adapted to follow complex surface shapes. For the past 40 years, inorganic silicon and gallium arsenide semiconductors, silicon dioxide insulators, and metals such as aluminum and copper have been the backbone of the semiconductor industry. However, there is a growing research effort in organic electronics to improve the semiconducting, conducting, and light-emitting properties of organics (polymers, oligomers) and hybrids (organic-inorganic composites) through novel synthesis and self-assembly techniques. Performance improvements, coupled with the ability to process these "active" materials at low temperatures over large areas on materials such as plastic or paper, may provide unique technologies, generate new applications, and provide opportunities to address the growing needs for pervasive computing.

The organic materials are often divided into two classes: small-molecular materials and polymers. The fundamental properties of both these classes are essentially the same, and the division mainly relates to the way thin films are prepared. Small molecules are typically thermally evaporated in vacuum whereas polymers are processed from solution. Yet, most small-molecular materials are soluble as well, or solubility can be increased by the synthetic addition of side chains. Polymers have been utilized for fabrication of organic electronic devices for decades. A report in 1977 on doped polyacetylene to achieve relatively high conductivity opened up important new vistas for physics and chemistry and for the technology in general. Early studies suggested that a key feature of electronic polymers is a backbone consisting of alternate single and double bonds resulting in a π-conjugated network. This in turn leads to a small energy gap, enabling the appearance of both semiconducting and metallic properties.[4] Conductive polymers are lighter, more flexible, and less expensive than inorganic conductors. This makes them a desirable alternative for many applications. It also creates the possibility of new applications that would otherwise be impossible using copper or silicon. Some of the other applications include smart windows and electronic paper. In this context, conductive polymers are likely to play an important role in the emerging science of molecular computers. In general, organic conductive polymers have a higher resistance and therefore have poor electrical conductivity compared to inorganic conductors.

The self-assembling or ordering of these organic and hybrid materials enhances the *p*-orbital overlap and is a key to improvements in carrier mobility. The recombination of charge carriers under an applied field can lead to the formation of an exciton that decays radiatively to produce light emission. For preparation of polymer thin films from solution, a number of techniques such as spin coating, sol-gels,[5] inkjet printing,[6] Langmuir Blodgett (LB) film deposition,[7] self-assembled technique,[8] EPD,[1] etc. have been utilized. The spin-coating technique, though cheap, requires excess of material solution as most of the solution gets blown away during film preparation. Also when thick films are required, this technique does not provide films of good uniformity and quality. The inkjet technique, though, gives a uniform pattern, but the production cost is quite high and it is a time-consuming process. The LB technique for film preparation requires the presence of amphiphilic groups in a material of interest, or the material of interest should be mixed with some amphiphilic materials for good-quality films. The use of an additional amphiphilic material hinders the materials of interest, decreases their surface exposure, and enhances the resistance of the film. And the stability of LB films is still a major problem. The use of the self-assembly technique for polymer layer formation is much restricted to the conditions required for self-assembled monolayer formation as it requires specific chemistry between the surface of the substrate and the functional group present in the polymer.

The EPD technique has recently gained much attention due to added advantages of high rate of deposition, controlled thickness, simple and easy method of fabrication, dense and uniform film preparation, etc.[9] Moreover, EPD from a colloidal suspension of desired conducting polymers provides the unique strategy to tailor nanostructured films[10–12] by a very simple method with unique morphology. Besides this, the use of EPD involves deposition from colloidal solution without any additives and thus ensures purity of the deposited polymer film, which is critical for the fabrication of optoelectronic and biosensor devices.

10.2 Electrophoretic Deposition

10.2.1 Definition

Electrophoretic deposition is a colloidal process in which charged powder particles, dispersed or suspended in a liquid medium, are attracted and deposited onto a conductive support of opposite charge on application of DC electric field. Despite being a wet process, EPD provides easy control over the film thickness and morphology by simple adjustment of deposition time and applied voltage. The term

Property	Electroplating	Electrophoretic Deposition
Moving species	Ions	Solid particles
Charge transfer on deposition	Ion reduction	None
Required conductance of liquid medium	High	Low
Preferred liquid	Water	Organic

Table 10.1 Comparison between EPD and Electroplating

electrodeposition is often ambiguously used to refer to either electroplating or EPD, although it more often refers to the former. Table 10.1 presents the difference between the two processes.

EPD can be of two types depending upon the electrode at which the deposition occurs. When the particles are positively charged, then the deposition occurs on the cathode (negatively biased electrode) and the process is called as cathodic EPD or cataphoresis. Deposition of negatively charged particles on the anode (positively biased electrode) is known as anodic EPD or anaphoresis. Either of the two modes of deposition can be used by suitable modification of the surface charge on the particles. Figure 10.1 shows a schematic of the EPD process. The arrow (Fig. 10.1) indicates the direction of movement of colloid particles.

10.2.2 Principle of EPD

EPD involves two processes: one is electrophoresis and the other is deposition. Electrophoresis is the phenomenon of motion of particles in a colloidal solution or suspension toward one of the electrodes under the influence of an electric field. It generally occurs when the distance over which the double-layer charge falls to zero is large compared to

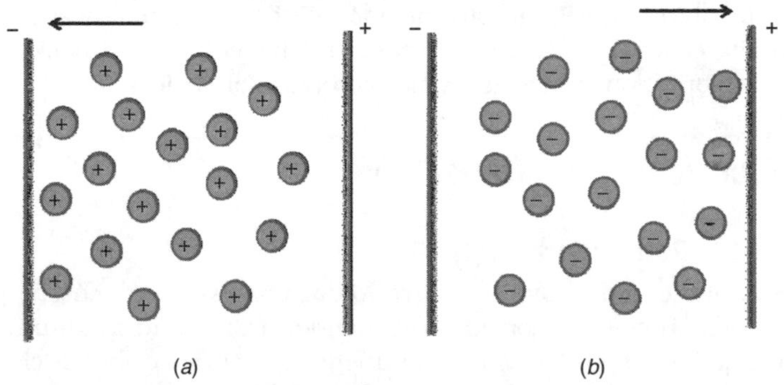

Figure 10.1 Schematic illustration of electrophoretic deposition process: (*a*) cathodic EPD and (*b*) anodic EPD.

the particle size. Under such a condition when the electric field is applied, the particles move relative to the liquid phase. In the second step (deposition), the particles are collected at one of the electrodes and form a coherent deposit on it. The deposit takes the shape imposed by the electrode. Zhitomirsky[13] has proposed a mechanism to explain the deposition of material on the electrode. The mechanism of deposition can be divided into three different categories depending upon the type of the material to be deposited and the nature of the suspension used for the deposition These include charge neutralization, zeta-potential lowering or electrochemical coagulation, and particle accumulation.

Particle Charge Neutralization

According to this concept, the particles undergo charge neutralization as they reach the electrode surface and become static.[14] This mechanism is important for single particles and monolayer deposits. It explains deposition of powders that get charged upon salt addition to the suspension, e.g., the deposition of aluminum.[15] But the limitation of this concept is that it explains the initial stage deposition from very dilute suspensions, but it is invalid under certain conditions such as (1) when EPD is performed for a longer time (thick deposits), (2) when particle-electrode processes are prevented (e.g., a semipermeable membrane induces deposition between the electrodes), and (3) when reactions occur at the electrode which alter the local pH.

Electrochemical Coagulation of Particles

This mechanism is based on the coagulation of the particles due to the reduction of repulsive forces between the particles near the electrode surface. The increase in electrolyte concentration around the particles near the depositing electrode lowers the zeta potential and thus induces flocculation.[1] But this mechanism is possible only when the deposition is in the aqueous phase where electrode reactions generate OH^- ions and is invalid when there is no increase of electrolyte concentration near the electrode. For such cases, Sarkar and Nicholson[16] gave an explanation by considering a positively charged oxide particle/lyosphere system moving toward the cathode in an EPD cell. According to this concept, the applied electric field and the fluid dynamics distort the double-layer envelope around the particle in such a manner that it will become wider behind and narrow in front. Moreover, the counter-ions on the wider end start reacting with the cations drifting toward the cathode and result in the thinning of the double layer. Under such conditions, the next particle with the thin, leading-edge double layer can now approach close enough for London van der Waals attractive forces to dominate and induce coagulation/deposition. The schematic of this mechanism is shown in Fig. 10.2.

This mechanism is, however, invalid when there is excessive concentration of cations near the cathode. Under such conditions, there is

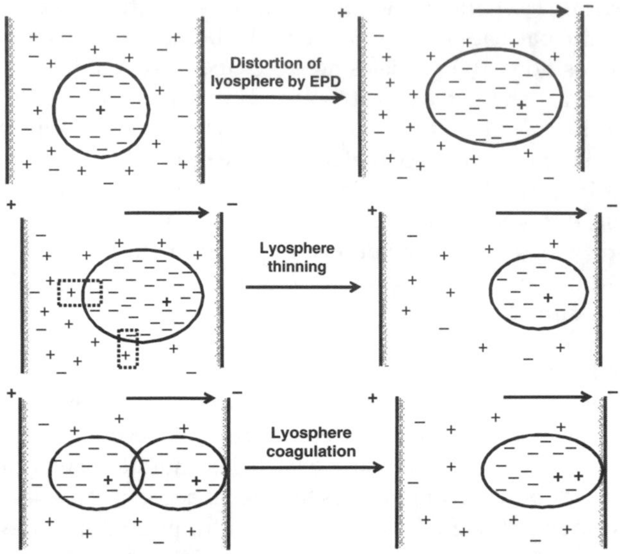

Figure 10.2 Schematic showing the deposition mechanism that involves lyosphere distortion, lyosphere thinning, and lyosphere coagulation.

depletion of H^+ by discharge at the cathode. This depletion results in an increase in the local pH toward the isoelectric point (pH_{iep}) and facilitates coagulation. The discharge of H^+ occurs as

$$H^+_{x=\infty} \xrightarrow{\text{Electrophoresis}} H^+_{x=0} + e^- \xrightarrow{\text{Charge transfer}} \tfrac{1}{2}H_2O \qquad (10.1)$$

When charge transfer at the electrode consumes H^+, $[H^+]$ ion concentration at the electrode/solution interface drops below the bulk value, creating a concentration gradient thereof. The concentration of H^+ as a function of distance and time is obtained by solving the classical diffusion equation with boundary conditions:

$$(J_{c,\text{total}})_{x=o} = -D_{\text{eff}}\left(\frac{\partial C}{\partial x}\right)_{x=0} \qquad (10.2)$$

$$= \lambda = \frac{I}{Z_c F} \qquad \text{at } x = 0 \qquad (10.3)$$

where I = current density
λ = constant
J = flux
D_{eff} = $D_c D_a (Z_a + Z_c)/(Z_c D_c + Z_a D_a)$, here subscripts c and a indicate cations and anions, respectively
Z = valency

Flocculation by Particle Accumulation

According to Hamaker and Verwey,[17] EPD is akin to sedimentation, and the primary function of the applied field is to move the particles toward the electrode. Accumulated particles then deposit due to the pressure exerted by those incoming and in the outer layers. This mechanism is feasible when deposition does not occur at the electrode, e.g., deposition on a dialysis membrane. It explains deposits on porous membranes that are not electrodes.

10.2.3 Theory of EPD

A successful EPD relies on the formation of well-stabilized, unagglomerated, and homogeneous colloidal suspension. A distinguishing feature of a colloidal system is that the contact area between the particle surface and the dispersing liquid is large. As a result, the interparticle forces strongly influence the suspension behavior. The dominating interparticle forces in most colloidal systems are (1) the van der Waals attractive force, (2) double-layer (electrostatic) repulsive force, and (3) steric (polymeric) forces. To obtain a well-stabilized suspension, particles dispersed in the suspending medium must exhibit sufficient repulsive forces to offset the van der Waals attraction. These forces are better understood in terms of electrical double layer and their interactions are discussed in the next section.

The Electrical Double Layer and Electrophoretic Mobility

EPD relies on the capability of powder to acquire an electric charge in the liquid in which it is dispersed. In general, when solid powder is dispersed in a polar liquid such as water, usually it results in the buildup of a charge at the solid-liquid interface. Sarkar and Nicholson[16] inserted a dialysis membrane between EPD electrodes in an Al_2O_3 suspension. The membrane is permeable to ions, but a dense deposit is formed thereon and the current is passed via ionic discharge at the cathode. They concluded that the majority of the charge is carried by ions that result in the passage of current. It is now well recognized that development of the electric charge on colloid particles dispersed in water is due to (1) surface group ionization (controlled by the pH of the dispersion media), (2) differential solubility of ions (e.g., silver iodide crystals are sparingly soluble in water and silver ions dissolve preferentially to leave a negatively charged surface), (3) isomorphous replacement/lattice substitution (e.g., in kaolinite, Si^{4+} is replaced by Al^{3+} to give negative charges), (4) charged crystal surface fracturing (crystals can reveal surfaces with different properties), and (5) specific ion adsorption (surfactant ions may be specifically adsorbed). This surface charge influences the distribution of nearby ions in the polar medium. The ions, which establish the surface charge, are termed *potential determining ions* (PDIs). These normally include ions of which the solid is composed; hydrogen and hydroxyl ions; and ions capable of forming complex or insoluble salts with the solid surface species. Ions of opposite charge (counter-ions) are attracted toward the surface, and ions of like charge (co-ions) are repelled from the surface. This leads to the formation of a net electric charge of one sign on one side of the interface and a charge of opposite sign on the other side, giving rise to what is called the *electrical double layer*.

10.2.4 Parameters Influencing EPD

There are two groups of parameters that determine characteristics of the electrophoretic process; one is related to the nature of the suspension and the other to the physical parameters, such as the electrical nature of the electrode and the amount and time for which the voltage is applied for the deposition. Hamaker[18] and Avgustnik et al.[19] were the first to describe the correlation between the amount of material deposited during EPD and the different influencing parameters. Hamaker's law relates the deposit yield (w) to the electric field strength E, the electrophoretic mobility μ, the surface area of the electrode A, and the particle mass concentration in the suspension C through the following equation:

$$w = \int_{t_1}^{t_2} \mu E A C \, dt \tag{10.4}$$

Avgustinik's law is based on cylindrical, co-axial electrodes, and the electrophoretic mobility has been expanded and is represented in terms of permittivity ε, the zeta potential ξ, and the viscosity of the suspension η:

$$w = \frac{l E \varepsilon \xi C t}{3 \ln(a/b) \eta} \tag{10.5}$$

where l = length of the deposition electrode
a = radius of the deposition electrode
b = radius of the co-axial counter electrode ($b > a$)

Biesheuval and Verweij[20] have considered three different phases during the deposition process, namely, solid phase, suspension phase, and phase having negligible or no solid particles. Both the deposit phase and the particle-free liquid phase grow at the expense of the suspension phase. By considering the movement of the boundary between the deposit and the suspension phases with time along with the continuity equation and expression for velocity of particles in the suspension, they[20] derived the following equation based on that of Avgustinik et al.[19]:

$$w = \frac{2\pi \mu l E C_d}{\ln(a/b)} \cdot \frac{\phi_s}{\phi_d - \phi_s} t \tag{10.6}$$

where ϕ_s = volumetric concentration of particles in suspension
ϕ_d = volumetric concentration of particles in deposit
C_d = mass concentration of particles in the deposit
μ = electrophoretic mobility = $\varepsilon\xi/6\pi\eta$

Ishihara et al.[21] and Chen et al.[22] used the following equation for the weight (w) of charged particles deposited per unit area of electrode, ignoring the charge carried by the free ions in the suspension

$$w = \frac{2}{3} C \varepsilon_0 \varepsilon_r \xi \left(\frac{1}{\eta}\right)\left(\frac{E}{L}\right) t \tag{10.7}$$

where C = concentration of the particles
ε_0 = permittivity of vacuum
ε_r = relative permittivity of the solvent
ξ = zeta potential of the particles
η = viscosity of the solvent
E = applied potential
L = distance between the electrodes
t = deposition time

Parameters Related to the Suspension

Size of the particle Successful EPD requires the development of a stable suspension composed of electrostatically charged particles homogeneously distributed in a suitable solvent. For larger particles, the main problem is that they tend to settle due to gravity. Under such conditions it is really difficult to get uniform deposition from sedimenting suspension of large particles. This will lead to a gradient in deposition, i.e., thinner deposit above and thicker deposit at the bottom, when the deposition electrode is placed vertical. Thus for the EPD of larger particles either a strong surface charge must exist, or the electrical double-layer region must increase in size. Particle size has also been found to have a prominent influence on controlling the cracking of the deposit during drying. Sato et al.[23] have investigated the effect of the size of $YBa_2Cu_3O_{7-\delta}$ particle on crack formation. The results reveal relatively less cracking in films deposited from the suspension consisting of smaller particles (0.06 μm) than the larger ones (3 μm).

Dielectric Constant of Liquid Powers[24] has investigated the effect of dielectric constant of the liquid on the EPD technique. Deposits could be obtained only with liquids for which the dielectric constant is in the range of 12 to 25. With too low a value of the dielectric constant, deposition does not occur because of insufficient dissociative power, while with a high dielectric constant, the high ionic concentration in the liquid reduces the size of the double-layer region and consequently the electrophoretic mobility. Thus, the ionic concentration of the liquid must remain low, a condition favored in liquids of low dielectric constant. Table 10.2 gives the values of dielectric constants of various organic solvents that can be used for EPD.

Solvent	Relative Dielectric Constant
Methanol	32.63
Ethanol	24.55
n-Propanol	20.33
Isopropanol	19.92
n-Butanol	17.51
Acetone	20.7
Acetylacetone	25.7
Acetonitrile	37.5

TABLE 10.2 Relative Dielectric Constants of Organic Solvents Used for EPD

Conductivity of Suspension Ferrari and Moreno[25] have proposed that conductivity of the suspension is a key factor and needs to be taken into account in EPD experiments. It has been pointed out that if the suspension is too conductive, particle motion is very low, and if the suspension is too resistive, the particles charge electronically and the stability is lost. They have found that the conductivity values are not useful for EPD, and there is only a narrow band of conductivity range at varying dispersant dosage and temperature in which the deposit is formed. This suitable region of conductivity is, however, expected to be different for different systems. The margin of conductivity region suitable for EPD can be increased by the applied current, ensuring success of the EPD process.[26]

Zeta Potential (ξ) Zeta potential measures the potential difference between the particle surface and the shear layer plane formed by the adsorbed ions and is a key factor in the EPD process. The value of the zeta potential is related to the stability of the colloidal dispersion. The zeta potential indicates the degree of repulsion between adjacent, similarly charged particles in dispersion. For molecules and particles that are small enough, and of low enough density to remain in suspension, a high zeta potential confers stability; i.e., the solution or dispersion resists aggregation. When the potential is low, attraction exceeds repulsion and the dispersion is likely to break and flocculate. And the colloids with high zeta potential (negative or positive) are electrically stabilized while colloids with low zeta potential tend to coagulate or flocculate as outlined in Table 10.3. Besides this, it plays an important role in determining the direction and migration velocity of particles during EPD and the green density of the deposit.

During EPD, particles come closer to one another with increasing attractive forces. Under such conditions if the particle charge is low,

Zeta Potential (mV)	Stability Behavior of Colloid
From 0 to ±5	Rapid coagulation or flocculation
From ±10 to ±30	Incipient instability
From ±30 to ±40	Moderate stability
From ±40 to ±60	Good stability
More than ±61	Excellent stability

TABLE 10.3 Effect of Zeta Potential on Colloid Stability

the particles coagulate even for relatively large interparticle distances, leading to porous, sponge-like deposits. On the contrary, if the particles have a high surface charge during deposition, they repel one another, occupying positions that may lead to high particle packing density.[27] It is therefore important to control the loading of desired material and concentration of solvents and additives in the EPD suspension in order to reach the highest possible green density of the deposit. The zeta potential can be controlled by a variety of charging agents such as acids, bases, and specifically adsorbed ions or polyelectrolytes, added to the suspension.[28]

There are a variety of additives that affect the magnitude of charge and its polarity. These additives act by different mechanisms. The main criteria for selection of a charging agent are the preferred polarity and the deposition rate of the particles. Chen et al.[29] have found that stability and the deposition rate of alumina from its alcoholic suspension are maximal at pH 2.2. At this pH there is a maximum positive zeta potential of alumina, but with the increase in the pH of the suspension the stability starts decreasing. This can be explained on the basis of the charging mechanism proposed by Wang et al.[30]

$$AlOH_2^+ \xleftarrow{H^+} AlOH \xrightarrow{OH^-} AlO^- + H_2O \tag{10.8}$$

Under basic conditions such as pH 11, AlOH tends to form AlO^-. However, in the presence of water it results in the formation of $AlOH_2^+$, and consequently there is lowering of the zeta potential at higher pH values than at pH 2. This may lead to high stability of the suspension at lower pH than at higher pH. Ma et al.[31] have demonstrated the effect of polymer additives on the zeta potential of the colloidal suspension, which is a measure of the colloid dispersion stability via the interaction strength of the colloid particles, and hence relate it with the stability of PZT ($PbZr_{0.52}Ti_{0.48}O_3$) colloidal suspension.

Stability of Suspension Electrophoresis is the phenomenon of motion of particles in a colloidal solution or suspension in an electric field, and it generally occurs when the distance over which the double-layer charge falls to zero is large compared to the particle size. In this condition, the particles will move relative to the liquid phase when the electric field is applied. Colloidal particles that are 1 μm or less in diameter tend to remain in suspension for long periods due to brownian motion. Particles larger than 1 μm require continuous hydrodynamic agitation to remain in suspension. The suspension stability is characterized by the settling rate and the tendency to undergo or avoid flocculation. Stable suspensions show no tendency to flocculate, settle slowly, and form dense and strongly adhering deposits at the bottom of the container. Flocculating suspensions settle rapidly and form low-density, weakly adhering deposits. If the suspension is too stable, the repulsive forces between the particles cannot be overcome by the electric field, and deposition does not occur. According to a model for EPD, the suspension should be unstable in the vicinity of the electrodes.[32] This local instability can be caused by the formation of ions from electrolysis or discharge of the particles; these ions then cause flocculation close to the electrode surface. It is desirable to find suitable physical/chemical parameters that characterize a suspension sufficiently so that its ability to deposit can be predicted.

Parameters Related to the EPD Process

Effect of Deposition Time At constant voltage, initially there is a linear relationship between the amount of material deposited and the deposition time, but later the deposition rate decreases with increased or prolonged deposition time. This is expected because while the potential difference between the electrodes is maintained constant, the electric field influencing electrophoresis decreases with deposition time due to formation of an insulating layer of ceramic particles on the electrode surface.[33]

Applied Voltage Normally the amount of deposit increases with increase in applied potential. Basu et al.[34] have found that there is more uniform deposition of films at moderate applied fields (25 to 100 V/cm) and the quality of the films deteriorates if relatively higher applied fields (>100 V/cm) are used. Since the formation of a particulate film on the electrode is a kinetic phenomenon, the accumulation rate of the particles influences their packing behavior in the coating. A higher applied field may cause turbulence in the suspension; the coating may be disturbed by flow in the surrounding medium, even during its deposition. In addition, particles can move so fast that they cannot find enough time to settle in their best positions to form a close-packed structure. Negishi et al.[35] have observed that the current density of *n*-propanol solvent in the absence of any powder is proportional to the applied voltage, and it tends to become unstable with increasing applied voltages (Fig. 10.3). Such stability data serve as a good guideline for deciding the deposition parameters and consequently the quality of

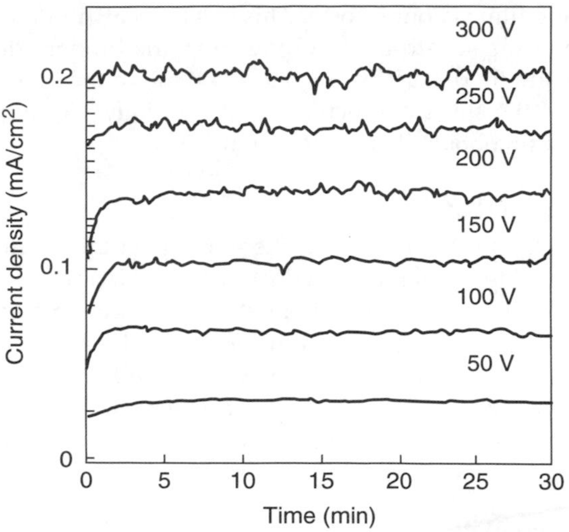

Figure 10.3 Stability of current density of *n*-propanol for different applied voltages. (*Reprinted from Ref. 1. Copyright 2007, with permission from Elsevier.*)

deposit formed by EPD. It is considered that the unstable current density influences the quality of deposition morphology. From the current density profile in Fig. 10.3, it is reasonable to suggest that the applied voltage should be less than 100 V in the case of *n*-propanol. It is observed that amount of YSZ (Y_2O_3-stabilized ZrO_2) deposition from the *n*-propanol bath increases with increasing applied voltage. However, the deposit surface morphologies are found to be flat at low voltages, and they became rougher with increasing applied voltage.

Concentration of Solid in Suspension Predominantly for multicomponent EPD, the volume fraction of the solid in the suspension plays an important role. In some cases, although each of the particle species has the same sign of surface charge, they could be deposited at different rates depending on the volume fraction of solids in the suspension. If the volume fraction of solids is high, the powders deposit at an equal rate. If, however, the volume fraction of solids is low, the particles can deposit at rates proportional to their individual electrophoretic mobility.[36]

Conductivity of Substrate The uniformity and conductivity of a substrate electrode are important parameters critical to the quality of the deposition of green films by EPD. Peng and Liu[37] have observed that low conductivity of the $La_{0.9}Sr_{0.1}MnO_3$ (LSM) substrate leads to nonuniform green film and slow deposition. Chen and Liu[22] have observed that when as-pressed LSM or LSM-YSZ composite pellets are used as substrates for EPD, the deposition rate of YSZ is slow and the resulting film is nonuniform. This has been attributed to the high resistance of the substrates resulting from the binder added. When the pellets are fired at 700°C for 0.5 h to remove the binder, the conductivity of the substrate increases substantially. Consequently, the green YSZ film obtained is of high quality.

10.2.5 Materials for EPD

The EPD technique has been used successfully for thick film of silica,[7, 38] nanosized zeolite membrane,[39] hydroxyapatite (HA) coating on a metal substrate for biomedical applications,[40–41] luminescent materials,[42] high-T_c superconducting films,[43] gas diffusion electrodes and sensors,[9] glass and ceramic matrix composites,[18] oxide nanorods,[44] carbon nanotube films,[45–48] layered ceramics,[49] piezoelectric materials, [50] etc. On the basis of the type of material used for electrophoretic deposition, these materials can be categorized as inorganic or organic.

Inorganic Materials

In the class of inorganic materials, EPD has been explored much for the processing and fabrication of a wide variety of ceramic materials. The EPD of ceramics was first studied by Hamaker[18] in 1940, and only in the 1980s did the process receive attention in the field of advanced ceramics. Abdollahi et al.[39] have applied the electrophoretic technique as the seeding method for the formation of zeolite ZSM-5 layers in order to achieve thin defect-free membranes with appropriate orientation. Using this technique, an oriented continuous layer of nanosized zeolite seeds is formed on the support, and the seeds act as nuclei for the next step, which is crystal growth under hydrothermal situation. Braun et al.[51] have reported the fabrication of transparent, homogeneous polycrystalline alumina ceramic with submicron microstructure by means of EPD. Recently, Besra et al.[52] have investigated EPD of YSZ particles from their suspension in acetylacetone onto a nonconducting NiO-YSZ substrate for solid oxide fuel cell applications. In principle, it is not possible to carry out EPD on nonconducting substrates. In this case, the EPD of YSZ particles on a NiO-YSZ substrate is made possible through the use of an adequately porous substrate. The continuous pores in the substrates, when saturated with the solvent, helped in establishing a "conductive path" between the electrode and the particles in suspension. Deposition rate is found to increase with increasing substrate porosity up to a certain value. EPD technique has also been explored for the nanocrystalline oxide coatings, e.g., MnO_2,[53] on the desired substrate.

Antonelli et al.[49] have recently reported the fabrication of dense, crack-free, and homogeneous thick films of BCT23 ($Ba_{0.77}Ca_{0.23}TiO_3$) by the electrophoretic technique. The BCT ceramic material is a ferroelectric material and has been reported as a promising multi-layer ceramic capacitor (MLCC). Trau et al.[54] and Bohmer[55] simultaneously reported the monolayer film formation of micron-sized latex particles by the electrophoretic technique. Highly ordered structures such as three-dimensional colloidal crystals (opals)[56–58] and ordered two-dimensional films of a binary mixture of colloids[59] have been described in the literature. The driving force for self-assembly in these systems has been investigated[54, 56, 59–61] and is generally ascribed to electrohydrodynamic flow.[62–63] There are less-frequent reports of electrophoretically deposited films of particles with dimensions in the low nanometer range. Bailey et al. have shown unordered films deposited by means of electrophoresis on substrates that were prepatterned using micro contact printing.[64] Another example is from Gao et al.,[65] who deposited CdTe nanoparticles on prepatterned indium tin oxide (ITO) electrodes. Inverse opals made by EPD of small particles into voids left between an ordered multilayer films of large particles have been shown by Gu et al.[58]

Generally, the assembled nanoparticulate films do not exhibit significant ordering. A number of factors make deposition of ordered films of nanoparticles and their investigation more challenging, as opposed to their micron-sized counterparts. First, for aqueous suspensions of nanocolloids, the thickness of the double layer is often comparable to the particle size, giving rise to considerable interparticle repulsion. Second, brownian motion for small particles is more important than for larger particles, thereby inhibiting ordering of the film.[66] Furthermore, electroosmotic flow arising from electrophoretic motion of ions in the aqueous liquid near charged surfaces, including the substrate, interferes with the well-defined motion of the charged particles. Kooij et al.[67] have reported nanocolloidal gold particle deposition from an aqueous benzoate/benzoic acid solution at metal-coated glass substrates, in the presence of an externally applied electric field. The spatial distribution of nanoparticles deposited in an applied field exhibits a higher degree of order compared to the random, irreversibly deposited nanocolloids at chemically functionalized surfaces. They have also explained electrohydrodynamic forces and capillary forces as the deriving forces for the ordering of the nanoparticles. Figure 10.4 shows the SEM images of nanocolloidal gold films deposited in the presence of external electric field.

EPD has been used to synthesize nickel-alumina, functionally graded materials from NiO, and alumina suspensions in ethanol by Nagarajan et al.[68] Functionally graded materials (FGMs) are composites with gradual transition of microstructure and/or composition. The importance of the gradual transition is to increase the strength of the bond between composites of dissimilar materials; e.g., ceramic/metal interfacial coherence can be increased by continuous gradation rather than a sharp discontinuity. Sarkar et al.[69] first obtained the EPD of functionally graded materials. Milczarek and Ciszewski[70] have used the electrophoretic technique for the easy and rapid deposition of Ni(II) and Co(II) phthalocyanines. Metal phthalocyanines (MPcs) are macrocyclic complexes and have been known to be excellent

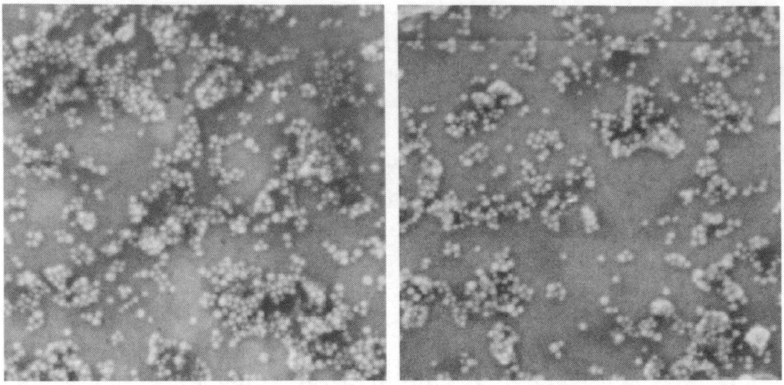

FIGURE 10.4 Scanning electron microscopy images of nanocolloidal gold films deposited in the presence of an external electric field. Depicted is the morphology after the deposition experiments shown of left and of right. The coverages amount to 41 and 9%, respectively. (*Reprinted from Ref. 67. Copyright 2007, with permission from Elsevier.*)

catalysts for many chemical reactions and also have application in the design of solid-state photovoltaic cells. Wang and Hu[40] have explored the use of EPD technique in patterning of HA[$Ca_{10}(PO_3)_6(OH)_2$, bioceramic] through the use of patterned metallic cathodes. Patterned bioceramic coatings of HA may find potential applications in orthopedic implants and biosensors. Bailey et al.[71] have reported a convenient approach for controlled fabrication of ultra-thin micropatterned colloidal gold films on conductive ITO-coated glass platforms using a combination of microtransfer molding and EPD techniques. These films readily diffract visible light and may prove useful as chemo- and electrochemically responsive optical diffraction gratings.

Limmer and Cao[44] have demonstrated a new method for the growth of oxide nanorods by combining sol-gel processing with EPD. They have named this technique sol gel electrophoresis, and this method has been reported to grow the nanorods of both single metal oxides (TiO_2, SiO_2) and complex oxides [$BaTiO_3$, $Sr_2Nb_2O_7$, and $Pb(Zr_{0.52}Ti_{0.48})O_3$] within the diameter range of 45 to 200 nm. The higher surface area of nanorods and their relatively shorter conduction path should combine to make solar cells that are more efficient. Another application is the use of the higher surface area of nanorods for sensors, detectors, and catalysts. Patterned, ordered arrays of unidirectionally aligned nanorods could serve as the foundation of two-dimensional photonic band gap crystals.

Organic Materials

EPD has been gaining increasing interest as an economical and versatile processing technique for the production of novel coatings or films of organic materials such as carbon nanotubes (CNTs), polymers, etc. on conductive substrates.

EPD of Carbon Nanotubes Since the discovery of CNT in 1991 by Iijima,[72] CNTs have been looked at extensively by researchers in various fields such as chemistry, physics, materials science, and electrical engineering. CNTs are unique nanostructured materials with remarkable physical and mechanical properties. Many of the remarkable properties of CNTs are now well established, and their exploitation in a wide range of applications forms a major part of current research and development efforts.[73] One of the challenges is to tackle the problem of manipulating CNTs, individually or collectively, to produce a particular arrangement needed for a given application. One very promising technique being developed for manipulating CNTs is EPD. For successful EPD, preparation of a stable dispersion of CNTs in a suitable solvent is necessary. The most common strategy is the production of an electrostatically stabilized dispersion, which in general requires the preparation of a solvent medium in which the particles have a high ξ potential, while keeping the ionic conductivity of the suspensions low. The stability of CNT suspensions, determined by ξ-potential measurements, has been studied mainly in aqueous and ethanol-based suspensions.[74]

The earliest investigations appear to be those of Du et al.,[45] who explored the possibility of using EPD to deposit multiwalled CNT (MWCNT) from ethanol/acetone suspensions on metallic substrates. They observed strong hydrogen evolution at the cathode, leading to a porous film of nanotubes with pore sizes ranging from 1 to 70 μm. Thomas et al.[46] have successfully deposited homogeneous MWCNT films onto stainless-steel substrates using EPD from aqueous suspensions of acid-oxidized nanotubes. No hydrogen evolution is observed during this deposition. This result contrasts with that of Du et al.[45] and it may be attributed to the lower electric field strength used by Thomas et al.[46] Du and Pan[75] have electrophoretically fabricated thin films of MWCNT using $(Mg(NO_3)_2)6H_2O$ as electrolyte, and they also reveal the application of these electrodes as super capacitors. These MWCNT electrodes exhibit a significantly small ESR and a high specific power density. The super capacitors also show superior frequency response, with a frequency "knee" more than 70 times higher than the highest reported knee frequency for super capacitors. In addition, this carbon nanotube thin film can act as a coating over an ordinary current collector to decrease the contact resistance between the active materials and the current collector for improved performance. EPD has been used to a limited extent to deposit single-wall CNTs (SWCNTs).[49] One report describes the production of SWCNT deposits from very dilute SWCNT suspensions in ethanol (1 mg SWCNT in 200 mL ethanol) after the addition of a suitable salt ($MgCl_2$).[50] Other solvents investigated for SWCNT deposition include dimethylformamide (DMF) and mixtures of distilled water and methanol.[76] Films containing long SWCNT bundles are obtained. Andrade et al.[77] have compared different techniques such as dip-coating, filtration, spray coating, and EPD for CNT deposition. The result suggests that dip-coating and EPD provide the smoothest CNTs and may be an interesting option for solar cell applications, among others.

Girishkumar et al.[78] used EPD to deposit a thin film of SWCNT modified with tetraoctylammonium bromide (TOAB) in tetrahydrofuran (THF) on aminopropyltriethoxysilane (APS) coated, optically transparent electrodes (OTEs) made of conductive glass. The TOAB binds to the surface of the CNT during sonication by hydrophobic interactions of its alkyl chain thereby preventing aggregation and settling of nanotubes. The same group has reported the fabrication of a membrane electrode assembly for hydrogen fuel cells by using EPD to deposit a SWCNT support and a Pt catalyst on carbon fiber electrodes.[79] Both the electrophoretically deposited nanotubes and platinum retained their nanostructured morphology on the carbon fiber surface. Kurnosov et al.[80] have suggested introducing a resistive material on top of the conductive cathode to improve the adhesion of the CNTs to the substrate and the uniformity of the deposited film. A suspension of SWCNT in $NiCl_2$/isopropyl alcohol was deposited on an ITO-coated aluminum cathode. Oh et al.[48] have performed a similar experiment in which functionalized SWCNTs are stabilized in $MgCl_2$/ethanol and are deposited on ITO-coated glass. In both cases, the nanotubes strongly adhere to the ITO coating. The adhesion has been attributed to two factors. The first is the interaction between the hydrophilic CNT and ITO surface.[48, 81] The second is the presence of the charger salt, $MgCl_2$, since Mg^{2+} ions form hydroxides at the surface of the negative electrodes that assist the interfacial bonding.[81] Girishkumar et al.[78] have introduced APS coating on OTE surface to obtain more uniform electrophoretically deposited films of CNTs.

Kamat et al.[82] have extensively investigated the assembly of solubilized SWCNTs into linear bundles at a high DC voltage (>100 V) and their deposition on OTE at relatively low DC voltage (~50 V). Purified SWCNTs are solubilized by mixing with TOAB in THF. The SWCNT films of varying thickness are obtained by adjusting the deposition time. At high DC voltage of >100 V, the CNTs do not deposit and become aligned perpendicular to the two electrodes (parallel to the field). The influence of electrode separation has been investigated by Kurnosov et al.[80] They have found that the uniformity of field emission depended significantly on the electrode separation. The best uniformity is obtained at the lower end of the separations tested (0.3 to 1.8 cm). These authors have observed that for larger electrode separations, the emission sites are concentrated at the edges of the electrodes due to nonuniformity of the electric field.

Once a porous CNT coating or film is obtained, EPD can be employed to deposit ceramic or metallic nanoparticles with the aim of infiltrating the CNT structure, or producing a layered structure. Alternatively, composite CNT/nanoparticulate coatings can be obtained by co-EPD from stable suspensions containing two or more components. The various components may be separately dispersed, coming together only during EPD, or may be preassembled to form a more complex building block. These opportunities have yet to be investigated systematically, but some indicative promising results have been obtained. Singh et al. have reported homogeneous and thick deposits of CNTs that can be coated and infiltrated with TiO_2 nanoparticles obtained by co-EPD. Co-EPD is carried out at a constant electric field of 20 V/cm.[83] Due to the complementary surface charge of CNT (negatively charged) and TiO_2 nanoparticles (positively charged), the two components attract each other in aqueous suspensions at the selected pH. These forces result in the deposition of TiO_2 nanoparticles on the surface of individual CNTs. Similarly, CNT/SiO_2 nanoparticle composite films have been obtained by EPD from aqueous suspensions, as discussed elsewhere.[84] The deposit is a three-dimensional network of interwoven CNTs coated and infiltrated by the SiO_2 nanoparticles. This type of porous CNT/titania and CNT/silica nanostructures may be useful for nanoelectronic devices.[85] More straightforwardly, the coating and infiltration of porous CNT assemblies with nanoparticles can be seen as a useful step toward homogeneous incorporation of CNTs in hard, structural, and functional matrices.[85] Kaya[41] has recently investigated EPD for coating of MWCNT reinforced ultrafine (20 nm) hydroxyapatite powders with Ti-6Al-4V for biomedical applications, such as total hip replacement.

EPD of Polymers EPD is an important commercial method of applying films to irregularly shaped metal articles, and adaptation of this method to polymers could increase their utility significantly. Li et al.[86] have presented a convenient approach for the formation of polyaniline (PANI) colloids with a size of ~100 nm. It has been demonstrated that the polyaniline colloidal suspensions have excellent processability when applied electrophoretically. More significantly, the method provides the means for delivering controlled amounts of materials to desired locations by manipulating the electric field. This makes possible the patterning of polyaniline, a technique that could be attractive, particularly for practical device applications. Li et al. have used this technique for the incorporation of polymeric coating on the MEMS platform, the NIST micro-hotplate, a conductometric gas sensor with an embedded microheater. The signal magnitude of this micro-hotplate device with an electrophoretically integrated polyaniline film correlates well with the gas concentration, with relatively short response and recovery times. The EPD of polyaniline colloids can be controlled with great flexibility by adjusting various process parameters such as the duration of the deposition, the colloid concentration, or the applied voltage. In addition, the electrophoretic

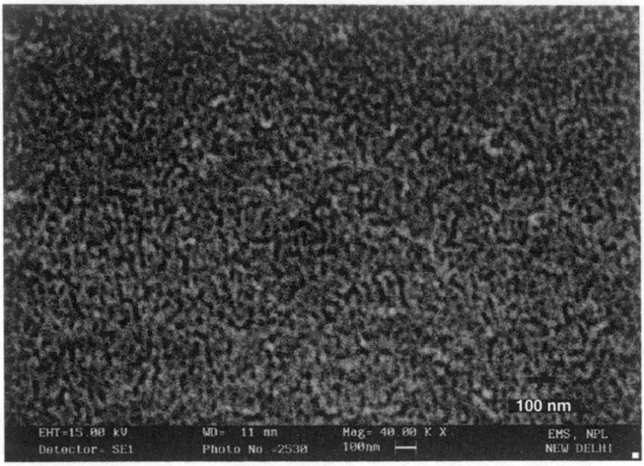

FIGURE 10.5 SEM of electrophoretically deposited nanostructured polyaniline film. (*Reprinted from Ref. 9. Copyright 2007, with permission from Elsevier.*)

patterning of polyaniline from its colloidal suspension has proved to be a scalable technique: it can be applied to macroscopic substrates as well as micro-fabricated device features. Recently, Dhand et al.[9] have reported the fabrication of nanostructured polyaniline films using this novel electrophoretic technique. A mechanism for the conformational changes in the PANI chain during EPD has also been proposed. Figure 10.5 shows the SEM of electrophoretically fabricated polyaniline film with nanostructured surface morphology.

Qariouh et al.[87] have carried out a systematic study to investigate the cataphoretic electrodeposition of polyetherimide from an aqueous medium onto an electrically conductive support (aluminum, steel mesh). The subsequent steps were also investigated. The yield and quality of deposited polyetherimide have been found to be strongly dependent upon a number of formulation variables that are closely related to the emulsion composition and electrodeposition conditions. Ma et al.[88] have applied classical colloidal theories that describe the particle behavior in suspension to polymer particulates/blocks. This suggests many possibilities in materials processing including coprocessing of composites via particulate or powder route to formulate complex microstructures. A polymeric material, polyetheretherketone (PEEK), has been used to examine the colloidal behavior of polymer particulates by Ma et al.[88] Zeta potential of the particulate suspension has been measured, and the electrostatic interaction of the particulates was examined by DLVO (Derjaguin, Landau, Verwey, and Overbeek) theory. The electrosteric effect between the charged particulates in suspension has also been studied. Based on the findings, an optimum condition for a dispersed stable suspension has been identified and applied to the EPD of PEEK particulates. Wang and Kuwabara[89] have demonstrated crack-free, relatively dense, and smooth $BaTiO_3$ film fabrication on the PANI layers-modified Si substrate. This method provides a potential route for fabricating ceramic films on the nonconductive Si or glass substrates using the EPD technique.

Bohmer[90] has studied the EPD of micron-sized polystyrene latex particles on an ITO electrode using in situ optical microscopy. Bohmer observed strong two-dimensional clustering of the particles on the electrode surface upon application of a potential. Clustering decreases somewhat with increasing salt concentration and breaks down upon reversal of the direction of the field. A different interpretation has been given to explain the clustering of polystyrene (PS) particles on the ITO electrode surface. Tada and Onoda[11,12] have demonstrated the nanostructured film fabrication of various conjugated polymers such as poly(3-alkylthiophene), polyfluorene derivative, and MEH-PPV {poly[2-(methoxy)-5-(2′-ethyl-hexyloxy)-1,4-phenylene vinylene]} etc. using electrophoretic technique. They have utilized this technique to prepare donor-acceptor nanocomposites consisting of conjugated conducting polymers as donor and C_{60} molecules as acceptor.[10] They have studied these electrophoretically deposited nanostructured polymer films for fabrication of organic light-emitting diodes, display devices, artificial fingerprint devices, and photovoltaic applications. Dhand et al.[91] have recently reported the preparation of PANI/MWCNT composite by electrophoretic route. Figure 10.6 shows the noncontact mode micrograph of electrophoretically deposited PANI and PANI/MWCNT composite. Pure polyaniline (Fig. 10.6*a*) reveals typical granular morphology with each granule having a diameter of 60 nm. Insignificant

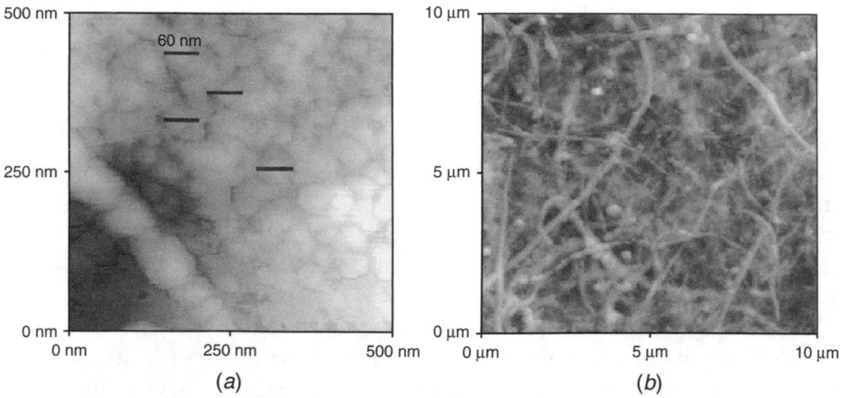

FIGURE 10.6 AFM images of electrophoretically deposited (*a*) polyaniline and (*b*) PANI/MWCNT-c films. (*Reprinted from Ref. 91. Copyright 2008, with permission from Elsevier.*) (See also color insert.)

agglomerate formation within the diameter range of 90 to 120 nm can be seen at certain places. The AFM image of the composite (Fig. 10.6*b*) clearly reveals uniform and homogeneous distribution of CNTs within the polymeric matrix poly[2-methoxy-5-(2′-ethyl-hexyloxy)-1,4-phenylene vinylene].

10.3 Applications of EPD

10.3.1 Photon Crystal Technology

The photon crystal technology (PCT) has attracted increasing interest for the past few years. The current focus of research lies in the search for a three-dimensional full photonic band gap (PBG).[52] The full PBG was first observed in the microwave regime; subsequent reductions in the wavelength were achieved. Recently, it has been developed in the 5 to 10 μm wavelength regions by wafer fusion techniques and in the 1.35 to 1.95 μm wavelength with silicon processing techniques.[92] The present goal is to achieve drastic reduction in the operating wavelength range because of the enormous number of applications of these materials when operated in the near-infrared/visible ranges and inhibition of the spontaneous emission of lucent materials embedded therein that can lead to a thresholdless laser. To achieve it, ordered arrays with micron and submicron parameters are needed. Various techniques based on the use of colloids have been developed to construct these solid arrays. Recently, a technique using a local electric field generated in the EPD suspension has been developed for a novel particle assembling. This technique is called the *micro-EPD* (μ-EPD) *process*. A microdot consisting of mono-dispersed polystyrene or silica spheres has been prepared, which works as photonic crystals.

The EPD technique produces single-crystal colloidal multilayers[93] on the time scale of minutes, which is a drastic acceleration in comparison with the most common gravity sedimentation technique. It is also quite simple to realize and yields three-dimensional photonic crystals of quality comparable to or higher than that with the other methods. Figure 10.7 shows SEM images of colloidal crystals made from the colloidal suspension of 300 nm PS latex spheres. This technique is applicable to a wide variety of particles; however, it is necessary to note that the colloids to be deposited should be tolerant to the addition of alcohol to the dispersion. Besides this, EPD offers the possibility of patterning[94] and impregnating the photonic crystals with luminescent materials. This technique has opened the possibility of preparation of uniform coatings of photonic crystals on curved substances such as spheres, which would be impossible by means of gravitational or centrifugal forces. These coatings can be the basis of unique diffraction devices. The overlap of the photonic stopband and

Figure 10.7 SEM images of colloidal crystal made from 300 nm PS latex spheres. (*Reprinted with permission from Ref. 93. Copyright 2008, American Chemical Society.*)

the spectrum of spontaneous emission of the impregnated species results in the redistribution of the emission energy in such a system. This phenomenon represents both fundamental and practical importance for the design of photonic devices.

Dziomkina et al.[95] have introduced a new method by combining electrophoresis and a lithographic approach for electrode patterning that can be used for the growth of either colloidal monolayers or colloidal crystals. The method controls and changes the colloidal crystal structure by introducing different patterns in a dielectric layer on top of the electrodes used as substrates. Achieving highly accurate growth and control of packing symmetry in colloidal crystals is of paramount importance for photonic applications.[96]

10.3.2 Light-Emitting Diodes

Applications of conjugated polymers possessing solubility like poly(3-alkylthiophene) and poly(2,5-dialkoxy-*p*-phenylene vinylene) to thin-film electronic devices, such as LEDs, have been widely studied. The solubility of these materials makes it possible to be processed into thin solid films through simple and cheap techniques such as spin coating. However, when thin films with submicron thickness are required, a dilute solution has to be used. In such cases, most of the polymer solution is blown away during spinning, and only a limited portion remains as film. Another problem is that the spin-coating technique is incompatible with patterning, which is necessary when one targets a full-color display. On the other hand, recent studies have shown that the morphology of the conjugated polymer film considerably affects the performance of devices such as photocells and field-effect transistors. In the case of field-effect transistors using spin-coated regioregular poly(3-hexylthiophene) films, e.g., different solvents yield different mobilities ranging over more than three orders of magnitude. These findings have encouraged the development of nanostructured conjugated polymer films.

Nanostructured conducting polymers can be obtained through the EPD process. Tada and Onoda[11] have reported the preparation of nanostructured conjugated polymer films through EPD from their colloidal suspension. The morphology of the films observed by an atomic-force microscope has indicated that the films consist of nanoparticles. It is suggested that the deposition as well as the drying of the film in the nonsolvent atmosphere is a key to obtain nanostructured film. A device fabricated from the nanostructured conjugated polymer film emitted light, indicating that pinhole-free nanostructured films can be obtained. On the other hand, this technique is another way to prepare dense films as used in ordinary polymer LEDs, if the optimal postdeposition treatment has been carried out. Tada and Onoda[12] have shown that a simple method like injection of a small amount of toluene solution of target material into a large amount of acetonitrile (nonsolvent for the relevant materials) can generate relatively stable colloidal suspensions of various materials such as C_{60}, poly(3-alkylthiophene), polyfluorene derivative, and MEH-PPV. The colloidal suspension of the mixtures of the materials can be similarly obtained. The colloidal particles can be easily collected through EPD to make nanostructured films, which are pinhole-free and uniform enough to make light-emitting devices as demonstrated. In comparison with the traditional approaches such as the spin-coating method, EPD requires less solubility and thus is applicable to a wider range of materials and composites. The key for this feature is the separation of solidification and film formation. Since the polymer films obtained through this process work as emission layers in a light emitting device, they are expected to work also in photocells as the active layer.

10.3.3 Organic Photocells

It is known that the use of donor-acceptor composite improves the photon/electron conversion performance of photocells with conjugated polymers.[97–98] If both components are incorporated at more than the percolation threshold concentration in the film, to make interpenetrating networks, both electrons and holes can be transported to and collected by the electrodes. Since most conjugated polymers have p-type nature, much effort has been devoted to synthesize acceptor materials with high solubility. Tada and Onoda[10] have proposed a novel route to obtain donor-acceptor nanocomposite by using EPD. The suspension required for the EPD consists of MEH-PPV as donor and C_{60} as acceptor. The film obtained by EPD shows notable quenching of photoluminescence indicating photo-induced charge transfer between the MEH-PPV donor and the C_{60} acceptor. This result suggests that the donor-acceptor nanocomposite obtained through the electrophoretic deposition in the mixture of suspensions is a promising material for organic photocells.

10.3.4 Biosensors

As mentioned, the electrophoretic technique is known to yield uniform, dense, and porous conducting polymer films.[9] In this context, EPD from a colloidal suspension of a conducting polymer has been shown to result in nanostructured conjugated polymer films.[10–12] Nanostructured polymer matrices have been found to provide increased surface area for high enzyme loading.[99] Furthermore, the high surface free energy of a nanostructured film strengthens binding and stabilizes the desired enzyme.[100] These unique properties of nanostructured conducting polymers offer excellent prospects for interfacing biological recognition events with electronic signal transduction and for designing new bioelectronic devices. Dhand et al.[9] have studied electrophoretically deposited nanostructured PANI film for application to a cholesterol biosensor. These nanostructured PANI derived bioelectrodes (ChOx/PANI/ITO) exhibit linearity up to 400 mg/dL of cholesterol (Fig. 10.8), sensitivity of 7.76×10^{-5} Abs $\times$ $(\text{mg/dL})^{-1}$ with negligible (0.1%) interference. Besides this, the value of the apparent Michaelis-Menten constant K_m^{app} indicative of enzyme-substrate interactions, has been found to be 0.62 m*M*. This low value K_m^{app} for ChOx/PANI/ITO bioelectrode reveals increased enzyme (cholesterol oxidase)-substrate (cholesterol) interactions, indicating distinct advantage of this matrix over other matrices used for cholesterol biosensor fabrication.

Dhand et al.[91] have also reported the preparation of a nanostructured composite film comprising emeraldine salt (ES) and carboxyl group

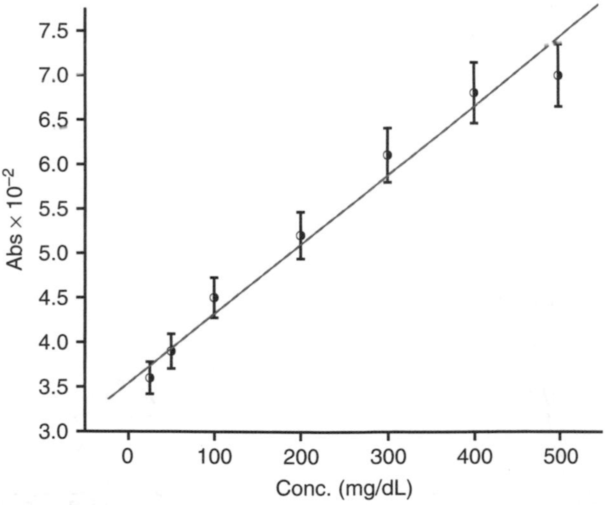

FIGURE 10.8 The calibration plot of ChOx/PANI/ITO bioelectrode: absorbance as a function of cholesterol concentration.

functionalized multiwalled carbon nanotubes (MWCNT-c) by using the electrophoretic technique. The results of the CV and EIS (electrochemical impedance spectroscopy) studies indicate enhanced electrochemical and charge transfer behavior of the composite as compared to pure polyaniline. This enhanced electrochemical response in electrophoretically deposited ES/MWCNT-c composite has been utilized to improve characteristics of biosensors. The application of ES/MWCNT-c/ITO electrode to biosensor for cholesterol indicates short response time (10 s) and high sensitivity (6800 nA/mM). This enhanced sensitivity is attributed to the incorporation of the MWCNT-c in the matrix and to the intimate association obtained between these two aromatic structures. Figure 10.9 shows the variation in the current measured at a fixed voltage of 0.28 V in LSV (linear sweep voltammetry) scans as a function of cholesterol concentration (1.3 to 13 m*M*).

In spite of many applications, the EPD technique has a number of limitations. In this context, it may be mentioned that the control of this process is difficult since the electrophoretic mobility of the colloidal particles is very sensitive to factors such as chemical environment, particle surface topography, suspension behavior, etc. It is presently impossible to predict whether suspensions will deposit electrophoretically. Another important area of concern is the cracking of the electrophoretically deposited materials such as ceramic coating during drying and sintering. Moreover, in aqueous EPD, there are a number of problems related to the aqueous suspension. Some of these are related to electrochemical reactions at the electrodes when current

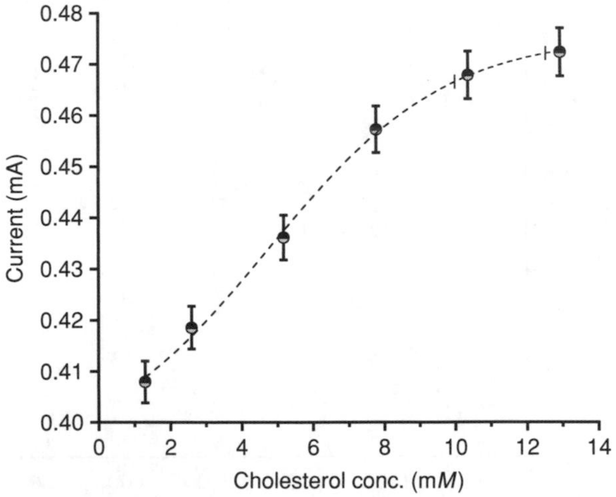

FIGURE 10.9 Variation in the amperometric current measured at 0.28 V as a function of cholesterol concentration. (*Reprinted from Ref. 91. Copyright 2008, with permission from Elsevier.*)

is passed through, which seriously affects the efficiency of the process and the uniformity of the deposit. Electrolysis of water occurs at low voltages, and gas evolution at the electrodes is inevitable at field strengths high enough to give reasonably short deposit times. This causes bubbles to be trapped within the deposit, unless special procedures are adopted, and damage the film topology. Another electrokinetic phenomenon occurring in an aqueous EPD is water electroosmosis, which consists of the movement of the liquid phase because of an external electric field. This could be helpful in EPD because it would accelerate drying of the deposit surface which is in contact with the electrode. So if the process is well controlled, demolding of the self-supported deposits can be easier. Contrarily, if the deposits were too thick, or the process was too fast, crack formation would occur as a consequence of the drying gradient.

10.4 Scope of Electrophoretically Deposited Polymers

It has been shown that EPD has gained considerable interest in recent years for fabrication of advanced materials. The process is simple, easy to use, and cost-effective, and it has found innumerable applications including thin- and thick-film deposition, layered ceramics, hybrid materials, fiber-reinforced composites as well as nanocomposites, nanoscale assembly of two- and three-dimensional ordered structures and micropatterned thin films. However, the process needs judicious choice of solvent media such that an appreciable magnitude of surface charge is developed on the powder surface in the suspension to ensure stability of the suspension as well as facilitate high electrophoretic mobility. EPD has a number of advantages over the usual filtration methods, including rapid but controlled deposition over a wide range of substrate materials. As long as the substrate is conducting, an adherent coating can be obtained. The EPD technique is a widely used industrial process that has been applied successfully for deposition of, e.g., phosphor for displays. It is an automated and high-throughput process that in general produces films with good homogeneity and packing density.

The EPD process is found to be the most promising technique for homogeneous, smooth, and rigid deposition of CNT with controlled thickness and morphology. In addition, the technique can be readily extended to allow the coating of CNTs onto large planar substrates, wires, individual fibers, fibrous structures, and porous components. These CNT-based films have numerous applications for these networks: antielectrostatic coatings; electrochromic or electrically heated windshields; field emitters; energy applications; displays; electromagnetic screening and touch panels; transistors for logic elements in macroelectronic systems or for optical elements with highly anisotropic properties; sensors; optoelectronics; and diodes. Moreover, electrophoretically deposited CNT films are robust and provide new ways to explore the electrochemical and electrocatalytic properties of carbon nanotube films. Polymer particulates/blocks can be electrophoretically moved under the action of an electric field. Hence this provides a broader selection of processing routes for composite material systems for both miniaturized and complex configurations using EPD.

The electrophoretic patterning of polyaniline from its colloidal suspension has proved to be a scalable technique: it can be applied to macroscopic substrates as well as microfabricated device features. Electrophoresis of charged polymer colloids on patterned electrode surfaces allows us to form colloidal monolayers and to grow colloidal crystals with different lattice structures, where the patterns of the electrode substrates determine the structures of the colloidal crystals grown on them. Achieving highly accurate growth and control of packing symmetry in colloidal crystals is of paramount importance for photonic applications. There is particular interest in this issue as the propagation of light in photonic crystals is highly dependent on their lattice structures. Colloidal monolayers fabricated by this method can be further used for the preparation of three-dimensional colloidal crystals with low packing symmetry, e.g., by means of a layer-by-layer deposition of oppositely charged colloidal particles. Furthermore, the monolayers can have additional applications, e.g., to study local reactions that take place in hollow spheres (nanocapsules) that are used as delivery vehicles for the controlled release of substances such as drugs, cosmetics, dyes, or inks, or for protection of sensitive agents such as enzymes and proteins. Electrophoretic deposition results in the formation of uniform, dense, pinhole-free nanostructured films of conducting polymers. These nanostructured polymeric films can be used as an effective matrix in light-emitting devices, photovoltaic cells, organic photocells, biosensors, etc.

In biosensor applications, nanostructured polymer matrices have been found to provide increased surface area for high enzyme loading. Furthermore, the high surface free energy of a nanostructure strengthens binding and stabilizes the desired enzyme. These unique properties of nanostructured conducting polymers offer excellent prospects for interfacing biological recognition events with electronic signal transduction and for designing new bioelectronic devices. In comparison with the traditional approaches such as the spin-coating method, the electrophoretic deposition requires less solubility and thus is applicable to a wider range of materials and composites. Electrophoretic deposition can be used for on-chip manipulation and assembly of colloidal particles. The structures assembled include cells, conductive microwires from metallic nanoparticles, and switchable two-dimensional crystals from polymer microspheres. The electric field-driven assembly of cells and nanoparticles can be used to create new types of biosensors, microbioassays, and bioelectronic circuits. This technique enables design of bacterial biofilms for biotechnological applications (e.g., biosensors and bioreactors) or as protective coatings of probiotic bacteria on, for instance, silver-impregnated urinary catheters or voice prostheses in biomedical applications.

In spite of these interesting developments, a lot remains to be done with regard to both fundamental understanding and the much needed improvement of the method of electrophoretic deposition for application to desired organic electronic devices.

References

1. L. Besra and M. Liu, A review on fundamentals and applications of electrophoretic deposition (EPD), *Progress Mater. Sci.* 52:1–61 (2007).
2. E. S. Kooij, E. A. M. Brouwer, and B. Poelsema, Electric field assisted nanocolloidal gold deposition, *J. Electroanal. Chem.* 611:208–216 (2007).
3. N. Koch, Organic electronic devices and their functional interfaces, *Chem. Phys. Chem.* 8:1438–1455 (2007).
4. M. Gerard, A. Chaubey, and B. D. Malhotra, Application of conducting polymers to biosensors, *Biosens. & Bioelectron.* 17:345–359 (2002).
5. A. Chaubey, K. K. Pande, and B. D. Malhotra, Application of polyaniline/sol-gel derived tetraethylorthosilicate films to an amperometric lactate biosensor, *Anal. Sci.*, 19:1477–1480 (2003).
6. A. F. Morgerac, B. Ballarinb, A. Filippinia, D. Frascaroa, C. Pianad, and L. Settia, An amperometric glucose biosensor prototype fabricated by thermal inkjet printing, *Biosens. & Bioelectron.* 20:2019–2026 (2005).
7. S. K. Sharma, R. Singhal, B. D. Malhotra, N. Sehgal, and A. Kumar, Langmuir–Blodgett film based biosensor for estimation of galactose in milk, *Electrochim. Acta* 49:2479–2485 (2004).
8. S. K. Arya, A. K. Prusty, S. P. Singh, P. R. Solanki, M. K. Pandey, M. Datta, and B. D. Malhotra, Cholesterol biosensor based on *N*-(2-aminoethyl)-3-aminopropyl-trimethoxysilane self-assembled monolayer, *Anal. Biochem.* 363:210–218 (2007).
9. C. Dhand, S. P. Singh, S. K. Arya, M. Datta, and B. D. Malhotra, Cholesterol biosensor based on electrophoretically deposited conducting polymer film derived from nano-structured polyaniline colloidal suspension, *Anal. Chim. Acta* 602:244–251 (2007).
10. K. Tada, and M. Onoda, Preparation of donor-acceptor nanocomposite through electrophoretic deposition, *Curr. Appl. Phys.* 5:5–8 (2005).
11. K. Tada, and M. Onoda, Preparation and application of nanostructured conjugated polymer film by electrophoretic deposition, *Thin Solid Films* 438–439:365–368 (2003).
12. K. Tada and M. Onoda, Nanostructured conjugated polymer films for electroluminescent and photovoltaic applications, *Thin Solid Films* 477:187–192 (2005).
13. I. Zhitomirsky, Cathodic electrodeposition of ceramic and organoceramic materials fundamental aspects, *Adv. in Coll. Int. Sci.* 97:279-317 (2002).
14. F. Grollion, D. Fayeulle, and M. Jeandin, Qualitative image analysis of electrophoretic coatings, *J. Mater. Sci. Lett.* 11:272–275 (1992).
15. D. R. Brown and F. W. Salt, The Mechanism of the electrophoretic deposition, *J. App. Chem.* 15:40–48 (1963).
16. P. Sarkar and P. S. Nicholson, Electrophoretic deposition (EPD): Mechanisms, kinetics, and application to ceramics, *J. Am. Ceram. Soc.* 79:1987–2002 (1996).
17. H. C. Hamaker and E. J. W. Verwey, Colloid stability: The role of the forces between the particles in electrodeposition and other phenomena, *Trans. Faraday Soc.* 35:180–185 (1940).

18. H. C. Hamaker, Formation of deposition by electrophoresis, *Trans. Farad Soc.* 35:279–283 (1940).
19. A. I. Avgustinik, V. S. Vigdergauz, and G. I. Zharavlev, Electrophoretic deposition of ceramic masses from suspension and calculation of deposit yields, *J. Appl. Chem. USSR* (English translation) 35:2175–2180 (1962).
20. P. M. Biesheuvel and H. Verweij, Theory of cast formation in electrophoretic deposition, *J. Am. Ceram. Soc.* 82:1451–1455 (1999).
21. T. Ishihara, K. Shimise, T. Kudo, H. Nishiguchi, T. Akbay, and Y. Takita, Preparation of yttria-stabilised zirconia thin-films on strontium doped $LaMnO_3$ cathode substrate via electrophoretic deposition for solid oxide fuel cells, *J. Am. Ceram. Soc.* 83:1921–1927 (2000).
22. F. Chen and M. Liu, Preparation of yttria-stabilised zirconia (YSZ) films on $La_{0.85}Sr_{0.15}MnO_3$ (LSM) and LSM-YSZ substrate using an electrophoretic deposition (EPD) process, *J. Eur. Ceram. Soc.* 21:127–134 (2001).
23. N. Sato, M. Kawachi, K. Noto, N. Yoshimoto, and M. Yoshizawa, Effect of particle size reduction on crack formation in electrophoretically deposited YBCO films, *Physica C* 357–360:1019–1022 (2001).
24. R. W. Powers, The electrophoretic forming of beta-alumina ceramic, *J. Electrochem. Soc.* 122:482–486 (1975).
25. B. Ferrari and R. Moreno, The conductivity of aqueous Al_2O_3 slips for electrophoretic deposition, *Mater. Lett.* 28:353–355 (1996).
26. B. Ferrari and R. Moreno, Electrophoretic deposition of aqueous alumina slip, *J. Eur. Ceram. Soc.* 17:549–556 (1997).
27. H. G. Krueger, A. Knote, U. Schindler, H. Kern, and A. Boccaccini, Composite ceramic metal coatings by means of combined electrophoretic deposition, *J. Mater. Sci.* 39:839–844 (2004).
28. M. Zarbov and I. Schuster, Methodology for selection of charging agents for electrophoretic deposition of ceramic particles, *Proceedings of the international symposium on electrophoretic deposition: fundamentals and applications*, The Electrochemical Society Inc, USA, Proc. 2002–21:39–46 (2002).
29. C. Y. Chen, S. Y. Chen, and D. M. Liu, Electrophoretic deposition forming of porous alumina membranes, *Acta Mater*, 47:2717–2726 (1999).
30. G. Wang, P. Sarkar, and P. S. Nicholson, Influence of acidity on the electrostatic stability of alumina suspensions in ethanol, *J. Am. Ceram. Soc.* 80:965–972 (1997).
31. J. Ma, R. Zhang, C. H. Liang, and L. Weng, Colloidal characterization and electrophoretic deposition of PZT, *Mater. Lett.* 57:4648–4654 (2003).
32. S. A. Troelstra, Applying coatings by electrophoresis, *Philips Tech. Rev.* 12:293–303 (1951).
33. I. Zhitomirsky, Electrophoretic deposition of hydroxyapatite, *J. Mater. Sci. Mater. Med.* 8:213–219 (1997).
34. R. N. Basu, C. A. Randall, and M. J. Mayo, Fabrication of dense zirconia electrolyte films for tubular solid oxide fuel cells by electrophoretic deposition, *J. Am. Ceram. Soc.* 84:33–40 (2001).
35. H. Negishi, H. Yanagishita, and H. Yokokawa, Electrophoretic deposition of solid oxide fuel cell material powders, *Proceedings of the electrochemical society on electrophoretic deposition: Fundamentals and applications*, Pennington, USA, 2002–21:214–221 (2002).
36. L. Vandeperre, O. Van Der Biest, and W. J. Clegg, Silicon carbide laminates with carbon interlayers by electrophoretic deposition. *Key. Eng. Mater.* (Pt. 1, Ceramic and Metal Matrix Composites) 127–131:567–573 (1997).
37. Z. Peng and M. Liu, Preparation of dense platinum-yttria stabilized zirconia and yttria stabilized zirconia films on porous $La_{0.9}Sr_{0.1}MnO_3$ (LSM) substrates. *J. Am. Ceram. Soc.* 84:283–288 (2001).
38. K. Hasegawa, S. Kunugi, M. Tatsumisago, and T. Minami, Preparation of thick films by electrophoretic deposition using modified silica particles derived by sol–gel method. *J. Sol–gel Sci. Technol.* 15:243–249 (1999).
39. M. Abdollahi, S. N. Ashrafizadeh, and A. Malekpour, Preparation of zeolite ZSM-5 membrane by electrophoretic deposition method, *Microporous & Mesoporous Mater* 106:192–200 (2007).

40. R. Wang and Y. X. Hu, Patterning hydroxyapatite biocoating by electrophoretic Deposition, *J. Biomed. Mater. Res. Part A*: 67A:270–275 (2003).
41. C. Kaya, Electrophoretic deposition of carbon nanotube-reinforced hydroxyapatite bioactive layers on Ti–6Al–4V alloys for biomedical applications, *Ceram. Intl.* 34:1843-1847 (2008).
42. M. J. Shane, J. B. Talbot, B. G. Kinney, E. Sluzky, and H. R. Hesse, Electrophoretic deposition of phosphors: II deposition experiments and analysis, *J. Colloid Interface Sci.* 165:334–340 (1994).
43. M. T. Ochsenkuehn-Petropoulou, A. F. Altzoumailis, R. Argyropoulou, and K. M. Ochsenkuehn, Superconducting coatings of MgB_2 prepared by electrophoretic deposition, *Anal. Bioanal. Chem.* 379:792–795 (2004).
44. S. J. Limmer and G. Cao, Sol-gel electrophoretic deposition for the growth of oxide nanorods, *Adv. Mater.* 15:427–431 (2003).
45. C. S. Du, D. Heldbrant, and N. Pan, Preparation and preliminary property study of carbon nanotubes films by electrophoretic deposition, *Mater. Lett.* 57:434–438 (2002).
46. B. J. C. Thomas, A. R. Boccaccini, and M. S. P. Shaffer, Multi-walled carbon nanotube coatings using electrophoretic deposition (EPD), *J. Am. Ceram. Soc.* 88:980–982 (2005).
47. O. Zhou, H. Shimoda, B. Gao, S. Oh, L. Fleming, and G. Yue, Material science of carbon nanotubes: Fabrication, integration, and properties of macroscopic structures of carbon nanotubes, *Acc. Chem. Res.* 35:1045–1053 (2002).
48. S. Oh, J. Zhang, Y. Cheng, H. Shimoda, and O. Zhou, Liquid-phase fabrication of patterned carbon nanotube field emission cathodes, *Appl. Phys. Lett.* 84:3738–3740 (2004).
49. E. Antonelli, R. S. da Silva, F. S. de Vicente, A. R. Zanatta, and A. C. Hernandes, Electrophoretic deposition of $Ba_{0.77}Ca_{0.23}TiO_3$ nanopowders, *J. Mater. Process. Technol.* 203:526-531 (2008).
50. J. Van Tassel and C. A. Randall, Electrophoretic deposition and sintering of thin/thick PZT film, *J. Eur. Ceram. Soc.* 19:955–958 (1999).
51. A. Braun, G. Falk, and R. Clasen, Transparent polycrystalline alumina ceramic with sub-micrometre microstructure by means of electrophoretic deposition, *Mat.-wiss. u. Werkstofftech.* 37:293-297 (2006).
52. L. Besra, C. Compson, and M. Liu, Electrophoretic deposition on nonconducting substrates: The case of YSZ film on NiO-YSZ composite substrates for solid oxide fuel cell application, *J. Power Sources* 173:130–136 (2007).
53. C. Y. Chen, Y. Ru Lyu, C. Y. Su, H. M. Lin, and C. K. Lin, Characterization of spray pyrolyzed manganese oxide powders deposited by electrophoretic deposition technique, *Surf. & Coatings Technol.* 202:1277–1281 (2007).
54. M. Trau, D. A. Saville, and I. A. Aksay, Field-induced layering of colloidal crystals, *Science* 272:706–709 (1996).
55. M. Bohmer, In situ observation of 2-dimensional clustering during electrophoretic deposition, *Langmuir.* 12:5747–5750 (1996).
56. M. Trau, D. A. Saville, and I. A. Aksay, Assembly of colloidal crystals at electrode interfaces, *Langmuir.* 13:6375–6381 (1997).
57. A. L. Rogach, N. A. Kotov, D. S. Koktysh, J. W. Ostrander, and G. A. Ragoisha, Electrophoretic deposition of latex-based 3D colloidal photonic crystals: A technique for rapid production of high-quality opals, *Chem. Mater.* 12:2721–2726 (2000).
58. Z. Z. Gu, S. Hayami, S. Kubo, Q. B. Meng, Y. Einaga, D. A. Tryk, A. Fujishima, et al., Porous film by electrophoresis, *J. Am. Chem. Soc.* 123:175–176 (2001).
59. W. D. Ristenpart, I. A. Aksay, and D. A. Saville, Electrically guided assembly of planar superlattices in binary colloidal suspensions, *Phys. Rev. Lett.* 90:128303 (2003).
60. P. J. Sides, Electrodynamically particle aggregation on an electrode driven by an alternating electric field normal to it, *Langmuir* 17:5791–5800 (2001).
61. W. Ristenpart, I. A. Aksay, and D. A. Saville, Electrohydrodynamic flow, kinetic experiments and scaling analysis, *Phys. Rev. E* 69:021405 (2004).

62. J. A. Fagan, P. J. Sides, and D. C. Prieve, Vertical motion of a charged colloidal particle near an ac polarized electrode with a nonuniform potential distribution: Theory and experimental evidence, *Langmuir* 20:4823–4834 (2004).
63. J. A. Fagan, P. J. Sides, and D. C. Prieve, Evidence of multiple electrohydrodynamic forces acting on a colloidal particle near an electrode due to an alternating current electric field, *Langmuir* 21:1784–1794 (2005).
64. R. C. Bailey, K. J. Stevenson, and J. T. Hupp, Assembly of micropatterned colloidal gold thin films via microtransfer molding and electrophoretic deposition, *Adv. Mater.* 12:1930–1934 (2000).
65. M. Gao, J. Sun, E. Dulkeith, N. Gaponik, U. Lemmer, and J. Feldmann, Lateral patterning of CdTe nanocrystal films by the electric field directed layer-by-layer assembly method, *Langmuir* 18:4098–4102 (2002).
66. O. D. Velev and K. H. Bhatt, On-chip micromanipulation and assembly of colloidal particles by electric fields, *Soft Matter* 2:738–750 (2006).
67. E. S. Kooij, E. A. M. Brouwer, and B. Poelsema, Electric field assisted nanocolloidal gold deposition, *J. Electroanal. Chem.* 611:208–216 (2007).
68. N. Nagarajan and P. S. Nicholson, Nickel-Alumina functionally graded materials by electrophoretic deposition, *J. Am. Ceram. Soc.*, 87:2053–2057 (2004).
69. P. Sarkar, S. Datta, and P. S. Nicholson, Functionally graded ceramic/ceramic and metal/ceramic composites by electrophoretic deposition, *Compos. Part B: Eng.* 28:49–56 (1997).
70. G. Milczarek and A. Ciszewski, Preparation of phthalocyanine modified electrodes. An electrophoretic approach, *Electroanalysis* 17:371–374 (2005).
71. R. C. Bailey, K. J. Stevenson, and J. T. Hupp, Assembly of micropatterned colloidal gold thin films via microtransfer molding and electrophoretic deposition, *Adv. Mater.* 12:1930–1934 (2000).
72. S. Iijima, Helical microtubules of graphitic carbon, *Nature* 354:56–58 (1991).
73. M. Trojanowicz, Analytical applications of carbon nanotubes: A review, *Trends in Anal. Chem.* 25:480–489 (2006).
74. C. Niu, E. K. Sichel, R. Hoch, D. Moy, and H. Tennent, High power electrochemical capacitors based on carbon nanotube electrodes, *Appl. Phys. Lett.* 70:1480–1482 (1997).
75. C. Du and N. Pan, High power density supercapacitor electrodes of carbon nanotube films by electrophoretic deposition, *Nanotechnology* 17:5314–5318 (2006).
76. H. Zhao, H. Song, Z. Li, G. Yuan, and Y. Jin, Electrophoretic deposition and field emission properties of patterned carbon nanotubes, *Appl. Surf. Sci.* 251:242–244 (2005).
77. M. J. Andrade, M. D. Lima, V. Skakalov, C. P. Bergmann, and S. Roth, Electrical properties of transparent carbon nanotube networks prepared through different techniques, *Phys. Stat. Sol.* (RRL) 5:178–180 (2007).
78. G. Girishkumar, K. Vinodgopal, and P. V. Kamat, Carbon nanostructures in portable fuel cells: Single-walled carbon nanotube electrodes for methanol oxidation and oxygen reduction, *J. Phys. Chem. B* 108:19960–19966 (2004).
79. G. Girishkumar, M. Rettker, R. Underhile, D. Binz, K. Vinodgopal, P. McGinn, and P. Kamat, Single-wall carbon nanotube-based proton exchange membrane assembly for hydrogen fuel cells, *Langmuir* 21:8487–8494 (2005).
80. D. Kurnosov, A. S. Bugaev, K. N. Nikolski, R. Tchesov, and E. Sheshin, Influence of the interelectrode distance in electrophoretic cold cathode fabrication on the emission uniformity, *Appl. Surf. Sci.* 215:232–236 (2003).
81. H. Shimoda, S. J. Oh, H. Z. Geng, R. J. Walker, X. B. Zhang, and L. E. McNeil, Self-assembly of carbon nanotubes, *Adv. Mater.* 14:899–901 (2002).
82. P. Kamat, K. Thomas, S. Barazzouk, G. Girishkumar, K. Vinodgopal, and D. Meisel, Self-assembled linear bundles of single wall carbon nanotubes and their alignment and deposition as a film in a dc field, *J. Am. Chem. Soc.* 126:10757–10762 (2004).
83. I. Singh, C. Kaya, M. S. P. Shaffer, B. J. C. Thomas, and A. R. Boccaccini, Bioactive ceramic coatings containing carbon nanotubes on metallic substrates by electrophoretic deposition (EPD), *J. Mater. Sci.* 41:8144–8151 (2006).

84. A. R. Boccaccini, J. A. Roether, B. J. C. Thomas, M. S. P. Shaffer, E. Chavez, and E. Stoll, The electrophoretic deposition of inorganic nanoscaled materials, *J. Ceram. Soc. Jpn.* 114:1–14 (2006).
85. R. Colorado and A. R. Barron, Silica-coated single-walled nanotubes: Nanostructure formation, *Chem. Mater.* 16:2691–2693 (2004).
86. G. Li, C. Martinez, and S. Semancik, Controlled electrophoretic patterning of polyaniline from a colloidal suspension, *J. Am. Chem. Soc.* 127:4903–4909 (2005).
87. H. Qariouh, N. Raklaoui, R. Schue, F. Schue, and C. Bailly, Electrophoretic deposition of polyetherimide from an aqueous emulsion: Optimisation of some deposition parameters, *Polym. Int.* 48:1183–1192 (1999).
88. J. Ma, C. Wang, and C. H. Liang, Colloidal and electrophoretic behavior of polymer particulates in suspension, *Mater. Sci. & Eng.* C 27:886–889 (2007).
89. J. Q. Wang and M. Kuwabara, Electrophoretic deposition of $BaTiO_3$ films on a Si substrate coated with conducting polyaniline layers, *J. Eur. Ceram. Soc.* 28:101–108 (2008).
90. Marcel Bohmer, In situ observation of 2-dimensional clustering during electrophoretic deposition, *Langmuir* 12:5747–5750 (1996).
91. C. Dhand, S. K. Arya, S. P. Singh, B. P. Singh, Monika Datta, and B. D. Malhotra, Preparation of polyaniline/multiwalled carbon nanotube composite by novel electrophoretic route *Carbon* 46:1727–1735 (2008).
92. J. G. Fleming and S. Y. Lin, Three-dimensional photonic crystal with a stop band from 1.35 to 1.95 μm, *Opt. Lett.* 24:49–51 (1999).
93. A. L. Rogach, N. A. Kotov, D. S. Koktysh, J. W. Ostrander, and G. A. Ragoisha, Electrophoretic deposition of latex-based 3D colloidal photonic crystals: A Technique for rapid production of high-quality opals, *Chem. Mater.* 12:2721–2726 (2000).
94. R. C. Hayward, D. A. Saville, and I. A. Aksay, Electrophoretic assembly of colloidal crystals with optically tunable micropatterns, *Nature* 404:56–59 (2000).
95. N. V. Dziomkina, M. A. Hempenius, and G. J. Vancso, Symmetry control of polymer colloidal monolayers and crystals by electrophoretic deposition onto patterned surfaces, *Adv. Mater.* 17:237–240 (2005).
96. E. Yablonovitch, Inhibited spontaneous emission in solid-state physics and electronics, *Phys. Rev. Lett.* 58:2059–2062 (1987).
97. N. S. Sariciftci, L. Smilowitz, A. J. Heeger, and F. Wudl, Photoinduced electron transfer from a conducting polymer to buckminsterfullerene, *Science* 258:1474-1476 (1992).
98. S. Morita, A. A. Zakhidov, and K. Yoshino, Doping effect of buckminsterfullerene in conducting polymer: Change of absorption spectrum and quenching of luminescence, *Solid State Commun* 82:249–252 (1992).
99. X. Luo, A. Morrin, A. J. Killard, and M. R. Smyth, Application of nanoparticles in electrochemical sensors and biosensors, *Electroanalysis* 18(4): 319–326 (2006).
100. J. Kim, J. W. Grate, and P. Wang, Nanostructures for enzyme stabilization chemical, *Eng. Sci.* 61:1017–1026 (2005).

CHAPTER 11

Electrochemical Surface Switches and Electronic Ion Pumps Based on Conjugated Polymers

Magnus Berggren* and Agneta Richter-Dahlfors†

OBOE—Strategic Center for Organic Bioelectronics

11.1 Electronic Control of Surface Properties

In a vast number of biotechnological applications, the interface between an aqueous fluid and a solid surface is of great importance. This interface may serve as the anchoring site for biomolecules in biosensor systems, act as a catalyst for chemical reactions, etc. In the technology area of microfluidics, major efforts have been devoted to design and engineer the chemistry and topography of solid surfaces in order to control the flow of liquids in dispensing and analysis systems of various complexity. In the cell biology field, surfaces have been shown to act as promoters for specific protein adsorption, and this effect has been utilized to dictate the specific proliferation and differentiation scheme of cultivated cells.[1] Numerous recent breakthroughs in material and surface science have provided a toolbox of technologies that today can be used to tailor-make the fluid-solid interface in order to express desired

*Organic Electronics, ITN, Linköpings Universitet, Norrköping, Sweden.
†Department of Neuroscience, Karolinska Institutet, Stockholm, Sweden.

static characteristics for the target application. However, for many bioapplications, it may be preferable to achieve dynamic control of the chemical and physical surface properties. Dynamic control of the surface tension, chemistry, and charge achieved by electric biasing will provide a novel technology with great potential to advance current research in cell biology.

11.1.1 Wettability Switches Based on Conducting Polymers

Conjugated polymers have been extensively explored in electrochemical (EC) devices. Their principle of operation is defined by the dynamic change of the fundamental chemical and/or physical properties of the conjugated polymer bulk upon EC switching. For instance, EC control of the volume, optical absorption, and impedance define the function of polymer actuators,[2] electrochromic displays,[3] and EC transistors.[4,5] As the oxidation state of a polymer film is altered, not only its bulk properties are switched but also its nature along its outermost surface. Several research groups have explored the use of EC switching of conjugated polymer thin films to achieve dynamic control of the surface tension.[6–8] The surface of an oxidized polymer film expresses a higher density of dipoles compared to the neutral polymer surface, and intuitively, it should therefore exhibit a relatively higher surface energy. However, wettability along conjugated polymer surfaces is somewhat more complex, since topography and the properties of doping ions must be taken into the account.

11.1.2 Surface Switches Based on P3AT, PPy, and PANI

Poly(3-alkylthiophenes)[9] (P3AT) can be processed and patterned from organic solvents. Films of poly(3-hexylthiophene) (P3HT) can be deposited on top of solid electrolytes to form a device configuration expressing its EC active surface toward the environment (Fig. 11.1). Along this surface, the water contact angle[8] is found to be $\theta = 101.8°$. By applying a positive voltage (~1 V) to the P3HT film and by grounding a counterelectrode, in contact with the common electrolyte, P3HT oxidizes according to Eq. (11.1):

$$\text{P3HT} + \text{X}^- \rightarrow \text{P3HT:X} + e^- \qquad (11.1)$$

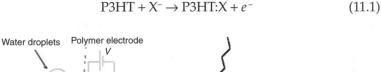

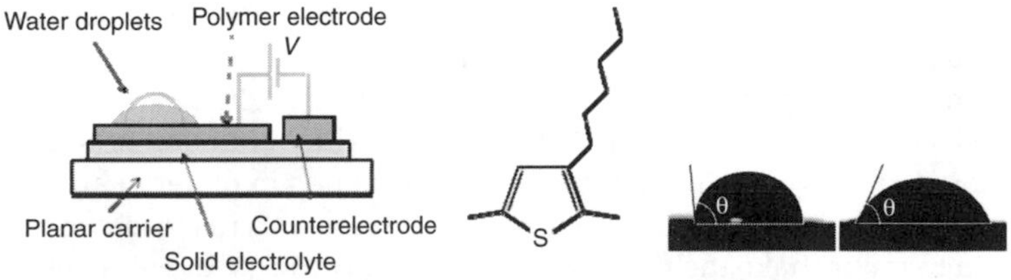

Figure 11.1 Left: Structure of an EC surface switch with a conjugated polymer as the active material. Middle: The chemical structure of P3HT. Right: Photographs of the water droplets added to the neutral (left) and oxidized (right) surfaces, respectively. θ = contact angle.

In the oxidized state, the P3HT film exhibits a water contact angle of $\theta = 89.1°$ (Fig. 11.1). In both oxidation states, the water contact angles are rather high and the associated difference is small. This is explained by the presence of hexyl side chains. Along the P3HT surface, the side groups point outward to a great extent. Therefore, they shield the net effect of the dynamic change of the dipole characteristics that occur along the core of the polythiophene backbone, upon EC switching.

Detergent acids such as DBSA (dodecylbenzene sulfonic acid)[6] are commonly used as the doping ion to obtain high conductivity of conjugated polymers, e.g., polyaniline (PANI) (Fig. 11.2). In the doped oxidized state, the acid group locks the doping ion to the conjugated backbone, leaving the nonpolar part of the molecule pointing away from the polymer chain and also from the surface. In this case, the oxidized film possesses a relatively lower surface energy, thus resulting in high water contact angles. In the reduced neutral state, the doping ions are decoupled from the conjugated polymer and can more freely rotate to expose the more polar acid groups away from the polymer main chain. This increases the surface tension and the water contact angle is therefore lowered (Fig. 11.2). The water contact angles along PANI:DBSA-based surface switches are found to be 9° and 37° for the reduced and oxidized states, respectively (Fig. 11.2).[10]

The relationship between the water contact angle θ for a liquid l droplet residing along a planar solid s surface also in contact with air or vapor v, and the contributing surface tension quantities is predicted by Young's equation, (11.2), where γ is the surface tension for the different interfaces.

$$\gamma_{sv} = \gamma_{sl} + \gamma_{lv} \cos\theta \tag{11.2}$$

To further increase the net difference of the water contact angle upon electrochemical switching of the conjugated polymer, one can

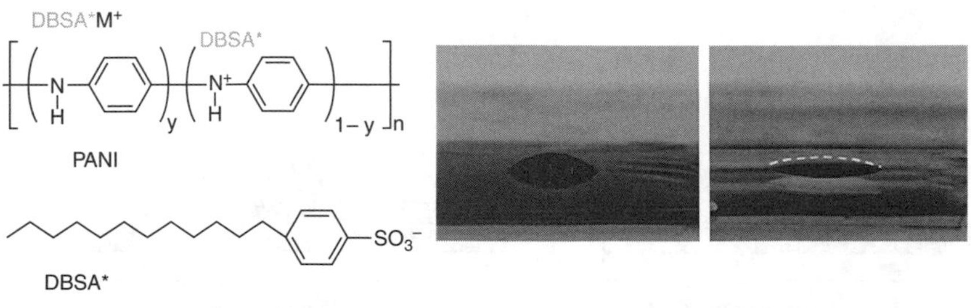

Figure 11.2 Left: The chemical structure of PANI and DBSA. Water droplets added to surface switches including PANI:DBSA, in which PANI is in its oxidized (middle) and reduced (right) state. (*From Ref. 10. Copyright 2004, Wiley-VCH Verlag GmbH & Co. KGaA. Reproduced with permission.*) (See also color insert.)

texture the solid surface. This increases the effective area between the liquid and the solid polymer surface, thus modifying Eq. (11.2). The wetting characteristics along textured surfaces have been treated and investigated in the past and can be predicted using the Cassie-Baxter or the Wenzel relationships.[11] If a droplet, applied onto a textured surface, stays standing on the pillar structures according to a fakir situation, and the surrounding cavities are filled with vapor or air, the Cassie-Baxter situation applies [see Eq. (11.3)]. In contrast, if the droplet wets the entire surface, Young's equation is simply modified by the roughness factor r, which equals the ratio between the real surface area and the projected area [Eq. (11.4)].

$$\cos\theta^* = f_s \cos\theta_s + f_v \cos\theta_v \tag{11.3}$$

$$\cos\theta^* = r\cos\theta \tag{11.4}$$

where θ^* = apparent water contact angle
f_s = area fractions of the solid
f_v = area fractions of vapor

Polypyrrole (PPy) has been electrochemically synthesized to form a mesh around pillars made of an insulating polymer (SU8), all made on top of a PEDOT:PSS electrode surface. As the PPy mesh is electrochemically switched from the oxidized to its neutral state, both the surface tension and its volume change. The result is that the effective SU8 pillar height decreases at the same time as the wetting character along the PPy phase is altered (Fig. 11.3). In its pristine oxidized state, the droplet is "hanging" on the SU8 pillars in accordance to the fakir situation (Cassie-Baxter's case). As the PPy mesh is reduced, the aspect ratio of the structured surface decreases, causing the water droplet to contact both the SU8 pillars and the PPy mesh (Wenzel's case). This has a major impact on the wettability characteristics. In the pristinely oxidized state, the apparent water contact angle is 129° while it decreases to 44° when PPy is reduced (Fig. 11.4).

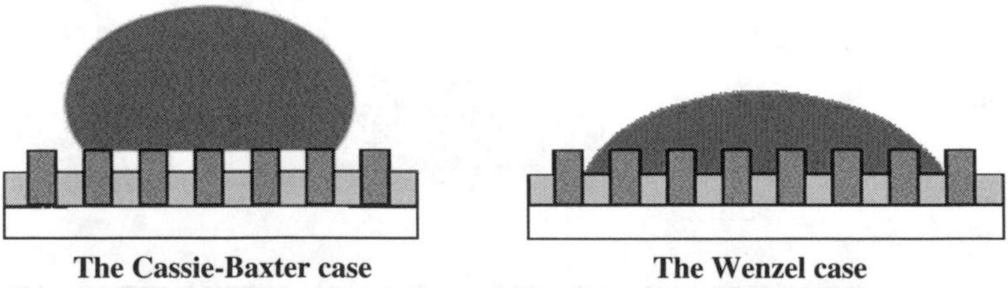

Figure 11.3 Illustrations of water droplets added to textured surfaces according to the Cassie-Baxter (left) and Wenzel (right) cases.

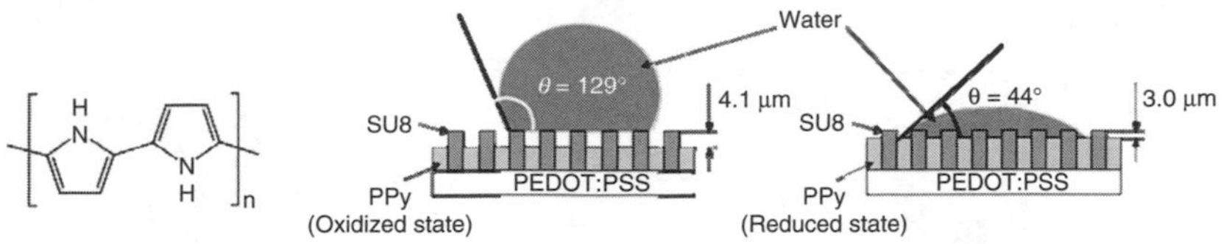

FIGURE 11.4 Left: The chemical structure of PPy. A water droplet added to a textured surface switch, composed of SU8 pillars and a PPy mesh with PPy in its oxidized state (middle) and after switching to its reduced state (right). (*Reproduced with permission from X. Wang, M. Berggren, O. Inganas, "Dynamically controlled surface energy and topography of microstructured conducting, polymer upon electrochemical reduction,* Langmuir *24(11):5942–5948 (2008) © 2008 American Chemical Society.*")

11.1.3 Integration of Wettability Switches in Microfluidic Systems

As dimensions of microfluidic systems are reduced, capillary forces become increasingly dominant in controlling the flow of liquids, such as aqueous samples. If the surface tension is changed along the ceiling, floor, or walls of the microfluidic channel, the aqueous flow can be controlled.[6, 12] EC P3HT-based surface tension switches were combined with microfluidic systems made from PDMS (polydimethylsiloxane) (Fig. 11.5).[13] P3HT surface switches, individually addressable, were defined under each "branch" of PDMS channel Y-junctions. The P3HT surfaces were then either switched to the oxidized state or left unswitched in the neutral state. Then a water sample was applied to the inlet at the channel "trunk" of the first Y-junction. The oxidized P3HT surface exhibits a relatively higher surface tension which then provides relatively higher capillary force acting on the aqueous sample, compared to channels including a reduced P3HT floor. We found that water samples were guided considerably faster through the channel branches including a P3HT floor switched to the oxidized state compared to branches including P3HT floors switched to the neutral state. Electronic gating of fluids might open for active dispensing of water samples possible to use in, e.g., lab-on-a-chip applications.

11.1.4 Electronic Control of Cell Seeding and Proliferation Using Surface Switches

Eukaryotic cells are commonly cultivated and propagated in cell culture dishes. Soon after cells are seeded into the wells, they adhere to the bottom surface of the well and eventually they start to proliferate.[14] It is important to notice that cells do not adhere directly to the plastic surface of the cell culture dish. This is so because cells are usually handled in a suspension of cell culturing medium containing large amounts of serum proteins. As these proteins rapidly diffuse in the medium, they immediately adhere to the bottom of the well, thus

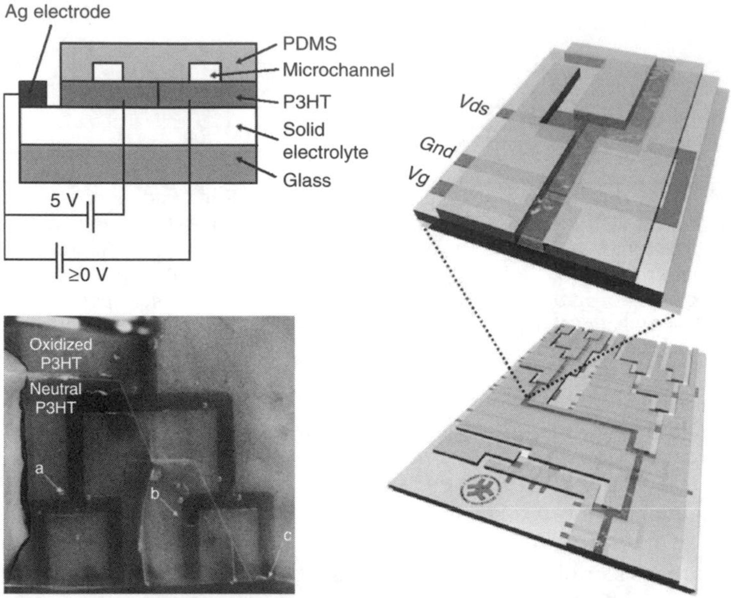

FIGURE 11.5 Top left: P3HT-based surface wettability switches used to control the flow of aqueous samples in microfluidic systems. Bottom left: Water is transported relatively faster along the microchannel of the Y-branches that include a floor of P3HT switched to the oxidized state (c). Once the water approaches the neutral P3HT (red color), it slows down considerably (a and b). Right: Electronic control of the gating of aqueous samples can be used in various lab-on-a-chip technologies to enable multiplexing of the sample analyte and different reagents. (*From Ref. 13. Reproduced by permission of the Royal Society of Chemistry, 2006.*) (See also color insert.)

providing a protein-coated plastic surface that cells adhere to. Various kinds of extracellular matrix (ECM) proteins, e.g., fibronectin and vitronectin, are among the serum proteins that coat the surface, and they act as cellular anchoring sites, because the cells express receptors that bind specifically to the ECM proteins. This can be exemplified by the interaction of integrins, located in the plasma membrane of the cells, and fibronectin present in the extracellular matrix. Depending on the amount of adherent proteins as well as their conformational state, the cell adhesion properties and the growth characteristics may differ. Major efforts have been devoted to design and manufacture cell culture dishes with specific protein adsorption characteristics, which are used to direct cell growth and differentiation. In some situations, it may be advantageous to alter the surface properties to enable control of the cell seeding characteristics in situ. Already in 1994, Wong et al. performed an experiment in which PPy thin films, electrochemically synthesized on ITO, were used as substrates for cell

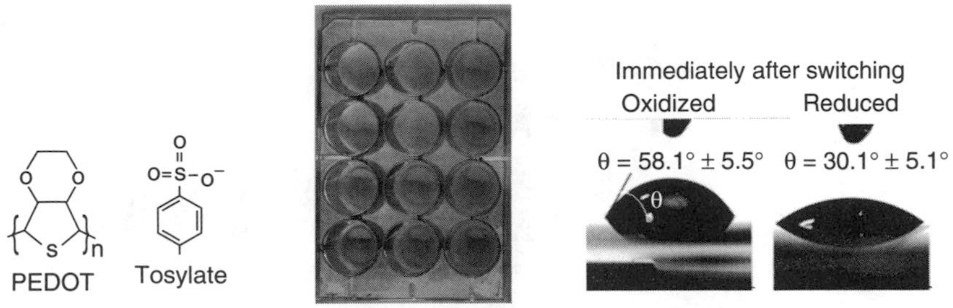

FIGURE 11.6 Left: The chemical structure of poly(3,4-ethylenedioxythiophene) (PEDOT) and Tosylate. Middle: PEDOT:Tosylate has been chemically synthesized along the surface of standard 12-well cell culture plates. Right: The water contact angle of planar PEDOT:Tosylate surfaces switched to the oxidized and reduced state, respectively.

growth.[15] They found that cells preferred to adhere to and spread along the fibronectin-coated PPy electrodes that were switched to the oxidized state. On the fibronectin-coated PPy electrodes switched to the neutral state, cells obtained a rounded morphology and were less prone to spread along the surface.

PEDOT can be chemically synthesized with Tosylate as the doping ion to form electrodes on various planar insulating carriers. These films can be manufactured along the surfaces of wells in standard cell culturing plates (Fig. 11.6). Electronically separated electrodes are achieved by cutting the electrode films using a plotter knife or an ordinary scalpel. The oxidation states of adjacent electronically isolated PEDOT:Tosylate electrodes are controlled by biasing addressing pads located outside the wells, while the cell medium serves as the electrolyte.

We used the PEDOT:Tosylate electrodes to analyze whether the electroactive surface could be used to direct stem cell differentiation. Neuronal stem cells c17.2 were seeded onto the reduced and oxidized electrodes. We found that the stem cells adhered to a much greater extent to the oxidized surface compared to the reduced surface. Surface spectroscopy studies and protein adsorption experiments revealed that the amount of serum protein albumin differs vastly between the reduced and oxidized surfaces. Such electronic control of cell seeding and density may open possibilities for novel devices that can be used to direct the differentiation of stem cells.

11.2 Electronic Ion Pumps Based on PEDOT:PSS

In PEDOT:PSS[16]-based (Fig. 11.7) thin-film EC devices, fast electrochemical switching throughout the bulk can occur because the polymer film electrodes allow transport of both ionic and electronic species. In the previously described surface switches, this may not be crucial

FIGURE 11.7 Left: Chemical structure of PEDOT and PSS. Middle: The conjugated PEDOT:PSS-based ion pump made of patterned, adjacent PEDOT:PSS electrodes (A through D). Right: By addressing the electrodes ions (M^+) migrate from the source to the target electrolytes through the nonconducting PEDOT:PSS channel (pink). (See also color insert.)

since all desired action specifically occurs along the outermost surface. But in other conjugated polymer-based EC devices (e.g., displays and transistors), switching of the bulk is crucial.[17] For instance, in a polymer EC transistor,[5] the electronic current between drain and source is controlled by the impedance state, which is further controlled by the EC state, within the channel. Conversely, it would be of great interest to enable electronic control of the migration of (charged) biosignaling molecules for a variety of bioapplications.

Electronic ion pumps were recently constructed using adjacent electrodes of PEDOT:PSS and separated source and target electrolytes[18] (Fig. 11.7). The two aqueous electrolytes are connected via a PEDOT:PSS channel, in which the PEDOT phase has been made permanently nonconducting (for electronic currents) by electrochemical over-oxidation.[19] Thus, the channel can only conduct ionic species, primarily throughout the polyelectrolyte PSS phase in the channel. As the electrodes are addressed, ions are first forced to enter the B electrode. Then the biased B-C electrode configuration drives ions from the source compartment toward the target electrolyte. Finally, the C-D electrode configuration promotes launching of ions away from the C-electrode surface to enter the target medium.

11.2.1 Electronic Control of Proton Oscillations

Initially, we studied the transport of positively charged ions, primarily protons and different alkali metal ions such as K^+. As protons are pumped from the A-B electrolyte to the C-D target electrolyte, they enter the target electrolyte along the border at which the PEDOT:PSS channel contacts the target electrolyte. To study the overall efficiency and the charge transport characteristics of this ion pump device, several different experiments were performed. First, proton migration was recorded using a HCl aqueous source electrolyte of pH = 0 and a target electrolyte based on a $CaCl_2$ aqueous solution of pH = 7. At $V_{BC} = 5$ V

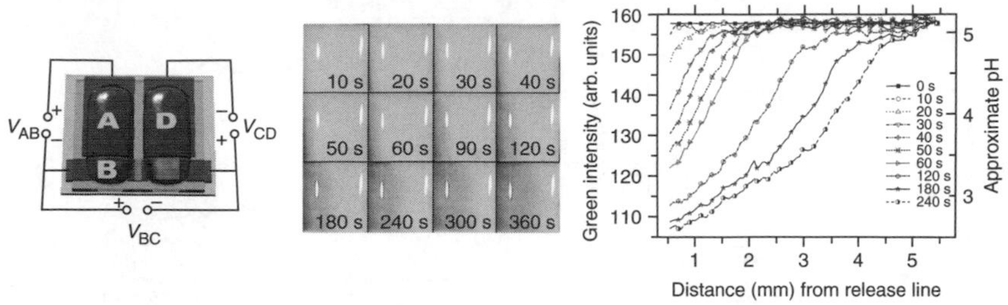

FIGURE 11.8 Left: The electronic ion pump with a pH paper on the C electrode as an indicator of the proton gradient. Middle: The proton gradient formed over time. Right: The associated measured pH gradients vs. time. (*Reprinted from Ref. 20. Copyright 2008, with permission from Elsevier.*) (See also color insert.)

the proton current was recorded at the same time as the pH gradient formed inside the target electrolyte was imaged using an ordinary pH paper (Fig. 11.8). As soon as 10 to 20 s after biasing the B-C electrode, the first delivered protons were observed. Then a dynamic gradient is established by diffusion, which evolves and spreads over time. We also demonstrated that proton oscillations can be induced by applying a pulsed voltage difference to the B and C electrodes.[20] Induction of proton gradients as well as proton oscillations, with the associated pH ranging from 3 to 5, is of great interest since pH is an important regulator for a vast array of cellular functions.

11.2.2 Electronic Ion Pumps to Regulate Intracellular Ca^{2+} Signaling

The Ca^{2+} ion is of major importance in a number of cell signaling processes in eukaryotic cells, e.g., to regulate metabolism, exocytosis, and gene transcription.[21] The specificity of a particular Ca^{2+} flux is represented by the spatial and temporal resolution of the Ca^{2+} signal, which is often oscillatory in nature. It is well established that excessively high extracellular concentrations of K^+ (≥ 50 m*M*) depolarize the cell membrane, whereby voltage-operated Ca^{2+} pumps located in the plasma membrane are activated. This leads to a controlled influx of Ca^{2+}, resulting in an increase of the intracellular [Ca^{2+}]. Using 0.1 *M* KCl as the aqueous electrolyte in the reservoir on top of the A-B electrodes, and 0.1 *M* Ca^{2+} acetate aqueous electrolyte in the receiving compartment on top of the C-D electrodes, we showed that K^+ ions could be pumped from the source to the target electrolyte in the ion pump at $V_{BC} = 10$ V, $V_{AB} = V_{CD} = 1$ V (Fig. 11.9). An almost constant delivery rate was observed during the first 600 s followed by a pinchoff caused by consumption of any of the A or D electrodes. Interestingly, we found that the ion pump operates at a very high efficiency, close to 100%. Knowing that every charge that is transported from the AB source to the CD compartment is a K^+ ion, experiments were performed in an attempt to mimic

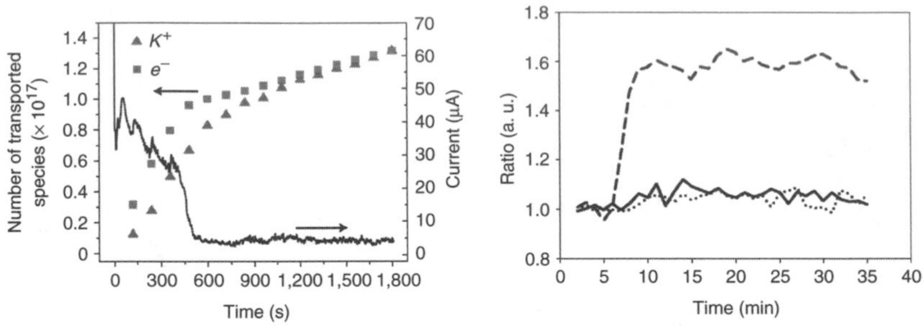

FIGURE 11.9 Left: The current and numbers of transported K^+ ions vs. time of biasing the ion pump. Right: Intracellular $[Ca^{2+}]$ as a response to pumping K^+ ions. Flat lines represent experiments where cells are treated with pharmacological agents that block voltage-gated ion channels. (*Reproduced with permission from Ref. 18. Copyright 2007, Nature Publishing Group.*)

physiological Ca^{2+} signaling using HCN-2 neuronal cells. Thus, cells were seeded in cell culturing medium on top of the CD electrode and were allowed to adhere to the electrode surface. As the ion pump was biased, transported K^+ reached a high, local concentration which depolarized the cells located in close proximity to the PEDOT:PSS channel. The concomitant increase in intracellular $[Ca^{2+}]$ was recorded using a fluorophore-based microscopy technique (Fig. 11.9). The physiological nature of the response to K^+ delivered by the ion pump device was demonstrated in experiments where the action of the voltage-operated Ca^{2+} channels was abrogated by pharmacological agents. In these experiments, no increase in the intracellular $[Ca^{2+}]$ was observed (Fig. 11.9).

Traditional techniques used to transfer bioactive substances from a reservoir to a desired point of delivery in close proximity to the target cell normally rely on the use of microfluidic systems or syringes. These are typically hard to downsize to the micron level, which is required for delivery at the single-cell level. Also, such devices are difficult to integrate into highly sophisticated delivery systems aiming to achieve complex signal patterns at the resolution of individual cell or subcellular level. Using a solid state organic ion pump technology, which is possible to manufacture using standard processing tools traditionally used for microelectronics, will pave the path for novel types of cell experiments where high spatial and temporal resolution is desired.

Acknowledgments

The authors wish to thank SSF (OBOE), VR, KAW, and KVA for financial support. In addition, the authors wish to thank Dr. Joakim Isaksson, now at Philips Research in Eindhoven (Netherlands), and David Nilsson, now at Acreo Insitute in Norrköping (Sweden), for providing data and illustrations.

References

1. Recknor, Jennifer B.; Recknor, Justin C.; Sakaguchi, Donald S.; and Mallapragada, Surya K., Oriented astroglial cell growth on micropatterned polystyrene substrates, *Biomater.* 25(14):2753 (2004); Stevens, Molly M.; and George, Julian H., Exploring and engineering the cell surface interface, *Science* 310(5751):1135 (2005); Tengvall, Pentti; Lundstrom, Ingemar; and Liedberg, Bo, Protein adsorption studies on model organic surfaces: An ellipsometric and infrared spectroscopic approach, *Biomater.* 19(4–5):407 (1998); Jia-Wei, Shen; Tao, Wu; Qi, Wang; and Hai-Hua, Pan, Molecular simulation of protein adsorption and desorption on hydroxyapatite surfaces, *Biomater.* 29(5):513 (2008).
2. Smela, Elisabeth, Conjugated polymer actuators for biomedical applications, *Adv. Mater.* 15(6):481 (2003); Jager, Edwin W. H.; Inganas, Olle; and Lundstrom, Ingemar, Microrobots for micrometer-size objects in aqueous media: Potential tools for single-cell manipulation, *Science* 288(5475):2335 (2000).
3. Argun, Avni A., et al., Multicolored electrochromism in polymers: Structures and devices. *Chem. Mater.* 16(23):4401 (2004).
4. Thackeray, James W.; White, Henry S.; and Wrighton., Mark S., Poly(3-methylthiophene)-coated electrodes: Optical and electrical properties as a function of redox potential and amplification of electrical and chemical signals using poly(3-methylthiophene)-based microelectrochemical transistors, *J. Phys. Chem.* 89:5133 (1985); Bernards, D. A., and Malliaras, G. G., Steady-state and transient behavior of organic electrochemical transistors. *Adv. Funct. Mater.* 17(17):3538 (2007).
5. Nilsson, David, et al., Bi-stable and dynamic current modulation in electrochemical organic transistors, *Adv. Mater.* 14(1 Jan. 4):51 (2002).
6. Causley, Jennifer, et al., Electrochemically-induced fluid movement using polypyrrole, *Synth. Met.* 151(1):60 (2005).
7. Sun, Taolei, et al., Reversible switching between superhydrophilicity and superhydrophobicity, *Angewandte Chem. — Int. Ed.* 43(3):357 (2004).
8. Isaksson, J.; Robinson, L.; Robinson, N. D.; and Berggren, M., Electrochemical control of surface wettability of poly(3-alkylthiophenes), *Surf. Sci.* 600(11):148 (2006).
9. Bao, Zhenan; Dodabalapur, Ananth; and Lovinger, Andrew J., Soluble and processable regioregular poly(3-hexylthiophene) for thin film field-effect transistor applications with high mobility, *Appl. Phys. Lett.* 69(26):4108 (1996).
10. Isaksson, Joakim, et al., A solid-state organic electronic wettability switch, *Adv. Mater.* 16(4):316 (2004).
11. Feng, X., and Jiang, L., Design and creation of superwetting/antiwetting surfaces, *Adv. Mater.* 18:3063 (2006).
12. Nadkarni, Suvid; Yoo, Byungwook; Basu, Debarshi; and Dodabalapur, Ananth, Actuation of water droplets driven by an organic transistor based inverter, *Appl. Phys. Lett.* 89(18):184105 (2006); Idota, Naokazu, et al., Microfluidic valves comprising nanolayered thermoresponsive polymer-grafted capillaries, *Adv. Mater.* 17(22):2723 (2005).
13. Robinson, L., et al., Electrochemical wettability switches gate aqueous liquids in microfluidic systems, *Lab on a Chip* 6(10):1277 (2006).
14. Yoshida, Mutsumi; Langer, Robert; Lendlein, Andreas; and Lahann, Joerg, From advanced biomedical coatings to multi-functionalized biomaterials, *Polymer Rev.* 46(4):347 (2006).
15. Wong, Joyce Y.; Langer, Robert; and Ingber, Donald E., Electronically conducting polymers can noninvasively control the shape and growth of mammalian cells, *Proc. Natl. Acad. of Sci.* 91:3201 (1994).
16. Groenendaal, L., et al., Poly(3,4-ethylenedioxythiophene) and its derivatives: Past, present, and future, *Adv. Mater.* 12(7):481 (2000).
17. Wang, X. Z.; Shapiro, B.; and Smela, E., Visualizing ion currents in conjugated polymers, *Adv. Mater.* 16(18):1605 (2004).
18. Isaksson, J., et al., Electronic control of Ca^{2+} signalling in neuronal cells using an organic electronic ion pump, *Nature Mater.* 6(9):673 (2007).

19. Tehrani, Payman, et al., Patterning polythiophene films using electrochemical overoxidation, *Smart Mater. & Struct.* 14(4):21 (2005).
20. Isaksson, J., et al., Electronically controlled pH gradients and proton oscillations, *Organic Electron.* 9(3):303 (2008).
21. Berridge, Michael J.; Bootman, Martin D.; and Lipp, Peter, Molecular biology: Calcium—a life and death signal, *Nature* 395(6703):645 (1998).

Index

A

B

C

D

P

Q

R

S

T